Student Solutions Manual

W9-BLK-096

for

Kotz/Treichel/Weaver's

Chemistry & Chemical Reactivity

Sixth Edition

Alton J. Banks
North Carolina State University

THOMSON

BROOKS/COLE

Australia • Canada • Mexico • Singapore • Spain • United Kingdom • United States

Printed in the United States of America
1 2 3 4 5 6 7 09 08 07 06 05

Printer: Thomson/West

ISBN: 0-534-99852-6

For more information about our products,
contact us at:
Thomson Learning Academic Resource Center
1-800-423-0563

For permission to use material from this text or
product, submit a request online at
http://www.thomsonrights.com.
Any additional questions about permissions can be
submitted by email to **thomsonrights@thomson.com.**

Cover Images: Motohiko Murakami

Thomson Higher Education
10 Davis Drive
Belmont, CA 94002-3098
USA

Asia (including India)
Thomson Learning
5 Shenton Way
#01-01 UIC Building
Singapore 068808

Australia/New Zealand
Thomson Learning Australia
102 Dodds Street
Southbank, Victoria 3006
Australia

Canada
Thomson Nelson
1120 Birchmount Road
Toronto, Ontario M1K 5G4
Canada

UK/Europe/Middle East/Africa
Thomson Learning
High Holborn House
50–51 Bedford Road
London WC1R 4LR
United Kingdom

Latin America
Thomson Learning
Seneca, 53
Colonia Polanco
11560 Mexico
D.F. Mexico

Spain (including Portugal)
Thomson Paraninfo
Calle Magallanes, 25
28015 Madrid, Spain

To the student:

The skills involved in solving chemistry problems are acquired only by discovering the conceptual paths which connect the available data to the desired piece(s) of information. These paths are discovered in different ways by different people. What is true is that those discoveries frequently require repetition. Working multiple problems is one very good approach to clarifying and solidifying the fundamental concepts. I would suggest that this **Student Solutions Manual** will provide maximum benefit if you consult it *after* you have attempted to solve a problem.

The selected Study Questions have been chosen by the authors of your text to allow you to discover the range and depth of your understanding of chemical concepts. The importance of mastering the "basics" cannot be overemphasized. You will find that the text, **Chemistry & Chemical Reactivity, 6th Edition**, has a wealth of study questions to assist you in your study of the science we call Chemistry.

Many of the questions contained in your book—and this solutions manual—have multiple parts. In many cases, comments have been added to aid you in the process of gathering available data and applicable conversion factors, and connecting them via fundamental concepts. In working these multiple-step questions, you may find an answer which differs slightly from those given here. This may be a result of "rounding" intermediate answers. The procedure followed in this manual was to report intermediate answers to the appropriate number of significant figures, and to calculate the "final" answer without any intermediate rounding. In cases involving atomic and molecular masses, those quantities were expressed with at least one digit more than the number of digits needed for the data provided.

A word of appreciation is due to several people. Thanks go to the authors, especially Dr. John C. Kotz, for conversations held during the development of this manual. Collaboration with Dr. Susan M. Young, author of the Instructor's Resource Manual, has also been helpful. The many fine folks at the publishers have been helpful as well. Thanks go to Peter McGahey, Annie Mac, and Ellen Bitter. Very special thanks are extended to my wife, Dr. Catherine Hamrick Banks, for her invaluable assistance in typing and proofreading this manuscript. One of my many blessings is to have such a person as a patient wife and a chemical colleague.

While we have worked diligently to remove all errors from this text, I am certain that some have escaped the many inspections. I accept responsibility for all those errors. Please contact me regarding any errors.

Alton J. Banks
Department of Chemistry
North Carolina State University
Raleigh, North Carolina 27695

Table of Contents

Chapter 1
Matter and Measurement

Practicing Skills

Matter: Elements and Atoms, Compounds and Molecules

1. Names for the following elements:

    (a) C - carbon   (c) Cl - chlorine   (e) Mg - magnesium

    (b) K - potassium   (d) P - phosphorus   (f) Ni - nickel

3. Symbols for the following elements:

    (a) barium - Ba   (c) chromium - Cr   (e) arsenic - As

    (b) titanium - Ti   (d) lead - Pb   (f) zinc - Zn

5. In each of the pairs, which is an element and which is a compound?

    Compounds contain **more than one element**, so for the pairs given:

    (a) NaCl is a compound while sodium (Na) is an element of that compound.

    (b) Sugar is a compound (with C,H,O) while carbon (C) is an element.

    (c) Gold chloride is a compound (with Au and Cl) while gold (Au) is an element.

Physical and Chemical Properties

7. Determine if the property is a physical or chemical property for the following:

    (a) color   a physical property

    (b) transformed into rust   a chemical property

    (c) explode   a chemical property

    (d) density   a physical property

    (e) melts   a physical property

    (f) green   a physical property

    Physical properties are those that can be observed or measured without changing the composition of the substance. Exploding or transforming into rust results in substances which are **different** from the original substances— and represent chemical properties.

9. Descriptors of physical versus chemical properties:

    (a) Color and physical state are physical properties. (colorless, liquid) while **burning** reflects a chemical property.

(b) Shiny, metal, orange, and liquid are physical properties while **reacts readily** describes a chemical property.

Using Density

11. What mass of ethylene glycol (in grams) possesses a volume of 500. mL of the liquid?

$$\frac{500.\ mL}{1} \cdot \frac{1\ cm^3}{1\ mL} \cdot \frac{1.11\ g}{cm^3} = 555\ g$$

13. What volume of a liquid compound with a density of 0.718 g/cm^3 has a mass of 2.00 g?

$$\frac{2.00\ g}{1} \cdot \frac{1\ cm^3}{0.718\ g} \cdot \frac{1\ mL}{1\ cm^3} = 2.79\ mL$$

15. The metal will displace a volume of water that is equal to the volume of the metal. Hence the difference in volumes of water (20.2-6.9) corresponds to the volume of metal. Since 1 mL = 1 cm^3, the density of the metal is then:

$$\frac{Mass}{Volume} = \frac{37.5\ g}{13.3\ cm^3} \text{ or } 2.82\ \frac{g}{cm^3}$$

From the list of metals provided, the metal with a density closest to this is **Aluminum**.

Temperature Scales

17. Express 25 °C in kelvins:

$$K = (25\ °C + 273) \text{ or } 298 \text{ kelvins}$$

19. Make the following temperature conversions:

°C	K
(a) 16	16 + 273.15 = 289
(b) 370 - 273 or 97	370
(c) 40	40 + 273.15 = 310

Note no decimal point after 40

Using Units

21. Express 42.195 km in meters; in miles:

$$\frac{42.195\ km}{1} \cdot \frac{1000\ m}{1\ km} = 42195\ m$$

$$\frac{42.195\ km}{1} \cdot \frac{0.62137\ miles}{1\ km} = 26,218.\ miles$$

The factor (0.62137 mi/km) is found inside the back cover of the text.

23. Express the area of a 2.5 cm x 2.1 cm stamp in cm^2 ; in m^2 :

$$2.5 \text{ cm} \cdot 2.1 \text{ cm} = 5.3 \text{ cm}^2$$

$$5.3 \text{ cm}^2 \cdot \left(\frac{1 \text{ m}}{100 \text{ cm}}\right)^2 = 5.3 \times 10^{-4} \text{ m}^2$$

25. Express 250. mL in cm^3; in liters (L); in m^3 ; in dm^3:

$$\frac{250.\text{cm}^3}{1\text{beaker}} \cdot \frac{1\text{cm}^3}{1\text{mL}} = \frac{250. \text{ cm}^3}{1 \text{ beaker}}$$

$$\frac{250.\text{cm}^3}{1\text{beaker}} \cdot \frac{1\text{L}}{1000\text{cm}^3} = \frac{0.250 \text{ L}}{1 \text{ beaker}}$$

$$\frac{250.\text{cm}^3}{1\text{beaker}} \cdot \frac{1\text{m}^3}{1\times10^6\text{cm}^3} = \frac{2.50 \times 10^{-4} \text{ m}^3}{1 \text{ beaker}}$$

$$\frac{250.\text{cm}^3}{1\text{beaker}} \cdot \frac{1\text{L}}{1000\text{cm}^3} \cdot \frac{1 \text{ dm}^3}{1 \text{ L}} = \frac{0.250 \text{ dm}^3}{1 \text{ beaker}}$$

27. Convert book's mass of 2.52 kg into grams:

$$\frac{2.52 \text{ kg}}{1 \text{ book}} \cdot \frac{1 \times 10^3 \text{ g}}{1 \text{ kg}} = \frac{2.52 \times 10^3 \text{ g}}{\text{book}}$$

Accuray, Precision, and Error

29. Using the data provided, the averages and their deviations are as follows:

Data point	Method A	deviation	Method B	deviation
1	2.2	0.2	2.703	0.777
2	2.3	0.1	2.701	0.779
3	2.7	0.3	2.705	0.775
4	2.4	0.0	5.811	2.331
Averages:	2.4	0.2	3.480	1.166

Note that the deviations for both methods are calculated by first determining the average of the four data points, and then subtracting the individual data points from the average (without regard to sign)

(a) The average density for method A is 2.4 ± 0.2 grams while the average density for method B is 3.480 ± 1.166 grams—if one includes all the data points. *Data point 4 in Method B has a large deviation, and should probably be excluded* from the calculation. If one omits data point 4, Method B gives a density of 2.703±0.001 g

(b) The percent error for each method :

Error = experimental value - accepted value

From Method A   error = (2.4 - 2.702) = 0.3

From Method B   error = (2.703 - 2.702) = 0.001 (omitting data point 4)

error = (3.480 - 2.702) = 0.778 (including all data points)

Percent error (Method A) = $\dfrac{0.3}{2.702} \cdot \dfrac{100}{1}$ = 11.1% (about 10% to 1 s.f.)

(Method B) = $\dfrac{0.001}{2.702} \cdot \dfrac{100}{1}$ = 0.037 % (about 0,04% to 1 s.f.)

(c) Precision and Accuracy of each method:

If one counts all data points, the deviations **for all data points** of Method A are less than those for **the data points of** Method B, Method A offers *better precision* . On the other hand, omitting data point 4, Method B offers both *better accuracy* (average closer to the accepted value) and *better precision* (since the value is known to a greater number of significant figures).

General Questions

31. For the gemstone turquoise:

(a) Qualitative: blue-green color;

Quantitative: density; mass

(b) Extensive: Mass

Intensive: Density; Color

(c) Volume: $\dfrac{2.5 \text{ g}}{1} \cdot \dfrac{1 \text{ cm}^3}{2.65 \text{ g}}$ = 0.94 cm^3

33. For the gemstone aquamarine:

(a) Elemental symbols: Aluminum: **Al**   ; Silicon: **Si**   ;Oxygen: **O**

(b) Physical properties of elements and mineral: Oxygen is a gas, while aluminum, silicon, and aquamarine are solids at room temperature. Regarding color, oxygen is—in the concentration shown here, colorless, while aluminum and silicon are gray. The gemstone is a bluish color.

35. Melting point and boiling point of Ne in kelvins:

MP = -248.6 °C + 273.15 = 24.6 K

BP = -246.1°C   + 273.15 = 27.1 K

37. Express the length 1.97 Angstroms in nanometers? In picometers?

$$\frac{1.97 \text{ Angstrom}}{1} \bullet \frac{1 \times 10^{-10} \text{m}}{1 \text{ Angstrom}} \bullet \frac{1 \times 10^{9} \text{ nm}}{1 \text{ m}} = 0.197 \text{ nm}$$

$$\frac{1.97 \text{ Angstrom}}{1} \bullet \frac{1 \times 10^{-10} \text{m}}{1 \text{ Angstrom}} \bullet \frac{1 \times 10^{12} \text{pm}}{1 \text{m}} = 197 \text{ pm}$$

39. Diameter of red blood cell = 7.5 μm

(a) In meters: $\dfrac{7.5 \text{ μm}}{1} \bullet \dfrac{1 \text{ m}}{1 \times 10^{6} \text{μm}} = 7.5 \times 10^{-6} \text{m}$

(b) In nanometers: $\dfrac{7.5 \text{ μm}}{1} \bullet \dfrac{1 \text{ m}}{1 \times 10^{6} \text{μm}} \bullet \dfrac{1 \times 10^{9} \text{nm}}{1 \text{m}} = 7.5 \times 10^{3} \text{nm}$

(c) In picometers: $\dfrac{7.5 \text{ μm}}{1} \bullet \dfrac{1 \text{ m}}{1 \times 10^{6} \text{μm}} \bullet \dfrac{1 \times 10^{12} \text{pm}}{1 \text{m}} = 7.5 \times 10^{6} \text{ pm}$

41. Mass of Pt contained in 1.53g of cisplatin (65.0% Pt)

$$\frac{1.53 \text{g cisplatin}}{1} \bullet \frac{65.0 \text{ g Pt}}{100 \text{ g cisplatin}} = 0.995 \text{ g Pt}$$

43. Mass of procaine hydrochloride (in mg) in 0.50 mL of solution

$$\frac{0.50 \text{ mL}}{1} \bullet \frac{1.0 \text{ g}}{1 \text{ mL}} \bullet \frac{10. \text{ g procaine HCl}}{100 \text{ g solution}} \bullet \frac{1 \times 10^{3} \text{ mg procaine HCl}}{1 \text{ g procaine HCl}} = 50. \text{ mg procaine HCl}$$

45. The volume of the brass is the major question, since its volume will add to the volume of water initally present (50.0 mL).

$$\frac{154 \text{ g brass}}{1} \bullet \frac{1 \text{ cm}^{3}}{8.56 \text{ g}} \bullet \frac{1 \text{ mL}}{1 \text{ cm}^{3}} = 18.0 \text{ mL}$$

The water in the graduated cylinder will rise an additional 18.0 mL for a final reading of 58.0 mL.

47. Mass of necklace = 67g. Volume = $(26.0 - 22.5)\text{cm}^3$ D = 67g/3.5 cm^3 = 19 g/cm^3. The accepted density for gold is 19.3 g/cm^3, so the necklace is probably gold.

The value of the gold is: $\dfrac{67 \text{ g}}{1} \bullet \dfrac{1 \text{ troy oz}}{31.1 \text{ g}} \bullet \dfrac{\$380}{1 \text{ troy oz}} = \$820$ (to 2sf).

So the price of $300 is a good one!!

Conceptual Questions

49. The non-uniform appearance of the mixture indicates that samples taken from different regions of that mixture would be different—a characteristic of a **heterogeneous mixture**. The components of a mixture retain their physical properties, Recalling that iron is attracted to a magnetic field, while sand is generally not attracted in this way suggests that passing a magnet through the mixture would separate the sand and iron, as the iron adhered to the magnet and left the sand behind.

51. Will aluminum (D = 2.70 g/cm^3) and plastic (D = 1.37 g/cm^3) float or sink in carbon tetrachloride (D = 1.58 g/cm^3)?
    Since substances in a fluid usually seek a density that is equal to their own, the plastic (with a much lower density than the carbon tetrachloride) will float. The aluminum (with a density that is greater) will sink. You probably most associate this phenomenon with helium-filled balloons "floating" in the atmosphere. The density of the helium is much less than the surrounding atmosphere, so the balloon "rises".

53. Since the material is found in the kitchen, place a bit of it in a sauce pan, and heat. Sugar melts at a much lower temperature (about 160 °C) than does NaCl(about 800 °C). Sugar is also much more soluble than NaCl.

55. The accepted value for a normal human temperature is 98.6 °F. On the Clasius scale this corresponds to:
$$°C = \frac{5}{9}(98.6 - 32) = 37°C$$

    Since the melting point of gallium is 29.8 °C, the gallium should melt in your hand.

57. Study Question 51 also addresses this phenomenon. HDPE with a density of 0.97 g/mL will float in any liquid whose density is greater than 0.97 g/mL and sink in any liquid whose density is less than that of HDPE. Of the liquids listed, HDPE should float in ethylene glycol, water, acetic acid, and glycerol.

59. Indicate the relative arrangements of the particles in each of the following:

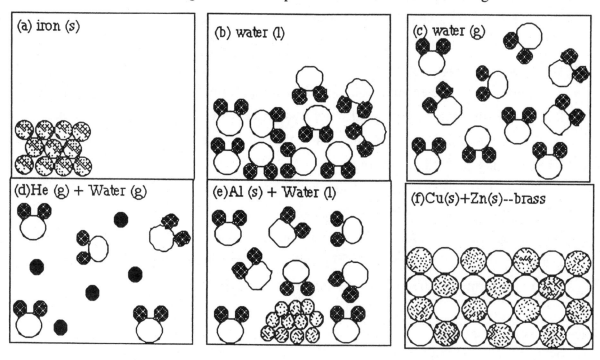

61. Experimental method to determine if a liquid is water:

There are several methods. One method would be to weigh an accurately known volume of the liquid. An empty dry weighed 10.0 mL graduated cylinder could be filled to the 10.0 mL mark with the liquid, and reweighed. The mass of liquid divided by the volume would provide the density of the liquid. That density could be compared with published values of the density of water *at that temperature*.

To determine if the water contains dissolved salts, test the electrical conductivity. Pure water is a very poor conductor. Water containing dissolved salts (and therefore the ions produced when that salt dissolves) would conduct an electric current.

To determine is there is salt dissolved in the water, one can boil the solution to dryness. Any dissolved salts would not boil (at water's normal boiling temperature) and would remain as a solid residue in the beaker.

63. Pouring three immiscible liquids into a test tube will result in three discrete layers in which the liquids arrange themselves from the most dense liquid (at bottom) to the least dense liquid (at top).

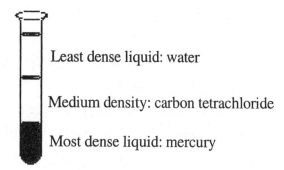

Least dense liquid: water

Medium density: carbon tetrachloride

Most dense liquid: mercury

65. For the reaction of elemental potassium reacting with water:
    (a) States of matter involved: **Solid** potassium reacts with **liquid** water to produce **gaseous** hydrogen and aqueous potassium hydroxide solution (a homogenous mixture).
    (b) The observed change is chemical. The products (hydrogen and potassium hydroxide) are quite different from elemental potassium and water. Litmus paper would also provide the information that while the original water was neither acidic nor basic, the solution produced would be basic. (It would change the color of red litmus paper to blue.)
    (c) The reactants: potassium and water
    The products: hydrogen and potassium hydroxide solution
    (d) Potassium reacts **vigorously** with water. Potassium is less dense than water, and floats atop the surface of the water. The reaction produces enough heat to ignite the hydrogen gas evolved. The flame observed is typically violet-purple in color. The potassium hydroxide formed is soluble in water (and therefore not visible).

67. Describe experiments to:
    (a) Separate salt (NaCl) from water:

    Heating a solution of salt dissolved in water so that the solution boiled, capturing, and condensing the water vapor would separate salt from water. This process is known as distillation.

    (b) Separate iron filings from small pieces of lead:

    Iron is diamagnetic, i.e. it is attracted to a magnetic field. Placing the mixture of metal pieces on a stiff sheet of paper, and drawing a bar magnet underneath the paper will cause the iron filings to follow the magnet, separating the two metals. Like Part (a) of this problem, this is an example of a physical technique

    (c) Separate elemental sulfur from sugar:
    Pour the two solids into a beaker of water, stir and allow the sugar to dissolve, while the sulfur remains in the solid state. Filter the mixture though a conical filter paper to separate the solid sulfur from the solution. Carefully distilling the water off the sugar (see Part (a)) would separate the sugar from the water.

69. To determine the mass of iron per gram of cereal, weigh and record the mass of the cereal in a box. Pour the dry cereal into a container with lots of water, and stir. Enclose a bar magnet (or other strong magnet) in a plastic bag (e.g. a Ziploc™ bag) and draw the magnet through the mixture of cereal and water. The iron chips will be attracted to the magnet.

Carefully remove the iron filings from the exterior of the plastic bag, dry and weigh them. Divide the mass of the iron filings by the mass of the cereal to obtain the ratio of iron per gram of cereal.

71. (a) Reactants: P_4 and Cl_2　　　Product: PCl_3

(b) Changes in structure from reactants to products:

P_4 +	$6Cl_2$ $\rightarrow$	$4PCl_3$
tetrahedral	linear	pyramidal

Mathematics of Chemistry

Exponential Notation

73. Express the following numbers in exponential notation:

(a) $0.054 = 5.4 \times 10^{-2}$ To locate the decimal behind the first non-zero digit, we move the decimal place to the right by 2 spaces (-2)

(b) $5462 = 5.462 \times 10^3$ To locate the decimal behind the first non-zero digit, we move the decimal place to the left by 3 spaces (+3)

(c) $0.000792 = 7.92 \times 10^{-4}$ To locate the decimal behind the first non-zero digit, we move the decimal place to the right by 4 spaces (-4)

Significant Figures

77. The number of significant figures in each of the following numbers:

The Rule numbers referred to are those found in the math interchapter of the text.

(a) 0.0123　　　　3sf; Rule 1 – 0 to the left of the 1 locate the decimal point

(b) 3.40×10^3　　　3sf; All digits to the left of the "x" are considered significant

(c) 1.6402　　　　5 sf; "trapped" zeros are considered significant

(d) 1.020　　　　　4 sf; Rule 1- when a number is greater than 1, all zeros to the right of the decimal point are significant.

79. Express the product of three numbers to the proper number of significant figures:

$(0.0546)(16.0000)(\dfrac{7.779}{55.85}) = 0.122$　　　(3 sf are allowed—owing to 0.0546)

Graphing

81. Calibration curve for spectrophotometer:

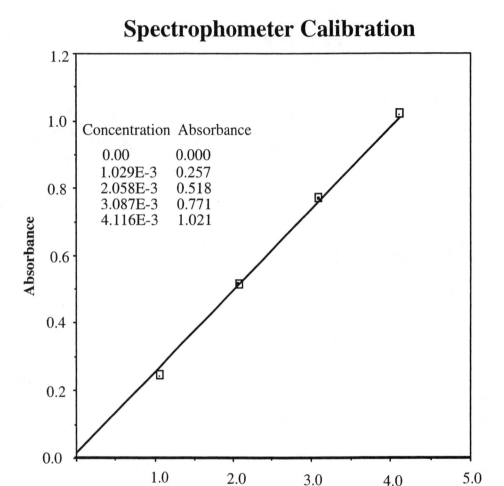

Spectrophometer Calibration

Concentration	Absorbance
0.00	0.000
1.029E-3	0.257
2.058E-3	0.518
3.087E-3	0.771
4.116E-3	1.021

Concentration $(\times 10^{3})$

Plotting the data results in the graph below:

Using Cricket Graph™ to plot the "best straight line" gives an equation: $y = 248.4x + 0.0022$

The concentration when absorbance = 0.635 is (from the graph) 2.55×10^{-3} M

The slope of the line is the coefficient of x or 248.4.

83. From the graph shown in your text:
    (a) The value of x when y = 4.0? Reading on the x axis at the point at which y = 4.0 gives an

    x value of about 0.21

    (b) The value of y when x = 0.30? Reading on the y axis at the point at which x 0.30 gives a

    value of about 5.6.

(c)The slope and intercept of the line: The line crosses the Y axis at about 0.20 units.
Calculating the slope is done by calculating how fast Y values change (rise) as a function of changes in X (run). Arbitrarily selecting two points on the graph we can calculate:

Slope = $(\Delta Y/\Delta X)$ = (7.50 – 2.00)/(0.40 – 0.10) = 5.50/0.30 = approximately 18

(d) The value of y when x = 1.0? Since this point is not on the graph, we can calculate it from the equation of the line. y = mx + b is the equation of a straight line, so we substitute for m (the slope) our value of 18, and for b (the intercept) our value of 0.20.

y = (18)(1.0) + 0.20 = 18.2 (and when adding 0.20 to 18, we report the sum as 18.

Using Equations

85. Solving the equation for "C":

(0.502)(123) = (750.)C   and rearranging the equation by dividing by 750. gives

$$\frac{(0.502)(123)}{750.} = C = 0.0823 \ (\text{to 3 sf})$$

87. Solve the following equation for T:

The equation: (4.184)(244)(T-292.0) + (0.449)(88.5)(T-369.0) = 0

Expanding the equation gives:

(4.184)(244)T - (4.184)(244)(292.0) + (0.449)(88.5)T - (0.449)(88.5)(369.0) = 0

1,020.896T - 298,101.632 + 39.7365T - 14,662.769 = 0

grouping the terms (1,020.896T + 39.7365T) gives 1,060.6325T likewise

(- 298,101.632 + - 14,662.769) gives -312,764.401.

The equation now is equivalent to 1,060.6325T -312,764.401= 0 . If we add the negative term to both sides: 1,060.6325T = 312,764.401and T = 312,764.401/1,060.6325 or 294.884797986.

This is clearly too many significant figures. Rounding to 3sf , gives a value of 295.

Problem Solving

89. Volume of a 1.50 carat diamond:

This problem is in essence (a) conversion of the unit "carat" to units of "grams", then using the density of diamond to convert from mass to volume.

$$\frac{1.50 \text{ carat}}{1} \cdot \frac{0.200 \text{ g}}{1 \text{ carat}} \cdot \frac{1 \text{ cm}}{3.513 \text{ g}} = 0.0854 \text{ cm}^3$$

91. Mass of a gold coin 2.2 cm in diameter and 3.0 mm thick:

To calculate the mass of the coin we'll need to determine the volume of the coin, and then use the density of gold.

Volume = $\pi r^2 \bullet$ thickness = $3.14159 \bullet (\frac{2.2 \text{ cm}}{2})^2 \bullet (\frac{3.0 \text{ mm}}{1} \bullet \frac{1 \text{ cm}}{10 \text{ mm}}) = 1.1 \text{ cm}^3$

Note that dividing the diameter (2.2cm) by 2 gives the radius (r). The last two terms are used to convert the thickness of the coin in millimeters to centimeters.

The product of $r^2 \bullet$ thickness gives the volume in cubic centimeters.

The mass of the coin is then: $V \bullet D = 1.1 \text{ cm}^3 \bullet \frac{19.3 \text{ g}}{\text{cm}^3} = 22 \text{ g}$ (to 2 sf)

93. Express the fuel density (1.77 lb/L) in units of kg/L:

$\frac{1.77 \text{ lb}}{1 \text{ L}} \bullet \frac{453.6 \text{ g}}{1 \text{ lb}} \bullet \frac{1 \text{ kg}}{1000 \text{ g}} = 0.803 \text{kg/L}$

Mass of fuel needed can be determined by determining the difference in mass between that required (22300 kg) – and that on-board (expressed presently as 7682 L).

Convert 7682 L to mass:

$\frac{7682 \text{ L}}{1} \bullet \frac{0.803 \text{ kg}}{1 \text{ L}} = 6,167.66 \text{ kg or } 6170 \text{ kg (to 3sf)}$

Knowing the mass of fuel on-board, we can calculate the mass required:

Mass needed: 22300kg – 6170 = 16130 kg needed

95. Calculate the mass of 12 ounces of aluminum in grams and from that mass, the volume:

$$\text{Volume} = \frac{\text{Mass}}{\text{Density}} = \frac{12 \text{ oz} \bullet \frac{454 \text{ g}}{16 \text{ oz}}}{2.70 \text{ g/cm}^3} = 130 \text{ cm}^3$$

Volume = Area • Thickness

Express the area in units of cm^2, then calculate the thickness by dividing the volume by

the area: Area = $75 \text{ ft}^2 \bullet (\frac{12 \text{ in}}{1 \text{ ft}})^2 \bullet (\frac{2.54 \text{ cm}}{1 \text{ in}})^2 = 7.0 \times 10^4 \text{ cm}^2$

then   Volume = Area   •   Thickness

$130 \text{ cm}^3 = 7.0 \times 10^4 \text{ cm}^2 \bullet$ Thickness

and   Thickness $= \frac{130 \text{ cm}^3}{7.0 \times 10^4 \text{ cm}^2} = 1.8 \times 10^{-3} \text{ cm or } 1.8 \times 10^{-2} \text{ mm}$

97. To determine the thickness of the oil layer, we can think about the oil layer as having a certain volume, (V = l x w x h), and that our "task" is to determine the "thickness" -- or h in our formula. The volume of oil is 1 teaspoon (5 cm³). The area covered (l x w) is 0.5 acres. So if we divide the volume by the area, we should have the **thickness**.

$$\frac{5 \text{ cm}^3}{1 \text{ teaspoon}} \bullet \frac{1 \text{ teaspoon}}{0.5 \text{ acre}} \bullet \frac{2.47 \text{ acres}}{1.0 \times 10^4 \text{ m}^2} \bullet (\frac{1 \text{ m}}{100 \text{ cm}})^2 = 2 \times 10^{-7} \text{ cm}$$

Volume / Area	Acre converted to square meters	Conversion of sq. m to sq. cm

The thickness is related to the size of the molecules in that the layer can be **no thinner** than the diameter of a molecule. Hence the greater the diameter of the molecule, the greater the thickness of the layer.

99. To determine the volume of Cu (in units of L), we need only use the density in g/cm³ and then express cm³ in units of L.

$$\frac{0.546 \text{ g Cu}}{1} \bullet \frac{1 \text{ cm}^3}{8.96 \text{ g Cu}} \bullet \frac{1 \text{ L}}{1000 \text{ cm}^3} = 6.09 \times 10^{-5} \text{ L}$$

101. To calculate the area of the Buddha, we need to determine the total volume of Au used. The density of Au (19.31 g/cm³) will be useful. The problem asks for area in m². One way to approach the problem is to express the density in g/m³ .

$$\frac{19.31 \text{ g Au}}{1 \text{ cm}^3} \bullet \frac{(100)^3 \text{cm}^3}{1 \text{ m}^3} = \frac{19.31 \times 10^6 \text{ g Au}}{1 \text{ m}^3}.$$ The second factor in the calculation is the conversion of length (in cm to m) expressed in three dimensions (recall that the SI system of volume is to consider volume (and its units) as length (and its units) cubed.

Using the density and the mass, we can calculate the volume of Au (in cubic meters).

$$\frac{279 \times 10^3 \text{ g Au}}{1} \bullet \frac{1 \text{ m}^3}{19.31 \times 10^6 \text{ g Au}} = 1.44 \times 10^{-2} \text{m}^3$$

Since V = area x thickness, the area may be determined:

$$1.44 \times 10^{-2} \text{ m}^3 = \text{area} \bullet (1.5 \times 10^{-3} \text{mm}) \bullet \frac{1 \text{ m}}{10^3 \text{ mm}} \text{ and}$$

area = (1.44 x 10⁻² m³)/(1.5 x 10⁻⁶ m) = 9.6 x 10³ m²

103. The room is 18 ft x 15 ft x 8.5 ft. Express the length (ft) in units of m:

$$\frac{1 \text{ ft}}{1} \bullet \frac{12 \text{ in}}{1 \text{ ft}} \bullet \frac{2.54 \text{ cm}}{1 \text{ in}} \bullet \frac{1 \text{ m}}{100 \text{ cm}} = 0.3048 \text{ m}.$$ Now we can express all lengths in units of m.

(a) Express the Volume in units of m^3 and L:

$$V = \left(\frac{18 \text{ ft}}{1} \bullet \frac{0.3048 \text{ m}}{1 \text{ ft}} \right) \bullet \left(\frac{15 \text{ ft}}{1} \bullet \frac{0.3048 \text{ m}}{1 \text{ ft}} \right) \bullet \left(\frac{8.5 \text{ ft}}{1} \bullet \frac{0.3048 \text{ m}}{1 \text{ ft}} \right) = 65 \text{ m}^3$$

The volume expressed in units of L is calculated by converting m^3 to L

The conversion factor may be calculated, but it is also at the table at the end of the text

$(1 \text{ L} = 1 \times 10^{-3} \text{ m}^3)$.

$V = (65 \text{ m}^3)(1 \text{ L}/1 \times 10^{-3} \text{ m}^3) = 6.5 \times 10^4 \text{ L}$

(b) Mass of air in the room in kg ; in lb, given that the density of air is 1.2 g/L

$$\frac{6.5 \times 10^4 \text{ L}}{1} \bullet \frac{1.2 \text{ g}}{1 \text{ L}} \bullet \frac{1 \text{ kg}}{1000 \text{ g}} = 78 \text{ kg}$$

$$\frac{6.5 \times 10^4 \text{ L}}{1} \bullet \frac{1.2 \text{ g}}{1 \text{ L}} \bullet \frac{1 \text{ lb}}{453.6 \text{ g}} = 170 \text{ lb } (2\text{sf })$$

105. (a) Calculate the density of and identify the unknown liquid.

	Mass
Beaker + liquid	16.08 g
Beaker (empty)	12.20 g
Liquid	3.88 g

Given that the volume of 3.50 mL had a mass of 3.88 g, the density of the liquid is:

$$D = \frac{3.88 \text{ g}}{3.50 \text{ mL}} = 1.11 \text{ g/mL (to 3 sf)}$$

Referring to the table of densities given, the liquid is most likely *ethylene glycol*.

(b) If the volume were known only to 2sf, would the results be sufficiently accurate to

identify the unknown?

If the volume were measured to only 2 sf, the resulting density for our calculation would

be 1.1 g/mL. For these 5 liquids, this accuracy would be sufficient, since Acetic acid and

water (two likely candidates) would have a density of ~1.0 g/mL (to 2 sf).

107. A 7.50×10^2 mL sample of gas has a mass of 0.9360 g,

(a) What is the density of the gas in units of g/L ?

$$\frac{0.9360 \text{ g}}{7.50 \times 10^2 \text{ mL}} \bullet \frac{1000 \text{ mL}}{1 \text{ L}} = 1.25 \text{ g/L}$$

(b) Of the gases shown in the table, three have densities that are equal to 1.25 g/L (expressed as 3 sf). One would not be able to exclusively identify the gas.

(c) With the more accurately determined volume, the density can be calculated as:

$$\frac{0.9360 \text{ g}}{7.496 \times 10^2 \text{ mL}} \cdot \frac{1000 \text{ mL}}{1 \text{ L}} = 1.249 \text{ g/L} . \text{ (4 sf)}$$

The more accurate density would certainly allow one to discard ethene (one of the 3), but neither CO nor N_2 could be eliminated.

109. Mass of Hg in the capillary:

Mass of capillary with Hg	3.416 g
Mass of capillary without Hg	3.263 g
Mass of Hg	0.153 g

To determine the volume of the capillary, calculate the volume of Hg that is filling it.

$$\frac{0.153 \text{ g Hg}}{1} \cdot \frac{1 \text{ cm}^3}{13.546 \text{ g Hg}} = 1.13 \times 10^{-2} \text{ cm}^3 (3 \text{ sf})$$

Now that we know the volume of the capillary, and the length of the tubing (given as 16.75 mm—or 1.675 cm), we can calculate the radius of the capillary using the equation:

Volume $= \pi r^2 l$.

1.13×10^{-2} cm^3 = (3.1416)r^2(1.675 cm), and solving for r^2

$$\frac{1.13 \times 10^{-2} \text{ cm}^3}{(3.1416)(1.675 \text{ cm})} = 2.15 \times 10^{-3} \text{ cm}^2 = r^2$$

So **r** is the square root of (2.15 x 10^{-3} cm^2) or 4.63 x 10^{-2} cm. The diameter would then be twice this value or 9.27 x 10^{-2} cm.

Chapter 2
Atoms and Elements

Practicing Skills

Atom:Their Composition and Structure

1.

Fundamental Particles	Protons	Electrons	Neutrons
Electrical Charges	+1	-1	0
Present in nucleus	Yes	No	Yes
Least Massive	1.007 u	**0.00055 u**	1.007 u

3. The discovery of radioactivity revealed the complex nature of that atom. While Dalton thought the atom to be indivisible—of one "body"—the phenomenon of radioactivty clearly pointed to "parts", which composed the "whole" of the atom.

5. The diameter of an atom if the nucleus of the atom were approximately 6 cm:
   Your text (Section 2.1—specifically Exercise 2.1) provides the relative sizes of the nuclear and atomic diameters, with the **nuclear radius** on the order of 0.001pm and the **atomic radius** approximately 100pm.

Nuclear diameter	Atomic diameter	Ratio
0.002 pm	200 pm	1:100,000
6 cm (orange)	?	1:100,000

   If the nuclear diameter is 6 cm, then the atomic diameter is 600,000 cm.
   Translating that number into larger units: 600,000 cm = 6,000 m or 6 km.

7. Element 86: Symbol: Rn ; Name: Radon

9. Mass number for:
   (a) Mg (at. no. 12) with 15 neutrons : 27
   (b) Ti (at. no. 22) with 26 neutrons : 48
   (c) Zn (at. no. 30) with 32 neutrons : 62
   The mass number represents the SUM of the protons + neutrons in the nucleus of an atom.
   The atomic number represents the # of protons, so (atomic no. + # neutrons)=mass number

11.   Mass number (A)   = no. of protons + no. of neutrons;

Atomic number (Z) = no. of protons

(a) $^{39}_{19}K$ (b) $^{84}_{36}Kr$ (c) $^{60}_{27}Co$

13.   substance   protons   neutrons   electrons

		protons	neutrons	electrons
(a)	magnesium-24	12	12	12
(b)	tin-119	50	69	50
(c)	thorium-232	90	142	90

Note that the number of protons and electrons are **equal** for any **neutral atom**. The number of protons is **always** equal to the atomic number. The mass number equals the sum of the numbers of protons and neutrons.

Isotopes

15. For technetium- 99 (at. no. 43) : # protons : 43

neutrons : (99 - 43) = 56

electrons : 43

17. Isotopes of cobalt (atomic number 27) with 30, 31, and 33 neutrons:

would have symbols of $^{57}_{27}Co$, $^{58}_{27}Co$, and $^{60}_{27}Co$ respectively.

Isotope Abundance and Atomic Mass

19. Thallium has two stable isotopes ^{203}Tl and ^{205}Tl. The more abundant isotope is:_____
The atomic weight of thallium is 204.4. The fact that this weight is closer to 205 than 203 indicates that the 205 isotope is the more abundant isotope. Recall that the atomic weight is the "weighted average" of all the isotopes of each element. Hence the more abundant isotope will have a "greater contribution" to the atomic weight than the less abundant one.

21. The atomic mass of lithium is:

$$(0.0750)(6.015121) + (0.9250)(7.016003) = 6.94 \, u$$

Recall that the atomic mass is a weighted average of all isotopes of an element,
and is obtained by **adding** the *product* of (relative abundance x mass) for all isotopes.

23. The two stable isotopes of silver are Ag-107 and Ag-109. The masses of the isotopes are, respectively: 106.9051 and 108.9047. The atomic weight of Ag on the periodic table is 107.868.

Since this weight is a weighted average, a 50:50 mixture of the two would have an atomic weight exactly mid-way between the two isotopic masses. Adding the masses of these two isotopes yields (108.9047 + 106.9051) = 215.8098. One-half this value is 107.9049. Given the proximity of this number to that of the published atomic weight indicates that the two stable isotopes of silver exist in a 50:50 mix.

25. The average atomic weight of gallium is 69.723 (from the periodic table). If we let **x** represent the abundance of the lighter isotope, and **(1-x)** the abundance of the heavier isotope, the expression to calculate the atomic weight of gallium may be written:

   (x)(68.9257) + (1 - **x**)(70.9249) = 69.723

[Note that the sum of all the isotopic abundances must add to 100% -- or 1 (in decimal notation).] Simplifying the equation gives:

$$68.9257 \text{ u } x \quad + \text{ } 70.9249 \text{ u } - 70.9249 \text{ u } x = \text{ } 69.723 \text{ u}$$
$$-1.9992 \text{ u } x = (69.723 \text{ u } - 70.9249)$$
$$-1.9992 \text{ u } x = -1.202 \text{ u}$$
$$x = 0.6012$$

So the relative abundance of isotope 69 is 60.12 % and that of isotope 71 is 39.88 %.

Atoms and the Mole

27. The mass, in grams of:

(a) 2.5 mol Al:

$$\frac{2.5 \text{ mol Al}}{1} \bullet \frac{27.0 \text{ g Al}}{1 \text{ mol Al}} = 67.5 \text{ g Al (68 g Al to 2sf)}$$

(b) 1.25×10^{-3} mol Fe:

$$\frac{1.25 \times 10^{-3} \text{mol Fe}}{1} \bullet \frac{55.85 \text{ g Fe}}{1 \text{ mol Fe}} = 0.0698 \text{ g Fe (3 sf)}$$

(c) 0.015 mol Ca:

$$\frac{0.015 \text{ mol Ca}}{1} \bullet \frac{40.1 \text{ g Ca}}{1 \text{ mol Ca}} = 0.60 \text{ g Ca (2 sf)}$$

(d) 653 mol Ne:

$$\frac{653 \text{ mol Ne}}{1} \bullet \frac{20.18 \text{ g Ne}}{1 \text{ mol Ne}} = 1.32 \times 10^4 \text{g Ne (3 sf)}$$

Note that, whenever possible, one should use a molar mass of the substance that contains **one more** significant figure than the data, to reduce round-off error.

29. The amount (moles) of substance represented by:

(a) 127.08 g Cu:

$$\frac{127.08 \text{ g Cu}}{1} \bullet \frac{1 \text{ mol Cu}}{63.546 \text{ g Cu}} = 1.9998 \text{ mol Cu(5 sf)}$$

(b) 0.012 g Li: $$\frac{0.012 \text{ g Li}}{1} \cdot \frac{1 \text{ mol Li}}{6.94 \text{ g Li}} = 1.7 \times 10^{-3} \text{ mol Li (2 sf)}$$

(c) 5.0 mg Am: $$\frac{5.0 \text{ mg Am}}{1} \cdot \frac{1 \text{ g Am}}{10^{3} \text{ mg Am}} \cdot \frac{1 \text{ mol Am}}{243 \text{ g Am}} = 2.1 \times 10^{-5} \text{ mol Am (2 sf)}$$

(d) 6.75 g Al $$\frac{6.75 \text{ g Al}}{1} \cdot \frac{1 \text{ mol Al}}{26.98 \text{ g Al}} = 0.250 \text{ mol Al}$$

31. 1-gram samples of He, Fe, Li, Si, C:

Which sample contains the **largest number** of atoms? ...the **smallest number** of atoms?

If we calculate the number of atoms of any one of these elements, say He, the process is:

$$\frac{1.0 \text{ g He}}{1} \cdot \frac{1 \text{ mol He}}{4.0026 \text{ g He}} \cdot \frac{6.0221 \times 10^{23} \text{ atoms He}}{1 \text{ mol He}} = 1.5 \times 10^{23} \text{ atoms He}$$

All the calculations proceed analogously, with the ONLY numerical difference attributable to the atomic mass of the element. Therefore the element with the **smallest** atomic weight(He) will have the **largest number** of atoms, while the element with the **largest** atomic weight(Fe) will have the **smallest number** of atoms.. This is a great question to answer by **thinking** rather than calculating.

33. The average mass of one copper atom:

One mole of copper (with a mass of 63.546 g) contains 6.0221×10^{23} atoms. So the average mass of **one** copper atom is:

$$\frac{63.546 \text{ g Cu}}{6.0221 \times 10^{23} \text{ atoms Cu}} = 1.0552 \times 10^{-22} \text{ g/Cu atom}$$

The Periodic Table

35. The elements in Group 5A are:

nitrogen (N - nonmetal),

phosphorus (P - nonmetal),

arsenic (As - metalloid),

antimony (Sb - metalloid),

bismuth (Bi - metal)

37. Periods with 8 elements: **2**; Periods 2 (at.no. 3-10) and 3 (at.no. 11-18)

    Periods with 18 elements: **2**; Periods 4 (at.no 19-36) and 5 (at.no. 37-54)

    Periods with 32 elements: **1**; Period 6 (at.no. 55-86)

39. Elements fitting the following descriptions:

	Description	Elements
(a)	Nonmetals	C, Cl
(b)	Main group elements	C, Ca, Cl, Cs
(c)	Lanthanides	Ce
(d)	Transition elements	Cr, Co, Cd, Cu, Ce, Cf, Cm
(e)	Actinides	Cf, Cm
(f)	Gases	Cl

41. Classify the elements as metals, metalloids, or nonmetals:

	Metals	Metalloids	Nonmetals
N			X
Na	X		
Ni	X		
Ne			X
Np	X		

43. Categorize elements as metals, main groups, transition metals

	Metals	Main groups	Transition metals
Sodium	X	X	
Silicon		X	
Sulfur		X	
Scandium	X		X
Selenium		X	
Strontium	X	X	
Silver	X		X
Samarium	X		X

General Questions

45.

Symbol	^{58}Ni	^{33}S	^{20}Ne	^{55}Mn
Number of protons	<u>28</u>	<u>16</u>	10	<u>25</u>
Number of neutrons	<u>30</u>	<u>17</u>	10	30
Number of electrons in the neutral atom	<u>28</u>	<u>16</u>	<u>10</u>	25
Name of element	<u>nickel</u>	<u>sulfur</u>	<u>neon</u>	<u>manganese</u>

47. Given that the average atomic mass for potassium is 39.0983 u, the lighter isotope, ^{39}K would be the more abundant of the **remaining** isotopes(ignoring K-40). Remember that atomic masses are **weighted averages**, that is the average atomic mass is closer to the mass of the most abundant isotope.

49. Regarding the elements:
   (a) the most abundant metal: Mg
   (b) the most abundant nonmetal: H
   (c) the most abundant metalloid: Si
   (d) most abundant transition element: Fe
   (e) halogens included (and most abundant): F, Cl, Br (Cl is most abundant)

51. Of the following, the one that is impossible:
   (a) Ag foil that is 1.2×10^{24} m thick. This is greater than the diameter of a silver atom and therefore possible.
   (b) A sample of K containing 1.784×10^{24} atoms. This is approximately 3 moles of potassium (113g) and therefore not impossible.
   (c) A gold coin of mass 1.23×10^{-3} kg (alternatively stated: 1.23 g)—not impossible
   (d) 3.43×10^{-27} mol S_8: This choice is impossible. If you multiply Avogadro's number (6.022×10^{23}) by the number of mol of sulfur given, the number of particles is: 0.00207 or LESS than one molecule of S_8.

53. Reviewing the periodic table:
   (a) An element in Group 2A : beryllium, magnesium, calcium, strontium, barium, radium
   (b) An element in the third period: sodium, magnesium, aluminum, silicon, phosphorus, sulfur, chlorine, argon
   (c) An element in the 2nd period in Group 4A: carbon
   (d) An element in the third period in Group 6A: sulfur

(e) A halogen in the fifth period: iodine

(f) An alkaline earth element in the third period: magnesium

(g) A noble gas element in the fourth period: krypton

(h) A nonmetal in Group 6A and the third period: sulfur

(i) A metalloid in the fourth period: germanium or arsenic

55. Using the furnished plot of density as a function of atomic number:

(a) Three elements in the series with the greatest density: At. no. 27 (Cobalt), At. no. 28 (Nickel), and At. no. 29 (Copper). The density of all three **metals** is approximately 9 g/cm^3.

(b) The element in the second period with the largest density is Boron (atomic number 5) while the element in the third period with the largest density is Aluminum (atomic number 13). Both of these elements belong to group 3A.

(c) Elements from the first 36 elements with very **low densities** are **gases**. These include Hydrogen, Helium, Nitrogen, Oxygen, Fluorine, Neon, Chlorine, Argon and Krypton.

57. Which of the following elements has the **largest number of atoms** in the mixture? .

$$\frac{52 \text{ g Ga}}{1} \cdot \frac{1 \text{ mol Ga}}{69.723 \text{ g Ga}} \cdot \frac{6.0221 \times 10^{23} \text{ atoms Ga}}{1 \text{ mol Ga}} = 4.5 \times 10^{23} \text{ atoms Ga}$$

$$\frac{9.5 \text{ g Al}}{1} \cdot \frac{1 \text{ mol Al}}{26.9815 \text{ g Al}} \cdot \frac{6.0221 \times 10^{23} \text{ atoms Al}}{1 \text{ mol Al}} = 2.1 \times 10^{23} \text{ atoms Al}$$

$$\frac{112 \text{ g As}}{1} \cdot \frac{1 \text{ mol As}}{74.9216 \text{ g As}} \cdot \frac{6.0221 \times 10^{23} \text{ atoms As}}{1 \text{ mol As}} = 9.00 \times 10^{23} \text{ atoms As}$$

The mixture has more As atoms than either Al or Ga.

59. Given 15 g of the atoms, Y, B, Cu, which has the largest number of atoms?

See SQ 31 for a similar problem. Like SQ-31, you do not necessarily have to perform the calculation. Since you are given an identical mass of each of the atoms, the element with the smallest atomic weight would represent the largest number of moles of atoms—B.

61. Number of moles of Kr in 0.00789 g Kr:

$$0.00789 \text{ g Kr} \cdot \frac{1 \text{ mol Kr}}{83.80 \text{ g Kr}} = 9.42 \times 10^{-5} \text{ mol Kr (3 sf)}$$

Number of atoms: 6.02217×10^{23} atoms Kr /mol Kr $\cdot$ 9.42×10^{-5} mol Kr= 5.67×10^{19}

63. Arrange the elements from least massive to most massive:

Calculate a common metric by which to compare the substances (say grams?)

(a) $\dfrac{3.79 \times 10^{24}\ \text{atoms Fe}}{1} \cdot \dfrac{1\ \text{mol Fe}}{6.0221 \times 10^{23}\ \text{atomFe}} \cdot \dfrac{55.845\ \text{g Fe}}{1\ \text{mol Fe}} = 351\ \text{g Fe}$

(b) $\dfrac{19.921\ \text{mol H}_2}{1} \cdot \dfrac{2.0158\ \text{g H}_2}{1\ \text{mol H}_2} = 40.157\ \text{g H}_2$

(c) $\dfrac{8.576\ \text{mol C}}{1} \cdot \dfrac{12.011\ \text{g C}}{1\ \text{mol C}} = 103.0\ \text{g C}$

(d) $\dfrac{7.4\ \text{mol Si}}{1} \cdot \dfrac{28.0855\ \text{g Si}}{1\ \text{mol Si}} = 210\ \text{g Si}$

(e) $\dfrac{9.221\ \text{mol Na}}{1} \cdot \dfrac{22.9898\ \text{g Na}}{1\ \text{mol Na}} = 212.0\ \text{g Na}$

(f) $\dfrac{4.07 \times 10^{24}\ \text{atoms Al}}{1} \cdot \dfrac{1\ \text{mol Al}}{6.0221 \times 10^{23}\ \text{atom Al}} \cdot \dfrac{26.9815\ \text{g Al}}{1\ \text{mol Al}} = 182\ \text{g Al}$

(g) $\dfrac{9.2\ \text{mol Cl}_2}{1} \cdot \dfrac{70.9054\ \text{g Cl}_2}{1\ \text{mol Cl}_2} = 650\ \text{g Cl}_2$

In ascending order of mass: H_2, C, Al, Si, Na, Fe, Cl_2

65. Data:

Oil Drop	Measured Charge on Drop(C)	Relative charge
1	1.59×10^{-19}	1
2	11.1×10^{-19}	7
3	9.54×10^{-19}	6
4	15.9×10^{-19}	10
5	6.36×10^{-19}	4

Millikan's thesis was that one couldn't be certain to have **only one** electron on an oil droplet, but that the **smallest** number of electrons (and hence the smallest charge) was **one**. Begin by dividing ALL of the data by the *smallest charge* (the relative charge).

(a) Hence, the smallest charge found was on **drop 1**, so one could conclude that the charge on the electron is 1.59×10^{-19} Coulombs.

(b) The number of electrons on each drop are found in the right-most column above. (1,7,6,10,4)

(c) To calculate the percent error, one would normally average the 5 data points. However since the oil droplets have varying numbers of electrons, we should first calculate the average charge for each droplet. Doing so reveals that the **average value** for the charge on the electron is **1.59×10^{-19} Coulombs,** and *to two significant values* the average deviation is zero.

The error (experimentally determined - accepted value) or
$(1.59 \times 10^{-19} - 1.60 \times 10^{-19})$ is -0.01×10^{-19}. The percent error is approximately 0.5%.

67. Suppose we had 68 mole of K atoms. That mass would be:

$\dfrac{68 \text{ mol K}}{1} \cdot \dfrac{39.0983\text{g K}}{1 \text{ mol K}} = 2{,}658 \text{ g K}$. Similarly, 32 mol of Na atoms would have a mass of:

$\dfrac{32 \text{ mol Na}}{1} \cdot \dfrac{22.9898\text{g Na}}{1 \text{ mol Na}} = 735 \text{ g Na}$

The total mass would then be approximately (2658+735) or 3393g. The weight percent of K would then be (2658/3393) or 0.783—or 78 (weight) % K.—to 2 significant figures.

Summary and Conceptual Questions

69. An atom of Helium-4, would have 2 protons and 2 electrons, and 2 neutrons. The nucleus would contain the protons and neutrons.

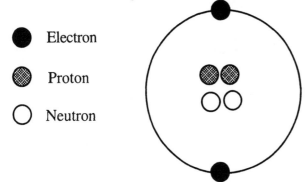

- ● Electron
- ▦ Proton
- ○ Neutron

71. Necessary information to calculate the number of atoms in one cm^3 of iron:

A sample calculation to arrive at an exact value is shown below:

$\dfrac{1.00 \text{ cm}^3}{1} \cdot \dfrac{7.86 \text{ g Fe}}{1 \text{ cm}^3} \cdot \dfrac{1 \text{ mol Fe}}{55.845 \text{ g Fe}} \cdot \dfrac{6.0221 \times 10^{23} \text{ atoms Fe}}{1 \text{ mol Fe}}$

Note that we needed: (b) atomic weight of Fe, (c) Avogadro's number, and (d) density.

73. Given the greater reactivity of Ca over Mg with water, one would anticipate that Ba would be even more reactive than Ca or Mg—with a more vigorous release of hydrogen gas. Reactivity of these metals increases down the group. Mg is in period 3, Ca in period 4, and Ba is in period 6. This trend is noted for Group IA as well.

75. Volume of Lithium to provide 256 moles of Li:

$\dfrac{256 \text{ mol Li}}{1} \cdot \dfrac{6.941 \text{ g Li}}{1 \text{ mol Li}} \cdot \dfrac{1 \text{ cm}^3 \text{ Li}}{0.534 \text{ g Li}} = 3{,}327 \text{cm}^3$ or 3330 cm^3 to 3 sf

The length of the edge:

Since the volume of a cube is (length)^3, the length is the cube root of the volume that we calculated: $(3330 \text{ cm}^3)^{1/3}$ or 14.9 cm.

77. Number of atoms in a cylinder of Na: 12.00 cm in length; Diameter = 4.5 cm; D= 0.971 g/cm3

Volume of cylinder $= \pi r^2$ x length = 190.85 cm^3 or 190 cm^3 to 2 sf

[HINT: Don't forget to convert the diameter of 4.5 cm to a radius of 2.25 cm]

Knowing the volume of the cylinder and the density of sodium, we can calculate the number of atoms of Na:

$$\frac{190.85 \text{ cm}^3 \text{ Na}}{1} \bullet \frac{0.971 \text{ g Na}}{1 \text{cm}^3 \text{ Na}} \bullet \frac{1 \text{ mol Na}}{22.9898 \text{ g Na}} \bullet \frac{6.0221 \times 10^{23} \text{ atoms Na}}{1 \text{ mol Na}} = 4.9 \times 10^{24} \text{ atoms Na}$$

79. Number of atoms of C in 2.0000 g sample of C.

$$2.0000 \text{ g C} \bullet \frac{1 \text{ mol C}}{12.011 \text{ g C}} \bullet \frac{6.0221 \times 10^{23} \text{ atoms C}}{1 \text{ mol C}} = 1.0028 \times 10^{23} \text{ atoms C}$$

If the accuracy of the balance is +/- 0.0001g then the mass could be 2.0001g C, so we would do a similar calculation to obtain 1.00281 x 10^{23} atoms C—a difference of 5 x 10^{18} atoms C.

81. Jar contains a fixed number of jellybeans. Estimate the number of jellybeans without counting each one. This is very similar to the problem which R.A. Millikan faced when determining the charge on one electron. One solution: Weigh the jar both empty and full of jellybeans. The difference is the mass of the jellybeans. Then weigh a fixed number of jellybeans (say 10). Calculate the mass of **one** of the jellybeans (by dividing by 10, if that's the number of jellybeans you weighed.) Dividing the **total** mass of the jellybeans by the mass of **one** jellybean will reveal the **number of jellybeans in the jar**.

Chapter 3
Molecules and Compounds

Practicing Skills

Molecular Formulas and Models

1. The formula for sulfuric acid is H_2SO_4. The molecule is **not flat**. The O atoms are arranged around the sulfur at the corners of a tetrahedron—that is the O-S-O angle would be about 109 degrees. The hydrogen atoms are connected to two of the oxygen atoms also with angles (H-O-S) of approximately 109 degrees.

3. The molecular formula for cis-Platin is : $Pt(NH_3)_2Cl_2$

    The structural formula is:

    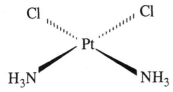

Ions and Ion Charges

5. Most commonly observed ion for:

    (a) Magnesium: 2+ —like all the alkaline earth metals

    (b) Zinc : 2+

    (c) Nickel: 2+

    (d) Gallium: 3+ (an analog of Aluminum)

7. The symbol and charge for the following ions:

    (a) barium ion Ba^{2+}

    (b) titanium(IV) ion Ti^{4+}

    (c) phosphate ion PO_4^{3-}

    (d) hydrogen carbonate ion HCO_3^-

    (e) sulfide ion S^{2-}

    (f) perchlorate ion ClO_4^-

    (g) cobalt(II) ion Co^{2+}

    (h) sulfate ion SO_4^{2-}

9. When potassium becomes a monatomic ion, potassium—like all alkali metals—**loses 1 electron.** The noble gas atom with the same number of electrons as the potassium ion is **argon.**

Ionic Compounds

11. Barium is in Group 2A, and is expected to form a 2+ ion while bromine is in group 7A and expected to form a 1- ion. Since the compound would have to have an **equal amount** of negative and positive charges, the formula would be $BaBr_2$.

13. Formula, Charge, and Number of ions in:

	cation	# of	anion	# of
(a) K_2S	K^+	2	S^{2-}	1
(b) $CoSO_4$	Co^{2+}	1	SO_4^{2-}	1
(c) $KMnO_4$	K^+	1	MnO_4^-	1
(d) $(NH_4)_3PO_4$	NH_4^+	3	PO_4^{3-}	1
(e) $Ca(ClO)_2$	Ca^{2+}	1	ClO^-	2

15. Cobalt oxide

Cobalt(II) oxide CoO cobalt ion : Co^{2+}

Cobalt(III) oxide Co_2O_3 Co^{3+}

17. Provide correct formulas for compounds:

(a) $AlCl_3$ The tripositive aluminum ion requires three chloride ions.

(b) KF Potassium is a monopositive cation. Fluoride is a mononegative anion.

(c) Ga_2O_3 is correct; Ga is a 3+ ion and O forms a 2- ion

(d) MgS is correct; Mg forms a 2+ ion and S forms a 2- ion

Naming Ionic Compounds

19. Names for the ionic compounds

(a) K_2S potassium sulfide

(b) $CoSO_4$ cobalt(II) sulfate

(c) $(NH_4)_3PO_4$ ammonium phosphate

(d) $Ca(ClO)_2$ calcium hypochlorite

21. Formulas for the ionic compounds

(a) ammonium carbonate $(NH_4)_2CO_3$

(b) calcium iodide CaI_2

(c) copper(II) bromide $CuBr_2$

(d) aluminum phosphate $AlPO_4$

(e) silver(I) acetate $AgCH_3CO_2$

23. Names and formulas for ionic compounds:

cation	anion CO_3^{2-}	anion I^-
Na^+	Na_2CO_3 sodium carbonate	NaI sodium iodide
Ba^{2+}	$BaCO_3$ barium carbonate	BaI_2 barium iodide

Coulomb's Law

25. The fluoride ion has a smaller radius than the iodide ion. Hence the distance between the sodium and fluoride ions will be less than the comparable distance between sodium and iodide. Coulomb's Law indicates that the attractive force becomes greater as the distance between the charges grows smaller—hence NaF will have stronger forces of attraction.

Naming Binary, Nonmetal Compounds

27. Names of binary nonionic compounds
    (a) NF_3 nitrogen trifluoride

    (b) HI   hydrogen iodide
    (c) BI_3 boron triiodide
    (d) PF_5 phosphorus pentafluoride

29. Formulas for:
    (a) sulfur dichloride SCl_2
    (b) dinitrogen pentaoxide N_2O_5
    (c) silicon tetrachloride $SiCl_4$
    (d) diboron trioxide B_2O_3

Molecules, Compounds, and the Mole

31. Molar mass of the following: (with atomic weights expressed to 4 significant figures)
    (a) Fe_2O_3 $(2)(55.85) + (3)(16.00) = 159.7$
    (b) BCl_3 $(1)(10.81) + (3)(35.45) = 117.2$
    (c) $C_6H_8O_6$ $(6)(12.01) + (8)(1.008) + (6)(16.00) = 176.1$

33. Molar mass of the following: (with atomic weights expressed to 4 significant figures)
    (a) $Ni(NO_3)_2 \cdot 6H_2O$ $(1)(58.69) + (2)(14.01) + 6(16.00) + (12)(1.008) + (6)(16.00)$
    $$= 290.8$$

    (b) $CuSO_4 \cdot 5H_2O$ $(1)(63.55) + (1)(32.07) + 4(16.00) + (10)(1.008) + (5)(16.00)$
    $$= 249.7$$

35. Moles represented by 1.00 g of the compounds:

Molar masses are calculated as before. To determine the # of moles in 1.00 g of the compound, we recall that 1 mol of a substance has a mass equal to the Molar mass expressed in units of grams. Hence we calculate for (a) the # of mol in 1.00 g:

$$\frac{1.00 \text{ g C}_3\text{H}_7\text{OH}}{1} \bullet \frac{1 \text{ mol C}_3\text{H}_7\text{OH}}{60.10 \text{ g C}_3\text{H}_7\text{OH}} = 0.0166 \text{ mol C}_3\text{H}_7\text{OH}$$

Compound	Molar mass	Moles in 1.00 g
(a) C_3H_7OH	60.10	0.0166
(b) $C_{11}H_{16}O_2$	180.2	0.00555
(c) $C_9H_8O_4$	180.2	0.00555

37. Moles of acetonitrile in 2.50 kg:
   1. Molar mass of CH_3CN:

   $$(2)(12.01) + (3)(1.008) + (1)(14.01) = 41.05 \text{ g/mol}$$

   2. Moles:

   $$2.50 \times 10^3 \text{g} \bullet \frac{1 \text{ mol CH}_3\text{CN}}{41.05 \text{ g}} = 60.9 \text{ mol CH}_3\text{CN}$$

39. Regarding sulfur trioxide:

   1. Amount of SO_3 in 1.00 kg: $\dfrac{1.00 \times 10^3 \text{g SO}_3}{1} \bullet \dfrac{1 \text{ mol SO}_3}{80.07 \text{ g SO}_3} = 12.5 \text{ mol SO}_3$

   2. Number of SO_3 molecules: $\dfrac{12.5 \text{ mol SO}_3}{1} \bullet \dfrac{6.022 \times 10^{23} \text{ molecules}}{1 \text{ mol SO}_3} = 7.52 \times 10^{24}$

   3. Number of S atoms: With 1 S atom per SO_3 molecule-- 7.52×10^{24} S atoms

   4. Number of O atoms: With 3 O atoms per SO_3 molecule—$3 \times 7.52 \times 10^{24}$ O atoms
   $$\text{or } 2.26 \times 10^{25} \text{ O atoms}$$

Percent Composition

41. Mass percent for: [4 significant figures]
   (a) PbS: $(1)(207.2) + (1)(32.06) = 239.3$ g/mol
   $$\%Pb = \frac{207.2 \text{ g Pb}}{239.3 \text{ g PbS}} \times 100 = 86.60 \%$$
   $$\%S = 100.00 - 86.60 = 13.40 \%$$

(b) C_3H_8: $(3)(12.01)+(8)(1.008) = 44.09$ g/mol

$$\%C = \frac{36.03 \text{ g C}}{44.09 \text{ g C}_3\text{H}_8} \times 100 = 81.71 \%$$

$\%H = 100.00 - 81.71 = 18.29 \%$

(c) $C_{10}H_{14}O$: $(10)(12.01) + (14)(1.008) + (1)(16.00) = 150.21$ g/mol

$$\%C = \frac{120.1 \text{ g C}}{150.21 \text{ g C}_{10}\text{H}_{14}\text{O}} \times 100 = 79.96 \%$$

$$\%H = \frac{14.112 \text{ g H}}{150.21 \text{ g C}_{10}\text{H}_{14}\text{O}} \times 100 = 9.394 \%$$

$\%O = 100.00 - (79.96 + 9.394) = 10.65 \%$

43. Mass of lead present in 10.0 g of PbS:

From SQ 41(a), the % of Pb in PbS is 86.60%, so in 10.0g of PbS there would be:

10.0 g PbS x 86.60% = 8.66 g of Pb

45. Mass of CuS to provide 10.0 g of Cu:

To calculate the weight percent of Cu in CuS, we need the respective atomic weights:

Cu = 63.546   S = 32.066   adding CuS = 95.612

The % of Cu in CuS is then: $\dfrac{63.546 \text{ g Cu}}{95.612 \text{ g CuS}} \times 100 = 66.46 \%$ Cu

Now we can calculate the mass of CuS that will provide 10.0 g of Cu:

$$\frac{10.0 \text{ g Cu}}{1} \cdot \frac{95.612 \text{ g CuS}}{63.546 \text{ g Cu}} = 15.0 \text{ g CuS}$$

Empirical and Molecular Formulas

47. The empirical formula ($C_2H_3O_2$) would have a mass of 59.04 g.

Since the molar mass is 118.1 g/mol we can write

$$\frac{1 \text{ empirical formula}}{59.04 \text{ g succinic acid}} \cdot \frac{118.1 \text{ g succinic acid}}{1 \text{ mol succinic acid}} = \frac{2.0 \text{ empirical formulas}}{1 \text{ mol succinic acid}}$$

So the molecular formula contains 2 empirical formulas (2 x $C_2H_3O_2$) or $C_4H_6O_4$.

49. Provide the empirical or molecular formula for the following, as requested:

	Empirical Formula	Molar Mass (g/mol)	Molecular Formula
(a)	CH	26.0	C_2H_2
(b)	CHO	116.1	$C_4H_4O_4$
(c)	CH_2	112.2	C_8H_{16}

Note that we can calculate the mass of an empirical formula by adding the respective atomic weights (13 for CH, for example). The molar mass (26.0 for part (a)) is obviously twice that for an empirical formula, so the molecular formula would be 2 x empirical formula (or C_2H_2 in part (a)).

51. Calculate the empirical formula of acetylene by calculating the atomic ratios of carbon and hydrogen in 100 g of the compound.

$$92.26 \text{ g C} \cdot \frac{1 \text{mol C}}{12.011 \text{g C}} = 7.681 \text{ mol C}$$

$$7.74 \text{ g H} \cdot \frac{1 \text{mol H}}{1.008 \text{g H}} = 7.678 \text{ mol H}$$

Calculate the atomic ratio: $\dfrac{7.68 \text{mol C}}{7.68 \text{mol H}} = \dfrac{1 \text{mol C}}{1 \text{mol H}}$

The atomic ratio indicates that there is 1 C atom for 1 H atom (1:1). The **empirical formula is then CH**. The formula mass is 13.01. Given that the molar mass of the compound is 26.02 g/mol, there are two formula units per molecular unit, hence the **molecular formula for acetylene is C_2H_2** .

53. Determine the empirical and molecular formulas of cumene:

The percentage composition of cumene is 89.94% C and (100.00-89.94) or 10.06%H.
We can calculate the ratio of mol C: mol H as done in SQ64.

$$89.94 \text{ g C} \cdot \frac{1 \text{mol C}}{12.011 \text{g C}} = 7.489 \text{ mol C}$$

$$10.06 \text{ g H} \cdot \frac{1 \text{mol H}}{1.008 \text{g H}} = 9.981 \text{ mol H}$$

Calculating the atomic ratio:
$$\frac{9.981 \text{ mol H}}{7.489 \text{ mol C}} = \frac{1.33 \text{ mol H}}{1.00 \text{ mol C}} \quad \text{or a ratio of 3C : 4H}$$
So the empirical formula for cumene is C_3H_4 , with a formula mass of 40.06

If the molar mass of 120.2 g/mol, then dividing the "empirical formula mass" into the molar mass gives: 120.2/40.06 or 3 empirical formulas **per** molar mass. The **molecular formula** is then 3 x C_3H_4 or C_9H_{12}.

55. Empirical and Molecular formula for Mandelic Acid:

$$63.15 \text{ g C} \cdot \frac{1 \text{ mol C}}{12.0115 \text{ g C}} = 5.258 \text{ mol C}$$

$$5.30 \text{ g H} \cdot \frac{1 \text{ mol H}}{1.0079 \text{ g H}} = 5.28 \text{ mol H}$$

$$31.55 \text{ g O} \cdot \frac{1 \text{ mol O}}{15.9994 \text{ g O}} = 1.972 \text{ mol O}$$

Using the smallest number of atoms, we calculate the ratio of atoms:

$$\frac{5.258 \text{ mol C}}{1.972 \text{ mol O}} = \frac{2.666 \text{ mol C}}{1 \text{ mol O}} \quad \text{or} \quad \frac{22/3 \text{ mol C}}{1 \text{mol O}} \quad \text{or} \quad \frac{8/3 \text{ mol C}}{1 \text{ mol O}}$$

So 3 mol O combine with 8 mol C and 8 mol H so the empirical formula is $C_8H_8O_3$.
The formula mass of $C_8H_8O_3$ is 152.15. Given the data that the molar mass is 152.15 g/mL, the molecular formula for mandelic acid is $C_8H_8O_3$.

Determining Formulas from Mass Data

57. Molecules of water per formula unit of $MgSO_4$:

From 1.687 g of the hydrate, only 0.824 g of the magnesium sulfate remain.

The mass of water contained in the solid is: (1.687-0.824) or 0.863 grams

Use the molar masses of the solid and water to calculate the number of moles of each substance present:

$$0.824 \text{g} \cdot \frac{1 \text{ mol MgSO}_4}{120.36 \text{ g MgSO}_4} = 6.85 \times 10^{-3} \text{ mol of magnesium sulfate}$$

$$0.863 \text{ g H}_2\text{O} \cdot \frac{1 \text{ mol H}_2\text{O}}{18.02 \text{ g H}_2\text{O}} = 4.79 \times 10^{-2} \text{ mol of water; the ratio of water to MgSO}_4 \text{ is:}$$

$$\frac{4.79 \times 10^{-2} \text{ mol water}}{6.85 \times 10^{-3} \text{ mol magnesium sulfate}} = 6.99$$

So we write the formula as $MgSO_4 \cdot 7 \text{ H}_2\text{O}$.

59. Given the masses of xenon involved, we can calculate the number of moles of the element:

$$0.526 \text{ g S} \cdot \frac{1 \text{mol Xe}}{131.29 \text{g Xe}} = 0.00401 \text{ mol Xe}$$

The mass of fluorine present is: 0.678 g compound - 0.526 g Xe = 0.152 g F

$$0.152 \text{ g F} \cdot \frac{1 \text{mol F}}{19.00 \text{g F}} = 0.00800 \text{ mol F}$$

Calculating atomic ratios:

$$\frac{0.00800 \text{ mol F}}{0.00401 \text{ mol Xe}} = \frac{2 \text{ mol F}}{1 \text{ mol Xe}} \text{ indicating that the empirical formula is } XeF_2$$

61. Formula of compound formed between zinc and iodine:

Calculate the amount of zinc and iodine present:

$$\frac{2.50 \text{ g Zn}}{1} \cdot \frac{1 \text{ mol Zn}}{65.39 \text{ g Zn}} = 3.82 \times 10^{-02} \text{ mol Zn and}$$

$$\frac{9.70 \text{ g I}_2}{1} \cdot \frac{1 \text{ mol I}_2}{253.8 \text{ g I}_2} = 3.82 \times 10^{-02} \text{ mol I}_2 \text{ (recall that the iodine is a diatomic specie, and}$$

would be the form of iodine reacting). Note that the amount of Zinc and I_2 combined are identical, making the formula for the compound ZnI_2. An **alternative** way of solving the problem would be to use the **atomic mass** of iodine (126.9g/ mol I) to represent 7.64×10^{-02} mol of I. The ratio of Zn:I would then be 1:2 with the formula being the same as that determined above.

General Questions on Molecules and Compounds

63. Possible compounds from ions:

	CO_3^{2-}	SO_4^{2-}
NH_4^+	$(NH_4)_2CO_3$	$(NH_4)_2SO_4$
Ni^{2+}	$NiCO_3$	$NiSO_4$

Compounds are electrically neutral—hence the total positive charge contributed by the cation (+ion) has to be equal to the total negative charge contributed by the anion (- ion). Since both carbonate and sulfate are di-negative anions, two ammonium ions are required, while only one nickel(II) ion is needed.

65. # of electrons in a Sr atom?   38 (same as the number of protons)

Sr atoms **lose** electrons when forming ions—a characteristic of metals!

Sr atoms have 2 electrons in the outer shell (in Group or Family 2), so Sr atoms lose 2 electrons when forming an ion—making the Sr^{2+} ion isoelectronic with the Kr atom.

67. Compound from the list with the highest weight percent of Cl: One way to answer this question is to calculate the %Cl in each of the five compounds. An observation that each compound has the same number of Cl atoms provides a "non-calculator" approach to answering the question. Since 3 Cl atoms will contribute the same TOTAL mass of Cl to the formula weights, the compound with the highest weight percent of Cl will also have the **lowest** weight percent of the other atom. Examining the atomic weights of the "other" atoms:

B	As	Ga	Al	P
10.81	74.92	69.72	26.98	30.97

B contributes the smallest mass of these five atoms, hence the smallest contribution to the atomic weights of the five compounds—hence BCl_3 has the highest weight percent of Cl.

69. Following the logic of SQ67, one can quickly answer this question—without the aid of a calculator.

N	C	Mg	Ca
14.01	12.01	24.31	40.08

Given the fact that each of the four compounds cited has the same number of O atoms, the compound with the highest weight percent of O will have the lowest weight percent of the "non-oxygen" atom. So CO will have the highest weight percent of oxygen.

71. Na_3BO_3 is the formula for sodium borate. The formula for the borate ion is: BO_3

Requiring 3 sodium ions (each with a +1 charge) tells us that the borate ion has a –3 charge—making borate an **anion**.

73. To determine the greater mass, let's first ask the question about the molar mass of Adenine. The formula for adenine is: $C_5H_5N_5$ with a molar mass of 135.13 g. The number of molecules requested is exactly 1/2 mole of adenine molecules. So 1/2 mol of adenine molecules would have a mass of 1/2(135.13g) or 67.57 g. So 1/2 mol of adenine has a greater mass than 40.0 g of adenine.

75. A drop of water has a volume of 0.05 mL. Assuming the density of water is 1.00 g/cm^3, the number of molecules of water may be calculated by first determining the mass of water present.

$$\frac{0.05 \text{ mL}}{1} \bullet \frac{1 \text{ cm}^3}{1 \text{ mL}} \bullet \frac{1.00 \text{ g}}{1 \text{ cm}^3} = 0.05 \text{ g water}$$

The molar mass of water is 18.02 g. The number of moles of water is then:

$$\frac{0.05 \text{ g water}}{1} \bullet \frac{1 \text{ mol water}}{18.02 \text{ g water}} = 2.77 \times 10^{-3} \text{ mol.}$$

The number of molecules is then:

2.77×10^{-3} mol $\times$ 6.02×10^{23} molecules/mol = 1.67×10^{21} molecules

(or 2×10^{21} molecules—to 1 sf.

77. Molar mass and mass percent of the elements in $Cu(NH_3)_4SO_4 \bullet H_2O$:

Molar Mass: (1)(Cu) + (4)(N) + 12(H) + (1)(S) + (4)(O) + (2)(H) + (1)(O).

Combining the hydrogens and oxygen from water with the compound:

(1)(Cu) +   (4)(N) +   14(H) +   (1)(S) +   (5)(O) =

(1)(63.546) +   (4)(14.0067) + 14(1.0079) + (1)(32.066) + (5)(15.9994) = 245.72 g/mol

The mass percents are:

$$Cu: (63.546/245.75) \times 100 = \quad 25.86\% \text{ Cu}$$
$$N: (56.027/245.75) \times 100 = \quad 22.80\% \text{ N}$$
$$H: (14.111/245.75) \times 100 = \quad 5.742\% \text{ H}$$
$$S: (32.066/245.75) \times 100 = \quad 13.05\% \text{ S}$$
$$O: (79.997/245.75) \times 100 = \quad 32.55\% \text{ O}$$

The mass of Copper and of Water in 10.5 g of the compound:

For Copper: $\dfrac{10.5 \text{ g compound}}{1} \bullet \dfrac{25.86 \text{ g Cu}}{100.00 \text{ g compound}} = 2.72 \text{ g Cu}$

For Water: $\dfrac{10.5 \text{ g compound}}{1} \bullet \dfrac{18.02 \text{ g H}_2\text{O}}{245.72 \text{ g compound}} = 0.770 \text{ g H}_2\text{O}$

79. The empirical formula of malic acid, if the ratio is: $C_1H_{1.50}O_{1.25}$

Since we prefer all subscripts to be integers, we ask what "multiplier" we can use to convert each of these subscripts to integers **while** retaining the ratio of C:H:O that we're given. Multiplying each subscript by 4 (we need to convert the 0.25 to an integer) gives a ratio of $C_4H_6O_5$.

81. The empirical formula for iron oxalate:

0.109 g of the compound contain 38.82% iron or (0.3882 x 0.109 = 0.0423) g iron

The law of mass action tells us that the rest is oxalate:

(0.109 g compound - 0.0423 g iron) = 0.0667 g oxalate ($C_2O_4{}^{2-}$)

Convert these masses into moles of iron and oxalate:

$$\frac{0.0423 \text{ g Fe}}{55.847 \text{ g/mol}} = 7.57 \times 10^{-4} \text{ moles of iron}$$

$$\frac{0.0667 \text{ g oxalate}}{88.02 \text{ g/mol}} = 7.57 \times 10^{-4} \text{ moles of } C_2O_4{}^{2-}$$

The ratio of oxalate to iron is: 1:1 and the empirical formula for the compound is FeC_2O_4

83. For ephedrine:

(a) The molecular formula and molar mass:

$C_{10}H_{15}NO$ with a corresponding molar mass of: 165.23 g

The structural formula of ephedrine:

Converting the picture shown in your text to a structural formula
gives:

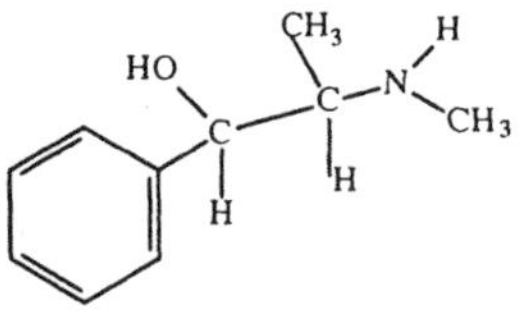

(b) The weight percent of Carbon is: $[(10 \times 12.011)/165.23] \times 100$ or
72.69% C

(c) The amount of ephedrine in 0.125 g sample:

$$\frac{0.125 \text{ g ephedrine}}{1} \bullet \frac{1 \text{ mol ephedrine}}{165.2 \text{ g ephedrine}} = 7.57 \times 10^{-4} \text{ mol}$$

(d) Molecules of ephedrine in 0.125 g sample:

$7.57 \times 10^{-4} \text{ mol} \bullet 6.022 \times 10^{23} \text{ molecules/mol} = 4.56 \times 10^{20} \text{ molecules}$

How many Carbon atoms?

Given that there are 10C atoms per molecule of ephedrine, the number of C atoms is then 10
times the number of ephedrine molecules or 4.56×10^{21} C atoms

85. Ionic compounds; formulas and names
   (c) Li_2S lithium sulfide
   (d) In_2O_3 indium oxide
   (g) CaF_2 calcium fluoride

Pair a & b & f: consist of two non-metals — a covalent compound is anticipated

Pair e : Argon doesn't typically form ionic compounds

87. Formulas for compounds; identify the ionic compounds

    (a) sodium hypochlorite   NaClO   ionic

    (b) boron triiodide BI_3

    (c) aluminum perchlorate $Al(ClO_4)_3$ ionic

    (d) calcium acetate $Ca(CH_3CO_2)_2$ ionic

    (e) potassium permanganate $KMnO_4$ ionic

    (f) ammonium sulfite $(NH_4)_2SO_3$ ionic

    (g) potassium dihydrogen phosphate KH_2PO_4 ionic

    (h) disulfur dichloride S_2Cl_2

    (i) chlorine trifluoride ClF_3

    (j) phosphorus trifluoride PF_3

89. Symbols, Formulas, and names of selected substances:

Supplied terms are emboldened.

Cation	Anion	Name	Formula
Li^+	ClO_4^-	lithium perchlorate	**$LiClO_4$**
Al^{3+}	PO_4^{3-}	**aluminum phosphate**	$AlPO_4$
Li^+	**Br^-**	**lithium bromide**	LiBr
Ba^{2+}	NO_3^-	barium nitrate	**$Ba(NO_3)_2$**
Al^{3+}	O^{2-}	**aluminum oxide**	Al_2O_3
Fe^{3+}	CO_3^{2-}	**iron(III) carbonate**	$Fe_2(CO_3)_3$

Names of ionic substances are composed by naming the cation (+ ion) and then the anion (- ion). Note in the case of the iron cation, the roman numeral III is needed to differentiate it from the other iron ion (2+) or II. Formulas for ionic compounds are written using sufficient numbers of cations to have a charge that is *equal in magnitude but opposite in sign* to the charge provided by the anion. In the case of aluminum oxide, for example, 2 Al ions contributes a (2 x 3+) 6+ charge. The oxide ion bears a charge of 2-, hence 3 such ions provide a total 6- charge, establishing electrical neutrality between the ion pairs.

91. Empirical and Molecular formula of Azulene:

Given the information that azulene is a hydrocarbon, if it is 93.71 % C, it is also (100.00 - 93.71) or 6.29 % H.

In a 100.00 g sample of azulene there are

$$93.71 \text{ g C} \cdot \frac{1 \text{ mol C}}{12.0115 \text{ g C}} = 7.802 \text{ mol C and}$$

$$6.29 \text{ g H} \cdot \frac{1 \text{ mol H}}{1.0079 \text{ g H}} = 6.241 \text{ mol H}$$

The ratio of C to H atoms is: 1.25 mol C : 1 mol H   or a ratio of 5 mol C: 4 mol H (C_5H_4).

The mass of such an empirical formula is ≈ 64. Given that the molar mass is ~128 g/mol, the molecular formula for azulene is $C_{10}H_8$.

93. Molecular formula of cadaverine:

Calculate the amount of each element in the compound (assuming that you have 100 g)

$$58.77 \text{ g C} \cdot \frac{1 \text{ mol C}}{12.0115 \text{ g C}} = 4.893 \text{ mol C}$$

$$13.81 \text{ g H} \cdot \frac{1 \text{ mol H}}{1.0079 \text{ g H}} = 13.70 \text{ mol H}$$

$$27.40 \text{ g N} \cdot \frac{1 \text{ mol N}}{14.0067 \text{ g N}} = 1.956 \text{ mol N}$$

The ratio of C:H:N can be found by dividing each by the smallest amount (1.956): to give $C_{2.50}H_7N_1$ and converting each subscript to an integer (multiplying by 2) $C_5H_{14}N_2$. The weight of this "empirical formula" would be approximately 102, hence the molecular formula is also $C_5H_{14}N_2$.

95. The empirical formula for MMT:

Moles of each atom present in 100. g of MMT:

$$49.5 \text{ g C} \cdot \frac{1 \text{ mol C}}{12.0115 \text{ g C}} = 4.13 \text{ mol C}$$

$$3.2 \text{ g H} \cdot \frac{1 \text{ mol H}}{1.0079 \text{ g H}} = 3.2 \text{ mol H}$$

$$22.0 \text{ g O} \cdot \frac{1 \text{ mol O}}{15.9994 \text{ g O}} = 1.38 \text{ mol O}$$

$$25.2 \text{ g Mn} \cdot \frac{1 \text{ mol Mn}}{54.938 \text{ g Mn}} = 0.459 \text{ mol Mn}$$

The ratio of C:H:O:Mn can be found by dividing each by the smallest amount (0.459): to give $MnC_9H_7O_3$.

97. Chromium oxide has the formula Cr_2O_3.

The weight percent of Cr in Cr_2O_3 is: [(2 x 52.00)/((2 x 52.00) + (3 x 16.00))] x 100 or

(104.00/152.00) x 100 or 68.42% Cr.

[The numerator is the sum of the mass of 2 atoms of Cr, while the denominator is the sum of the mass of 2 atoms of Cr and 3 atoms of O.]

The weight of chromium oxide necessary to produce 850 kg Cr:

$$\frac{850 kg Cr}{1} \cdot \frac{100 \text{ kg } Cr_2O_3}{68.42 \text{ kg Cr}} = 1,200 \text{ kg } Cr_2O_3 \text{ to 2sf}$$

The second fraction represents the %Cr in the oxide. Dividing the desired mass of Cr by the percent Cr (or multiplying by the reciprocal of that percentage) gives the mass of oxide needed.

<u>99.</u> I_2 + Cl_2 $\rightarrow$ I_xCl_y

0.678 g   (1.246 - 0.678)   1.246 g

Calculate the ratio of I : Cl atoms

0.678 g I $\cdot \dfrac{1 \text{ mol I}}{126.9 \text{ g I}}$ = 5.34 x 10^{-3} mol I atoms

0.568 g Cl $\cdot \dfrac{1 \text{ mol Cl}}{35.45 \text{ g Cl}}$ = 1.6 x 10^{-2} mol Cl atoms

The ratio of Cl : I is : $\dfrac{1.6 \times 10^{-2} \text{mol Cl atoms}}{5.34 \times 10^{-3} \text{mol I atoms}} = 3.00 \dfrac{\text{Cl atoms}}{\text{I atoms}}$

The empirical formula is ICl_3 (FW = 233.3)

Given that the molar mass of I_xCl_y was 467 g/mol, we can calculate the number of empirical formulas per mole:

$$\frac{467 g/mol}{233.3 g/empirical formula} = 2 \frac{\text{empirical formula}}{\text{mol}}$$

for a molecular formula of I_2Cl_6.

<u>101.</u> Mass of Fe in 15.8 kg of FeS_2 :

% Fe in FeS_2 = $\dfrac{55.85 g Fe}{119.97 g FeS_2}$ x 100 = 46.55 % Fe

and in 15.8 kg FeS_2

15.8 kg FeS_2 $\cdot \dfrac{46.55 \text{ kg Fe}}{100.00 \text{ kg } FeS_2}$ = 7.35 kg Fe

103. The formula of barium molybdate is $BaMoO_4$. What is the formula for sodium molybdate?

This question is easily answered by observing that the compound indicates ONE barium ion. Since the barium ion has a 2+ charge, 2 Na+ cations would be needed, making the formula for sodium molybdate Na_2MoO_4 or choice (d).

105. Mass of Bi in two tablets of Pepto-Bismol™ ($C_{21}H_{15}Bi_3O_{12}$):

Moles of the active ingredient:

$$\frac{2 \text{ tablets}}{1} \bullet \frac{300. \times 10^{-3} \text{g } C_{21}H_{15}Bi_3O_{12}}{1 \text{ tablet}} \bullet \frac{1 \text{ mol } C_{21}H_{15}Bi_3O_{12}}{1086 \text{ g } C_{21}H_{15}Bi_3O_{12}} = 5.52 \times 10^{-4} \text{ mol } C_{21}H_{15}Bi_3O_{12}$$

$$\frac{2 \text{ tablets}}{1} \bullet \frac{300. \times 10^{-3} \text{g } C_{21}H_{15}Bi_3O_{12}}{1 \text{ tablet}} \bullet \frac{1 \text{ mol } C_{21}H_{15}Bi_3O_{12}}{1086 \text{ g } C_{21}H_{15}Bi_3O_{12}} \bullet \frac{3 \text{ mol Bi}}{1 \text{ mol } C_{21}H_{15}Bi_3O_{12}} \bullet \frac{208.98 \text{ g Bi}}{1 \text{ mol Bi}}$$

$$= 0.346 \text{ g Bi}$$

107. What is the molar mass of ECl_4 and the identity of E?

2.50 mol of ECl_4 has a mass of 385 grams. The molar mass of ECl_4 would be:

$$\frac{385 \text{ g } ECl_4}{2.50 \text{ mol } ECl_4} = 154 \text{ g/mol } ECl_4.$$

Since the molar mass is 154, and we know that there are 4 chlorine atoms per mole of the compound, we can subtract the mass of 4 chlorine atoms to determine the mass of E. $154 - 4(35.5) = 12$. The element with an atomic mass of 12 g/mol is **carbon**.

109. For what value of n, will Br compose 10.46% of the mass of the polymer, $Br_3C_6H_3(C_8H_8)_n$?

Knowing that the 3Br atoms comprises 10.46% of the formula weight of the polymer, we can write the fraction:

$$\frac{3 \bullet Br}{\text{formula weight}} = 0.1046 \text{ (where 3*Br) is 3 times the atomic weight of Br. Substituting we}$$

get $\dfrac{239.7}{\text{formula weight}} = 0.1046$ or $\dfrac{239.7}{0.1046} = $ formula weight $= 2,292$ (to 4sf)

Noting that the "fixed" part of the formula contains 3Br atoms, 6 C atoms, and 3 H atoms, we can calculate the mass associated with this part of the molecule. ($239.7 + 75.09 = 314.79$). Subtracting this mass from the total (2292) gives 1,977 as the mass corresponding to the "C_8H_8" units. Since each such unit has a mass of 104.15 (8C + 8H), we can divide the mass of *one* C_8H_8 unit into the 1977 mass remaining:

$$\frac{1977 \text{ u}}{104.15 \text{u}/C_8H_8} = 18.98 \; C_8H_8 \text{ units to give a value for } n \text{ of 19.}$$

Summary and Conceptual Questions:

111. Calculate:

(a) moles of nickel—found by density **once** the volume of foil is calculated.

$$V = 1.25 \text{ cm} \times 1.25 \text{ cm} \times 0.0550 \text{ cm} = 8.59 \times 10^{-2} \text{ cm}^3$$

$$\text{Mass} = \frac{8.908 \text{ g}}{1 \text{ cm}^3} \cdot 8.59 \times 10^{-2} \text{ cm}^3 = 0.766 \text{ g Ni}$$

$$0.766 \text{ g Ni} \cdot \frac{1 \text{ mol Ni}}{58.69 \text{ g Ni}} = 1.30 \times 10^{-2} \text{ mol Ni}$$

(b) Formula for the fluoride salt:

$$\text{Mass F} = (1.261 \text{ g salt} - 0.766 \text{ g Ni}) = 0.495 \text{ g F}$$

$$\text{Moles F} = 0.495 \text{ g F} \cdot \frac{1 \text{ mol F}}{19.00 \text{ g F}} = 2.60 \times 10^{-2} \text{ mol F},$$

so 1.30×10^{-2} mol Ni combines with 2.60×10^{-2} mol F , indicating a formula of NiF_2

(c) Name: Nickel(II) fluoride

113. Compound composition is C 54.0%   H 6.00%   O 40.0%

For C atoms: $\dfrac{54.0 \text{ g C}}{1} \cdot \dfrac{1 \text{ mol C}}{12.01 \text{ g C}} = 4.496$ mol C atoms

For H atoms: $\dfrac{6.00 \text{ g H}}{1} \cdot \dfrac{1 \text{ mol H}}{1.008 \text{ g H}} = 5.952$ mol H atoms

For O atoms: $\dfrac{40.0 \text{ g O}}{1} \cdot \dfrac{1 \text{ mol O}}{16.00 \text{ g O}} = 2.500$ mol O atoms

Typically, to establish the ratio of C:H:O atoms, we divide all values by the smallest of the 3 (2.500). Noting however that the # of O atoms can be made integral by multiplying by 2, we can establish the following:

C: (4.496 x 2) =8.99   H: (5.952 x 2) = 11.904 and O (2.500 x 2) = 5.000

Rounding to integral values gives an empirical formula of $C_9H_{12}O_5$. (or answer choice (d))

Erroneous empirical formulas are frequently arrived at when one rounds the moles of atoms—instead of attempting to find a whole number ratio. For each, answer choice (a), can be arrived at by rounding 4.496 C atoms to 4; 5.952 H atoms to 5; and 2.5 O atoms to 2. Another source of error arises when one truncates the number of significant figures prematurely, resulting in erroneous atomic ratios.

115. For Uranium metal being heated:

(a) U metal: 0.169 g –moles of U: $\dfrac{0.169 \text{ g U}}{1} \cdot \dfrac{1 \text{ mol U}}{238.029 \text{ g U}} = 7.10 \times 10^{-4}$ mol U

Oxide: 0.199 g- 0.169 g U = 0.030 g O (according to the Law of Conservation of Mass)

Moles of O: $\dfrac{0.030 \text{ g O}}{1} \cdot \dfrac{1 \text{ mol O}}{16.00 \text{ g O}} = 1.875 \times 10^{-3}$ mol O

The empirical formula of the oxide is arrived at by dividing all mol values by the smallest (in this case the moles of U): U:1.00;   O:2.64

Noting that the value for O may also be expressed as $2^2/_3$ or 8/3, one can multiply both the 1 (value of U) and the 8/3 (value of O) by 3 to obtain an empirical formula of U_3O_8. This oxide must be a mix of uranium (IV) oxide and uranium(VI) oxide.

The number of moles of oxide:

Note that the empirical formula indicates that for 1 mol of the oxide, 3 mol of uranium are needed. Earlier in this part of the question, we calculated the # of moles of U to be: 7.10×10^{-4} mol U.

$$\frac{7.10 \times 10^{-4} \text{ mol U}}{1} \cdot \frac{1 \text{ mol U}_3\text{O}_8}{3 \text{mol U}} = 2.37 \times 10^{-4} \text{ mol U}_3\text{O}_8$$

(b) Since atomic weights are weighted averages, isotopes of greater abundance have a "greater effect" on the atomic weight than those of lesser abundance. Noting that the reported atomic weight of U is 238.02 g/mol, the isotope ^{238}U will be in greater abundance than the two lighter isotopes noted.

(c) The number of waters of hydration in the compound $UO_2(NO_3)_2 \cdot z\ H_2O$:

Mass of hydrated compound: 0.865g

Mass of anydrous compound: <u>0.679g</u>

Mass of water: 0.186g which would correspond to :

$$\frac{0.186 \text{ g H}_2\text{O}}{1} \cdot \frac{1 \text{mol H}_2\text{O}}{18.02 \text{ g H}_2\text{O}} = 1.03 \times 10^{-2} \text{ mol H}_2\text{O}$$

The number of moles of the anhydrous compound is:

$$\frac{0.679 \text{ g UO}_2(\text{NO}_3)_2}{1} \cdot \frac{1 \text{ mol UO}_2(\text{NO}_3)_2}{394.04 \text{ g UO}_2(\text{NO}_3)_2} = 1.72 \times 10^{-3} \text{ mol UO}_2(\text{NO}_3)_2$$

The ratio of moles of water to moles of compound is 5.99 or 6:1, giving 6 waters of hydration.

117. Question: How would one calculate the number of Al atoms in a cube of alum with a linear dimension of 3.00 cm?

Several concepts provide the solution to this question:

1) The volume of the cube: 3.00 cm x 3.00 cm x 3.00 cm = 27.0 cm^3

2) The density of alum (1.757 g/cm^3)—from a reference source—e.g. Handbook of Chemistry and Physics. The product of the volume x the density gives the mass of alum present:

$$\frac{27.0 \text{ cm}^3}{1} \cdot \frac{1.757 \text{ g alum}}{1 \text{ cm}^3} = 47.4 \text{ g alum}$$

3) The weight percent of Al in the alum provides the mass of Al present:

$$\frac{47.4 \text{ g alum}}{1} \bullet \frac{26.98 \text{ g Al}}{474.38 \text{ g alum}} = 2.70 \text{ g Al}$$

4) Avogadro's number and the atomic weight of Al get to the "final" answer:

$$\frac{2.70 \text{ g Al}}{1} \bullet \frac{1 \text{ mol Al}}{26.98 \text{ g Al}} \bullet \frac{6.022 \times 10^{23} \text{ atoms Al}}{1 \text{ mol Al}} = 6.02 \times 10^{22} \text{ atoms Al}$$

Chapter 4
Chemical Equations and Stoichiometry

Practicing Skills
Balancing Equations

Balancing equations can be a matter of "running in circles" if a reasonable methodology is not employed. While there isn't one "right place" to begin, generally you will suffer fewer complications if you begin the balancing process using a substance that contains the **greatest number** of elements **or** the **largest subscript** values. Noting that you must have at least that many atoms of each element involved, coefficients can be used to increase the "atomic inventory". In the next few questions, you will see one **emboldened** substance in each equation. This emboldened substance is the one that I judge to be a "good" starting place. One last hint--modify the coefficients of uncombined elements, i.e. those not in compounds, <u>after</u> you modify the coefficients for compounds containing those elements -- <u>not before</u>!

1. Balanced equation for combustion of liquid pentane:
   $$\textbf{C}_5\textbf{H}_{12}\,(l) + 8\,O_2\,(g) \rightarrow 6\,H_2O\,(l) + 5\,CO_2\,(g)$$
   1. A minimum of 5 C and 12 H (a C_5H_{12} molecule) suggests coefficients of 5 for CO_2 and 6 for H_2O.
   2. Coefficients of 6 for H_2O and 5 for CO_2 will indicate a **total** of 16 O atoms or 8 molecules of the diatomic element, O_2.

3. (a) $4\,Cr\,(s) + 3\,O_2\,(g) \rightarrow \textbf{2}\,\textbf{Cr}_2\textbf{O}_3\,(s)$
   1. Note the need for <u>at least</u> 2 Cr and 3 O atoms.
   2. Oxygen is diatomic -- we'll need an <u>even</u> number of oxygen atoms, so
      try : $2\,Cr_2O_3$.
   3. $3\,O_2$ would give 6 O atoms on both sides of the equation.
   4. 4 Cr would give 4 Cr atoms on both sides of the equation.
   (b) $Cu_2S\,(s) + O_2\,(g) \rightarrow 2\,Cu(s) + \textbf{SO}_2\,(g)$
   1. A minimum of 2 O in SO_2 is required, and is provided with one molecule of elemental oxygen.
   2. 2 Cu atoms (on the right) indicates 2 Cu (on the left).

   (c) $\textbf{C}_6\textbf{H}_5\textbf{CH}_3\,(l) + 9\,O_2\,(g) \rightarrow 4\,H_2O\,(l) + 7\,CO_2\,(g)$
   1. A minimum of 7 C and 8 H is required.
   2. $7\,CO_2$ furnishes 7 C and $4\,H_2O$ furnishes 8 H atoms.
   3. $4\,H_2O$ and $7\,CO_2$ furnish a total of 18 O atoms, making the coefficient of $O_2 = 9$.

5. Balance and name the reactants and products:
   (a) **Fe$_2$O$_3$** (s) + 3 Mg(s) → 3 MgO (s) + 2 Fe (s)
   1. Note the need for <u>at least</u> 2 Fe and 3 O atoms.
   2. 2 Fe atoms would provide the proper iron atom inventory.
   3. 3 MgO would give 3 O atoms on both sides of the equation.
   4. 3 Mg would give 3 Mg atoms on both sides of the equation.

   Reactants: iron(III) oxide and magnesium

   Products: magnesium oxide and iron

   (b) **AlCl$_3$** (s) + 3 NaOH(aq) → Al(OH)$_3$ (s) + 3 NaCl (aq)
   1. Note the need for <u>at least</u> 1 Al and 3 Cl atoms.
   2. 3 NaCl molecules would provide the proper Cl atom inventory.
   3. 3NaCl would require 3 Na atoms on the left side—a coefficient of 3 for NaOH is needed.
   4. 3 OH groups (from Al(OH)$_3$) would give 3 OH groups needed on both sides of the
      equation—so a coefficient of 3 for NaOH is needed to provide that balance.

   Reactants: aluminum chloride and sodium hydroxide

   Products: aluminum hydroxide and sodium chloride.

   (c) 2 NaNO$_3$ (s) + H$_2$SO$_4$ (l) → **Na$_2$SO$_4$** (s) + 2 HNO$_3$ (l)
   1. Note the need for <u>at least</u> 2 Na and 1 S and 4 O atoms.
   2. 2 NaNO$_3$ will provide the proper Na atom inventory.
   3. The coefficient of 2 in front of NaNO$_3$ requires a coefficient of 2 for HNO$_3$ —providing a
      balance for N atoms.
   4. The implied coefficient of 1 for Na$_2$SO$_4$ suggests a similar coefficient for H$_2$SO$_4$ —to
      balance the S atom inventory.
   5. O atom inventory is done "automatically" when we balanced N and S inventories.

   Reactants: sodium nitrate and sulfuric acid

   Products: sodium sulfate and nitric acid

   [....although nitric acid typically exists as an aqueous solution.]

   (d) **NiCO$_3$** (s) + 2 HNO$_3$ (aq) → Ni(NO$_3$)$_2$ (aq) + CO$_2$ (g) + H$_2$O (l)
   1. Note the need for <u>at least</u> 1 Ni atom on both sides. This inventory will mandate 2 NO$_3$
      groups on the right —and also on the left. Since these come from HNO$_3$ molecules, we'll
      need 2 HNO$_3$ on the left.
   2. The 2 H from the acid and the CO$_3$ from nickel carbonate, provide 2H, 1 C and 3 O atoms.
      1 H$_2$O takes care of the 2H, and **one** of the O atoms, 1 CO$_2$ consumes the 1 C and the
      remaining 2 O atoms.

   Reactants: nickel(II) carbonate and nitric acid

Products: nickel(II) nitrate, carbon dioxide, and water

Mass Relationships in Chemical Reactions: Basic Stoichiometry

7. Moles of oxygen needed to react with 6.0 mol of Al:

$$4 \text{ Al (s)} + 3 \text{ O}_2 \text{ (g)} \rightarrow 2 \text{ Al}_2\text{O}_3 \text{ (s)}$$

$$6.0 \text{ mol Al} \cdot \frac{3 \text{ mol O}_2}{4 \text{ mol Al}} = 4.5 \text{ mol O}_2$$

What mass of Al_2O_3 should be produced?

$$6.0 \text{ mol Al} \cdot \frac{2 \text{mol Al}_2\text{O}_3}{4 \text{mol Al}} \cdot \frac{102 \text{ g Al}_2\text{O}_3}{1 \text{ mol Al}_2\text{O}_3} = 310 \text{ g Al}_2\text{O}_3 \text{ (to 2 sf)}$$

9. Quantity of Br_2 to react with 2.56 g of Al:

According to the balanced equation 2 mol of Al react with 3 mol of Br_2 .

Calculate the # of moles of Al, then multiply by 3/2 to obtain # mol of Br_2 required.

$$2.56 \text{ g Al} \cdot \frac{1 \text{ mol Al}}{26.98 \text{ g Al}} \cdot \frac{3 \text{ mol Br}_2}{2 \text{ mol Al}} \cdot \frac{159.8 \text{ g Br}_2}{1 \text{ mol Br}_2} = 22.7 \text{ g Br}_2$$

Mass of Al_2Br_6 expected:

This could be solved in several ways. The simplest is to recognize that—according to the Law of Conservation of Matter, mass is conserved in a reaction. If 22.7 g of bromine react with exactly 2.56 g of aluminum, the total products would also have a mass of (22.7 g + 2.56 g) or 25.3 g Al_2Br_6.

11. The reaction of iron with oxygen to given iron(III) oxide:

   (a) The balanced equation for the reaction:

   $$4 \text{ Fe (s)} + 3 \text{ O}_2 \text{ (g)} \rightarrow 2 \text{ Fe}_2\text{O}_3 \text{ (s)}$$

   (b) Mass of Fe_2O_3 produced when 2.68 g Fe react:

   $$2.68 \text{ g Fe} \cdot \frac{1 \text{ mol Fe}}{55.85 \text{ g Fe}} \cdot \frac{2 \text{ mol Fe}_2\text{O}_3}{4 \text{ mol Fe}} \cdot \frac{159.7 \text{ g Fe}_2\text{O}_3}{1 \text{ mol Fe}_2\text{O}_3} = 3.83 \text{ g Fe}_2\text{O}_3$$

   (c) Mass of oxygen required:

   This could be solved in several ways. The simplest is to recognize that—according to the Law of Conservation of Matter, mass is conserved in a reaction. If 3.83 g Fe_2O_3 are produced when 2.68 g of iron react, the total oxygen required would have a mass of (3.83 g - 2.68 g) or 1.15 g O_2.

13. Removal of SO_2 by $CaCO_3$:

   (a) Mass of $CaCO_3$ required to remove 155 g of SO_2:

   $$155 \text{ g SO}_2 \cdot \frac{1 \text{ mol SO}_2}{64.06 \text{ g SO}_2} \cdot \frac{2 \text{ mol CaCO}_3}{2 \text{ mol SO}_2} \cdot \frac{100.1 \text{ g CaCO}_3}{1 \text{ mol CaCO}_3} = 242 \text{ g CaCO}_3$$

(b) Mass of $CaSO_4$ when 155 g of SO_2 is consumed completely:

$$155 \text{ g SO}_2 \bullet \frac{1 \text{ mol SO}_2}{64.06 \text{ g SO}_2} \bullet \frac{2 \text{ mol CaSO}_4}{2 \text{ mol SO}_2} \bullet \frac{136.1 \text{ g CaSO}_4}{1 \text{ mol CaSO}_4} = 329 \text{ g CaSO}_4$$

Amounts Tables and Chemical Stoichiometry

15. Emboldened quantities indicate given data

Equation	2 PbS(s) +	3 O₂(g) →	2 PbO(s) +	2SO₂(g)
Initial amount (mol)	**2.5**			
Change in amount upon reaction (mol)	-2.5	-3/2(2.5) = 3.75	+2.5	+2.5
Amount after complete reaction (mol)	0	0	2.5	2.5

The amount of O_2 required is 1.5 times the amount of PbS (note the 3:2 ratio). With the required amount of oxygen present, the amount of PbO and SO_2 formed is equal to the amount of PbS consumed (note the ratios of 2:2 for both PbS:PbO and PbS:SO$_2$)

17. For the reaction of elemental Cr with oxygen to produce chromium(III) oxide:

Equation	4 Cr(s) +	3 O₂(g) →	2 Cr₂O₃(s)
Initial amount (g)	**0.175**		
Initial amount (mol)	3.37×10^{-3}		
Change in amount upon reaction (mol)	-3.37×10^{-3}	$-3/4(3.37 \times 10^{-3})$ = 2.52×10^{-4}	$+1/2(3.37 \times 10^{-3})$ = 1.68×10^{-3}
Amount after complete reaction (mol)	0	0	$+2.40 \times 10^{-2}$
Amount after complete reaction (g)	0	0	0.256

$$\frac{0.175 \text{ g Cr}}{1} \bullet \frac{1 \text{ mol Cr}}{52.00 \text{ g Cr}} = 3.37 \times 10^{-3} \text{ mol Cr}$$

The balanced equation indicates that for each mol of elemental Cr, 1/2 mol of Cr_2O_3 is produced (the 4:2 ratio), so 3.37×10^{-3} mol Cr produces 1.68×10^{-3} mol Cr_2O_3

$$\frac{1.68 \times 10^{-3} \text{ mol Cr}_2\text{O}_3}{1} \bullet \frac{152.0 \text{ g Cr}_2\text{O}_3}{1 \text{ mol Cr}_2\text{O}_3} = 0.256 \text{ g Cr}_2\text{O}_3$$

The amount of oxygen required is 3/4 of the amount of elemental Cr (4:3 ratio), with each mole of oxygen having a mass of 32.00 g.

$$\frac{3.37 \times 10^{-3} \text{ mol Cr}}{1} \bullet \frac{3 \text{ mol O}_2}{4 \text{ mol Cr}} \bullet \frac{32.00 \text{ g O}_2}{1 \text{ mol O}_2} = 0.0808 \text{ g O}_2$$

Limiting Reactants and Amounts Tables

19. The reaction to produce sodium sulfide: $Na_2SO_4(aq) + 4\ C(s) \rightarrow Na_2S(aq) + 4\ CO(g)$

With 15 g of Na_2SO_4 and 7.5 g of C, what is the limiting reactant?

The amounts table:

Equation	$Na_2SO_4(aq)$	$+ 4\ C(s)\rightarrow$	$Na_2S(aq) +$	$4CO(g)$
Initial amount (g)	**15**	**7.5**		
Initial amount (mol)	0.106	0.625		
Change in amount upon reaction (mol)	-0.106	-4(0.106) = 0.424	+0.106	+0.424
Amount after complete reaction (mol)	0	0.201	+0.106	+0.424

$$\frac{15\ g\ Na_2SO_4}{1} \cdot \frac{1\ mol\ Na_2SO_4}{142\ g\ Na_2SO_4} = 0.106\ mol\ Na_2SO_4 \qquad \frac{7.5\ g\ C}{1} \cdot \frac{1\ mol\ C}{12.00\ g\ C} = 0.625\ mol\ C$$

The coefficients of the balanced equation indicate the need for 4 mol of C for 1 mol of Na_2SO_4. The actual ratio of C: Na_2SO_4 is 0.625:0.106 (or 5.89:1), indicating that Na_2SO_4 is the limiting reagent. With Na_2SO_4 as the limiting reagent, the amount of Na_2S produced is identical (on a mol:mol basis). The mass of Na_2S produced is:

$$\frac{0.106\ mol\ Na_2S}{1} \cdot \frac{78.06\ g\ Na_2S}{1\ mol\ Na_2S} = 8.27\ g\ Na_2S\ or\ 8.3\ g\ (to\ 2\ sf)$$

21. Identify the limiting reactant when 1.6 mol of S_8 and 35 mol of F_2 react:

The balanced equation is: $S_8 + 24\ F_2 \rightarrow 8\ SF_6$

Determine the required ratio (from the balanced equation): $\dfrac{24\ mol\ F_2}{1\ mol\ S_8}$

Determine the ratio of actual amounts: $\dfrac{35\ mol\ F_2}{1.6\ mol\ S_8} = \dfrac{21.9\ mol F_2}{1\ mol\ S_8}$

Since the ratio of actual fluorine available is **less than** the required ratio, **fluorine is the limiting reactant.**

23. For the reaction of methane with water:

The amounts table:

Equation	$CH_4(g)$ +	$H_2O(g)$ →	$CO_2(g)$ +	3 $H_2(g)$
Initial amount (g)	**995**	**2510**		
Initial amount (mol)	62.0	139.3		
Change in amount upon reaction (mol)	-62.0	-62.0	+62.0	+3(62.0)
Amount after complete reaction (mol)	0	139.3 - 62.0 = 77.3		+186

(a) Limiting reagent:

$$995 \text{ g } CH_4 \cdot \frac{1 \text{ mol } CH_4}{16.04 \text{ g } CH_4} = 62.0 \text{ mol } CH_4$$

$$2510 \text{ g } H_2O \cdot \frac{1 \text{ mol } H_2O}{18.02 \text{ g } H_2O} = 139.3 \text{ mol } H_2O$$

The required ratio of methane to water: $\dfrac{1 \text{ mol } H_2O}{1 \text{ mol } CH_4}$ and

the actual ratio: $\dfrac{139.3 \text{ mol } H_2O}{62.0 \text{ mol } CH_4} = 2.25$

The actual ratio indicates that **methane is the limiting reactant.**

(b) Maximum mass of H_2 possible:

$$62.0 \text{ mol } CH_4 \cdot \frac{3 \text{ mol } H_2}{1 \text{ mol } CH_4} \cdot \frac{2.016 \text{ g } H_2}{1 \text{ mol } H_2} = 375 \text{ g } H_2$$

(c) Mass of water remaining:

Since 1 mol of methane reacts with 1 mol of water, we know that 62.0 mol of CH_4 reacts with 62.0 mol of water. The amount of water remaining is: (139.3 – 62.0) = 77.3 mol H_2O. The mass of this amount of water is:

$$77.2 \text{ mol } H_2O \cdot \frac{18.02 \text{ g } H_2O}{1 \text{ mol } H_2O} = 1,393 \text{ g } H_2O \text{ or } 1390g \text{ (to 3 sf)}$$

25. Hexane gas burns in air (O_2) to give CO_2 and H_2O.

The amounts table:

Equation	2 $C_6H_{14}(g)$ +	19 $O_2(g)$→	12 $CO_2(g)$	+ 14 $H_2O(g)$
Initial amount (g)	**215**	**215**		
Initial amount (mol)	2.49	6.72		
Change in amount	-2/19(6.72)	-6.72	+12/19(6.72)	+14/19(6.72)

upon reaction (mol)	= -0.707		= 4.24	= 4.95
Amount after complete reaction (mol)	2.49-0.707 = 1.78	0	4.24	4.95

(a) The balanced equation for the reaction.

$2C_6H_{14}(g) + 19\ O_2(g) \rightarrow 12\ CO_2(g) + 14\ H_2O(g)$. A coefficient of 6 for CO_2 seems a reasonable start, with a coefficient of 7 for H_2O providing the balance for H. However, a quick examination of the count of O atoms on the right gives an **odd** number. This indicates that doubling the coefficient for hexane will provide 12 C (on the right), and 14 as a coefficient for H_2O. With a total of 38 O atoms (on the right), 19 O_2 molecules provides the balance.

(b) If 215 g of C_6H_{14} is mixed with 215 g of O_2, what masses of CO_2 and H_2O are produced in the reaction?

Moles of each reactant: $\dfrac{215\ g\ C_6H_{14}}{1} \bullet \dfrac{1\ mol\ C_6H_{14}}{86.18\ g\ C_6H_{14}} = 2.49\ mol\ C_6H_{14}$ and

$\dfrac{215\ g\ O_2}{1} \bullet \dfrac{1\ mol\ O_2}{32.0\ g\ O_2} = 6.72\ mol\ O_2$

Determining the required and actual ratios:

$\dfrac{19\ mol\ O_2}{2\ mol\ C_6H_{14}} = \dfrac{9.5\ mol\ O_2}{1\ mol\ C_6H_{14}}$ and the actual ratio: $\dfrac{6.72\ mol\ O_2}{2.49\ mol\ C_6H_{14}} = \dfrac{2.69\ mol\ O_2}{1\ mol\ C_6H_{14}}$

indicating that **oxygen is the limiting reagent.**

The stoichiometric ratios provide:

$6.72\ mol\ O_2 \bullet \dfrac{12\ mol\ CO_2}{19\ mol\ O_2} \bullet \dfrac{44.01\ g\ CO_2}{1\ mol\ CO_2} = 187\ g\ CO_2$, and for the 4.95 mol of water:

$\dfrac{4.95\ mol\ H_2O}{1} \bullet \dfrac{18.02\ g\ H_2O}{1\ mol\ H_2O} = 89.2\ g\ H_2O$

(c) What mass of the excess reactant remains after the hexane has been burned?

With 1.78 mol of C_6H_{14} remaining, the mass is:

$\dfrac{1.78\ mol\ C_6H_{14}}{1} \bullet \dfrac{86.18\ g\ C_6H_{14}}{1\ mol\ C_6H_{14}} = 154\ g\ C_6H_{14}$

Percent Yield

27. Percent yield of CH_3OH:

$\dfrac{actual}{theoretical} = \dfrac{332\ g\ CH_3OH}{407\ g\ CH_3OH} = 81.6\%$ yield

29. In the formation of $Cu(NH_3)_4SO_4$:

(a) The theoretical yield of $Cu(NH_3)_4SO_4$ from 10.0 g of $CuSO_4$:

$$10.0 \text{ g CuSO}_4 \bullet \frac{1 \text{ mol CuSO}_4}{159.6 \text{ g CuSO}_4} \bullet \frac{1 \text{ mol Cu(NH}_3)_4\text{SO}_4}{1 \text{ mol CuSO}_4}$$

$$\bullet \frac{227.7 \text{ g Cu(NH}_3)_4\text{SO}_4}{1 \text{ mol Cu(NH}_3)_4\text{SO}_4} = 14.3 \text{ g Cu(NH}_3)_4\text{SO}_4$$

(b) Percentage yield of the compound:

$$\frac{12.6 \text{ g compound}}{14.3 \text{ g compound}} \bullet 100 = 88.3\%$$

Analysis of Mixtures

31. Mass percent of $CuSO_4 \bullet 5 H_2O$ in the mixture:

Mass of H_2O = 1.245 g - 0.832 g = 0.413 g H_2O

Since this water was a part of the hydrated salt, let's calculate the mass of that salt present:
In 1 mol of $CuSO_4 \bullet 5 H_2O$ there are 90.10 g H_2O and 159.61 g $CuSO_4$ or
249.71 g $CuSO_4 \bullet 5 H_2O$. These masses correspond to the molar masses of anhydrous
$CuSO_4$ and 5$\bullet$ mol H_2O. So:

$$0.413 \text{ g H}_2\text{O} \bullet \frac{249.71 \text{ g CuSO}_4 \bullet 5\text{H}_2\text{O}}{90.10 \text{ gH}_2\text{O}} = 1.14 \text{ g hydrated salt}$$

$$\% \text{ hydrated salt} = \frac{1.14 \text{ g hydrated salt}}{1.245 \text{ g mixture}} \times 100 = 91.9\%$$

33. Mass percent of $CaCO_3$ in a limestone sample:

Note that the ratio of carbon dioxide to calcium carbonate is 1:1 (from balanced equation).
Calculate the amount of carbon dioxide represented by 0.558 g CO_2

$$0.558 \text{ g CO}_2 \bullet \frac{1 \text{ mol CO}_2}{44.01 \text{ g CO}_2} = 0.01268 \text{ mol CO}_2$$

Since there would have been 0.01268 mol of $CaCO_3$, the mass is :

$$0.01268 \text{ mol CaCO}_3 \bullet \frac{100.1 \text{ g CaCO}_3}{1 \text{ mol CaCO}_3} = 1.269 \text{ g CaCO}_3.$$

The percent of $CaCO_3$ in the sample is:

$$\frac{1.269 \text{ g CaCO}_3}{1.506 \text{ g sample}} \bullet 100 = 84.3\%$$

35. Mass percent of Tl_2SO_4 in 10.20 g sample:

    Begin by balancing the equation:

$$Tl_2SO_4 + 2NaI(aq) \rightarrow 2TlI(s) + Na_2SO_4(aq)$$

This equation tells us that for each mol of Tl_2SO_4, we expect **two** moles of TlI.

Determine the amount of TlI: $0.1964 \text{ g TlI} \bullet \dfrac{1 \text{ mol TlI}}{331.29 \text{ g TlI}} = 5.928 \times 10^{-4} \text{ mol TlI}$

Use the ratio described by the balanced equation:

$5.928 \times 10^{-4} \text{ mol TlI} \bullet \dfrac{1 \text{ mol Tl}_2SO_4}{2 \text{ mol TlI}} = 2.964 \times 10^{-4} \text{ mol Tl}_2SO_4$

Using the molar mass of thallium(I) sulfate gives the mass of the sulfate present in the sample:

$2.964 \times 10^{-4} \text{ mol Tl}_2SO_4 \bullet \dfrac{504.83 \text{ g Tl}_2SO_4}{1 \text{ mol Tl}_2SO_4} = 0.1496 \text{ g Tl}_2SO_4$

The mass percent of Tl_2SO_4 in the sample is $\dfrac{0.1496 \text{ g Tl}_2SO_4}{10.20 \text{ g sample}} \bullet 100 = 1.467\%$

Using Stoichiometry to determine Empirical and Molecular Formulas

37. The basic equation is:

$$C_xH_y + O_2 \rightarrow x\, CO_2 + \frac{y}{2}\, H_2O$$

Without balancing the equation, one can see that all the C in CO_2 comes from the styrene as does all the H in H_2O. Let's use the percentage of C in CO_2 to determine the mass of C in styrene, and the percentage of H in H_2O to provide the mass of H in styrene.

$1.481 \text{ g CO}_2 \bullet \dfrac{12.01 \text{ g C}}{44.01 \text{ g CO}_2} = 0.404 \text{ g C}$

Similarly:

$0.303 \text{ g H}_2O \bullet \dfrac{2.02 \text{ g H}}{18.02 \text{ g H}_2O} = 0.0340 \text{ g H}$

Alternatively, the mass of H could be determined by subtracting the mass of C from the 0.438 g styrene

Mass H = 0.438 g styrene - 0.404 g C

Establish the ratio of C atoms to H atoms

$0.0340 \text{ g H} \bullet \dfrac{1 \text{ mol H}}{1.008 \text{ g H}} = 0.0337 \text{ mol H}$

$0.404 \text{ g C} \bullet \dfrac{1 \text{ mol C}}{12.011 \text{ g C}} = 0.0336 \text{ mol C}$

This number of H and C atoms indicates an empirical formula for styrene of 1:1 or C_1H_1.

39. Combustion of Cyclopentane:

The approach is similar to that for study question 37.

The general equation is $C_xH_y + O_2 \rightarrow x\,CO_2 + \frac{y}{2}\,H_2O$

All the carbon from cyclopentane eventually resides in the CO_2 produced.

Calculate the amount of C in cyclopentane:

$$0.300 \text{ g } CO_2 \cdot \frac{1 \text{ mol } CO_2}{44.01 \text{ g } CO_2} \cdot \frac{1 \text{ mol C}}{1 \text{ mol } CO_2} = 0.00682 \text{ mol C}$$

Similarly the amount of H in cyclopentane:

$$0.123 \text{ g } H_2O \cdot \frac{1 \text{ mol } H_2O}{18.01 \text{ g } H_2O} \cdot \frac{2 \text{ mol H}}{1 \text{ mol } H_2O} = 0.0137 \text{ mol H}$$

(a) The empirical formula is then: $\dfrac{0.0137 \text{ mol H}}{0.00682 \text{ mol C}} = \dfrac{2.00 \text{ mol H}}{1.00 \text{ mol C}}$

C_1H_2

(b) If the molar mass is 70.1 g/mol, the molecular formula is:

Since the molecular formula represents some **multiple** of the empirical formula,

calculate the "empirical formula mass".

For C_1H_2 that mass is 14.03 g/empirical formula [(1 x 12.0115) + (2 x 1.0079)]

Calculate the # of empirical formulas in a molecular formula:

$$\frac{70.1 \text{ g/molecular formula}}{14.03 \text{ g/empirical formula}} = 5 \text{ , giving a molecular formula of } C_5H_{10}.$$

41. An unknown compound has the formula $C_xH_yO_z$.

When 0.0956 g of the compound burns in air, 0.1356 g of CO_2 and 0.0833 g of H_2O result.

What is the empirical formula of the compound?

All the C that originated in the compound resides in the CO2 produced, and all the H in the water produced originated in the compound. The same can not be said for the O. While some of the O in both carbon dioxide and water originated in the compound, some was incorporated from the atmosphere during the combustion process. For that reason, we can calculate the mass of C and H that resides in the two products, and obtain **by difference** [total mass of compound – (mass of C + mass of H)]the mass of O originally contained in the compound.

$$\frac{0.1356 \text{ g } CO_2}{1} \cdot \frac{12.01 \text{ g } CO_2}{44.01 \text{ g } CO_2} = 0.03700 \text{ g C and similarly for H}$$

$$\frac{0.0833 \text{ g } H_2O}{1} \cdot \frac{2.02 \text{ g H}}{18.02 \text{ g } H_2O} = 0.00934 \text{ g H;}$$

Now we can determine the mass of O originating

in the compound: 0.0956 g C,H, and O (in the compound)- 0.03700 g C – 0.00934 g H = 0.04926 g O

Determine the number of moles of each of the 3 elements:

$$0.03700 \text{ g C} \cdot \frac{1 \text{ mol C}}{12.011 \text{ g C}} = 3.081 \times 10^{-3} \text{ mol C}$$

$$0.00934 \text{ g H} \cdot \frac{1 \text{ mol H}}{1.008 \text{ g H}} = 9.26 \times 10^{-3} \text{ mol H}$$

$$0.04926 \text{ g O} \cdot \frac{1 \text{ mol O}}{16.00 \text{ g O}} = 3.079 \times 10^{-3} \text{ mol O}$$

Establish a small whole-number ratio of the amounts of C,H,O:

Note that C and O are present in equi-molar amounts (3.08×10^{-3}). Dividing the number of moles of H by the moles of C (or O), gives a ratio of 3 H: 1 C or 3 H:1 O—for an empirical formula of $C_1H_3O_1$, more commonly written CH_3O,

If the molar mass if 62.1 g/mol, what is the molecular formula?

Adding the atomic weights of 1C, 3H, and 1 O gives approximately (12+3+16)=31 for an "empirical formula" weight.

The molecular formula is then : $\dfrac{62.1 \text{ g}}{1 \text{ mol}} \cdot \dfrac{1 \text{ empirical formula}}{31 \text{ g}} = \dfrac{2 \text{ empirical formulas}}{1 \text{ mol}}$ so the

molecular formula is $C_2H_6O_2$.

43. Formula of the carbonyl compound formed with nickel:

$$Ni_x(CO)_y(s) + O_2 \rightarrow x \, NiO \, (s) + y \, CO_2(g)$$

Given the mass of NiO formed, we can calculate the mass of Ni in NiO (and also in the nickel carbonyl compound)

Quantity of Ni :

$$0.0426 \text{ g NiO} \cdot \frac{1 \text{ mol NiO}}{74.6924 \text{ g NiO}} = 5.70 \times 10^{-4} \text{ mol NiO} \quad \text{(and an equal number of moles of}$$

Nickel since the oxide has 1mol Ni:1mol NiO). That number of moles of Ni would have a mass

of: $5.70 \times 10^{-4} \text{ mol Ni} \cdot \dfrac{58.693 \text{g Ni}}{1 \text{ mol Ni}} = 0.03347 \text{ g Ni.}$

Quantity of CO contained in the nickel carbonyl compound:

0.0973 g compound – 0.03347 g Ni = 0.06383 g CO

Amount of CO contained in the compound:

$$0.06383 \text{ g CO} \bullet \frac{1 \text{ mol CO}}{28.010 \text{ g CO}} = 2.279 \times 10^{-3} \text{ mol CO}$$

The ratio of Ni: CO is: $\dfrac{2.279 \times 10^{-3} \text{ mol CO}}{5.70 \times 10^{-4} \text{ mol Ni}} = 4$ and the empirical formula is $Ni(CO)_4$.

General Questions on Stoichiometry

45. Balance:

   (a) synthesis of urea:

   $$CO_2(g) + 2\,NH_3(g) \rightarrow CO(NH_2)_2(s) + H_2O(l)$$

   1. Note the need for two NH_3 in each molecule of urea, so multiply NH_3 by 2.
   2. $2\,NH_3$ provides the two H atoms for a molecule of H_2O.
   3. Each CO_2 provides the O atom for a molecule of H_2O.

   (b) synthesis of uranium(VI) fluoride

   $$UO_2(s) + 4\,HF(aq) \rightarrow UF_4(s) + H_2O(l)$$

   $$UF_4(s) + F_2(g) \rightarrow UF_6(s)$$

   1. The 4 F atoms in UF_4 requires 4 F atoms from HF.   (equation 1)
   2. The H atoms in HF produce 2 molecules of H_2O.   (equation 1)
   3. The 1:1 stoichiometry of UF_6 : UF_4 provides a simple balance.   (equation 2)

   (c) synthesis of titanium metal from TiO_2:

   $$TiO_2(s) + 2\,Cl_2(g) + 2\,C(s) \rightarrow TiCl_4(l) + 2\,CO(g)$$

   $$TiCl_4(l) + 2\,Mg(s) \rightarrow Ti(s) + 2\,MgCl_2(s)$$

   1. The O balance mandates 2 CO for each TiO_2.   (equation 1)
   2. A coefficient of 2 for C provides C balance.   (equation 1)
   3. The Ti balance (TiO_2 : $TiCl_4$) requires 4 Cl atoms, hence 2 Cl_2 (equation 1)
   4. The Cl balance requires 2 $MgCl_2$, hence 2 Mg.   (equation 2)

47. For the reaction of benzene with oxygen:

   (a) the products of the reaction:

   Combination of C_6H_6 with O_2 gives the oxide of C and the oxygen of H:

   $$CO_2(g) + H_2O(g)$$

   (b) the balanced equation for the reaction is:

   $$2\,C_6H_6(l) + 15\,O_2(g) \rightarrow 12\,CO_2(g) + 6\,H_2O(g)$$

   (c) mass of oxygen, in grams, needed to completely consume the 16.04 g C_6H_6.

   $$16.04 \text{ g } C_6H_6 \bullet \frac{1 \text{ mol } C_6H_6}{78.11 \text{ g } C_6H_6} \bullet \frac{15 \text{ mol } O_2}{2 \text{ mol } C_6H_6} \bullet \frac{31.999 \text{ g } O_2}{1 \text{ mol } O_2} = 49.28 \text{ g } O_2 \,.$$

(d) total mass of products expected:

   This could be solved in several ways. Perhaps the simplest is to recognize that—according to the Law of Conservation of Matter, mass is conserved in a reaction. If 49.28 g of oxygen react with exactly 16.04 g of benzene, the total products would also have a mass of (49.28 +16.04) g or 65.32 g.

49. Mass of acetone available from decomposition of 125 mg of acetoacetic acid(AA)

$$CH_3COCH_2CO_2H \rightarrow CH_3COCH_3 + CO_2$$

   molar mass: 102.09 g/mol   58.08 g/mol

$$125 \text{ mg AA} \cdot \frac{1 \text{mol AA}}{102.09 \text{gAA}} \cdot \frac{1 \text{mol acetone}}{1 \text{mol AA}} \cdot \frac{58.08 \text{ g acetone}}{1 \text{ mol acetone}} = 71.1 \text{ mg acetone}$$

   Note that it **is not necessary** to convert milligrams of acetoacetic acid into grams of acetoacetic acid, since the mass of product (acetone) is also expressed in milligrams.

51. (a) Balanced equation: $2 \text{ Fe(s)} + 3 \text{ Cl}_2(g) \rightarrow 2 \text{ FeCl}_3(s)$

   (b) 1. Mass of Cl_2 to react with 10.0 g iron:

$$10.0 \text{ g Fe} \cdot \frac{1 \text{ mol Fe}}{55.85 \text{ g Fe}} \cdot \frac{3 \text{ mol Cl}_2}{2 \text{ mol Fe}} \cdot \frac{70.91 \text{ g Cl}_2}{1 \text{ mol Cl}_2} = 19.0 \text{ g Cl}_2$$

   2. Amount of $FeCl_3$ produced:

$$10.0 \text{ g Fe} \cdot \frac{1 \text{ mol Fe}}{55.85 \text{ g Fe}} \cdot \frac{2 \text{ mol FeCl}_3}{2 \text{ mol Fe}} = 0.179 \text{ mol FeCl}_3$$

$$0.179 \text{ mol FeCl}_3 \cdot \frac{162.2 \text{ g FeCl}_3}{1 \text{ mol FeCl}_3} = 29.0 \text{ g FeCl}_3$$

   (c) Percent yield of $FeCl_3$:

$$\frac{\text{Actual}}{\text{Theoretical}} \text{ x } 100 \text{ or } \frac{18.5 \text{ g FeCl}_3}{29.0 \text{ g FeCl}_3} \text{ x } 100 = 63.7 \%$$

   (d) With 10.0 g of iron and 10.0 g of Cl_2, the theoretical yield of $FeCl_3$:

   Part (b) tells us that 10.0 g of Fe would require 19.0 g of Cl_2. This indicates that Cl_2 would be the limiting reagent.

$$\frac{10.0 \text{ g Cl}_2}{1} \cdot \frac{1 \text{ mol Cl}_2}{70.91 \text{ g Cl}_2} \cdot \frac{2 \text{ mol FeCl}_3}{3 \text{ mol Cl}_2} \cdot \frac{162.2 \text{ g FeCl}_3}{1 \text{ mol FeCl}_3} = 15.3 \text{ g FeCl}_3$$

53. For the reaction of $TiCl_4(l) + 2 H_2O(l) \rightarrow TiO_2(s) + 4 HCl(g)$:
   (a) Names of the compounds:

   $TiCl_4(l)$ –titanium(IV) chloride—also called titanium tetrachloride

$H_2O(l)$ – water (non-systematic name)

$TiO_2(s)$ – titanium(IV) oxide—also known as titanium dioxide

$HCl(g)$ – hydrogen chloride (the aqueous form is known as hydrochloric acid)

(b) Mass of water to react with 14.0 mL of (d = 1.73 g/mL)

$$\frac{14.0 \text{ mL TiCl}_4}{1} \cdot \frac{1.73 \text{ g TiCl}_4}{1 \text{ mL TiCl}_4} \cdot \frac{1 \text{ mol TiCl}_4}{189.7 \text{g TiCl}_4} \cdot \frac{2 \text{ mol H}_2\text{O}}{1 \text{ mol TiCl}_4} \cdot \frac{18.02 \text{ g H}_2\text{O}}{1 \text{ mol H}_2\text{O}} = 4.60 \text{ g H}_2\text{O}$$

(c) Mass of products expected:

$$\frac{14.0 \text{ mL TiCl}_4}{1} \cdot \frac{1.73 \text{ g TiCl}_4}{1 \text{ mL TiCl}_4} \cdot \frac{1 \text{ mol TiCl}_4}{189.7 \text{ g TiCl}_4} \cdot \frac{1 \text{ mol TiO}_2}{1 \text{ mol TiCl}_4} \cdot \frac{79.866 \text{ g TiO}_2}{1 \text{ mol TiO}_2} = 10.2 \text{ g TiO}_2$$

$$\frac{14.0 \text{ mL TiCl}_4}{1} \cdot \frac{1.73 \text{ g TiCl}_4}{1 \text{ mL TiCl}_4} \cdot \frac{1 \text{ mol TiCl}_4}{189.7 \text{ g TiCl}_4} \cdot \frac{4 \text{ mol HCl}}{1 \text{ mol TiCl}_4} \cdot \frac{36.46 \text{ g HCl}}{1 \text{ mol HCl}} = 18.6 \text{ g HCl}$$

55. Production of Sodium azide: $NaNO_3 + 3 \text{ NaNH}_2 \rightarrow NaN_3 + 3 \text{ NaOH} + NH_3$

Mass of NaN_3 produced when 15.0 g of $NaNO_3$ (85.0 g/mol) reacts with 15.0 g of $NaNH_2$:

$$\frac{15.0 \text{ g NaNO}_3}{1} \cdot \frac{1 \text{ mol NaNO}_3}{85.0 \text{ g NaNO}_3} = 0.176 \text{ mol NaNO}_3$$

$$\frac{15.0 \text{ g NaNH}_2}{1} \cdot \frac{1 \text{ mol NaNH}_2}{39.0 \text{ g NaNH}_2} = 0.385 \text{ mol NaNH}_2$$

Determine the limiting reagent:

Required ratio: $\dfrac{3 \text{ mol NaNH}_2}{1 \text{ mol NaNO}_3} = 3$ Available ratio: $\dfrac{0.385 \text{ mol NaNH}_2}{0.176 \text{ mol NaNO}_3} = 2.19$

The ratios indicate that $NaNH_2$ is the limiting reagent.

$$\frac{0.385 \text{ mol NaNH}_2}{1} \cdot \frac{1 \text{ mol NaN}_3}{3 \text{ mol NaNH}_2} \cdot \frac{65.0 \text{ g NaN}_3}{1 \text{ mol NaN}_3} = 8.33 \text{ g NaN}_3$$

57. Copper(I) sulfide reacts with O_2 upon heating to give copper metal and sulfur dioxide.

(a) Write a balanced equation for the reaction.

$$Cu_2S + O_2 \rightarrow 2 \text{ Cu}(s) + SO_2(g)$$

(b) Mass of copper metal that can be obtained from exactly 500 g of copper(I) sulfide?

$$\frac{500. \text{ g Cu}_2\text{S}}{1} \cdot \frac{1 \text{ mol Cu}_2\text{S}}{159.14 \text{ g Cu}_2\text{S}} \cdot \frac{2 \text{ mol Cu}}{1 \text{ mol Cu}_2\text{S}} \cdot \frac{63.54 \text{ g Cu}}{1 \text{ mol Cu}} = 399 \text{ g Cu}$$

59. Empirical formula of B_xH_y

Using the molar mass of B_2O_3, calculate the amount of the compound formed and the amount of B contained in that compound.

$$0.422 \text{ g B}_2\text{O}_3 \cdot \frac{21.62 \text{ g B}}{69.62 \text{ g B}_2\text{O}_3} = 0.131 \text{ g B}$$

The Law of Conservation of matter tells us that since 0.148 g of the compound B_xH_y has

0.131 g B, then it must also contain (0.148 – 0.131) or 0.017g H.

Now calculate the amount of B and H, using the respective atomic masses:

$0.131 \text{ g B} \bullet \dfrac{1 \text{ mol B}}{10.811 \text{ g B}} = 1.21 \times 10^{-2}$ mol B and

$0.017 \text{ g H} \bullet \dfrac{1 \text{ mol H}}{1.0079 \text{ g H}} = 1.7 \times 10^{-2}$ mol H (actually 1.69×10^{-2} to 3 sf)

The ratios of B:H are then: $\dfrac{1.7 \times 10^{-2} \text{ mol H}}{1.21 \times 10^{-2} \text{ mol B}} = \dfrac{1.39 \text{ mol H}}{1 \text{ mol B}}$

Noting that the numerator is almost 1.4, a multiplier of 5 would provide a formula of B_5H_7.

61. The combustion of menthol can be represented: $C_xH_yO_z + O_2 \rightarrow CO_2 + H_2O$

It's important to note that while **all** the C in CO_2 originated in the menthol and **all** the H in H_2O originated in the menthol—**not all** of the oxygen originates in the menthol. So we begin by calculating the masses of C and H , and subtracting those masses from the 95.6 mg of compound to determine the mass of O present in menthol.

$269 \text{ mg } CO_2 \bullet \dfrac{12.01 \text{g C}}{44.02 \text{g } CO_2} = 73.41$ mg C

$110 \text{ mg } H_2O \bullet \dfrac{2.02 \text{ g H}}{18.02 \text{ g } H_2O} = 12.33$ mg H

Mass O = 95.6 mg - (73.41 mg C + 12.33 mg H) = 9.86 mg O

Now we can calculate the # of moles of each of these atoms:

$73.41 \times 10^{-3} \text{ g C} \bullet \dfrac{1 \text{ mol C}}{12.01 \text{ g C}} = 6.11 \times 10^{-3}$ mol C

$12.33 \times 10^{-3} \text{ g H} \bullet \dfrac{1 \text{ mol H}}{1.008 \text{ g H}} = 12.23 \times 10^{-3}$ mol H

$9.86 \times 10^{-3} \text{ g O} \bullet \dfrac{1 \text{ mol O}}{16.00 \text{ g O}} = 0.616 \times 10^{-3}$ mol O

We can express the **ratio** of C,H, and O present in menthol by dividing these three amounts by the smallest (O content).

$\dfrac{6.11 \times 10^{-3} \text{molC}}{0.616 \times 10^{-3} \text{molO}} = 10$ $\dfrac{12.23 \times 10^{-3} \text{molH}}{0.616 \times 10^{-3} \text{molO}} = 20$

The empirical formula is then $C_{10}H_{20}O$.

63. (a) The balanced equation for the reaction: $FeCl_2 + Na_2S \rightarrow FeS + 2 NaCl$

(b) The limiting reagent when 40. g of $FeCl_2$ react with 40. g of Na_2S:

$\dfrac{40 \text{ g } FeCl_2}{1} \bullet \dfrac{1 \text{ mol } FeCl_2}{127 \text{ g } FeCl_2} = 0.32$ mol $FeCl_2$

$$\frac{40 \text{ g Na}_2\text{S}}{1} \cdot \frac{1 \text{ mol Na}_2\text{S}}{78.0 \text{ g Na}_2\text{S}} = 0.51 \text{ mol Na}_2\text{S}$$

The 1:1 mole ratio indicated by the balanced equation tells us that $FeCl_2$ will be the limiting reagent.

(c) Mass of FeS produced:

$$\frac{0.32 \text{ mol FeCl}_2}{1} \cdot \frac{1 \text{ mol FeS}}{1 \text{ mol FeCl}_2} \cdot \frac{87.9 \text{ g FeS}}{1 \text{ mol FeS}} = 28 \text{ g FeS}$$

(d) Mass of the reagent present in excess after the reaction:

The required ratio is 1:1, so 0.32 mol Na_2S are required, leaving (0.51-0.32) mol Na_2S:

$$\frac{0.19 \text{ mol Na}_2\text{S}}{1} \cdot \frac{78 \text{ g Na}_2\text{S}}{1 \text{ mol Na}_2\text{S}} = 15 \text{ g Na}_2\text{S}$$

(e) Mass of $FeCl_2$ needed to completely react with 40 g of Na_2S:

From part (a) we note that 40 g corresponds to 0.51 mol Na_2S:

$$\frac{0.51 \text{ mol Na}_2\text{S}}{1} \cdot \frac{1 \text{ mol FeCl}_2}{1 \text{ mol Na}_2\text{S}} \cdot \frac{127 \text{ g FeCl}_2}{1 \text{ mol FeCl}_2} = 65 \text{ g FeCl}_2$$

<u>65</u>. 1.056 g MCO_3 produced MO + 0.376 g CO_2

The MO had a mass of (1.056 - 0.376) 0.680 g

According to the equation given, CO_2 and MO are produced in equimolar amounts.

$$0.376 \text{ g CO}_2 \cdot \frac{1 \text{ mol CO}_2}{44.0 \text{ g CO}_2} = 8.54 \times 10^{-3} \text{ mol CO}_2$$

So the metal oxide (0.680 g) must correspond to 8.54×10^{-3} mol of metal oxide. The molar mass of metal oxide is then: $\dfrac{0.680 \text{ g}}{8.54 \times 10^{-3} \text{ mol}} = 79.6 \text{ g/ mol}$

If the oxide contains one mol of O atoms per mol of M atoms, we can deduce the molar mass of M   MO = 79.6 g/mol

$$79.6 = M \text{ g/mol} + 16.0 \text{ g O/mol}$$

$$63.6 \text{ g/mol} = M$$

The metal with the atomic weight close to 63.6 is (b) Cu.

<u>67</u>. $TiO_2 + H_2 \rightarrow H_2O + Ti_xO_y$

$\qquad$ 1.598 g $\qquad\qquad\qquad\qquad$ 1.438 g

The moles of TiO_2 initially present:

$$1.598 \text{ g TiO}_2 \cdot \frac{1 \text{ mol TiO}_2}{79.879 \text{ g TiO}_2} = 0.02000 \text{ mol TiO}_2 \text{ (and Ti)}$$

The mass of Ti present in the TiO_2 :

$$1.598 \text{ g TiO}_2 \cdot \frac{47.88 \text{ g Ti}}{79.879 \text{ g TiO}_2} = 0.9579 \text{ g Ti}$$

Since all the Ti in the unknown compound, Ti_xO_y, originates in the TiO_2, 0.9579 g of the

1.438 g of the new oxide is Ti, leaving (1.438 - 0.9579) g of O.

The number of moles of O is:

$$0.480 \text{ g O} \cdot \frac{1 \text{ mol O}}{16.00 \text{ g O}} = 0.03000 \text{ mol O}$$

The new oxide is then Ti $_{0.02000}$ O$_{0.03000}$ or Ti_2O_3

69. A 1.236-g sample of the herbicide liberated the chlorine as Cl ion, and the ion precipitated as

AgCl, with a mass of 0.1840 g. The mass percent of 2,4-D in the sample:

First, determine the moles of chlorine present. Noting that each molecule of 2,4D has 2 atoms

of Cl, allows you to calculate the number of moles of 2,4D (each with a molar mass of 221.04

g/mol)

$$\frac{0.1840 \text{ g AgCl}}{1} \cdot \frac{1 \text{ mol AgCl}}{143.32 \text{ g AgCl}} \cdot \frac{1 \text{ mol Cl}}{1 \text{ mol AgCl}} \cdot \frac{1 \text{ mol 2,4D}}{2 \text{ mol Cl}} \cdot \frac{221.04 \text{ g 2,4D}}{1 \text{ mol 2,4D}} = 0.1419 \text{ g 2,4D}$$

The mass percent of 2,4D in the sample is:

$$\frac{0.1419 \text{ g 2,4D}}{1.236 \text{ g sample}} \cdot 100 = 11.48 \text{ % 2,4D}$$

71. Commercial sodium "hydrosulfite" is 90.1% pure $Na_2S_2O_4$. The sequence of reactions used

to prepare the compound is:

$$Zn(s) + 2 SO_2(g) \rightarrow ZnS_2O_4(s)$$

$$ZnS_2O_4(s) + Na_2CO_3(aq) \rightarrow ZnCO_3(s) + Na_2S_2O_4(aq)$$

(a) What mass of pure $Na_2S_2O_4$ can be prepared from 125 kg of Zn, 500 kg of SO_2, and an

excess of Na_2CO_3?

Since we know that sodium carbonate is present in excess, determine the limiting reagent (of

Zn or SO_2)

$$\frac{125000 \text{ g Zn}}{1} \cdot \frac{1 \text{ mol Zn}}{65.39 \text{ g Zn}} = 1910 \text{ mol Zn (3 sf)}$$

$$\frac{500000 \text{ g SO}_2}{1} \cdot \frac{1 \text{ mol SO}_2}{64.06 \text{ g SO}_2} = 7810 \text{ mol SO}_2 \text{ A quick observation of the amounts}$$

present tells us that Zn is the limiting reagent. (1910 mol Zn would require 3820 mol

SO_2—and we have 7810!)

To determine the mass of pure $Na_2S_2O_4$ that can be prepared from the amounts of substances given, observe two relationships: that of Zn to ZnS_2O_4 (1:1) and that of ZnS_2O_4 to $Na_2S_2O_4$ (1:1). With excess SO_2, and excess Na_2CO_3, examine the amount of $Na_2S_2O_4$ that can be formed.

$$\frac{1910 \text{ mol Zn}}{1} \cdot \frac{1 \text{ mol } ZnS_2O_4}{1 \text{ mol Zn}} \cdot \frac{1 \text{ mol } Na_2S_2O_4}{1 \text{ mol } ZnS_2O_4} \cdot \frac{174.11 \text{ g } Na_2S_2O_4}{1 \text{ mol } Na_2S_2O_4} = 333,000 \text{ g } Na_2S_2O_4$$

or 333 kg (to 3sf)

(b) Mass of the commercial product that contains the $Na_2S_2O_4$ produced using the amounts of reactants in part (a):

$$\frac{333 \text{ kg } Na_2S_2O_4}{1} \cdot \frac{100 \text{ kg commercial product}}{90.1 \text{ kg } Na_2S_2O_4} = 369 \text{ kg commercial product}$$

73. Mass of H_2SO_4 that can be prepared from 525 kg of FeS_2:

Examine the 3 equations given. Note that in the first equation, 4 mol FeS_2 give 8 mol SO_2. Carrying the SO_2 forward, 8 mol SO_2 would give 8 mol SO_3. In the 3rd equation, 8 mol SO_3 would give rise to 8 mol H_2SO_4.

Calculate the # of mol of pyrite present:

$$525 \times 10^3 \text{ g of } FeS_2 \cdot \frac{1 \text{ mol } FeS_2}{119.98 \text{ g } FeS_2} = 4376 \text{ mol } FeS_2$$

From this we can determine the mol H_2SO_4 possible:

$$\frac{4376 \text{ mol } FeS_2}{1} \cdot \frac{8 \text{ mol } H_2SO_4}{4 \text{ mol } FeS_2} \cdot \frac{98.08 \text{ g } H_2SO_4}{1 \text{mol } H_2SO_4} = 858,000 \text{ g } H_2SO_4 \text{ or } 858 \text{ kg } H_2SO_4.$$

75. A mixture of butene, C_4H_8, and butane, C_4H_{10}, has a mass if 2.86 g.

8.80 g of CO_2 and 4.14 g of H_2O result upon combustion.

What is the weight percents of butene and butane in the mixture?

Write the balanced equations for the combustions:

Butene: $C_4H_8 + 6 O_2 \rightarrow 4 CO_2(g) + 4 H_2O(g)$

Butane: $C_4H_{10} + 13/2 O_2 \rightarrow 4 CO_2(g) + 5 H_2O(g)$

Note that both butene and butane provide 4 mol of C for each mol of the hydrocarbon. Butene also produces 4 mol of H_2O (8 mol of H), while butane produces 5 mol of H_2O (10 mol of H).

Consider the amounts of CO_2 and H_2O produced:

$$\frac{8.80 \text{ g } CO_2}{1} \cdot \frac{1 \text{ mol } CO_2}{44.01 \text{ g } CO_2} \cdot \frac{1 \text{ mol C}}{1 \text{ mol } CO_2} = 0.200 \text{ mol C}$$

$$\frac{4.14 \text{ g } H_2O}{1} \cdot \frac{1 \text{ mol } H_2O}{18.02 \text{ g } H_2O} \cdot \frac{2 \text{ mol H}}{1 \text{ mol } H_2O} = 0.459 \text{ mol H}$$

Examining the ratios of C:H that exist (from the calculation above), recall that a 100% butene sample would have, for 0.200 mol C, 0.400 mol H (two mol H for one mol C). A 100% butane sample would have, for 0.200 mol C, 0.500 mol H (2.5 mol H for 1 mol C).

Examining the differences between the amounts of H (0.500 – 0.400) in these two samples, and the difference we find in our mixture (0.459-0.400) we have a percent difference of:

(0.059)/0.100 x 100 = 59% of the mixture that is butane (and 41% butene).

These percentages correspond to 1.687 g butane, and 1.172 g butene.

[As a double check, calculate the mass of CO_2 and H_2O that these two amounts of butane and butene would generate.]

77. $NaHCO_3$ can be decomposed quantitatively by heating:

$$2 \text{ } NaHCO_3(s) \rightarrow Na_2CO_3(s) + CO_2(g) + H_2O(g)$$

A 0.682 g sample of impure $NaHCO_3$ yielded a solid residue (consisting of Na_2CO_3 and other solids) with a mass of 0.467 g. What was the mass percent of $NaHCO_3$ in the sample?

We solve by assuming that the only specie in the impure sample to liberate CO_2 is $NaHCO_3$. Further, the heating will drive off any water formed as gas. The balanced equation indicates that for each mole of CO_2 we get one mole of H_2O. The difference in mass between the solid residue and the original sample is then water and carbon dioxide. (0.682 g – 0.467 g) = 0.215 g. Treat the loss of mass as a molecule of $CO_2 + H_2O$ combined—with a molar mass of that is equal to the **combined** molar masses of $CO_2 + H_2O$ or 62.02 g/mol.

Moles of "water and carbon dioxide" =
$$\frac{0.215 \text{ g}}{1} \cdot \frac{1 \text{ mol "water & carbon dioxide"}}{62.02 \text{ g}} = 3.47 \times 10^{-3} \text{mol "water and carbon dioxide"}$$
The balanced equation indicates that one gets 1 mol of carbon dioxide (and one of water) for each 2 mol of $NaHCO_3$ that decomposes. So the mixture originally contained $2 \times 3.47 \times 10^{-3}$ mol of $NaHCO_3$. The mass of $NaHCO_3$ to which this corresponds is:

$$\frac{6.94 \times 10^{-3} \text{ mol } NaHCO_3}{1} \cdot \frac{84.01 \text{ g } NaHCO_3}{1 \text{ mol } NaHCO_3} = 0.582 \text{ g } NaHCO_3$$

The mass percent of $NaHCO_3$ is : (0.582 g $NaHCO_3$/ 0.682 g sample) x 100 =

$$85.4 \% \ NaHCO_3$$

Summary and Conceptual Questions

<u>79.</u> The graph representing the mass of iron consumed with a **fixed** mass of bromine shows that for the mass of bromine used, the amount of iron consumed maximizes at 2 g of Fe.

(a) According to the graph, the mass of Fe consumed maximizes at 2.0 g Fe with a fixed amount of bromine(about 8.6 g). Any excess iron that is added does not result in additional product.

$$2.0 \text{ g Fe} \bullet \frac{1 \text{ mol Fe}}{55.85 \text{ g Fe}} = 0.036 \text{ mol Fe}$$

The amount of bromine would be:

$$8.6 \text{ g Br} \bullet \frac{1 \text{ mol Br}}{79.9 \text{ g Br}} = 0.108 \text{ mol Br}$$

(b) The ratio of $\dfrac{Br}{Fe}$ is $\dfrac{0.108}{0.036} = 2.99$ or 3:1 (to 1 sf)

(c) The empirical formula is $FeBr_3$.

(d) The balanced equation: $2 \text{ Fe} + 3 \text{ Br}_2 \rightarrow 2 \text{ FeBr}_3$

(e) The product is $FeBr_3$: *iron(III) bromide*

(f) When 1.00 g of Fe is added to the Br_2, *Fe is the limiting reactant* (the addition of more Fe results in the formation of more product).

81. (a) The weight percent of N, Pt, and Cl in cis-platin $[Pt(NH_3)_2Cl_2]$:

Adding the atomic weights of 1 Pt, 2N, 6H, and 2Cl gives a molar mass of: 300.05 g/mol.

The weight percent of N:

$$\frac{2 \times 14.01 \text{ g N}}{300.05 \text{ g cisplatin}} \bullet 100 = 9.34 \ \%N$$

The weight percent of Pt:

$$\frac{1 \times 195.08 \text{ g Pt}}{300.05 \text{ g cisplatin}} \bullet 100 = 65.016 \ \% \text{ Pt and the weight percent of Cl:}$$

$$\frac{2 \times 35.453 \text{ g Cl}}{300.05 \text{ g cisplatin}} \bullet 100 = 23.631 \ \% \text{ Cl}$$

(b) Cisplatin is made by reaction K_2PtCl_4 with ammonia.

$$K_2PtCl_4(aq) + 2\ NH_3(aq) \rightarrow Pt(NH_3)_2Cl_2(aq) + 2\ KCl\ (aq)$$

If you begin with 16.0 g of K_2PtCl_4, what mass of ammonia should be used to completely consume the K_2PtCl_4?

Convert mass of K_2PtCl_4 to mol K_2PtCl_4. Note that 2 mole of NH_3 are required for each mole of K_2PtCl_4.

$$\frac{16.0\ g\ of\ K_2PtCl_4}{1} \cdot \frac{1\ mol\ K_2PtCl_4}{415.09\ g\ of\ K_2PtCl_4} \cdot \frac{2\ mol\ NH_3}{1\ mol\ K_2PtCl_4} \cdot \frac{17.03\ g\ NH_3}{1 mol\ NH_3} = 1.31\ g\ NH_3$$

What mass of cisplatin will be produced?

$$\frac{16.0\ g\ K_2PtCl_4}{1} \cdot \frac{1\ mol\ K_2PtCl_4}{415.09\ g\ K_2PtCl_4} \cdot \frac{1\ mol\ Pt(NH_3)_2Cl_2}{1\ mol\ K_2PtCl_4} \cdot \frac{300.05\ g\ Pt(NH_3)_2Cl_2}{1\ mol\ Pt(NH_3)_2Cl_2} = 11.6\ gPt(NH_3)_2Cl_2$$

83. Each of the 3 flasks contains 0.100 mol of HCl. The mass of Zn differs. Calculate the # of mol Zn in each flask.

Flask 1: 7.00 g Zn $\cdot \dfrac{1\ mol\ Zn}{65.39\ g\ Zn}$ =0.107 mol Zn

Flask 2: 3.27 g Zn $\cdot \dfrac{1\ mol\ Zn}{65.39\ g\ Zn}$ =0.0500 mol Zn

Flask 3: 1.31 g Zn $\cdot \dfrac{1\ mol\ Zn}{65.39\ g\ Zn}$ =0.0200 mol Zn

The balanced equation tells us that for each mol of Zn, we need 2 mol of HCl. In flask 1, 0.107 mol Zn exceeds the amount of HCl available. The reaction consumes all the HCl, and leaves unreacted Zn metal. In flask 2, the 0.0500 mol Zn reacts **exactly** with the 0.100 mol of HCl, leaving **no** unreacted Zn or HCl. In flask 3, the 0.0200 mol of Zn react with 0.0400 mol of HCl—completely consuming **all the Zn**, and leaving unreacted HCl. The smaller amount of Zn present (0.0200 mol) produces a smaller amount of H_2 (0.0200 mol)—thus not totally inflating the balloon.

Chapter 5
Reactions in Aqueous Solution

Practicing Skills

Electrolytes and Solubility of Compounds

1. What is an electrolyte? What are experimental means for discriminating between weak and strong electrolytes?

   An electrolyte is a substance whose aqueous solution conducts an electric current.

   As to experimental means for discriminating between weak and strong electrolytes, refer to the apparatus in the Active Figure5.2. NaCl is a strong electrolyte and would cause the bulb to glow brightly—reflecting a large number of ions in solution while aqueous ammonia or vinegar (an aqueous solution of acetic acid) would cause the bulb to glow only dimly—indicating a smaller number of ions in solution.

3. Predict water solubility:
   (a) $CuCl_2$ is expected to be soluble, while CuO and $FeCO_3$ are not. Chlorides are generally water soluble, while oxides and carbonates are not.
   (b) $AgNO_3$ is soluble. AgI and Ag_3PO_4 are not soluble. Nitrate salts are soluble. Phosphate salts are generally insoluble. While halides are generally soluble, those of Ag^+ are not.
   (c) K_2CO_3, KI and $KMnO_4$ are soluble. In general, salts of the alkali metals are soluble.

5. Ions produced when the compounds dissolve in water.

Compound	Cation	Anion
(a) KOH	K^+	OH^-
(b) K_2SO_4	$2 K^+$	SO_4^{2-}
(c) $LiNO_3$	Li^+	NO_3^-
(d) $(NH_4)_2SO_4$	$2 NH_4^+$	SO_4^{2-}

7.

Compound	Water Soluble	Cation	Anion
(a) Na_2CO_3	yes	$2 Na^+$	CO_3^{2-}
(b) $CuSO_4$	yes	Cu^{2+}	SO_4^{2-}
(c) NiS	no		
(d) $BaBr_2$	yes	Ba^{2+}	$2 Br^-$

Precipitation Reactions and Net Ionic Equations

9. $CdCl_2(aq) + 2\,NaOH(aq) \rightarrow Cd(OH)_2(s) + 2\,NaCl(aq)$

   Net ionic equation: $Cd^{2+}(aq) + 2\,OH^-(aq) \rightarrow Cd(OH)_2(s)$

11. Balanced equations for precipitation reactions:

   (a) $NiCl_2(aq) + (NH_4)_2S(aq) \rightarrow NiS(s) + 2\,NH_4Cl(aq)$

   Net ionic equation: $Ni^{2+}(aq) + S^{2-}(aq) \rightarrow NiS(s)$

   (b) $3\,Mn(NO_3)_2(aq) + 2\,Na_3PO_4(aq) \rightarrow Mn_3(PO_4)_2(s) + 6\,NaNO_3(aq)$

   Net ionic equation: $3\,Mn^{2+}(aq) + 2\,PO_4^{3-}(aq) \rightarrow Mn_3(PO_4)_2(s)$

Acids and Bases

13. $HNO_3(aq) + H_2O(\ell) \rightarrow H_3O^+(aq) + NO_3^-(aq)$

   alternatively: $HNO_3(aq) \rightarrow H^+(aq) + NO_3^-(aq)$

15. $H_2C_2O_4(aq) \rightarrow H^+(aq) + HC_2O_4^-(aq)$

   $HC_2O_4^-(aq) \rightarrow H^+(aq) + C_2O_4^{2-}(aq)$

17. $MgO(s) + H_2O(\ell) \rightarrow Mg(OH)_2(s)$

Reactions of Acids and Bases

19. Complete and Balance

   (a) $2\,CH_3CO_2H(aq) + Mg(OH)_2(s) \rightarrow Mg(CH_3CO_2)_2(aq) + 2\,H_2O(\ell)$
       acetic   magnesium   magnesium   water
       acid   hydroxide   acetate

   (b) $HClO_4(aq) + NH_3(aq) \rightarrow NH_4ClO_4(aq)$
       perchloric   ammonia   ammonium
       acid   perchlorate

21. Write and balance the equation for barium hydroxide reacting with nitric acid:

   $Ba(OH)_2(s) + 2\,HNO_3(aq) \rightarrow Ba(NO_3)_2(aq) + 2\,H_2O(\ell)$
       barium   nitric   barium   water
       hydroxide   acid   nitrate

Writing Net Ionic Equations

23. (a) $(NH_4)_2CO_3(aq) + Cu(NO_3)_2(aq) \rightarrow CuCO_3(s) + 2\,NH_4NO_3(aq)$

   (net) $CO_3^{2-}(aq) + Cu^{2+}(aq) \rightarrow CuCO_3(s)$

(b) $Pb(OH)_2(s) + 2\ HCl(aq) \rightarrow\ PbCl_2(s) + 2\ H_2O(\ell)$

(net) $Pb(OH)_2(s) + 2\ H^+(aq) + 2\ Cl^-(aq) \rightarrow PbCl_2(s) + 2\ H_2O(\ell)$

(c) $BaCO_3(s)\ + 2\ HCl(aq) \rightarrow BaCl_2(aq) + H_2O(\ell) + CO_2(g)$

(net) $BaCO_3(s)\ + 2\ H^+(aq) \rightarrow Ba^{2+}(aq) + H_2O(\ell) + CO_2(g)$

25. (a) $AgNO_3(aq) + KI(aq)\ \rightarrow AgI(s) + KNO_3(aq)$

(net) $Ag^+(aq)\ +\ I^-(aq)\ \rightarrow AgI(s)$

(b) $Ba(OH)_2(aq) + 2\ HNO_3(aq)\ \rightarrow\ 2\ H_2O(\ell)\ + Ba(NO_3)_2(aq)$

(net) $OH^-(aq) +\ H^+(aq) \rightarrow H_2O(\ell)$

(c) $2\ Na_3PO_4(aq) + 3\ Ni(NO_3)_2(aq)\ \rightarrow Ni_3(PO_4)_2(s) + 6\ NaNO_3(aq)$

(net) $2\ PO_4^{3-}(aq) + 3\ Ni^{2+}(aq) \rightarrow Ni_3(PO_4)_2(s)$

Gas-Forming Reactions

27. Write and balance the equation for iron(II) carbonate reacting with nitric acid:

$$FeCO_3(s)\ +\ 2\ HNO_3(aq)\ \rightarrow\ Fe(NO_3)_2(aq)\ +\ H_2O(\ell)\ +\ CO_2(g)$$
| iron(II) carbonate | nitric acid | iron(II) nitrate | water | carbon dioxide |

Types of Reactions in Aqueous Solution

29. Acid-Base (AB) , Precipitation (PR), or Gas-Forming (GF)

(a) $Ba(OH)_2(s) + 2\ HCl(aq) \rightarrow\ BaCl_2(aq) + 2\ H_2O(\ell)$ AB

(b) $2\ HNO_3(aq) + CoCO_3(s) \rightarrow Co(NO_3)_2(aq) + H_2O(\ell) + CO_2(g)$ GF

(c) $2\ Na_3PO_4(aq) + 3\ Cu(NO_3)_2(aq) \rightarrow Cu_3(PO_4)_2(s) + 6\ NaNO_3(aq)$ PR

31. Acid-Base (AB) , Precipitation (PR), or Gas-Forming (GF)

(a) $MnCl_2(aq)\ +\ Na_2S(aq)\ \rightarrow\ MnS(s)\ +\ 2\ NaCl(aq)$ PR

(net) $Mn^{2+}(aq)\ +\ S^{2-}(aq)\ \rightarrow\ MnS(s)$

(b) $K_2CO_3(aq) + ZnCl_2(aq)\ \rightarrow\ ZnCO_3(s) + 2\ KCl(aq)$ PR

(net) $CO_3^{2-}(aq)\ +\ Zn^{2+}(aq)\ \rightarrow\ ZnCO_3(s)$

Product- or Reactant-Favored Reactions

33. Feature causing the reaction to be product-favored:

(a) The formation of the solid CuS "drives" the reaction.

(b) The formation of the liquid H_2O "drives" the reaction

Oxidation Numbers

35. For questions on oxidation number, read the symbol (x) as "the oxidation number of x."

(a) BrO_3^- (Br) + 3(O)   = -1

Since oxygen almost always has an oxidation number of -2, we can substitute this value and solve for the oxidation number of Br.

(Br) + 3(-2)   = -1

(Br)   = +5

(b) $C_2O_4^{2-}$ 2 (C) + 4 (O) = -2

2 (C) + 4 (-2) = -2

2 (C) + -8   = -2

2 (C)   = +6

(C)   = +3

(c) F^- The oxidation number for any monatomic ion is the charge on the ion. So (F) = -1

(d) CaH_2 (Ca) + 2 (H) = 0

(Ca) + 2 (-1) = 0

(Ca)   = +2

(e) H_4SiO_4 4(H) + (Si) + 4(O) = 0

4(+1) + (Si) + 4(-2) = 0

(Si)   = +4

(f) HSO_4^- (H) + (S) + 4(O) = -1

(+1) + (S) + 4(-2) = -1

(S)   = +6

Oxidation-Reduction Reactions

37. (a) Oxidation-Reduction: $Zn(s)$ has an oxidation number of 0, while $Zn^{2+}(aq)$ has an oxidation number of +2—hence Zn is being oxidized.
N in NO_3^- has an oxidation number of +5, while N in NO_2 has an oxidation number of +4—hence N is being reduced.

(b) Acid-Base reaction: There is no change in oxidation number for any of the elements in this reaction—hence it is NOT an oxidation-reduction reaction.
H_2SO_4 is an acid, and $Zn(OH)_2$ acts as a base.

(c) Oxidation-Reduction: $Ca(s)$ has an oxidation number of 0, while $Ca^{2+}(aq)$ has an oxidation number of +2—hence Ca is being oxidized.

H in H_2O has an oxidation number of +1, while H in H_2

has an oxidation number of 0—hence H is being reduced.

39. Determine which reactant is oxidized and which is reduced:

(a) $C_2H_4(g) + 3 O_2(g) \rightarrow 2 CO_2(g) + 2 H_2O(g)$

ox. number

specie	before	after	has experienced	functions as the
C	-2	+4	oxidation	(C_2H_4) reducing agent
H	+1	+1	no change	
O	0	-2	reduction	(O_2) oxidizing agent

(b) $Si(s) + 2 Cl_2(g) \rightarrow SiCl_4(\ell)$

ox. number

specie	before	after	has experienced	functions as the
Si	0	+4	oxidation	(Si) reducing agent
Cl	0	-1	reduction	(Cl_2) oxidizing agent

Solution Concentration

41. Molarity of Na_2CO_3 solution:

$$6.73 \text{ g } Na_2CO_3 \cdot \frac{1 \text{ mol } Na_2CO_3}{106.0 \text{ g } Na_2CO_3} = 0.0635 \text{ mol } Na_2CO_3$$

$$\text{Molarity} \equiv \frac{\# \text{ mol}}{L} = \frac{0.0635 \text{ mol } Na_2CO_3}{0.250 \text{ L}} = 0.254 \text{ M } Na_2CO_3$$

Concentration of Na^+ and CO_3^{2-} ions:

$$\frac{0.254 \text{ mol } Na_2CO_3}{L} \cdot \frac{2 \text{ mol } Na^+}{1 \text{ mol } Na_2CO_3} = 0.508 \text{ M } Na^+$$

$$\frac{0.254 \text{ mol } Na_2CO_3}{L} \cdot \frac{1 \text{ mol } CO_3^{2-}}{1 \text{ mol } Na_2CO_3} = 0.254 \text{ M } CO_3^{2-}$$

43. Mass of $KMnO_4$:

$$\frac{0.0125 \text{ mol } KMnO_4}{L} \cdot \frac{0.250 \text{ L}}{1} \cdot \frac{158.0 \text{ g } KMnO_4}{1 \text{ mol } KMnO_4} = 0.494 \text{ g } KMnO_4$$

45. Volume of 0.123 M NaOH to contain 25.0 g NaOH:

   Calculate moles of NaOH in 25.0 g:

   $$\frac{25.0 \text{ g NaOH}}{1} \cdot \frac{1 \text{ mol NaOH}}{40.00 \text{ g NaOH}} = 0.625 \text{ mol NaOH}$$

   The volume of 0.123 M NaOH that contains 0.625 mol NaOH:

   $$0.625 \text{ mol NaOH} \cdot \frac{1 \text{ L}}{0.123 \text{ mol NaOH}} \cdot \frac{1 \times 10^3 \text{ mL}}{1 \text{L}} = 5.08 \times 10^3 \text{ mL}$$

47. Identity and concentration of ions in each of the following solutions:

   (a) 0.25 M $(NH_4)_2SO_4$ gives rise to (2 x 0.25) 0.50 M NH_4^+ ions & 0.25 M SO_4^{2-} ions

   (b) 0.123 M Na_2CO_3 gives rise to (2 x 0.123) 0.246 M Na^+ ions & 0.123 M CO_3^{2-}.

   (c) 0.056 M HNO_3 gives rise to 0.056 M H^+ ions & 0.056 M NO_3^- ions.

Preparing Solutions

49. Prepare 500. mL of 0.0200 M solution of Na_2CO_3:

   Decide what amount (#moles) of sodium carbonate are needed:

   $$\frac{0.0200 \text{ M Na}_2\text{CO}_3}{1 \text{ L}} \cdot \frac{0.500 \text{ L}}{1} = 0.0100 \text{ mol Na}_2\text{CO}_3$$

   What mass does 0.0100 mol sodium carbonate have?

   MW = 106.0 g Na_2CO_3/ mol Na_2CO_3 and using 1/100 mol would require:

   0.0100 mol Na_2CO_3 • 106.06 g/mol Na_2CO_3 or *1.06 g Na_2CO_3*.

   To prepare the solution, take 1.06 g of Na_2CO_3, and transfer it to the volumetric flaks. Add a

   bit of distilled water and stir carefully until all the solid Na_2CO_3 has dissolved. Once the solid

   has dissolved, and distilled water to the calibrated mark on the neck of the volumetric flask.

   Stopper, stir to assure complete mixing!

51. Molarity of HCl in the diluted solution:

   We can calculate the molarity if we know the number of moles of HCl in the 25.0 mL solution

   1. Moles of HCl in 25.0 mL of 1.50 M HCl :

   $$M \times V = \frac{1.50 \text{ mol HCl}}{L} \cdot \frac{25.00 \times 10^{-3} L}{1} = 0.0375 \text{ mol HCl}$$

   2. When that number of moles is distributed in 500. mL:

   $$\frac{0.0375 \text{ mol HCl}}{0.500 \text{ L}} = 0.0750 \text{ M HCl}$$

Perhaps a shorter way to solve this problem is to note the number of moles (found by multiplying the original molarity times the volume) is distributed in a given volume, resulting in the diluted molarity. Mathematically: $M_1 \times V_1 = M_2 \times V_2$

53. Using the definition of molarity we can calculate the amount of H_2SO_4 in the following:

$$\frac{0.125 \, mol \, H_2SO_4}{L} \bullet 1.00 \, L = 0.125 \text{ moles of } H_2SO_4$$

If we dilute 20.8 mL of 6.00 M H_2SO_4 to one liter we obtain:

$$\frac{6.00 \, M \, H_2SO_4}{1 \, L} \bullet 0.0208 L = 0.1248 \text{ mol } H_2SO_4 \text{ or a 0.125 M solution of } H_2SO_4$$

Method (a) is correct!

pH

55. The hydrogen ion concentration of a wine whose pH = 3.40:

Since pH is defined as $-\log[H^+]$; $[H^+] = 10^{-3.40} = 4.0 \times 10^{-4} \, M \, H^+$

The solution has a pH < 7.00, so it is **acidic**.

57. The $[H^+]$ and pH of a solution of 0.0013 M HNO_3:

Since nitric acid is a strong acid (and strong electrolyte), we can state that:

$[H^+] = 0.0013$ M; the pH = - log (0.0013) or 2.89

59. Make the interconversions and decide if the solution is acidic or basic:

	pH	$[H^+]$	Acidic/Basic
(a)	**1.00**	0.10	Acidic
(b)	**10.50**	3.2×10^{-11} M	Basic
(c)	4.89	**1.3×10^{-5} M**	Acidic
(d)	7.64	**2.3×10^{-8} M**	Basic

pH values less than 7 indicate an acidic solution while those greater than 7 indicate a basic solution. Similarly solutions for which the $[H^+]$ is greater than 1.0×10^{-7} are acidic while those with hydrogen ion concentrations LESS THAN 1.0×10^{-7} are basic.

Stoichiometry of Reactions in Solution

61. Volume of 0.109 M HNO_3 to react with 2.50 g of $Ba(OH)_2$:

Need several steps:

1. calculate mol of barium hydroxide in 2.50 g

$$2.50 \text{ g Ba(OH)}_2 \cdot \frac{1 \text{ mol Ba(OH)}_2}{171.3 \text{ g Ba(OH)}_2}$$

2. calculate mole of HNO_3 needed to react with that # of mol of barium hydroxide

$$2.50 \text{ g Ba(OH)}_2 \cdot \frac{1 \text{ mol Ba(OH)}_2}{171.3 \text{ g Ba(OH)}_2} \cdot \frac{2 \text{ mol HNO}_3}{1 \text{ mol Ba(OH)}_2}$$

3. calculate volume of 0.109 M HNO_3 that contains that # of mol of nitric acid.

$$2.50 \text{ g Ba(OH)}_2 \cdot \frac{1 \text{ mol Ba(OH)}_2}{171.3 \text{ g Ba(OH)}_2} \cdot \frac{2 \text{ mol HNO}_3}{1 \text{ mol Ba(OH)}_2} \cdot \frac{1 \text{ L}}{0.109 \text{ M HNO}_3} = 0.268 \text{ L}$$

or 268 mL.

63. Mass of NaOH formed from 15.0 L of 0.35 M NaCl:

$$\frac{0.35 \text{ mol NaCl}}{1\text{L}} \cdot \frac{15.0 \text{ L}}{1} \cdot \frac{2 \text{ mol NaOH}}{2 \text{ mol NaCl}} \cdot \frac{40.0 \text{ g NaOH}}{1 \text{ mol NaOH}} = 210 \text{ g NaOH}$$

Mass of Cl_2 obtainable:

$$\frac{0.35 \text{ mol NaCl}}{1 \text{ L}} \cdot \frac{15.0 \text{ L}}{1} \cdot \frac{1 \text{ mol Cl}_2}{2 \text{ mol NaCl}} \cdot \frac{70.9 \text{ g Cl}_2}{1 \text{ mol Cl}_2} = 186 \text{ g Cl}_2 \text{ or } 190 \text{ mL (2 sf)}$$

65. Volume of 0.0138M $Na_2S_2O_3$ to dissolve 0.225 g of AgBr:

Calculate: 1. mol of AgBr

2. mol of $Na_2S_2O_3$ needed to react (balanced equation)

3. volume of 0.0138 M $Na_2S_2O_3$ containing that number of moles.

$$0.225 \text{ g AgBr} \cdot \frac{1 \text{ mol AgBr}}{187.8 \text{ g AgBr}} \cdot \frac{2 \text{ mol Na}_2\text{S}_2\text{O}_3}{1 \text{ mol AgBr}} \cdot \frac{1 \text{ L}}{0.0138 \text{ mol Na}_2\text{S}_2\text{O}_3}$$
$$\cdot \frac{1000 \text{ mL}}{1 \text{ L}} = 174 \text{ mL Na}_2\text{S}_2\text{O}_3$$

67. The balanced equation:

$$Pb(NO_3)_2(aq) + 2 NaCl(aq) \rightarrow PbCl_2(s) + 2 NaNO_3(aq)$$

Volume of 0.750 M $Pb(NO_3)_2$ needed:

$$\frac{2.25 \text{ mol NaCl}}{1 \text{ L}} \cdot \frac{1.00 \text{ L}}{1} \cdot \frac{1 \text{ mol Pb(NO}_3)_2}{2 \text{ mol NaCl}} \cdot \frac{1 \text{ L}}{0.750 \text{ mol Pb(NO}_3)_2} \cdot \frac{1000 \text{ mL}}{1 \text{ L}}$$
$$= 1500 \text{ mL or } 1.50 \times 10^3 \text{ mL}$$

Titrations

69. To calculate the volume of 0.812 M HCl needed, we calculate the moles of NaOH in 1.45 g, then use the stoichiometry of the balanced equation:

$$HCl(aq) + NaOH(aq) \rightarrow NaCl(aq) + H_2O(\ell)$$

$$1.45 \text{ g NaOH} \cdot \frac{1 \text{ mol NaOH}}{40.00 \text{ g NaOH}} \cdot \frac{1 \text{ mol HCl}}{1 \text{ mol NaOH}} \cdot \frac{1 \text{ L}}{0.812 \text{ mol HCl}} \cdot \frac{1000 \text{ mL}}{1 \text{ L}}$$

$$= 44.6 \text{ mL HCl}$$

71. Calculate:

1. moles of Na_2CO_3 corresponding to 2.150 g Na_2CO_3

2. moles of HCl that react with that number of moles (using the balanced equation)

3. the molarity of HCl containing that number of moles of HCl in 38.55 mL.

The balanced equation is:

$$Na_2CO_3(aq) + 2\,HCl(aq) \rightarrow 2\,NaCl(aq) + H_2O(\ell) + CO_2\,(g)$$

$$2.150 \text{ g } Na_2CO_3 \bullet \frac{1 \text{ mol } Na_2CO_3}{106.0 \text{ g } Na_2CO_3} \bullet \frac{2 \text{ mol HCl}}{1 \text{ mol } Na_2CO_3} \bullet \frac{1}{0.03855 \text{ L sol}}$$

$$= 1.052 \text{ M HCl}$$

73. Molar Mass of an acid if 36.04 mL of 0.509 M NaOH will titrate 0.954 g of an acid H_2A:

Note that the acid is a diprotic acid, and will require 2 mol of NaOH for each mol of the acid.

Calculate the # moles of NaOH: $\dfrac{0.509 \text{ mol NaOH}}{1 \text{ L}} \bullet 0.03604 \text{ L} = 0.0183 \text{ mol NaOH}$

The number of moles of acid will be **half** of the number of moles of NaOH:

$$0.0183 \text{ mol NaOH} \bullet \frac{1 \text{ mol } H_2A}{2 \text{ mol NaOH}} = 0.00917 \text{ mol } H_2A$$

Since we know the mass corresponding to this number of moles of acid, we can calculate the

molar mass (# g/mol): $\dfrac{0.954 \text{ g } H_2A}{0.00917 \text{ mol } H_2A} = 104 \text{ g/mol}$

75. Mass percent of iron in a 0.598 gram sample that requires 22.25 mL of 0.0123 M $KMnO_4$:

Note from the balanced, net ionic equation that 1 mol MnO_4^- requires 5 mol of Fe^{2+}.

The # mol of MnO_4^- is: $\dfrac{0.0123 \text{ M } MnO_4^-}{1 \text{ L}} \bullet 0.02225 \text{ L} = 2.74 \times 10^{-4} \text{ mol } MnO_4^-$

Using the ratio of permanganate ion to iron(II) ion we obtain:

$$2.74 \times 10^{-4} \text{ mol } MnO_4^- \bullet \frac{5 \text{ mol } Fe^{2+}}{1 \text{ mol } MnO_4^-} = 1.37 \times 10^{-3} \text{ mol } Fe^{2+}$$

Using the mass of the iron(II) ion, we can calculate the mass of iron present:

$$1.37 \times 10^{-3} \text{ mol } Fe^{2+} \bullet \frac{55.847 \text{ g } Fe^{2+}}{1 \text{ mol } Fe^{2+}} = 0.0764 \text{ g } Fe^{2+}$$

The mass percent of iron is then: $\dfrac{0.0764 \text{ g } Fe^{2+}}{0.598 \text{ g sample}} \bullet 100 = 12.8 \text{ \% } Fe^{2+}$

General Questions

77. Formula for the following compounds:

   (a) soluble compound with Br^- ion: almost any bromide compound with the exception of Ag^+, Hg_2^{2+} and Pb^{2+}

   (b) insoluble hydroxide: almost any hydroxide except the salts of NH_4^+ and the alkali metal ions

   (c) insoluble carbonate: almost any carbonate except the salts of NH_4^+ and the alkali metal ions

   (d) soluble nitrate-containing compound: all nitrate-containing compounds are soluble

   The listing of soluble and insoluble compounds in Section 5.1 of your text will provide general guidelines for predicting the solubility of compounds.

79. For the following copper salts:

   Water soluble: $Cu(NO_3)_2$, $CuCl_2$ — nitrates and chlorides are soluble

   Water insoluble: $CuCO_3$, $Cu_3(PO_4)_2$ — carbonates and phosphates are insoluble

81. **Spectator ions** in the following equation and the net ionic equation:

   $$2\ H^+(aq) + 2\ \mathbf{NO_3^-(aq)} + Mg(OH)_2(s) \rightarrow\ 2\ H_2O(\ell) + Mg^{2+}(aq) + 2\ \mathbf{NO_3^-(aq)}$$

   The emboldened nitrate ions are the spectator ions. The net ionic equation would be the first equation shown above without the spectator ions:

   $$2\ H^+(aq) +\ Mg(OH)_2(s) \rightarrow 2\ H_2O(\ell) + Mg^{2+}(aq)\ \text{[An acid-base exchange]}$$

83. For the reaction of chlorine with NaBr:

   $$Cl_2(g) + 2\ NaBr(aq) \rightarrow 2\ NaCl(aq) + Br_2(\ell)$$

   (a) Oxidized: **bromine's** oxidation number is changed from -1 to 0

   Reduced: **chlorine's** oxidation number is changed from 0 to –1

   (b) Oxidizing agent: $\mathbf{Cl_2}$ removes the electrons from NaBr

   Reducing agent: **NaBr** provides the electrons to the chlorine.

   (c) Mass of Cl_2 to react with 125 mL of 0.153 M NaBr:

   $$\frac{0.125\ L}{1} \bullet \frac{0.153\ mol\ NaBr}{1\ L} \bullet \frac{1\ mol\ Cl_2}{2\ mol\ NaBr} \bullet \frac{70.91\ g\ Cl_2}{1\ mol\ Cl_2} = 0.678\ g\ Cl_2$$

85. Greater mass of solute in 1 L of 0.1 M NaCl or 1 L of 0.06 M Na_2CO_3

   First determine the number of moles of each solute in the amount of solution indicated.

   Since Molarity is defined as #moles/Liter solution, the product of Molarity • Volume (in L) gives the # of moles. An alternate, but equivalent, question to ask is which has the greater mass: 0.1 mol of NaCl or 0.06 mol of Na_2CO_3. The molar mass of NaCl is approximately (23 + 35.5) or 58.8 grams, while that for Na_2CO_3 is [2(23) + 12 + 3(16)] or 106 grams. One-tenth mol of NaCl would have a mass of approximately (0.1 • 58.4) or 5.84 g while 0.06 mol of sodium carbonate would have a mass of (0.06 • 106) or 6.36 grams. To one sf the mass of NaCl and Na_2CO_3 would be 6 g.

87. A 500.0 mL solution of 0.20 M Na_2CO_3 will require 0.10 mol of Na_2CO_3:

Since 0.20 mol/L Na_2CO_3 • 0.5000 L = 0.10 mol Na_2CO_3.

The molar mass of Na_2CO_3 is 105.99 g/mol, so we need 10.6 g of the salt (11 g to 2 sf).

Weigh 10.6 g of solid into a beaker, add water until the solid dissolves, then quantitatively (no spills allowed!) transfer the liquid into the volumetric flask. Add water to the flask until the meniscus of the liquid rests atop the graduation marking the 500.0 mL mark for that flask.

89. Greater concentration of H^+ ion: 0.015 M HCl or solution with pH of 1.2

With HCl as a strong acid, the $[H^+]$ for 0.015 M HCl is 0.015 M H^+ or 1.5 x 10^{-2} M. The solution with pH = 1.2 has a $[H^+]$ of $10^{-1.2}$ (recall that pH is the negative logarithm of hydrogen ion concentration) or 6.3 x 10^{-2} M—so this solution has the greater concentration of hydrogen ion.

91. Reaction: $MgCO_3(s) + 2\ HCl(aq) \rightarrow CO_2(g) + MgCl_2(aq) + H_2O(\ell)$

(a) The net ionic equation: $MgCO_3(s) + 2\ H^+(aq) \rightarrow CO_2(g) + Mg^{2+}(aq) + H_2O(\ell)$

The spectator ion is the chloride ion (Cl^-).

(b) The production of $CO_2(g)$ characterizes this as a gas-forming reaction.

(c) Mass of $MgCO_3(s)$ to react with 125 mL of HCl whose pH = 1.56:

The solution with pH = 1.56 has a $[H^+]$ of $10^{-1.56}$ or 2.75 x 10^{-2} M.

$$\frac{0.0275\ mol\ H^+}{1\ L} \bullet \frac{0.125\ L}{1} \bullet \frac{1\ mol\ MgCO_3}{2\ mol\ H^+} \bullet \frac{84.31g\ MgCO_3}{1\ mol\ MgCO_3} = 0.14\ g\ MgCO_3$$

Since the pH given only has 2 sf, our answer has but 2.

93. Species present in aqueous solutions of:

compound	types of species	species present
(a) NH_3	molecules (weak base)	NH_3, NH_4^+, OH^-
(b) CH_3CO_2H	molecules (weak acid)	CH_3CO_2H, $CH_3CO_2^-$, H^+
(c) NaOH	ions (strong base)	Na^+ and OH^-
(d) HBr	ions (strong acid)	H^+ and Br^-

In every case, H_2O will be present (but omitted in this list)

95. Limiting reagent between 125 mL of 0.15 M CH_3CO_2H and 15.0 g $NaHCO_3$:

   moles CH_3CO_2H: $\dfrac{0.15 \text{ mol } CH_3CO_2H}{1L} \bullet 0.125 \text{ L} = 0.01875 \text{ mol } CH_3CO_2H$

   moles $NaHCO_3$: $15.0 \text{ g } NaHCO_3 \bullet \dfrac{1 \text{ mol } NaHCO_3}{84.01 \text{ g } NaHCO_3} = 0.179 \text{ mol } NaHCO_3$

Since the reaction proceeds with a 1:1 stoichiometry, *CH_3CO_2H is the limiting reagent.*

The reaction will also produce 1 mol of $NaCH_3CO_2$ for each mol of CH_3CO_2H.

$$\dfrac{0.01875 \text{ mol } CH_3CO_2H}{1} \bullet \dfrac{1 \text{ mol } NaCH_3CO_2}{1 \text{ mol } CH_3CO_2H} \bullet \dfrac{82.03 \text{ g } NaCH_3CO_2}{1 \text{ mol } NaCH_3CO_2} = 1.5 \text{ g } NaCH_3CO_2$$

97. Weight percent of $Na_2S_2O_3$ in 3.232 g sample of material:

$$\dfrac{0.246 \text{ mol } I_2}{1 \text{ L}} \bullet \dfrac{0.04021 \text{ L}}{1} \bullet \dfrac{2 \text{ mol } Na_2S_2O_3}{1 \text{ mol } I_2} \bullet \dfrac{158.11 \text{ g } Na_2S_2O_3}{1 \text{ mol } Na_2S_2O_3} = 3.13 \text{ g } Na_2S_2O_3$$

Now that we know the amount of sodium thiosulfate in the sample, we can calculate the percent of the compound in the impure mixture.

$$\dfrac{3.13 \text{ g } Na_2S_2O_3}{3.232 \text{ g mixture}} \bullet 100 = 96.8\%$$

99. Solubility: (See the Solubility guidelines in Section 5.1 of your text)
   (a) Two soluble copper(II) salts:

   copper(II) nitrate: $Cu(NO_3)_2$ and copper(II) halides: CuF_2, $CuCl_2$, $CuBr_2$, CuI_2

   Two insoluble copper(II) salts:

   copper(II) sulfide, CuS and copper(II) carbonate: $CuCO_3$

   (b) Two soluble barium salts:

   barium nitrate: $Ba(NO_3)_2$ and barium chloride: $BaCl_2$

   Two insoluble barium salts:

   barium sulfate: $BaSO_4$ and barium phosphate: $Ba_3(PO_4)_2$

101. Overall balanced and net equations:

   (a) aqueous lead(II) nitrate + aqueous potassium hydroxide:
   $Pb(NO_3)_2(aq) + 2 \text{ KOH}(aq) \rightarrow Pb(OH)_2(s) + 2 \text{ KNO}_3(aq)$
   (net) $Pb^{2+}(aq) + 2 \text{ OH}^-(aq) \rightarrow Pb(OH)_2(s)$

   Note that neither the nitrate or potassium ions appear in the net equation, since lead(II) nitrate and potassium hydroxide both exist as the four separate ions (as reactants); the potassium nitrate exists as separate ions (as product).

(b) aqueous copper(II) nitrate + aqueous sodium carbonate:

$Cu(NO_3)_2(aq) + Na_2CO_3(aq) \rightarrow CuCO_3(s) + 2\,NaNO_3(aq)$

(net) $Cu^{2+}(aq) + CO_3^{2-}(aq) \rightarrow CuCO_3(s)$

Both reactants are water soluble, and exist as the four independent ions.

Sodium nitrate (product) is also water soluble, and exists as the independent sodium and nitrate ions. Copper(II) carbonate is **not** water soluble, and exists as the entity, $CuCO_3(s)$.

The result is that nitrate and sodium ions do not appear in the net equation.

103. Mass of $NaHCO_3$ needed to exactly neutralize the H^+ in a solution of HCl (125 mL ; pH=2.56). The solution with pH = 2.56 has a $[H^+]$ of $10^{-2.56}$ or 2.75×10^{-3} M.

and 125 mL will contain: $\dfrac{2.75 \times 10^{-3}\ \text{mol } H^+}{1\ L} \cdot \dfrac{0.125\ L}{1} = 3.44 \times 10^{-4}$ mol H^+.

The equation is: $HCl\,(aq) + NaHCO_3(s) \rightarrow NaCl(aq) + H_2O(\ell) + CO_2\,(g)$

Knowing the amount of hydrogen ion is present in the 125 mL of solution, we can calculate the moles and the mass of $NaHCO_3$.

$$\dfrac{3.44 \times 10^{-4}\ \text{mol } H^+}{1} \cdot \dfrac{1\,\text{mol } NaHCO_3}{1\ \text{mol } H^+} \cdot \dfrac{84.01\ g\ NaHCO_3}{1\,\text{mol } NaHCO_3} = 0.029\ g\ NaHCO_3$$

105. pH of solution resulting when 250. mL of 0.0105 M NaOH is added to 250. mL of HCl solution with pH = 1.92:

HCl will react with NaOH on a 1 mol:1 mol basis. First determine the # mol of HCl in the original solution.

pH = 1.92 so $[H^+] = 10^{-1.92} = 1.20 \times 10^{-2}$ M HCl and since the volume is 250. mL:

The product of M • V = 3.01×10^{-3} mol HCl.

The number of moles of HCl that will react is equal to the # mol NaOH:

$$\dfrac{0.0105\ \text{mol NaOH}}{1\ L} \bullet 0.250\ L = 0.00263\ \text{mol NaOH} = 0.00263\ \text{mol HCl react}$$

The amount of HCl remaining is then:

3.01×10^{-3} mol HCl - 2.63×10^{-3} mol HCl = 3.8×10^{-4} mol HCl

and the concentration is: $\dfrac{3.8 \times 10^{-4}\,\text{mol}}{0.500\ L} = 7.6 \times 10^{-4}$ M

and the pH = -log[7.6×10^{-4}] = 3.13

107. For the reaction of $NaHCO_3$ with acetic acid (CH_3COOH):

$$\frac{200.\ mL\ vinegar}{1} \cdot \frac{50.0\ g\ vinegar}{1000\ mL\ vinegar} \cdot \frac{1\ mol\ vinegar}{60.05\ g\ vinegar} \cdot \frac{1\ mol\ NaHCO_3}{1\ mol\ vinegar} \cdot \frac{84.01\ g\ NaHCO_3}{1\ mol\ NaHCO_3}$$

$$\cdot \frac{1\ spoonful}{3.8\ g\ NaHCO_3} = 3.7\ spoonfuls$$

So 4 spoonfuls of sodium bicarbonate will be required.

109. 2.56 g $CaCO_3$ in a beaker containing 250. mL 0.125 M HCl. After reaction, does any $CaCO_3$ remain?

The amount of HCl present is: $\dfrac{0.125\ mol\ HCl}{1\ L} \cdot \dfrac{0.250\ L}{1} = 0.0313\ mol\ HCl$

The amount of HCl that reacts with the $CaCO_3$ corresponding to 2.56 g is:

$$\frac{2.56\ g\ CaCO_3}{1} \cdot \frac{1\ mol\ CaCO_3}{100.09\ g\ CaCO_3} \cdot \frac{2\ mol\ HCl}{1\ mol\ CaCO_3} = 0.0512\ mol\ HCl$$

This calculation shows that the amount of HCl present is NOT SUFFICIENT to consume the $CaCO_3$ corresponding to 2.56 g of the solid.

Mass of $CaCl_2$ produced:

Since HCl is the limiting reagent, the amount of $CaCl_2$ that can be produced is:

$$\frac{0.0313\ mol\ HCl}{1} \cdot \frac{1\ mol\ CaCl_2}{2\ mol\ HCl} \cdot \frac{110.98\ g\ CaCl_2}{1\ mol\ CaCl_2} = 1.74\ g\ CaCl_2$$

<u>111.</u> The equation: $Cr(NH_3)_xCl_3(aq) + x\ HCl\ (aq) \rightarrow x\ NH_4^+(aq) + Cr^{3+}(aq) + (x+3)Cl^-(aq)$

The amount of HCl: 23.63 mL of 1.500 M HCl (1.500 mol/L x 0.02363L) = 0.03545 mol

We don't know the ratio between moles of salt and moles of HCl. X can be no smaller than 1, but we don't know an upper bound for its value.

Assume that there is only 1 NH_3 in the chromium salt. The molar mass of that salt would be 175.38 g/mol. The 1.580 g of salt would correspond to 1.580 g/175.38 g/mol or 9.01×10^{-3} mol. We know, from the calculation of HCl that we have **at least** 0.03545 mol of HCl (and mol of NH_3. This tells us that there is more than 1 ammonia unit/unit of chromium salt.

Calculate the MM of a complex containing ever increasing ammonia units per formula unit of chromium salt. Those formulas and MM are contained in the 2 right columns of data below. If we divide those MM into 1.580g of salt, we can calculate the # of moles of salt present. This calculation is the third column of data in the chart. If we divide the number of moles of HCl by increasing number of ammonia units (data in 1st and 2nd column), we get a proposed number of moles of salt. Compare the number of moles of salt arrived at (in column 2) with the number of moles of salt arrived at in column 3. The graph is a plot of #NH_3 molecules vs (mol HCl/mol NH_3) and #NH_3 molecules vs #mol salt calculated from the assumed formulas (in column5). The intersection (at #NH$_3$ = 6) tells us that x has a value of 6, and the salt has the formula Cr(NH$_3$)$_6$Cl$_3$.

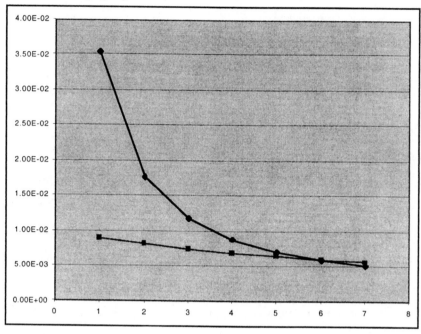

# NH3	# mol HCl/mol NH3	=1.580/MM	MM	Formula
1	3.55E-02	9.01E-03	175.38	CrCl3(NH3)1
2	1.77E-02	8.21E-03	192.41	CrCl3(NH3)2
3	1.18E-02	7.54E-03	209.45	CrCl3(NH3)3
4	8.86E-03	6.98E-03	226.48	CrCl3(NH3)4
5	7.09E-03	6.49E-03	243.51	CrCl3(NH3)5
6	5.91E-03	6.06E-03	260.54	CrCl3(NH3)6
7	5.06E-03	5.69E-03	277.57	CrCl3(NH3)7

113. Determine the # mol of methylene blue:

$$\frac{1.0 \text{ g } C_{16}H_{18}ClN_3S}{1} \bullet \frac{1 \text{ mol } C_{16}H_{18}ClN_3S}{319.86 \text{ g } C_{16}H_{18}ClN_3S} = 0.0031 \text{ mol } C_{16}H_{18}ClN_3S$$

The measurement indicates that the molar concentration of methylene blue is 4.1×10^{-8} M. The concentration of methylene blue in the measured sample will be equal to that of methylene blue in the pool--assuming that the methylene blue has been uniformly distributed throughout the pool water before the sample was taken. Recalling that M = # mol/V, and knowing that we know BOTH the molarity and the # mol, we can calculate the volume.

$$\frac{0.0031 \text{ mol methylene blue}}{1} \bullet \frac{1 \text{ L}}{4.1 \times 10^{-8} \text{ mol methylene blue}} = 76,000 \text{ L (2 sf)}$$

Note that we ignored the 50 mL of solution that originally contained the 1.0 g of methylene blue.

114. Determine the amount of Na_2CO_3 in the original sample:

$$\frac{0.2200 \text{ g impure Na}_2CO_3}{1} \cdot \frac{99.50 \text{ g Na}_2CO_3}{100.00 \text{ g impure Na}_2CO_3} \cdot \frac{1 \text{ mol Na}_2CO_3}{105.99 \text{ gNa}_2CO_3}$$

$$= 2.065 \times 10^{-3} \text{ mol Na}_2CO_3$$

The equation for the reaction of Na_2CO_3 with HCl shows a ratio of 2 mol HCl:1 mol Na_2CO_3. Hence the number of mol of HCl present is 4.131×10^{-3} mol.

In the separate experiment, 38.63 mL HCl was equal to 33.27 mL NaOH. Since we know that NaOH and HCl react on a 1mol:1mol basis, we know that the #mol of HCl and # mol NaOH in these two amounts are equal. Express the " HCl equivalence of 1 mL of NaOH"

$$1 \text{ mL NaOH} = \frac{38.63 \text{ mL HCl}}{33.27 \text{ mL NaOH}} = 1.161 \text{ mL HCl},$$ so the backtitration (which required 2.75

mL of NaOH) is equal to (2.75×1.161)mL HCl or 3.19 mL HCl.

This volume of HCl is present **in excess** of that needed to exactly react with the Na_2CO_3.

So the volume of HCl needed to react with the Na_2CO_3 is (43.50 –3.19) or 40.31 mL.

The concentration of HCl is then: 4.131×10^{-3} mol HCl/0.04031L = 0.1025M HCl

The concentration of NaOH is then calculated from the second experiment:

0.1025 mol HCl/L x 0.03863 L = **B** M NaOH x 0.03327 L NaOH and solving for **B**.

B = 0.1190 M NaOH.

115. Weight percent of Cu in 0.251 g of a copper containing alloy:
(a) Oxidizing and reducing agents in the two equations:

$$2 \text{ Cu}^{2+}(aq) + 5 \text{ I}^- (aq) \rightarrow 2 \text{ CuI}(s) + I_3^-(aq)$$

I^- reduces the Cu(II) ion to the Cu(I) ion—so I^- acts as the reducing agent.

Cu^{2+} oxidizes I^- to I_3^- (I_2 and I^-)–and acts as the oxidizing agent.

$$I_3^-(aq) + 2 \text{ S}_2O_3^{2-}(aq) \rightarrow S_4O_6^{2-}(aq) + 3I^-(aq)$$

$S_2O_3^{2-}$ reduces the I_3^-ion(I_2 and I^-)–to the I^- ion—so $S_2O_3^{2-}$acts as the reducing agent.

I_3^- oxidizes $S_2O_3^{2-}$ to $S_4O_6^{2-}$ –and acts as the oxidizing agent.

In $S_4O_6^{2-}$, the oxidation state of S can be thought of as (+2.5), and in $S_2O_3^{2-}$ as (+2)

(b) Several steps in this problem:
   (1) Note the relationship between Cu (Cu^{2+}) and I_3^- formed (2 mol Cu^{2+} : 1 mol I_3^-)
   (2) Excess I_3^- reacts with $S_2O_3^{2-}$ in a 1 mol I_3^- to 2 mol $S_2O_3^{2-}$ ratio)
   (3) Calculate the amount of thiosulfate in 26.32 mL of 0.101 M solution:

$$\frac{0.101 \text{ mol } S_2O_3{}^{2-}}{1 \text{ L}} \bullet 0.02632 \text{ L} = 0.00266 \text{ mol } S_2O_3{}^{2-} \text{ (to 3 sf)}$$

(4) This amount can now be used to calculate $I_3{}^-$ produced (step 2), and the amount of $I_3{}^-$ related to the amount of Cu present (step 1):

$$0.00266 \text{ mol } S_2O_3{}^{2-} \bullet \frac{1 \text{ mol } I_3{}^-}{2 \text{ mol } S_2O_3{}^{2-}} \bullet \frac{2 \text{ mol } Cu^{2+}}{1 \text{ mol } I_3{}^-} \bullet \frac{1 \text{ mol } Cu}{1 \text{ mol } Cu^{2+}} \bullet \frac{63.546 \text{ g } Cu}{1 \text{ mol } Cu} = 0.169 \text{ g } Cu$$

The weight percent is then: $\dfrac{0.169 \text{ g } Cu}{0.251 \text{ g alloy}} \bullet 100 = 67.3 \text{ % } Cu$

117. For the reaction in which Au is dissolved by treatment with sodium cyanide:
   (a) Oxidizing agent: elemental O_2 Reducing agent: elemental Au

   Substance oxidized: elemental Au (oxidation state changes from 0 to +1)
   Substance reduced: elemental O_2 (oxidation state changes from 0 to -2)

   (b) Volume of 0.075 M NaCN to extract gold from 1000 kg of rock:

   Mass of gold in the rock: $\dfrac{0.019 \text{ g gold}}{100 \text{ g rock}} \bullet \dfrac{10^3 \text{ g rock}}{1 \text{ kg rock}} \bullet 10^3 \text{ kg rock} = 190 \text{ g gold}$

   Amount of NaCN needed (from the balanced equation):

   $$190 \text{ g } Au \bullet \frac{1 \text{ mol } Au}{196.9 \text{ g } Au} \bullet \frac{8 \text{ mol NaCN}}{4 \text{ mol } Au} = 1.9 \text{ mol NaCN}$$

   Volume of 0.075 M NaCN that contains that amount of NaCN

   $$1.9 \text{ mol NaCN} \bullet \frac{1 \text{ L}}{0.075 \text{ mol NaCN}} = 26 \text{ L}$$

Summary and Conceptual Questions

<u>119.</u> To decide the relative concentrations, calculate the dilutions. Let's assume that the HCl has a concentration of 0.100 M. Then calculate the diluted concentrations in each case.

Student 1 :

20.0 mL of 0.100 M HCl is diluted to 40.0 mL total

The diluted molarity is then : $\dfrac{0.100 \text{ mol HCl}}{1 \text{ L}} \bullet 0.0200 \text{ L} = 0.400 \text{ L} \bullet \text{M}$

$$0.050 \frac{\text{mol HCl}}{\text{L}} = \text{M}$$

Student 2 :

20.0 mL of 0.100 M HCl is diluted to 80.0 mL total

The diluted molarity is : $\dfrac{0.100 \text{ mol HCl}}{1 \text{L}} \bullet 0.0200 \text{ L} = 0.0800 \text{ L} \bullet \text{M}$

$$0.025 \frac{\text{mol HCl}}{\text{L}} = \text{M}$$

So the second student's prepared HCl solution is **(c) half the concentration of the first student**'s. **However**, since the total number of moles of HCl in both solutions *is identical,* they will calculate **the same concentration for the original HCl solution.**

<u>121</u>. The amount of HCl contained in all 3 flasks is identical. The amount of Zn in flask 2 is a stoichiometric amount (1 mol Zn: 2 mol HCl). Flask 1 has excess Zn—but HCl limits the amount of hydrogen gas. Flask 3 contains an insufficient amount of Zn to react with all the HCl present—and a smaller volume of hydrogen gas results.

123. Using the reagents, $BaCl_2$, $BaCO_3$, $Ba(OH)_2$, H_2SO_4, Na_2SO_4,

Prepare barium sulfate by:

(a) a precipitation reaction
The reaction of $BaCl_2$ with Na_2SO_4 will perform this task:
$$BaCl_2 \text{ (aq) } + Na_2SO_4\text{(aq)} \rightarrow BaSO_4\text{(s)} + 2 \text{ NaCl (aq)}$$

(b) a gas-forming reaction
$$BaCO_3 + H_2SO_4 \rightarrow BaSO_4\text{(s)} + H_2O\text{(l)} + CO_2\text{(g)}$$
One would think about using $Ba(OH)_2$ as one reactant for part (a), but the substance isn't very water-soluble. $BaCl_2$ is much more water-soluble.

125. If the BAL is found to be 100 mg ethanol (C_2H_5OH) per deciliter of blood, a person will receive a DWI ticket. Suppose a person is found to have 0.033 mol of ethanol per liter of blood. Will the person be arrested?
Express the concentration of ethanol in g (or mg) per liter:

$$\frac{0.033 \text{ mol } C_2H_5OH}{1 \text{ L blood}} \cdot \frac{46.07 \text{ g } C_2H_5OH}{1 \text{ mol } C_2H_5OH} \cdot \frac{1000 \text{ mg } C_2H_5OH}{1 \text{ g } C_2H_5OH} \cdot \frac{1 \text{ L blood}}{10 \text{ dL blood}}$$
$$= 150 \text{ mg/dL blood}$$

Given these data, the person **should be arrested**.

Chapter 6
Principles of Reactivity:
Energy and Chemical Reactions

Practicing Skills

Energy

1. To move the lever, one uses mechanical energy. The energy resulting is manifest in electrical energy (which produces light); thermal energy would be released as the bulb in the flashlight glows.

Energy Units

3. Express 1200 Calories/day in Joules:

$$\frac{1200 \text{ Cal}}{\text{day}} \cdot \frac{1000 \text{ calorie}}{1 \text{ Cal}} \cdot \frac{4.184 \text{ J}}{1 \text{ cal}} = 5.0 \times 10^6 \text{ Joules/day}$$

5. Compare 170 kcal/serving and 280 kJ/serving.

$$\frac{170 \text{ kcal}}{1 \text{ serving}} \cdot \frac{1000 \text{ calorie}}{1 \text{ kcal}} \cdot \frac{4.184 \text{ J}}{1 \text{ cal}} \cdot \frac{1 \text{ kJ}}{1000 \text{ J}} = 710 \text{ kJoules/serving}$$

So 170 kcal/serving has a greater energy content.

Specific Heat Capacity

7. What is the specific heat capacity of mercury, if the molar heat capacity is 28.1 J/mol • K?
   Note that the difference in **units** of these two quantities is in the amount of substance. In one case, moles, while in the other grams.

$$28.1 \frac{\text{J}}{\text{mol} \cdot \text{K}} \cdot \frac{1 \text{ mol}}{200.59 \text{ g}} = 0.140 \frac{\text{J}}{\text{g} \cdot \text{K}}$$

9. Heat energy to warm 168 g copper from -12.2 °C to 25.6 °C:

Heat = mass x heat capacity x ΔT

For copper = $(168 \text{ g})(\frac{0.385 \text{ J}}{\text{g} \cdot K})[25.6°C - (-12.2) °C] \cdot \frac{1 \text{ K}}{1 \text{ °C}}$

= 2.44×10^3 J or 2.44 kJ

11. The final temperature of a 344 g sample of iron when 2.25 kJ of heat are added to a sample originally at 18.2 °C. The energy added is:

q_{Fe} = (mass)(heat capacity)(ΔT)

2.25×10^3 J = $(344 \text{ g})(0.449 \frac{\text{J}}{\text{g} \cdot \text{K}})(x)$ and solving for x we get:

14.57 K = x. and since 1K = 1°C, ΔT = 14.57 °C.

The final temperature is (14.57 + 18.2)°C or 32.8°C.

13. Final T of copper-water mixture:

We must **assume** that **no energy** will be transferred to or from the beaker containing the water. Then the **magnitude** of energy lost by the hot copper and the energy gained by the cold water will be equal (but opposite in sign).

$$q_{copper} = -q_{water}$$

Using the heat capacities of H_2O and copper, and expressing the temperatures in Kelvin $(K = °C + 273.15)$ we can write:

$$(45.5 \text{ g})(0.385 \frac{J}{g \cdot K})(T_{final} - 372.95 \text{ K}) = -(152. \text{ g})(4.184 \frac{J}{g \cdot K})(T_{final} - 291.65 \text{K})$$

Simplifying each side gives:

$$17.52 \frac{J}{K} \cdot T_{final} - 6533 \text{ J} = -636.0 \frac{J}{K} \cdot T_{final} + 185,480 \text{ J}$$

$$653.52 \frac{J}{K} \cdot T_{final} = 192013 \text{ J}$$

$$T_{final} = 293.81 \text{ K or } (293.81 - 273.15) \text{ or } 20.7 \text{ °C}$$

Don't forget: **Round numbers only at the end**.

15. Final temperature of water mixture:

This problem is solved almost exactly like question 13. The difference is that both samples are samples of water. From a mechanical standpoint, the heat capacity of both samples will be identical—and can be omitted from both sides of the equation:

$$q_{water} \text{ (at 95 °C)} = -q_{water} \text{ (at 22 °C)}$$

$$(85.2 \text{ g})(4.184 \frac{J}{g \cdot K})(T_{final} - 368.15 \text{ K}) = -(156 \text{ g})(4.184 \frac{J}{g \cdot K})(T_{final} - 295.15 \text{ K}) \text{ or}$$

$$(85.2 \text{ g})(T_{final} - 368.15 \text{ K}) = -(156 \text{ g})(T_{final} - 295.15 \text{ K}) \text{ or}$$

$$85.2 \text{ J/K } T_{final} - 31366.38 \text{ J} = -156 \text{ J/K } T_{final} + 46043.4 \text{ J}$$

rearranging: $241.2 \text{ J/K} \cdot T_{final} = 77409.78 \text{ J}$

$$321.0 \text{ K} = T_{final} \text{ or } 47.8 \text{ °C}$$

17. Exercise 6.3 in your text is another good example for this problem.

$$q_{metal} = -q_{water}$$

Remembering that $\Delta T = T_{final} - T_{initial}$, we can calculate the change in temperature for the water and the metal. Further, since we know the final and initial for both the metal and the water, we can calculate the temperature difference in units of Celsius degrees, since the **change** in temperature on the **Kelvin** scale would be numerically identical.

For the metal : $\Delta T = T_{final} - T_{initial} = (27.1 - 98.8)$ or $-71.7°C$ or -71.7 K.

For the water: $\Delta T = T_{final} - T_{initial} = (27.1 - 25.0)$ or $2.1°C$ or 2.1 K (recalling that a Celsius degrees and a Kelvin are the same "size".

$$(13.8 \text{ g})(C_{metal})(-71.7 \text{ K}) = -(45.0 \text{ g})(4.184 \tfrac{J}{g \cdot K})(2.1 \text{ K})$$

$$-989.46 \text{ g} \cdot K(C_{metal}) = -395 \text{ J}$$

$$C_{metal} = 0.40 \tfrac{J}{g \cdot K} \quad \text{(to 2 significant figures)}$$

Change of State

19. Quantity of energy evolved when 1.0 L of water at 0 °C solidifies to ice:

The mass of water involved: If we assume a density of liquid water of 1.000 g/mL,

1.0 L of water (1000 mL) would have a mass of 1000 g.

To freeze 1000 g water: $1000 \text{ g ice} \cdot \dfrac{333 \text{ J}}{1.000 \text{ g ice}} = 333. \text{ x } 10^3 \text{ J}$ or 330 kJ (to 2sf)

21. Heat required to vaporize (convert liquid to solid) 125 g C_6H_6:

The heat of vaporization of benzene is 30.8 kJ/mol.

Convert mass of benzene to moles of benzene: $125 \text{ g} \cdot 78.11 \text{ g/mol} = 1.60 \text{ mol}$
Heat required: $1.60 \text{ mol } C_6H_6 \cdot 30.8 \text{ kJ/mol} = 49.3 \text{ kJ}$

NOTE: As in question 19, no sign has been attached to the amount of heat, since we wanted to know the **amount**. If we want to assign a **direction** of heat flow in this question, then we would add a (+) to 49.3 kJ to indicate that heat is being **added** to the liquid benzene.

23. To calculate the quantity of heat for the process described, think of the problem in two steps:

   1) cool liquid from 23.0 °C to liquid at – 38.8 °C
   2) freeze the liquid at its freezing point (– 38.8 °C)

Note that the specific heat capacity is expressed in units of mass, so convert the volume of liquid mercury to **mass**. $1.00 \text{ mL} \cdot 13.6 \text{ g/mL} = 13.6 \text{ g Hg}$ (Recall:1 cm^3 = 1 mL)

1) The energy to cool 13.6 g of Hg from 23.0 °C to liquid at – 38.8 °C is:

$\Delta T = (234.35 \text{ K} - 296.15 \text{ K})$ or - 61.8 K

$13.6 \text{ g Hg} \cdot 0.140 \dfrac{J}{g \cdot K} \cdot - 61.8 \text{ K} = -118 \text{ J}$

2) To convert liquid mercury to solid Hg at this temperature:

$- 11.4 \text{ J/g} \cdot 13.6 \text{ g} = -155 \text{ J}$ (The (-) sign indicates that heat is being removed from the Hg.

The total energy released by the Hg is: [- 118 J + - 155 J] = - 273 J and since $q_{mercury} = - q_{surroundings}$ the amount released to the surroundings is 273 J.

25. To accomplish the process, one must:

   1) heat the ethanol from 20.0 °C to 78.29 °C (ΔT = 58.29 K)

   2) boil the ethanol (convert from liquid to gas) at 78.29 °C

   Using the specific heat for ethanol, the energy for the first step is:

   $(2.44 \frac{J}{g \cdot K})(1000 \text{ g})(58.29 \text{ K})$ = 142,227.6 J (142,000 J to 3 sf)

   To boil the ethanol at 78.29 °C, we need:

   $$855 \frac{J}{g} \cdot 1000 \text{ g} = 855,000 \text{ J}$$

   The total heat energy needed (in J) is (142,000 + 855,000) = 997,000 or 9.97×10^5 J

Enthalpy

   Note that in this chapter, I have left negative signs with the value for heat released

   (heat released = - ; heat absorbed = +)

27. For a process in which the $\Delta H°$ is negative, that process is **exothermic**.

   To calculate heat released when 1.25 g NO react, note that the energy shown (-114.1 kJ) is

   released when **2** moles of NO react, so we'll need to account for that:

   $1.25 \text{ g NO} \cdot \frac{1 \text{ mol NO}}{30.01 \text{ g NO}} \cdot \frac{-114.1 \text{ kJ}}{2 \text{ mol NO}}$ = -2.38 kJ

29. The combustion of isooctane (IO) is **exothermic**. The molar mass of IO is: 114.2 g/mol.

   The heat evolved is:

   $1.00 \text{ L of IO} \cdot \frac{0.69 \text{ g IO}}{1 \text{mL}} \cdot \frac{1 \times 10^3 \text{ mL}}{1 \text{ L}} \cdot \frac{1 \text{ mol IO}}{114.2 \text{ g IO}} \cdot \frac{-10922 \text{ kJ}}{2 \text{ mol IO}}$ = -3.3×10^4 kJ

Calorimetry

31. 100.0 mL of 0.200 M CsOH and 50.0 mL of 0.400 M HCl each supply 0.0200 moles of base

   and acid respectively. If we assume the specific heat capacities of the solutions are

   4.2 J/g • K, the **heat evolved** for 0.200 moles of CsOH is:

   q = (4.2 J/g • K)(150. g)(24.28 °C - 22.50°C) [and since 1.78°C = 1.78 K]

   q = (4.2 J/g • K)(150. g)(1.78 K)

   q = 1120 J

   The molar enthalpy of neutralization is: $\frac{-1120 \text{ J}}{0.0200 \text{ mol CsOH}}$ = - 56000 J/mol (to 2 sf)

   or -56 kJ/mol

33. For the problem, we'll assume that the coffee-cup calorimeter absorbs **no** heat.

    Since $q_{metal} = -q_{water}$

    Remembering that $\Delta T = T_{final} - T_{initial}$, we can calculate the change in temperature for the water and the metal. Further, since we know the final and initial for both the metal and the water, we can calculate the temperature difference in units of Celsius degrees, since the **change** in temperature on the **Kelvin** scale would be numerically identical.

    For the metal : $\Delta T = T_{final} - T_{initial} = (24.3 - 99.5)$ or $-75.2°C$ or $-75.2K$.

    For the water: $\Delta T = T_{final} - T_{initial} = (24.3 - 21.7)$ or $2.6°C$ or 2.6 K (recalling that a Celsius degrees and a Kelvin are the same "size".

    $(20.8 \text{ g})(C_{metal})(-75.2 \text{ K}) = -(75.0 \text{ g})(4.184 \frac{J}{g \cdot K})(2.6 \text{ K})$

    $-1564.16 \text{ g} \cdot \text{K}(C_{metal}) = -816 \text{ J}$

    $$C_{metal} = 0.52 \frac{J}{g \cdot K} \text{ (to 2 significant figures)}$$

35. Enthalpy change when 5.44 g of NH_4NO_3 is dissolved in 150.0 g water at 18.6 °C..

    Calculate the heat released by the solution: $\Delta T = (16.2 - 18.6)$ or -2.4 °C or -2.4 K

    $(155.4 \text{ g})(4.2 \frac{J}{g \cdot K})(-2.4 \text{ K}) = -1566 \text{ J}$ or -1600 J(to 2 sf)

    Calculate the amount of NH_4NO_3: 5.44 g $NH_4NO_3 \cdot \dfrac{1 \text{ mol } NH_4NO_3}{80.04 \text{ g } NH_4NO_3} = 0.0680$ mol

    Recall that the energy that was released by the solution is **absorbed** by the ammonium nitrate, so we change the sign from (-) to (+). The enthalpy change has been requested in units of kJ, so divide the energy (in J) by 1000:

    Enthalpy of dissolving $= \dfrac{1.566 \text{ kJ}}{0.0680 \text{ mol}} = 23.0$ kJ/mol or 23 kJ/mol (to 2 sf)

37. Calculate the heat evolved (per mol SO_2) for the reaction of sulfur with oxygen to form SO_2
    There are several steps:

    1) Calculate the heat transferred to the water:

       815 g $\cdot 4.184 \frac{J}{g \cdot K} \cdot (26.72 - 21.25)°C \cdot 1K/1°C = 18,700$ J

    2) Calculate the heat transferred to the bomb calorimeter

       923 J/K $\cdot (26.72 - 21.25)°C \cdot 1K/1°C = 5,050.$J

    3) Amount of sulfur present: 2.56 g $\cdot \dfrac{1 \text{ mol } S_8}{256.536 \text{ g } S_8} = 0.010$ mol S_8

    Note from the equation that 8 mol of SO_2 form from each mole of S_8

4) Calculating the quantity of heat related per mol of SO_2 yields:

$$\frac{(18,700\ J\ +\ 5,050\ J)}{0.08\ mol\ SO_2} = 297,000\ J/mol\ SO_2\ \text{or } 297\ kJ/mol\ SO_2$$

39. Quantity of heat evolved in the combustion of benzoic acid:

As in study question 37, we can approach this in several steps:

1) Calculate the heat transferred to the water:

$$775\ g \cdot 4.184\ \frac{J}{g \cdot K} \cdot (31.69 - 22.50\ )°C \cdot 1K/1°C = 29,800\ J$$

2) Calculate the heat transferred to the bomb calorimeter

$$893\ J/K \cdot (31.69 - 22.50\ )°C \cdot 1K/1°C = 8,210\ J$$

3) Amount of benzoic acid:

$$1.500\ g\ \text{benzoic acid} \cdot \frac{1\ mol\ \text{benzoic acid}}{122.1\ g\ \text{benzoic acid}} = 1.229 \times 10^{-2}\ mol\ \text{benzoic acid}$$

4) Heat evolved per mol of benzoic acid is:

$$\frac{(29,800\ J\ +\ 8,210\ J)}{1.229 \times 10^{-2}\ mol} = 3.09 \times 10^{6}\ J/mol\ \text{or } 3.09 \times 10^{3}\ kJ/mol$$

41. Heat absorbed by the ice : $\dfrac{333\ J}{1.00\ g\ ice} \cdot 3.54\ g\ ice = 1,180\ J$ (to 3 sf)

Since this energy (1180 J) is released by the metal, we can calculate the heat capacity of the

metal: heat   = heat capacity x mass x ΔT

$\quad$ -1180 J   = C x 50.0 g x (273.2 K - 373 K) [Note that ΔT is negative!]

$0.236\ \dfrac{J}{g \cdot K}$ = C

Note that the heat released (left side of equation) has a negative sign to indicate the **directional flow** of the energy.

Hess's Law

43 (a) Hess' Law allows us to calculate the overall enthalpy change by the appropriate combination of several equations. In this case we add the two equations, reversing the second one (with the concommitant reversal of sign)

$$CH_4\ (g)\ +\ 2\ O_2\ (g)\ \rightarrow CO_2\ (g)\ +\ 2\ H_2O\ (g) \qquad \Delta H° = -\ 802.4\ kJ$$

$$CO_2\ (g)\ +\ 2\ H_2O\ (g) \rightarrow CH_3OH(g) + 3/2\ O_2\ (g) \qquad \Delta H° = +\ 676\ kJ$$

$$\overline{CH_4\ (g)\ +\ 1/2\ O_2\ (g)\ \rightarrow CH_3OH(g)} \qquad\qquad \Delta H° = -\ 126\ kJ$$

(b) A graphic description of the energy change:

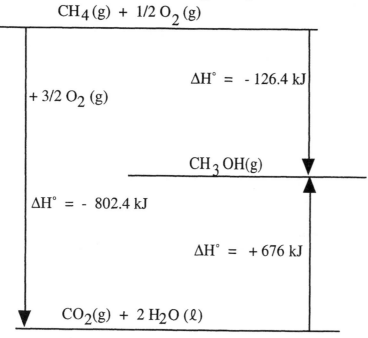

$$CH_4 (g) + 1/2\ O_2 (g)$$

$+ 3/2\ O_2 (g)$

$\Delta H° = -126.4\ kJ$

$CH_3 OH(g)$

$\Delta H° = -802.4\ kJ$

$\Delta H° = +676\ kJ$

$$CO_2(g) + 2\ H_2O\ (\ell)$$

45. The overall enthalpy change for $1/2\ N_2 (g) + 1/2\ O_2 (g) \rightarrow NO\ (g)$

For the overall equation, note that elemental nitrogen and oxygen are on the "left" side of the equation, and NO on the "right" side of the equation. Noting that equation 2 has 4 ammonia molecules consumed, let's multiply equation 1 by 2:

$2\ N_2 (g) + 6\ H_2 (g) \rightarrow 4\ NH_3 (g)$ $\Delta H = (2)(-91.8\ kJ)$

The second equation has NO on the right side :

$4\ NH_3 (g) + 5\ O_2 (g) \rightarrow 4\ NO\ (g) + 6\ H_2O\ (g)$ $\Delta H = -906.2 kJ$

The third equation has water as a product, and we need to "consume" the water formed in equation two, so let's reverse equation 3—changing the sign—AND multiply it by 6

$6\ H_2O\ (g) \rightarrow 6\ H_2 (g) + 3\ O_2 (g)$ $\Delta H = (+241.8)(6)\ kJ$

Adding these 3 equations gives

$2\ N_2 (g) + 2\ O_2 (g) \rightarrow 4\ NO\ (g)$ $\Delta H = 361\ kJ$

So, dividing all the coefficients by 4 provides the desired equation with a $\Delta H = +361 \cdot 0.25\ kJ$ or $90.3\ kJ$

Standard Enthalpies of Formation

47. The equation requested requires that we form **one** mol of product liquid CH_3OH from its elements—each in their standard state.

Begin by writing a balanced equation:

2 C (s,graphite) + O_2(g) + 4 H_2(g) → 2 CH_3OH(ℓ)

Now express the reaction so that you form one mole of CH_3OH—divide coefficients by 2

C (s) + 1/2 O_2(g) + 2 H_2(g) → CH_3OH(ℓ)

And from Appendix L, the $\Delta H_f°$ is reported as –238.4 kJ/mol

49. (a) The equation of the formation of Cr_2O_3 (s) from the elements:

2 Cr (s) + 3/2 O_2 (g) → Cr_2O_3 (s)

from Appendix L $\Delta H_f°$ is reported as: -1134.7 kJ/mol for the oxide.

(b) The enthalpy change if 2.4 g of Cr is oxidized to Cr_2O_3 (g) is:

$$2.4 \text{ g Cr} \cdot \frac{1 \text{ mol Cr}}{52.0 \text{ g Cr}} \cdot \frac{-1134.7 \text{ kJ}}{2 \text{ mol Cr}} = -26 \text{ kJ} \quad (\text{to 2 sf})$$

51. Calculate $\Delta H_{rx}°$ for the following processes:

(a) 1.0 g of white phosphorus burns:

P_4 (s) + 5 O_2 (g) → P_4O_{10} (s)   from Appendix L: $\Delta H_f°$ -2984.0 kJ/mol

$$1.0 \text{ g } P_4 \cdot \frac{1.0 \text{ mol } P_4}{123.89 \text{ g } P_4} \cdot \frac{-2984.0 \text{ kJ}}{1 \text{ mol } P_4} = -24 \text{ kJ}$$

(b) 0.20 mol NO (g) decomposes to N_2 (g) and O_2 (g):

From Appendix L: $\Delta H_f°$ for NO = 90.29 kJ/mol

Since the reaction requested is the **reverse** of $\Delta H_f°$, we change the sign to – 90.29 kJ/mol.

The enthalpy change is then $\frac{-90.29 \text{ kJ}}{1 \text{ mol}} \cdot 0.20 \text{ mol} = -18 \text{ kJ}$

(c) 2.40 g NaCl is formed from elemental Na and elemental Cl_2:

From Appendix L: $\Delta H_f°$ for NaCl (s) = - 411.12 kJ/mol

The amount of NaCl is: $2.40 \text{ g NaCl} \cdot \frac{1 \text{ mol NaCl}}{58.44 \text{ g NaCl}} = 0.0411 \text{ mol}$

The overall energy change is: -411.12 kJ/mol • 0.0411 mol = - 16.9 kJ

(d) 250 g of Fe oxidized to Fe_2O_3 (s):

From Appendix L: $\Delta H_f°$ for Fe_2O_3 (s) = - 824.2 kJ/mol

The overall energy change is:

$$250 \text{ g Fe} \cdot \frac{1 \text{ mol Fe}}{55.8847 \text{ g Fe}} \cdot \frac{1 \text{ mol } Fe_2O_3}{2 \text{ mol Fe}} \cdot \frac{-824.2 \text{ kJ}}{1 \text{ mol } Fe_2O_3} = -1.8 \times 10^3 \text{ kJ}$$

53. (a) The enthalpy change for the reaction:

$$4 NH_3(g) + 5 O_2(g) \rightarrow 4 NO(g) + 6 H_2O(g)$$

$\Delta H^\circ_f(kJ/mol)$ -45.90   0   +90.29   -241.83

$$\Delta H^\circ_{rxn} = [(4 \text{ mol})(+90.29 \frac{kJ}{mol}) + (6 \text{ mol})(-241.83 \frac{kJ}{mol})] -$$

$$[(4 \text{ mol})(-45.90 \frac{kJ}{mol}) + (5 \text{ mol})(0)]$$

= (-1089.82 kJ) - (-183.6 kJ)

= -906.2 kJ   **The reaction is exothermic.**

(b) Heat evolved when 10.0 g NH_3 react:

The balanced equation shows that 4 mol NH_3 result in the release of 906.2 kJ.

$$10.0 \text{ g } NH_3 \cdot \frac{1 \text{ mol } NH_3}{17.03 \text{ g } NH_3} \cdot \frac{-906.2 \text{ kJ}}{4 \text{ mol } NH_3} = -133 \text{ kJ}$$

55. (a) The enthalpy change for the reaction:

$$BaO_2 (s) \rightarrow BaO (s) + 1/2 O_2 (g)$$

Given ΔH°_f for BaO is: - 553.5 kJ/mol and ΔH°_f for BaO_2 is: -634.3 kJ/mol

This equation can be seen as the summation of the two equations:

(1) $Ba (s) + 1/2 O_2 (g) \rightarrow BaO (s)$

(2) $\underline{BaO_2 (s) \rightarrow Ba (s) + O_2 (g)}$

Equation (1) corresponds to the formation of BaO while equation(2) corresponds to the **reverse** of the formation of BaO_2

$$\Delta H^\circ_{rxn} = (\Delta H^\circ_f \text{ for BaO}) + -1(\Delta H^\circ_f \text{ for BaO}_2) =$$

$$\Delta H^\circ_{rxn} = 80.8 \text{ kJ} \quad \text{and the reaction is } \textbf{endothermic.}$$

(b) Energy level diagram for the equations in question:

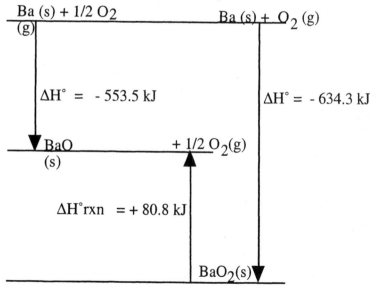

57. The molar enthalpy of formation of naphthalene can be calculated since we're given the enthalpic change for the reaction:

$$C_{10}H_8 (s) + 12 O_2(g) \rightarrow 10 CO_2(g) + 4 H_2O(\ell)$$

$\Delta H°_f(kJ/mol)$?   0   -393.509   -285.83

$$\Delta H°_{rxn} = \sum \Delta H°_f \text{ products } - \sum \Delta H°_f \text{ reactants}$$

$$-5156.1 \text{ kJ} = [(10 \text{ mol})(-393.509 \tfrac{kJ}{mol}) + (4 \text{ mol})(-285.83 \tfrac{kJ}{mol})] - [\Delta H°_f C_{10}H_8]$$

$$-5156.1 \text{ kJ} = (-5078.41 \text{ kJ}) - \Delta H°_f C_{10}H_8$$

$$-77.7 \text{ kJ} = -\Delta H°_f C_{10}H_8$$

$$77.7 \text{ kJ} = \Delta H°_f C_{10}H_8$$

Product- and Reactant-Favored Reactions

59. Calculate $\Delta H°_{rxn}$ for the equations and draw an energy level diagram for:

(a) Aluminum and chlorine to form $AlCl_3$ (s): $Al (s) + 3/2 Cl_2 (g) \rightarrow AlCl_3 (s)$

This equation represents the formation of $AlCl_3$ from it's elements, so we can calculate the $\Delta H°_{rxn}$ by looking up the $\Delta H°_f$ for $AlCl_3$ in Appendix L which is: -705.63 kJ/mol

(Reaction is product-favored.)

(b) Decomposition of HgO to form elemental liquid mercury and oxygen gas:

$$HgO (s) \rightarrow Hg (\ell) + 1/2 O_2 (g)$$

This equation represents the decomposition of HgO into its component elements, so we can calculate the $\Delta H°_{rxn}$ by looking up the $\Delta H°_f$ for HgO in Appendix L (which is: -90.83 kJ/mol) and **changing the sign to +90.83 kJ/mol.**

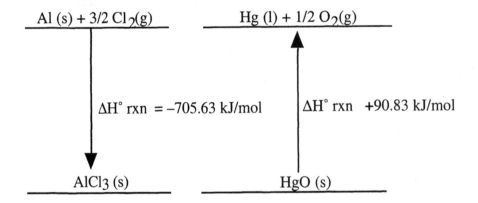

General Questions on Thermochemistry

61. Define and give an example of:

(a) Exothermic and Endothermic—the suffix "thermic" talks about heat, and the prefixes "exo" and "endo" tell us whether heat is ADDED to the surroundings from the system (exo) or REMOVED from the surroundings to the system (endo). Combustion reactions (e.g. gasoline burning in your automobile) are EXOthermic reactions, while ice melting is an ENDOthermic reaction.

(b) System and Surroundings—The "system" is the reactant(s) and product(s) of a reaction, while the "surroundings" is EVERYTHING else. Suppose we burn gasoline in an internal combustion engine. The gasoline (and air) in the cylinder(s) composes the "system", while the engine, and the air contacting the engine is the "surroundings". Together the "system" and "surroundings" compose the "universe"—at least in thermodynamics.

(c) Specific heat capacity—is the quantity of heat required to change the temperature of 1g of a substance by 1 degree Celsius. Water has a specific heat capacity of about 4.2 J/g•K, meaning that 1 gram of water at 15 degrees C, to which 4.2 J of energy is added, will have a temperature of 16 degrees C. (or 14 degrees C—if 4.2 J of energy is removed).

(d) State function—Any parameter which is dependent ONLY on the initial and final states. Chemists typically use CAPITAL letters to indicate state functions (e.g. H, S,) while non-state functions are indicated with LOWER CASE letters (e.g. q, w). Your checking account balance is a state function!

(e) Standard state—Defined as the MOST STABLE (PHYSICAL) STATE for a substance at at a pressure of 1 bar and at a specified temperature. (Typically 298K) The standard state for elemental nitrogen at 25 °C (298K) is **gas**.

(f) Enthalpy change—the heat transferred in a process that is carried out under constant pressure conditions is the enthalpy change, ΔH. The enthalpy change upon the formation of 1 mol of water(ℓ) is –285.8 kJ, meaning that 285.8 kJ is released upon the formation of 1 mol of liquid water from 1 mol of hydrogen (g) and 1/2 mol oxygen (g).

(g) Standard Enthalpy of Formation—the enthalpy change for the formation of 1 mol of a compound directly from its component elements, each in their standard states. The standard enthalpy of formation of nitrogen gas (N_2) = 0 kJ/mol .

63. Define system and surroundings for each of the following, and give direction of heat transfer:

(a) Methane is burning in a gas furnace in your home:

(System) methane + oxygen (Surroundings) components of furnace and the air in your home. The heat flows from the methane + oxygen to the furnace and air.

(b) Water drops on your skin evaporate:

(System) water droplets (Surroundings) your skin and the surrounding air. The heat flows from your skin and the air to the water droplet.

(c) Liquid water at 25 °C is placed in freezer:

(System) water (Surroundings) freezer. The heat flows from the water to the freezer.

(d) Aluminum and FeO react in a flask on a lab bench:

(System) Al and FeO (Surroundings) flask, lab bench, and air around flask. The heat flows from the reaction of Al and FeO into the surroundings.

65. First Law of Thermodynamics: $\Delta E = q + w$. This law says that the change in internal energy (ΔE) is a combination of heat (q) and work (w).

67. Standard Enthalpies of Formation for O(g), O_2 (g), O_3 (g).

Substance	ΔH_f (at 298K) kJ/mol
O(g)	249.170
O_2 (g)	0
O_3 (g)	142.67

What is the standard state of O_2? The standard state of oxygen (O_2) is as a gas.

Is the formation of O from O_2 exothermic?

$\Delta H_{reaction} = \Sigma \Delta H_f$ products $- \Sigma \Delta H_f$ reactants ($O_2 \rightarrow 2O$)

$\Delta H_{reaction} = (2 \text{ mol})(249.170) - (1 \text{ mol})(0) = 498.340$ kJ (endothermic)

What is the ΔH for 3/2 O_2(g) $\rightarrow O_3$ (g)

$\Delta H_{reaction} = \Sigma \Delta H_f$ products $- \Sigma \Delta H_f$ reactants

$\Delta H_{reaction} = (1 \text{ mol})(142.67 \text{ kJ/mol}) - (3/2 \text{ mol})(0) = 142.67$ kJ

69. To calculate the formation of $SnCl_4(\ell)$ and $TiBr_2(s)$ from $SnBr_2(s) + TiCl_4(\ell)$ we can use Hess' Law to add a series of appropriate equations. A screen capture from the CD-ROM shows the overall process:

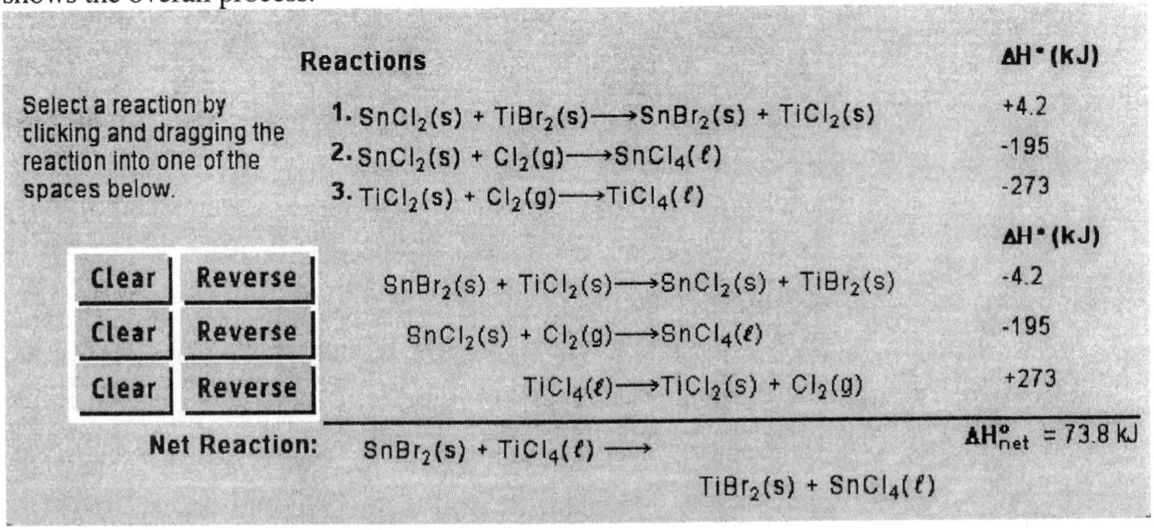

Given the integral values for "Reaction 2", $\Delta H° = 74$ kJ.

71. Releases more heat on cooling from 50 °C to 10 °C—50.0 g water or 100. g of ethanol?

[Specific heat of ethanol = 2.46 J/g•K; of water = 4.18 J/g •K]

There are at least two ways to solve this problem. The mechanical is to solve the equation: q = m•C•Δt for each of the two substances. The other way is to recognize that Δt is the same for both substances (40 °C). Hence the answer depends upon the product m•C. For ethanol the product is (100. x 2.46 J/g•K) for water the product is (50.0 x 4.18 J/g •K). The product for ethanol is GREATER, hence 100. g ETHANOL releases more heat than 50.0 g of water.

73. If 187 J raises the temperature of 93.45 g of Ag from 18.5 to 27.0°C, what is the specific heat capacity of silver?

Recall that q = m•c•Δt; so 187 J = 93.45 g • c • (27.0 – 18.5)°C. and

$$c = \frac{187 \text{ J}}{93.45 \text{ g} \cdot (27.0 - 18.5)°C} = 0.24 \text{ J/g} \cdot K$$

75. Addition of 100.0 g of water at 60 °C to 100.0 g of ice at 0.00 °C. The water cools to 0.00 °C. How much ice has melted?

As the ice absorbs heat from the water, two processes occur: (1) the ice melts and (2) the water cools. We can express this with the equation $q_{water} = -q_{ice}$

The melting of ice can be expressed with the heat of fusion of ice,333 J/g, as q = m • 333 J/g. The cooling of the water may be expressed: q = m • c • ΔT. Setting these quantities equal gives:

m • c • ΔT = -= m • 333 J/g or 100.0 g • (4.184 J/g •K) • (0 – 60)K = -x g • 333 J/g

[x = quantity of ice that melts. Note that since Celsius degrees and kelvin are the same "size", Δt is –60°C or –60 K]

100.0 g • (4.184 J/g •K) • (0 – 60)K = -x • 333 J/g

- 25104 J = - x • 333 J/g or $\frac{-25104 \text{ J}}{- 333 \text{ J/g}}$ = x or 75.4 g of ice.

77. 90 g (two 45 g cubes)of ice cubes (at 0 °C) are dropped into 500. mL tea at 20.0 °C (Assume a density of 1.00 g/mL for tea). What is the final temperature of the mixture if all the ice melts?

As in Study Question 75, $q_{water} = -q_{ice}$ and we can set up the expression.

m • c • ΔT = - m • 333 J/g NOTE however, that not only does all the ice melt (as in SQ75), but the melted ice warms to a temperature above 0 °C. We add a term to account for that:

$m_{tea} • c • \Delta T_{tea} = - [m_{ice} • 333 \text{ J/g} + m_{ice} • c • \Delta T_{ice}]$

500. g • (4.184 J/g•K) • (F – 293.2 K) = -[(90 g • 333 J/g) +

(90 g •(4.184 J/g•K) • (F -273.2 K))]

where F is the final temperature of the tea and melted ice.

2092F J - 613,374 J = - [29970 J + 377F J - 102,876 J] Simplifying:

2092F J - 613,374 J + 29970 J + 377F J - 102,876 J = 0

(2092F J + 377F J) + (- 613,374 J + 29970 J - 102,876 J) = 0

2469 F J + -686,280 = 0 or 2469 F J = 686,280 and F = (686,280/2469) = 278 K

and noting that 45 g of ice cube has 2 sf, we report a final temperature of 280 K.

79. Calculate the enthalpy change for the precipitation of AgCl (in kJ/mol):

1) How much AgCl is being formed?

250. mL of 0.16 M $AgNO_3$ will contain (0.250L • 0.16 mol/L) 0.040 mol of $AgNO_3$

125 mL of 0.32 M NaCl will contain (0.125L • 0.32 mol/L) 0.040 mol of NaCl.

Given the stoichiometry of the process, we anticipate the formation of 0.040 mol of AgCl.

2) How much energy is evolved?

$$375 \text{ g} \cdot 4.2 \frac{\text{J}}{\text{g} \cdot \text{K}} \cdot (296.05 \text{ K} - 294.30) \text{ K} = 2,800 \text{ J (to 2 sf)}$$

The enthalpy change is then - 2800 J (since the reaction **releases** heat).

The change in kJ/mol is $\dfrac{2800 \text{ J}}{0.040 \text{ mol}} \cdot \dfrac{1 \text{ kJ}}{1000 \text{ J}} = - 69$ kJ/mol

81. Heat evolved when ammonium nitrate is decomposed:

$\Delta T = (20.72-18.90) = 1.82 \,°C$ (or 1.82 K).

Heat absorbed by the calorimeter: 155 J/K • 1.82K = 282 J

Heat absorbed by the water : $415 \text{ g} \cdot 4.18 \dfrac{\text{J}}{\text{g} \cdot \text{K}} \cdot 1.82 \text{ K} = 3160$ J

Total heat **released** by the decomposition: 3160 J + 282 J = 3,440 J (to 3 sf)

With 7.647 g NH_4NO_3 = 0.09554 mol, heat released $= \dfrac{3440 \text{ J}}{0.09554 \text{ mol}} = 36.0$ kJ/mol

<u>83</u>. One can arrive at the desired answer if you recall the **definition** of $\Delta H°_f$. The definition is the
enthalpy change associated with the formation of **one mole** of the substance (in this case
B_2H_6) from its elements—each in their standard state (s for boron and g for hydrogen).

(a) Note that the 1st equation given uses **four** moles of B as a reactant — and we'll need only 2,
so divide the first equation by 2 to give:

2 B (s) + 3/2 O_2 (g)   → B_2O_3 (s) $\Delta H° = 1/2(-2543.8 \text{ kJ}) = -1271.9$ kJ

The formation of 1 mole of B_2H_6 will require the use of 6 moles of H (or 3 moles of H_2), so multiply the second equation by 3 to give:

$3 H_2 (g) + 3/2 O_2 (g) \rightarrow 3 H_2O (g)$ $\Delta H° = 3(-241.8 \text{ kJ}) = -725.4 \text{ kJ}$

Finally the third equation given has B_2H_6 as a **reactant and not a product**. So reverse

the third equation to give:

$B_2O_3 (s) + 3 H_2O (g) \rightarrow B_2H_6 (g) + 3 O_2 (g)$ $\Delta H° = -(-2032.9 \text{ kJ}) = +2032.9\text{kJ}$

(b) Adding the three equations gives the equation:

$2 B (s) + 3 H_2 (g) \rightarrow B_2H_6 (g)$ with a $\Delta H° = (-1271.9 + -725.4 + 2032.9)\text{kJ}$

or a $\Delta H°$ for B_2H_6 (g) of + 35.6 kJ

(c) Energy level diagram for the reactions:

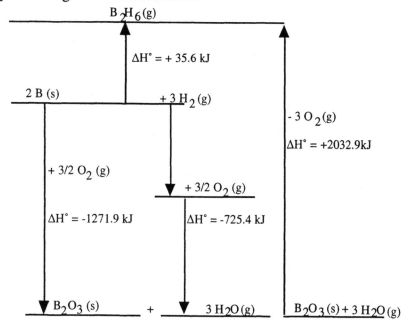

(d) Formation of B_2H_6 (g) is **reactant-favored**

85. The enthalpy change for the reaction:

$$Mg(s) + 2 H_2O(\ell) \rightarrow Mg(OH)_2 (s) + H_2(g)$$
$\Delta H°_f(\text{kJ/mol})$ 0   -285.83   -924.54   0

$$\Delta H°_{rxn} = (1 \text{ mol})(-924.54 \frac{\text{kJ}}{\text{mol}}) - (2 \text{ mol})(-285.83 \frac{\text{kJ}}{\text{mol}})$$

$$= -352.88 \text{ kJ or } -3.5288 \times 10^5 \text{ J}$$

Each mole of magnesium releases 352.88 kJ of heat energy.

Calculate the heat required to warm 25 mL of water from 25 to 85 °C.

heat   = heat capacity x mass x ∆T

$$= (4.184 \ \frac{kJ}{mol})(25 \ mL)(\frac{1.00 \ g}{1 \ mL})(60 \ K)$$

$$= 6276 \text{ or } 6300 \text{ J} \text{ or } 6.3 \text{ kJ} \text{ (to 2 sf)}$$

Magnesium required:

$$6.3 \text{ kJ} \cdot \frac{1 \text{ mol Mg}}{352.88 \text{ kJ}} \cdot \frac{24.3 \text{ g Mg}}{1 \text{ mol Mg}} = 0.43 \text{ g Mg}$$

87. (a) Enthalpy change for:

$$C(s) + \quad H_2O(g) \quad \rightarrow \quad CO(g) + \quad H_2(g)$$

∆H°f(kJ/mol)   0   -241.83   -110.525   0

$$\Delta H°_{rxn} \quad = \ [(1 \text{ mol})(-110.525 \ \frac{kJ}{mol}) + 0] - [\ 0 \ + \ (1 \text{ mol})(-241.83 \ \frac{kJ}{mol})]$$

$$= +131.31 \text{ kJ}$$

(b) The process is **endothermic,** so the reaction is **reactant-favored**.

(c) Heat involved when 1.0 metric ton (1000.0 kg) of C is converted to coal gas:

$$1000.0 \text{ kg C} \cdot \frac{1000 \text{ g C}}{1 \text{ kg C}} \cdot \frac{1 \text{ mol C}}{12.011 \text{ g C}} \cdot \frac{+131.31 \text{ kJ}}{1 \text{ mol C}} = 1.0932 \times 10^7 \text{ kJ}$$

89. For the combustion of C_8H_{18}:

$$C_8H_{18}(\ell) + 25/2 \ O_2(g) \rightarrow 8 \ CO_2(g) + 9 \ H_2O(\ell)$$

$$\Delta H°_{rxn} = [(8 \text{ mol})(-393.509 \ \frac{kJ}{mol}) + (9 \text{ mol})(-285.83 \ \frac{kJ}{mol})] - [(1 \text{ mol})(-259.2 \ \frac{kJ}{mol}) + 0]$$

$$\Delta H°_{rxn} = -5461.3 \text{ kJ}$$

Expressed on a gram basis: $-5461.3 \ \frac{kJ}{mol} \cdot \frac{1 \text{ mol } C_8H_{18}}{114.2 \text{ g } C_8H_{18}} = -47.81 \text{ kJ/g}$

For the combustion of CH_3OH:

$$2 \ CH_3OH(\ell) + 3 \ O_2 (g) \rightarrow 2 \ CO_2 (g) + 4 \ H_2O(\ell)$$

$$\Delta H°_{rxn} = [(2mol)(- 393.509 \text{ kJ/mol}) + (4 \text{ mol})(- 285.83 \text{ kJ/mol})]$$
$$- [(2mol)(- 238.4 \text{ kJ/mol}) + 0]$$

$$= [(- 787.0) + (- 967.2)] + 477.4 \text{ kJ}$$

$$= - 1453.5 \text{ kJ or } –726.77 \text{ kJ/mol}$$

Express this on a per mol and per gram basis:

$$\frac{-1453.5 \text{ kJ}}{2 \text{ mol } CH_3OH} \cdot \frac{1 \text{ mol } CH_3OH}{32.04 \text{ g } CH_3OH} = - 22.682 \text{ kJ/g}$$

On a per gram basis, **octane liberates the greater amount** of heat energy.

91. (a) Enthalpy change for formation of 1.00 mol of $SrCO_3$

$Sr (s) + C (graphite) + 3/2\ O_2 (g) \rightarrow SrCO_3 (s)$ using the data:

$$Sr(s) + 1/2\ O_2 (g) \rightarrow SrO(s) \qquad \Delta H°_f = -592\ kJ$$
$$SrO(s) + CO_2 (g) \rightarrow SrCO_3 (s) \qquad \Delta H°_{rxn} = -234\ kJ$$
$$C (graphite) + O_2 (g) \rightarrow CO_2 (g) \qquad \Delta H°_f = -394\ kJ$$

Let's add the equations to give our desired overall equation.

$$Sr(s) + 1/2\ O_2 (g) \rightarrow SrO(s) \qquad \Delta H°_f = -592\ kJ$$
$$\cancel{SrO(s)} + \cancel{CO_2(g)} \rightarrow \cancel{SrCO_3 (s)} \qquad \Delta H°_{rxn} = -234\ kJ$$
$$\underline{C (graphite) + O_2 (g) \rightarrow \cancel{CO_2} (g) \qquad \Delta H°_f = -394\ kJ}$$
$$Sr (s) + C (graphite) + 3/2\ O_2 (g) \rightarrow SrCO_3 (s) \qquad \Delta H°_{rxn} = -1220.\ kJ$$

(b) Energy diagram relating the energy quantities:

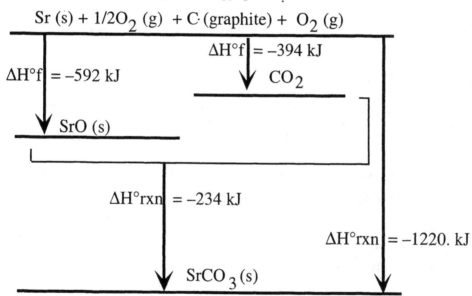

93. The desired equation is: $CH_4 (g) + 3\ Cl_2 (g) \rightarrow 3\ HCl (g) + CHCl_3 (g)$

Begin with equation 1 (the combustion of methane)

$$CH_4 (g) + 2\ O_2 (g) \rightarrow 2\ H_2O (\ell) + CO_2 (g) \qquad \Delta H = -890.4\ kJ = -890.4\ kJ$$

Noting that we form HCl as one of the products, using the second equation, we need to **reverse** it and (to adjust the coefficient of HCl to 3), multiply by 3/2 to give:

$$3/2\ H_2 (g) + 3/2\ Cl_2 (g) \rightarrow 3\ HCl \qquad \Delta H = -3/2(+184.6)\ kJ = -276.9\ kJ$$

Note that CO_2 formed in equation 1 doesn't appear in the overall equation so let's use equation 3 (reversed) to "consume" the CO_2:

$$CO_2 (g) \rightarrow C (graphite) + O_2 (g) \qquad \Delta H = -1(-393.5)\ kJ = +393.5\ kJ$$

Noting also that equation 1 produces 2 water molecules, let's "consume" them by using equation 4 (reversed) multiplied by 2:

$$2 H_2O (\ell) \rightarrow 2 H_2 (g) + O_2 (g) \qquad \Delta H = -2(-285.8) \text{ kJ} = + 571.6 \text{ kJ}$$

and finally we need to produce $CHCl_3$ which we can do with the equation that represents the ΔH_f for $CHCl_3$:

$$C(graphite) + 1/2 H_2 (g) + 3/2 Cl_2 (g) \rightarrow CHCl_3 (g) \quad \Delta H = -103.1 \text{ kJ}$$

The overall enthalpy change would then be:

$$\Delta H = -890.4 \text{ kJ} - 276.9 \text{ kJ} + 393.5 \text{ kJ} + 571.6 \text{ kJ} - 103.1 \text{ kJ} = -305.3 \text{ kJ}$$

95. Calculate the heat evolved by burning 1.00 kg of C to $CO_2(g)$,

$$C(s) + O_2 (g) \rightarrow CO_2 (g)$$

$\Delta H°_f(\text{kJ/mol}) \qquad 0 \qquad 0 \qquad - 393.509$

$$\Delta H°_{rxn} = (1 \text{ mol})(- 393.509 \tfrac{kJ}{mol}) - [(1 \text{ mol})(0 \tfrac{kJ}{mol}) + (1 \text{ mol})(0 \tfrac{kJ}{mol})]$$

$$= -393.509 \text{ kJ}$$

Now for the combustion of 1.00 kg:

$$\frac{1.00 \text{ kg C}}{1} \bullet \frac{1000 \text{ g C}}{1.00 \text{ kg C}} \bullet \frac{1 \text{ mol C}}{12.01 \text{ g C}} \bullet \frac{- 393.509 \text{ kJ}}{1 \text{ mol C}} = -3.28 \times 10^4 \text{ kJ}$$

Now calculate the energy associated with producing water gas ($CO + H_2$) from 1.00 kg C, and the subsequent combustion of that water gas to CO_2 and H_2O.

Two processes occur:
(1) $C(s) + H_2O (g) \rightarrow CO (g) + H_2(g)$ then combustion of that water gas
(2) $CO (g) + H_2(g) + O_2 (g) \rightarrow CO_2 (g) + H_2O(g)$

For the production of water gas:

$$C(s) + H_2O (g) \rightarrow CO (g) + H_2(g)$$

$\Delta H°_f(\text{kJ/mol}) \qquad 0 \qquad -241.83 \qquad\quad - 110.525 \qquad 0$

$$\Delta H°_{rxn} = [(1 \text{ mol})(- 110.525 \tfrac{kJ}{mol}) + (1 \text{ mol})(0 \tfrac{kJ}{mol})]-$$

$$[(1 \text{ mol})(0 \tfrac{kJ}{mol}) + (1 \text{ mol})(-241.83 \tfrac{kJ}{mol})]$$

$$= 131.305 \text{ kJ}$$

For 1.00 kg of C, this amount of energy would be: 83.3 mol C $\bullet$ 131.305 kJ/mol = 10,900 kJ

[Recall that 1.00 kg of C = $\dfrac{1000 \text{ g C}}{1} \bullet \dfrac{1 \text{ mol C}}{12.01 \text{ g C}} = 83.3 \text{ mol C}$]

Calculate the energy change for process (2):

$$CO (g) + H_2(g) + O_2 (g) \rightarrow CO_2 (g) + H_2O(g)$$

$\Delta H°_f(\text{kJ/mol}) \qquad -110.525 \quad 0 \qquad 0 \qquad\qquad - 393.509 \qquad -241.83$

$$\Delta H^\circ_{rxn} = (1 \text{ mol})(-393.509 \frac{kJ}{mol}) + (1 \text{ mol})(-241.83\frac{kJ}{mol})$$

$$- [(1 \text{ mol})(-110.525\frac{kJ}{mol}) + (1 \text{ mol})(0 \frac{kJ}{mol}) + (1 \text{ mol})(0 \frac{kJ}{mol})]$$

$$= -524.81 \text{ kJ per mol of CO.}$$

Given that we have 83.3 mol of CO, and 83.3 mol of H_2 (formed by process (1)), we can

calculate the amount of energy released in this process:

$$\frac{83.3 \text{ mol CO}}{1} \bullet \frac{-524.81 \text{ kJ}}{1 \text{ mol CO}} = -43,717. \text{ kJ} \quad (\text{or } -43,700 \text{ kJ to 3 sf})$$

If we compare the amount of energy released by the combustion of water gas (4.37×10^4 kJ)

with the energy released by the combustion of C (-3.28×10^4 kJ) we see that MORE energy

is released by the combustion of water gas. Recall however that the FORMATION of the water

gas consumed energy (1.09×10^4 kJ), so the NET amount of energy released is:

[4.37×10^4 kJ - 1.09×10^4 kJ = 3.28×10^4 kJ], so the amounts of energy produced by the

combustion of C to CO_2 and the combustion of water gas to CO_2 and H_2O are comparable.

Summary and Conceptual Questions

97. The first law of thermodynamics is another way of stating the Law of Conservation of Energy.

    (Section 6.1 of your textbook addresses this as well.)

99. Without doing calculations, decide whether each is product- or reactant- favored:

    (a) combustion of natural gas—as noted in Section 6.9 of your textbook, oxidation reactions

        of carbon and hydrogen are typically product-favored. This process is exothermic as well.

    (b) Decomposition of sugar to carbon and water- Sugar does not naturally decompose into

        carbon and water, hence we predict the reaction to be reactant-favored.

101. (a) When a hotter object comes into contact with a cooler one, the energy of the hotter object

        flows to the cooler object. The result is that the energy of the hotter object (and the motion

        of the particles) is reduced, the energy of the cooler object (and the motion of the

        particles) is increased.

    (b) When two objects reach thermal equilibrium, the energies of the two objects (and the

        motion of the particles) have become equal—and the objects have the same temperature.

103. Determine the value of ΔH for the reaction:

$$Ca(s) + 1/8 \, S_8 \, (s) + 2 \, O_2(g) \rightarrow CaSO_4(s)$$

Imagine this as the sum of several processes:

1) $Ca(s) + 1/2\ O_2(g) \rightarrow CaO(s)$

2) $1/8\ S_8\ (s) + 3/2\ O_2(g) \rightarrow SO_3(g)$

3) $CaO(s) + SO_3(g) \rightarrow CaSO_4(s)$ $\Delta H = -402.7\ kJ$

Note that the SUM of the three processes is the DESIRED equation (the formation of $CaSO_4(s)$). The OVERALL ΔH is then the SUM of the ΔH for process (1) and ΔH for process (2).

We know that the ΔH_{rxn} for (3) = -402.7 kJ or

$\Delta H_{rxn} = \Delta H_f\ CaSO_4(s) - [\Delta H_f\ CaO(s) + \Delta H_f\ SO_3(g)]$. Since we know the ΔH_{rxn} and BOTH the ΔH_f for CaO(s) and $SO_3(g)$, we can calculate the $\Delta H_f\ CaSO_4(s)$.

From Table 6.2, ΔH_f for CaO(s) = -635.09 kJ/mol and ΔH_f for $SO_3(g)$ = - 395.77 kJ/mol

$\Delta H_{rxn} = \Delta H_f\ CaSO_4(s) - [\Delta H_f\ CaO(s) + \Delta H_f\ SO_3(g)]$

$- 402.7\ kJ = \Delta H_f\ CaSO_4(s) - [-635.09\ kJ/mol + -\ 395.77\ kJ/mol]$

$- 1,433.6\ kJ = \Delta H_f\ CaSO_4(s)$

105. The molar heat capacities for Al, Fe, Cu, and Au are:

$0.897\ \dfrac{J}{g\bullet K}$ • $\dfrac{26.98\ g\ Al}{1\ mol\ Al} = 24.2\ \dfrac{J}{mol\bullet K}$

$0.449\ \dfrac{J}{g\bullet K}$ • $\dfrac{55.85\ gFe}{1\ mol\ Fe} = 25.1\ \dfrac{J}{mol\bullet K}$

$0.385\ \dfrac{J}{g\bullet K}$ • $\dfrac{63.55\ g\ Cu}{1\ mol\ Cu} = 24.5\ \dfrac{J}{mol\bullet K}$

$0.129\ \dfrac{J}{g\bullet K}$ • $\dfrac{197.0\ g\ Au}{1\ mol\ Au} = 25.4\ \dfrac{J}{mol\bullet K}$

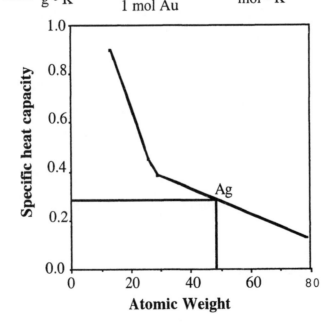

The graph shown is a plot of specific heat capacity versus atomic weight. As you can see, no simple linear relationship exists for these metals. The plot of the specific heat of Cu (atomic weight 63.55) and Au (atomic weight 197) does show a **decreasing** value of specific heat capacity as the atomic weight of the element increases. If you estimate the atomic weight to be about 100 (exact value is about 108), one could **estimate** a value of approximately 0.28 as the specific heat (compared to the experimental value of 0.236.

Alternatively, a quick examination of the values for the four metals above indicates that they are **quite similar**, with an average of 24.8 J/mol • K. This translates into:

$$24.8 \frac{J}{mol \cdot K} \cdot \frac{1 \text{ mol Au}}{107.9 \text{ g Au}} = 0.230 \text{ J/g} \cdot K$$

107. Mass of methane needed to heat the air from 15.0 to 22.0 °C:

Calculate the volume of air, then with the density and average molar mass, the moles of air present:

$$275 \text{ m}^2 \cdot 2.50 \text{ m} \cdot \frac{1000 \text{ L}}{1 \text{ m}^3} \cdot \frac{1.22 \text{ g air}}{1 \text{ L air}} \cdot \frac{1 \text{ mol air}}{28.9 \text{ g air}} = 2.90 \text{ x } 10^4 \text{ mol air}$$

The energy needed to change the temperature of that amount of air by (22.0 – 15.0)°C:

$$2.90 \text{ x } 10^4 \text{ mol air} \cdot 29.1 \frac{J}{mol \cdot K} \cdot 7.0 \text{ K} = 5.9 \text{ x } 10^6 \text{ J}$$

What quantity of energy does the combustion of methane provide?
The reaction may be written: $CH_4 (g) + 2 O_2 (g) \rightarrow 2 H_2O (g) + CO_2 (g)$

Using data from Appendix L:
$\Delta H_{rxn} = [(2 \text{ mol})(-241.83 \text{ kJ/mol}) + (1 \text{mol})(-393.509 \text{ kJ/mol})]$

$$- [(1 \text{ mol})(-74.87 \text{ kJ/mol}) + (2 \text{ mol})(0)] = -802.3 \text{ kJ}$$

The amount of methane necessary is:

$$5.9 \text{ x } 10^6 \text{ J} \cdot \frac{1 \text{ kJ}}{1000 \text{ J}} \cdot \frac{1 \text{ mol } CH_4}{802.3 \text{ kJ}} \cdot \frac{16.0 \text{ g } CH_4}{1 \text{ mol } CH_4} = 120 \text{ g } CH_4 \text{ (to 2 sf)}$$

109. Calculate the quantity of heat transferred to the surroundings from the water vapor condensation as rain falls.

Calculate the volume of water that falls, and then the mass of that water:
From the conversion factors listed in your textbook, calculate the area of $1 mi^2$ in cm^2
[1 km = 0.62137 mi and 1 km = 10^5 cm.]

$$\frac{1 \text{ mi}^2}{1} \cdot \frac{(1 \text{ km})^2}{(0.62137 \text{ mi})^2} \cdot \frac{(10^5 \text{ cm})^2}{(1 \text{ km})^2} = \frac{10^{10} \text{cm}^2}{0.38610} = 2.59 \text{ x } 10^{10} \text{ cm}^2$$

1 in = 2.54 cm so the VOLUME of water is 2.59 x 10^{10} cm^2 x 2.54 cm = 6.6 x 10^{10} cm^3.

The mass of water is: 6.6×10^{10} cm^3 x 1.0 g/cm^3 or 6.6×10^{10} g of water.

The amount of heat: $\dfrac{6.6 \times 10^{10}\text{g water}}{1} \bullet \dfrac{1 \text{ mol water}}{18.02 \text{ g water}} \bullet \dfrac{44.0 \text{ kJ}}{1 \text{ mol water}} = 1.6 \times 10^{11}$ kJ

Note the much larger energy for this process than for the detonation of a ton of dynamite.

Chapter 7
Atomic Structure

Practicing Skills
Electromagnetic Radiation

1. Using Figure 7.3:

   (a) Microwave radiation is less energetic than X-ray radiation.

   (b) Red light uses higher frequency light than Radar.

   (c) Infrared radiation is of longer wavelength than ultraviolet.

3. (a) The higher frequency light is the green 500 nm light.

   Recall that frequency and wavelength are inversely related.

   (b) The frequency of amber light (595 nm) is:

   $$\text{frequency} = \frac{\text{speed of light}}{\text{wavelength}} = \frac{2.9979 \times 10^8 \text{m/s}}{595 \text{ nm}} \cdot \frac{1.00 \times 10^9 \text{nm}}{1.00 \text{ m}}$$

   $$= 5.04 \times 10^{14} \text{ s}^{-1}$$

Electromagnetic Radiation and Planck's Equation

5. To calculate the energy of one photon of light with 500 nm wavelength, we need to first calculate the frequency of the radiation:

   $$\text{frequency} = \frac{\text{speed of light}}{\text{wavelength}} = \frac{2.9979 \times 10^8 \text{ m/s}}{5.0 \times 10^2 \text{ nm}} \cdot \frac{1.00 \times 10^9 \text{ nm}}{1.00 \text{ m}}$$

   $$= 6.0 \times 10^{14} \text{ s}^{-1}$$

   And the energy is then $E = h\upsilon$ or $(6.626 \times 10^{-34} \text{ J} \cdot \text{s} \cdot \text{photons}^{-1})(6.0 \times 10^{14} \text{ s}^{-1})$

   $$= 4.0 \times 10^{-19} \text{ J} \cdot \text{photons}^{-1}$$

   Energy of 1.00 mol of photons $= 4.0 \times 10^{-19} \text{ J} \cdot \text{photon}^{-1} \cdot \frac{6.0221 \times 10^{23} \text{ photons}}{1.00 \text{ mol photons}}$

   $$= 2.4 \times 10^5 \text{ J/mol photon}$$

7. The frequency of the line at 396.15 nm:

   $$\text{frequency} = \frac{\text{speed of light}}{\text{wavelength}} = \frac{2.9979 \times 10^8 \text{ m/s}}{3.9615 \times 10^2 \text{ nm}} \cdot \frac{1.00 \times 10^9 \text{ nm}}{1.00 \text{ m}}$$

   $$= 7.5676 \times 10^{14} \text{ s}^{-1}$$

   The energy of a photon of this light may be determined : $E = h\upsilon$

   Planck's constant, h, has a value of $6.626 \times 10^{-34} \text{ J} \cdot \text{s} \cdot \text{photons}^{-1}$

$$E = (6.626 \times 10^{-34} \text{ J} \cdot \text{s} \cdot \text{photons}^{-1})(7.5676 \times 10^{14} \text{ s}^{-1})$$
$$= 5.0143 \times 10^{-19} \text{ J} \cdot \text{photon}^{-1}$$

Energy of 1.00 mol of photons $= 5.0143 \times 10^{-19} \text{ J} \cdot \text{photon}^{-1} \cdot \dfrac{6.0221 \times 10^{23} \text{ photons}}{1.00 \text{ mol photons}}$

$$= 3.02 \times 10^{5} \text{ J/mol photon or } 302 \text{ kJ/mol photon.}$$

9. Since energy is proportional to frequency ($E = h\upsilon$), we can arrange the radiation in order of increasing energy per photon by listing the types of radiation in increasing frequency (or decreasing wavelength).

$\rightarrow$ Energy increasing $\rightarrow$

FM music   microwave   yellow light   x-rays

$\rightarrow$ Frequency (υ) increasing $\rightarrow$

$\leftarrow$ Wavelength (λ) increasing $\leftarrow$

Photoelectric Effect

11. Energy $= 2.0 \times 10^{2} \text{ kJ/mol} \cdot \dfrac{1 \text{ mol}}{6.0221 \times 10^{23} \text{ photons}} \cdot \dfrac{1.00 \times 10^{3} \text{ J}}{1.00 \text{ kJ}}$

$$= 3.3 \times 10^{-19} \text{ J} \cdot \text{photons}^{-1}$$

What wavelength of light would provide this energy ?

$$E = h\upsilon = \frac{hc}{\lambda} \quad \text{or} \quad \lambda = \frac{hc}{E} = \frac{(6.626 \times 10^{-34} \text{ J} \cdot \text{s} \cdot \text{photons}^{-1})(2.9979 \times 10^{8} \text{ m} \cdot \text{s}^{-1})}{3.3 \times 10^{-19} \text{ J} \cdot \text{photons}^{-1}}$$

$$= 6.0 \times 10^{-7} \text{ m or } 6.0 \times 10^{2} \text{ nm}$$

Radiation of this wavelength--in the **visible** region of the electromagnetic spectrum-- would appear **orange**.

Atomic Spectra and the Bohr Atom

13. (a) The **most energetic light** would be represented by the light of **shortest wavelength** (253.652 nm).

(b) The frequency of this light is :

$$\frac{2.9979 \times 10^{8} \text{ m/s}}{253.652 \text{ nm}} \cdot \frac{1.00 \times 10^{9} \text{ nm}}{1.00 \text{ m}} = 1.18190 \times 10^{15} \text{ s}^{-1}$$

The energy of 1 photon with this wavelength is:

$$E = h\upsilon = (6.62608 \times 10^{-34} \frac{\text{J} \cdot \text{s}}{\text{photon}})(1.18190 \times 10^{15} \text{ s}^{-1})$$

$$= 7.83139 \times 10^{-19} \frac{\text{J}}{\text{photon}}$$

(c) The line emission spectrum of mercury shows the visible region between $\approx$ 400 and 750 nm. The lines at 404 and 436 nm are present while the lines at 253 nm, 365 nm and 1013 nm lie outside the visible region. The 404 nm line is violet, while the 436 nm line is blue.

15. The Balmer series of lines terminates with $n_f = 2$. According to Figure 7.12, the transition originates at $n_i = 6$. Light of wavelength 410.2 nm would be violet.

17. (a) <u>Transitions from</u>   <u>to</u>

         n = 5   n = 4, 3, 2, or 1   (4 transitions)

         n = 4   n = 3, 2, or 1   (3 transitions)

         n = 3   n = 2 or 1   (2 transition)

         n = 2   n = 1   (1 transition)

Ten transitions are possible from these five quantum levels, providing 10 emission lines.

(b) Photons of the highest **frequency** are emitted in a transition from level of n = _5_ to a level with n = _1_. Recalling that energy is directly proportional to frequency ($E = h\upsilon$), the highest frequency will correspond to the highest energy—a transition from the two levels that differ most in energy.

(c) Emission line having the longest wavelength corresponds to a transition from level of n = _5_ to a level with n = _4_. Levels 4 and 5 are closer in energy than other possibilities given here, so a transition between the two would be of longest wavelength (lowest E).

19. (a) Photons of the lowest energy will be emitted in a transition from the level with **n = 3** to the level **n = 2**. This is easily seen with the aid of the equation

$$\Delta E = Rhc\left(\frac{1}{n^2_f} - \frac{1}{n^2_i}\right).$$

Since R, h, and c are constant for any transition, inspection shows that the smaller change in energy results if $n_f = 3$ and $n_i = 2$. (The fractions in the equation above would correspond to $\left(\frac{1}{4} - \frac{1}{16}\right)$ for the 4→2 transition or $\left(\frac{1}{4} - \frac{1}{9}\right)$ for the 3→2 transition.

(b) Once again, using the equation above, the fraction in parenthesis (and hence ΔE) will be greater for the transition from $n_i = 4$ to $n_f = 1$ than for the transition from 5 →2.

21. The wavelength of emitted light for the transition n = 3 to n = 1.

$$\Delta E = -Rhc\left(\frac{1}{1^2} - \frac{1}{3^2}\right) \text{ and the value of Rhc = 1312 kJ/mol, so}$$

$$\Delta E = -1312 \text{ kJ/mol} \left(\frac{1}{1^2} - \frac{1}{3^2}\right) \text{ or } = -1312 \text{ kJ/mol (8/9) = } -1166 \text{ kJ/mol}$$

To calculate the frequency and wavelength, we use $E = h\upsilon$. Recall that we must first express the energy **per photon** (as opposed to a mole of photons).

$$\Delta E = \frac{-1166 \text{ kJ/mol photons}}{6.022 x 10^{23} \text{ photons/1 mol photons}} \cdot \frac{10^3 J}{1 \text{ kJ}} = 1.936 \text{ x } 10^{-18} \text{J/photon}$$

and solving for frequency, $\upsilon = \dfrac{1.936 \text{ x } 10^{-18} \text{ J/photon}}{6.626 \text{ x } 10^{-34} \text{ J} \cdot \text{s /photon}} = 2.923 \text{ x } 10^{15} \text{ s}^{-1}$

Substituting into the relationship $\lambda\upsilon = c$ we get $\dfrac{2.998 \text{ x} 10^8 \text{ m}}{2.923 \text{ x } 10^{15} \text{ s}^{-1}} = 1.0257 \text{ x } 10^{-7} \text{m}$

$\lambda = 1.0257 \text{ x } 10^{-7}$ m or 102.6 nm (far ultraviolet)

DeBroglie and Matter Waves

23. Mass of an electron: $9.11 \text{ x } 10^{-31}$ kg

    Planck's constant: $6.626 \text{ x } 10^{-34}$ J • s • photon^{-1}

    Velocity of the electron: $2.5 \text{ x } 10^8$ cm • s^{-1} or $2.5 \text{ x } 10^6$ m • s^{-1}

$$\lambda = \frac{h}{m \cdot v} = \frac{6.626 \text{ x } 10^{-34} \text{ J} \cdot \text{s}}{(9.11 \text{ x } 10^{-31} \text{ kg} \cdot 2.5 \text{ x } 10^6 \text{ m} \cdot \text{s}^{-1})}$$

$$= 2.9 \text{ x } 10^{-10} \text{ m} = 2.9 \text{ Angstroms} = 0.29 \text{ nm}$$

25. The wavelength can be determined exactly as in Question 23:

$$\lambda = \frac{h}{m \cdot v} = \frac{6.626 \text{ x } 10^{-34} \text{ J} \cdot \text{s}}{(1.0 \text{ x } 10^{-1} \text{ kg} \cdot 30. \text{ m} \cdot \text{s}^{-1})}$$

$$= 2.2 \text{ x } 10^{-34} \text{ m} \quad \text{or} \quad 2.2 \text{ x } 10^{-25} \text{ nm}$$

Velocity to have a wavelength of $5.6 \text{ x } 10^{-3}$ nm

(First convert the wavelength to units of meters)

$$5.6 \text{ x } 10^{-3} \text{ nm} \cdot \frac{1 \text{ m}}{1 \text{ x } 10^9 \text{ nm}} = 5.6 \text{ x } 10^{-12} \text{ m}$$

Then rewriting the above equation:

$$v = \frac{h}{m \cdot \lambda} = \frac{6.626 \text{ x } 10^{-34} \text{ J} \cdot \text{s}}{(1.0 \text{ x } 10^{-1} \text{ kg} \cdot 5.6 \text{ x } 10^{-12} \text{ m})} = 1.2 \text{ x } 10^{-21} \frac{\text{m}}{\text{s}}$$

Quantum Mechanics

27. (a) n = 4   possible ℓ values = 0,1,2,3   (ℓ = 0,1,... (n - 1))

    (b) $\ell = 2$ possible m_ℓ values = -2,-1,0,+1,+2   (-ℓ ..., 0,....+ ℓ)

    (c) orbital = 4s   possible values for n = 4; $\ell = 0$; $m_\ell = 0$

    (d) orbital = 4f   possible values for n = 4; $\ell = 3$; m_ℓ = -3,-2,-1,0,+1,+2,+3

29. An electron in a 4p orbital must have n = 4 and ℓ = 1. The possible m_ℓ values give rise to the following sets of n, ℓ, and m_ℓ

n	ℓ	m_ℓ	
4	1	-1	Note that the **three values** of m describe
4	1	0	**three orbital orientations**.
4	1	+1	

31. Subshells in the electron shell with n = 4:

There are 4: s, p, d, and f sublevels corresponding to ℓ = 0, 1, 2, and 3 respectively.

Recall that values of ℓ from 0 to a maximum of (n-1) are possible.

33. Explain why each of the following is not a possible set of quantum numbers for an electron in an atom.

(a) n = 2, ℓ = 2, m_ℓ = 0   For n = 2, maximum value of ℓ is one (1).

(b) n = 3, ℓ = 0, m_ℓ = -2 For ℓ = 0, possible value of m_ℓ is 0.

(c) n = 6, ℓ = 0, m_ℓ = 1   For ℓ = 0, possible value of m_ℓ is 0.

35.
quantum number designation	maximum number of orbitals
(a) n = 3; ℓ = 0; m_ℓ = +1	none; for ℓ = 0, the only possible value of m_ℓ = 0
(b) n = 5; ℓ = 1	3 ("**p**" orbitals)
(c) n = 7; ℓ = 5	eleven; the # of orbitals is "2 ℓ +1"
(d) n = 4; ℓ = 2; m_ℓ = -2	1 (one of the three 4 "**p**" orbitals)

37. Which of the following orbitals cannot exist and why:

2s exists	n = 2	permits ℓ values as large as 1 (ℓ = 0 is an s sublevel)
2d cannot exist		ℓ = 2 is not permitted for n < 3 (ℓ = 2 is a d sublevel)
3p exists	n = 3	permits ℓ values as large as 2 (ℓ = 1 is a p sublevel)
3f cannot exist		ℓ = 3 is not permitted for n < 4 (ℓ = 3 is an f sublevel)
4f exists	ℓ = 4	permits ℓ values as large as 3 (ℓ = 3 is an f sublevel)
5s exists	n = 5	permits ℓ values as large as 4 (ℓ = 0 is an s sublevel)

39. The complete set of quantum numbers for :

	n	ℓ	m_ℓ	
(a)2p	2	1	-1, 0, +1	(3 orbitals)
(b)3d	3	2	-2, -1, 0, +1, +2	(5 orbitals)
(c)4f	4	3	-3, -2, -1, 0, +1, +2, +3	(7 orbitals)

41. With an n = 4, and ℓ = 2, this orbital belongs in the 4th level (n = 4) and with an ℓ = 2, this must be a "d" type orbital—so (d) 4d.

43. The number of nodal surfaces possessed by an orbital is equal to the value of the "ℓ" quantum number, so for each of the following:

orbital		number of planar nodes
(a) 2s	($\ell = 0$)	0
(b) 5d	($\ell = 2$)	2
(c) 5f	($\ell = 3$)	3

General Questions on Atomic Structure

45. Concerning the photoelectric effect:
    (a) Light is electromagnetic radiation- correct
    (b) Intensity of light beam related to frequency—incorrect. The intensity is related to the number of photons of light.
    (c) Light can be thought of as mass-less particles-correct.

47. Number of nodal surfaces for the following orbital types:

Orbital type	Nodal surfaces
s	0 (because $\ell = 0$)
p	1 (because $\ell = 1$)
d	2 (because $\ell = 2$)
f	3 (because $\ell = 3$)

49. Orbital types associated with values of "ℓ"

Orbital type	Values of "ℓ"
f	$\ell = 3$
s	$\ell = 0$
p	$\ell = 1$
d	$\ell = 2$

51.

Orbital Type	Number of orbitals in a Given Subshell	Number of Nodal Surfaces
s	1	0
p	3	1
d	5	2
f	7	3

110

53. Regarding red and green light emitted by a sign:

(a) Green light has the shorter wavelength and therefore higher energy photons.

(b) Green light has higher energy photons than red light, the shorter wavelength (500 nm) is green

(c) Since frequency and wavelength are inversely related, the shorter wavelength (green) must have the higher frequency.

55. For radiation of 850 MHz:

(a) The wavelength: $\dfrac{3.00 \times 10^8 \text{ m/s}}{850 \times 10^6 \text{ Hz}} \bullet \dfrac{1 \text{ Hz}}{1 \text{ s}^{-1}} = 0.35 \text{ m}$

(b) The energy of 1.0 mol of photons with $\upsilon = 850$ MHz:

$$E = \frac{hc}{\lambda} = \frac{(6.626 \times 10^{-34} \text{ J} \bullet \text{s} \bullet \text{photons}^{-1})(2.9979 \times 10^8 \text{ m} \bullet \text{s}^{-1})}{0.35 \text{ m}} \bullet \frac{6.0221 \times 10^{23} \text{photons}}{1 \text{ mol photons}}$$

$$= 0.34 \text{ J/mol}$$

(c) The energy of a mole of photons of 420 nm light:

$$E = \frac{hc}{\lambda} = \frac{(6.626 \times 10^{-34} \text{ J} \bullet \text{s} \bullet \text{photons}^{-1})(2.9979 \times 10^8 \text{ m} \bullet \text{s}^{-1})}{420 \times 10^{-9} \text{ m}} \bullet \frac{6.0221 \times 10^{23} \text{photons}}{1 \text{ mol photons}}$$

$$= 2.9 \times 10^5 \text{ J/mol}$$

(d) The energy of a mole of photons of blue light is much greater than that of the corresponding photons from a cell phone.

57.   "A Closer Look" following Example 7.3 in your text illustrates the calculation of the ionization energy for H's electron

$$E = \frac{-Z^2 Rhc}{n^2} = -2.179 \times 10^{-18} \text{ J/atom} \Rightarrow -1312 \text{ kJ/mol}$$

For He$^+$ the calculation yields

$$E = \frac{-(2)^2 (1.097 \times 10^7 \text{ m}^{-1})(6.626 \times 10^{-34} \text{ J} \bullet \text{s})(2.998 \times 10^8 \text{ m} \bullet \text{s}^{-1})}{(1)^2}$$

$$= -8.717 \times 10^{-18} \text{ J/ion and expressing this energy for a mol of ions}$$

$$= \frac{-8.717 \times 10^{-18} \text{ J}}{\text{ion}} \bullet \frac{6.0221 \times 10^{23} \text{ atoms}}{\text{mol}} \bullet \frac{1 \text{ kJ}}{1000 \text{ J}} = -5248 \frac{\text{kJ}}{\text{mol}}$$

The energy to remove the electron is then 5248 kJ/mol of ions. Note that this energy is four times that for H.

59. Orbitals in a H atom in order of increasing energy: 1s < 2s = 2p < 3s = 3p =3d < 4s
    For the hydrogen atom, energy levels increase with increasing values of "n" (Equation 7.4).

61. Wavelength and frequency for a photon with E = 1.173MeV:

    Using the energy relationship: E = hυ, we solve for frequency:

    $$\upsilon = \frac{1.172 \times 10^6 \text{ ev}}{6.626 \times 10^{-34} \text{ J} \bullet \text{s}} \bullet \frac{9.6485 \times 10^4 \text{ J/mol}}{1 \text{ ev}} \bullet \frac{1 \text{ mol}}{6.022 \times 10^{23} \text{ photons}}$$

    $$= 2.836 \times 10^{20} \text{ s}^{-1}$$

    Substituting into the wavelength-frequency relationship, we have:

    $$\lambda = c/\upsilon = \frac{2.998 \times 10^8 \text{ m/s}}{2.836 \times 10^{20} \text{ s}^{-1}} = 1.057 \times 10^{-12} \text{m}$$

63. Time for Sojourner's signal to travel 7.8×10^7 km:

    If light travels at 2.9979×10^8 m $\bullet$ s^{-1}, we can calculate the time:

    $$\frac{7.8 \times 10^7 \text{ km}}{1} \bullet \frac{1 \times 10^3 \text{ m}}{1 \text{ km}} \bullet \frac{1 \text{ s}}{2.9979 \times 10^8 \text{ m}} = 260 \text{ s}$$

    Given that there are 60 s in 1 minute: $260 \text{ s} \bullet \dfrac{1\text{min}}{60\text{s}} = 4.3$ minutes

65. (a) The quantum number n describes the **size (and energy)** of an atomic orbital .

    (b) The shape of an atomic orbitals is given by the quantum number **ℓ**.

    (c) A photon of green light has **more energy** than a photon of orange light.

    (d) The maximum number of orbitals that may be associated with the quantum numbers
        n= **4**, ℓ = **3** is **seven.** (corresponding to **m**ℓ values of ± 3,± 2, ± 1, and 0)

    (e) The maximum number of orbitals that may be associated with the quantum numbers
        n= **3,** ℓ = **2,** and **m**ℓ = -2 is **one.**

    (f) The orbital on the left is a **d orbital** and the one on the right is a **p orbital**, while the
        orbital in the middle is an **s orbital.**

    (g) When n = 5, the possible values of **ℓ are 0,1,2,3, and 4**. (Range is 0 ..... (n-1))

    (h) The maximum number of orbitals that can be assigned to the n = 4 shell is **16.**

n = 4	ℓ = 0	mℓ = 0	1 orbital
n = 4	ℓ = 1	mℓ = -1,0,+1	3 orbitals
n = 4	ℓ = 2	mℓ = -2,-1,0,+1,+2	5 orbitals
n = 4	ℓ = 3	mℓ = -3,-2,-1,0,+1,+2,+3	7 orbitals
			16 orbitals

Summary and Conceptual Questions

67. Two major assumption of Bohr's theory of atomic structure:

    (a) Electrons orbit nucleus in circular quantized (fixed energy) orbits.

    (b) An electron in an atom would remain in its lowest energy level unless disturbed.

69. The light visible from a sodium or mercury street light arises owing to (c) electrons are moving from a given energy level to one of lower n (and hence of lower energy).

71. The "wave-particle duality" describes the behavior of electrons—in which the particulate nature of the electron (e.g. as observed in the photoelectric effect) and the undulatory nature of the electron (e.g. the observation of diffraction patterns in the Davisson-Germer experiment) are both important to our explanation of natural phenomena. The implications of this duality arise in our understanding of electrons in atoms as waves existing in regions of high probability.

73. The Heisenberg Uncertainty Principle answers this question for us. This principles says that it is impossible to **simultaneously determine** both the position and energy of an electron. Reducing the uncertainty about one of these parameters increases the uncertainty of the measurement of the other parameter.

75. A photon with a wavelength of 93.8 nm can excite the electron up to level n = 6. (Figure 7.12). Transitions from n = 6 to lower values of n are possible. Transitions from level 6-> 5,4,3,2,1—**5** transitions;from level 5-> 4,3,2,1—**4** transitions; from level 4->3,2,1—**3** transitions; from level 3–>2,1— **2** transitions; from level 2->1—**1** transition. Total transitions: 5+4+3+2+1= 15 transitions. Wavelengths of most of these transitions are shown in Figure 7.12.

77. The Heisenberg Uncertainty Principle states that for objects on the subatomic scale (read electron) it is impossible to **accurately determine both** the velocity (read kinetic energy) and position of an electron **simultaneously.** Louis de Broglie proposed the matter wave equation ($\lambda = h/mv$) to calculate wavelengths of objects. The Rydberg constant, h, is relatively small. For objects as massive as a golf ball, (mv) would be very large—in comparison, making the wavelength so small as to be immeasurable. So—we **cannot measure** the wavelength of a golf ball in flight.

79. The pickle glows since the materials in the pickle are being "excited" by the addition of the energy (electric current). Since the pickle has been soaked in brine (NaCl), the electrons in the sodium atom are excited and release energy as they "return" to lower energy states, providing "yellow" light. The same kind of light is visible in many street lamps.

Chapter 8
Atomic Electron Configurations and Chemical Periodicity

Practicing Skills
Writing Electron Configurations of Atoms

1. The orbital box and spdf notation for P and Cl

	Orbital box notation	spdf notation

(a) P $\boxed{\uparrow\downarrow}$ $\boxed{\uparrow\downarrow}$ $\boxed{\uparrow\downarrow}\boxed{\uparrow\downarrow}\boxed{\uparrow\downarrow}$ $\boxed{\uparrow\downarrow}$ $\boxed{\uparrow}\boxed{\uparrow}\boxed{\uparrow}$ $1s^2 2s^2 2p^6 3s^2 3p^3$

(b) Cl $\boxed{\uparrow\downarrow}$ $\boxed{\uparrow\downarrow}$ $\boxed{\uparrow\downarrow}\boxed{\uparrow\downarrow}\boxed{\uparrow\downarrow}$ $\boxed{\uparrow\downarrow}$ $\boxed{\uparrow\downarrow}\boxed{\uparrow\downarrow}\boxed{\uparrow}$ $1s^2 2s^2 2p^6 3s^2 3p^5$

Note that Cl is in group 7A (17) indicating that there are SEVEN electrons in the outer shell, while P is in group 5A (15) indicating that there are FIVE electrons in the outer shell. Both Cl and P are on the "right side" of the periodic table—where elements have their "outermost" electrons in p subshells.

3. Electron configuration of chromium and iron:

(a) Cr: $1s^2 2s^2 2p^6 3s^2 3p^6 3d^5 4s^1$

(b) Fe: $1s^2 2s^2 2p^6 3s^2 3p^6 3d^6 4s^2$

The "surprising" configuration of elemental chromium—compared to the electron configuration of the preceding element, vanadium, arises from the stability of the "half-filled 3d sublevel" (which can be visualized as having a 4s electron occupy a 3d orbital—with the resultant $4s^1$ configuration.)

5. (a) Arsenic's electron configuration: (33 electrons)

spdf notation: $1s^2 2s^2 2p^6 3s^2 3p^6 3d^{10} 4s^2 4p^3$

noble gas notation: $[Ar] 3d^{10} 4s^2 4p^3$

(b) Krypton's electron configuration: (36 electrons)

spdf notation: $1s^2 2s^2 2p^6 3s^2 3p^6 3d^{10} 4s^2 4p^6$

noble gas notation: $[Kr]$

7. (a) Tantalum's noble gas and spdf notation (73 electrons)

spdf notation: $1s^2 2s^2 2p^6 3s^2 3p^6 3d^{10} 4s^2 4p^6 4d^{10} 4f^{14} 5s^2 5p^6 5d^3 6s^2$

noble gas notation: $[Xe] 4f^{14} 5d^3 6s^2$

(b) Platinum's noble gas and spdf notation (78 electrons)

spdf notation: $1s^2\ 2s^2\ 2p^6\ 3s^2\ 3p^6\ 3d^{10}\ 4s^2\ 4p^6 4d^{10}\ 4f^{14}\ 5s^2\ 5p^6\ 5d^9 6s^1$

noble gas notation: $[Xe]\ 4f^{14}\ 5d^9\ 6s^1$

Tantalum's configuration is expected, with Ta in period 5. and in group 5B. Platinum's configuration is a bit unexpected, but like other transition metals, it attempts to fill that d sublevel, resulting in a $5d^9 6s^1$ configuration rather than the expected $5d^8 6s^2$. Transition elements in the levels past period 4, do have several exceptions to the Aufbau principle.

9. Americium's noble gas and spdf notation (95 electrons)

spdf notation: $1s^2\ 2s^2\ 2p^6\ 3s^2\ 3p^6\ 3d^{10}\ 4s^2\ 4p^6 4d^{10}\ 4f^{14}\ 5s^2\ 5p^6 5d^{10} 5f^7 6s^2 6p^6 7s^2$

noble gas notation: $[Rn]\ 5f^7 7s^2$

Electron Configurations of Atoms and Ions and Magnetic Behavior

11. The orbital box representations for the following ions:

Orbital box notation

(a) Mg^{2+}

(b) K^+

(c) Cl^-

(d) O^{2-}

13. Electron configurations of:

(a) V   [Ar] $[Ar]\ 3d^3 4s^2$

(b) V^{2+} [Ar] $[Ar]\ 3d^3$

(c) V^{5+} [Ar]

Note that the V^{2+} ion contains unpaired electrons, and is therefore paramagnetic.

15. Manganese's orbital box and noble gas diagrams:

$$3d \qquad\qquad 4s$$

(a) Mn   [Ar] $\boxed{\uparrow}\,\boxed{\uparrow}\,\boxed{\uparrow}\,\boxed{\uparrow}\,\boxed{\uparrow}$ $\boxed{\uparrow\downarrow}$ [Ar] $3d^5 4s^2$

(b) Mn^{2+} [Ar] $\boxed{\uparrow}\,\boxed{\uparrow}\,\boxed{\uparrow}\,\boxed{\uparrow}\,\boxed{\uparrow}$ $\boxed{}$ [Ar] $3d^5$

(c) Having unpaired electrons, Mn^{2+} is **paramagnetic.**

(d) Mn^{2+} has five (5) unpaired electrons.

Quantum Numbers and Electron Configurations

17. Explain why the following sets of quantum numbers are not valid:

(a) $n = 4$, $\ell = 2$, $m_\ell = 0$, $m_s = 0$:

The possible values of m_s can only be $+1/2$ or $-1/2$

(b) $n = 3$, $\ell = 1$, $m_\ell = -3$, $m_s = -1/2$:

The possible values for m_ℓ is $-\ell$......0.....$+\ell$. Changing m_ℓ to $-1,0,+1$ would give a valid set of quantum numbers.

(c) $n = 3$, $\ell = 3$, $m_\ell = -1$, $m_s = +1/2$

The maximum value of ℓ is $(n-1)$. Changing ℓ to 2 would provide a valid set of quantum numbers.

19. Maximum number of electrons associated with the following sets of quantum numbers:

	Characterized as	Maximum number of electrons
(a) $n = 4$ and $\ell = 3$	4f electrons	14 (an "f" sublevel)
(b) $n = 6$, $\ell = 1$, $m_\ell = -1$	6p electrons	2 (a "p" orbital)
(c) $n = 3$, $\ell = 3$, $m_\ell = -3$,	NONE	With $n = 3$, the maximum value of ℓ can be 2 (i.e. n-1)

21. The electron configuration for Mg using the orbital box method:

$$1s \qquad 2s \qquad\quad 2p \qquad\quad 3s$$

Mg: $\boxed{\uparrow\downarrow}$ $\boxed{\uparrow\downarrow}$ $\boxed{\uparrow\downarrow}\,\boxed{\uparrow\downarrow}\,\boxed{\uparrow\downarrow}$ $\boxed{\uparrow\downarrow}$

Electron number   11 12

The noble gas notation: $[Ne]3s^2$

Electron

number:	n	ℓ	m_ℓ	m_s
11	3	0	0	+ 1/2
12	3	0	0	- 1/2

23. The electron configuration for Gallium using the orbital box method:

Noble gas configuration

3d   4s   4p

Ga   [Ar] $\uparrow\downarrow$ $\uparrow\downarrow$ $\uparrow\downarrow$ $\uparrow\downarrow$ $\uparrow\downarrow$ $\uparrow\downarrow$ $\uparrow$ ☐ ☐   [Ar] $3d^{10} 4s^2 4p^1$

A possible set of quantum numbers for the highest energy electron:

n	ℓ	m_ℓ	m_s
4	1	-1	+ 1/2

Periodic Properties

25. Elements arranged in order of increasing size: $C < B < Al < Na < K$

Radii from Figure 8.11 (in pm) $77 < 83 < 143 < 186 < 227$

Since K is in period 4, we anticipate it being larger than Na, its analog in 1A (period 3).

Al is to the right of Na, so we expect it to be smaller than Na.

B and C are in period 2, with B to the right of C, and therefore larger than C.

27. The specie in each pair with the larger radius:

(a) Cl^- is larger than Cl -- The ion has more electrons/proton than the atom.

(b) Al is larger than O -- Al is in period 3, while O is in period 2.

(c) In is larger than I -- Atomic radii decrease, in general, across a period.

29. The group of elements with correctly ordered increasing ionization energy (IE):

(c) $Li < Si < C < Ne$.

Neon would have the greatest IE. Silicon, being slightly larger in atomic radius than carbon, has a lesser IE. Lithium, the largest atom of this group, would have the smallest IE.

31. For the elements Na, Mg, O, and P:

(a) The largest atomic radius: Na

The greater the period number, the larger the atom.

Radius also decreases to the right in a given period.

(b) The largest (most negative) electron affinity: O

Nonmetals have a more negative EA than metals.

Down a group, the EA becomes more positive (as the "metallic" character increases).

(c) Increasing ionization energy: $Na < Mg < P < O$

The ionization energy varies inversely with the atomic radius.

The smaller the atom, the greater the IE.

33. (a) Increasing ionization energy: S < O < F

Ionization energy is inversely proportional to atomic size.

(b) Largest ionization energy of O, S, or Se: O

Oxygen is the smallest of these Group 6A elements, and hence has the largest IE.

(c) Most negative electron affinity of Se, Cl, or Br: Cl

Chlorine is the smallest of these three elements. EA tends to increase on a diagonal from the lower left of the periodic table to the upper right.(See Figure 8.12).

(d) Largest radius of O^{2-}, F^-, F: O^{2-}

The oxide ion has the largest electron : proton ratio. If one considers the attraction of the nuclear species (protons) for the extranuclear species (electrons), the greater the number of protons/electron the smaller the specie—owing to an increased electron-proton attraction. So the oxide ion (with 10 electrons and 8 protons) has the fewest electrons (of these three species) per proton.

General Questions

35. The diagram below, indicates that in (b) all the electrons are paired—giving rise to a diamagnetic substance—which is NOT strongly attracted to a magnetic field. In (a), the unpaired electron spins are aligned within the region (also known as **ferromagnetic**); in (c) the unpaired spins are not aligned—a paramagnetic material. The ferromagnetic material (a), will be **strongly attracted** to the magnetic field.

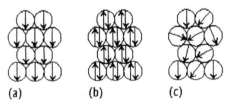

(a)   (b)   (c)

37. The electron configuration for U and U^{4+} :

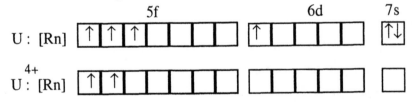

Both species have unpaired electrons and are therefore **paramagnetic**.

39. Using the spectroscopic notation the atom described would have an electron configuration:

$$1s^2 \ 2s^2 \ 2p^6 \ 3s^2 \ 3p^6 \ 4s^2$$

(a) Adding the electrons gives a sum of 20. Since this is a **neutral** atom, the # of protons and electrons wouldbe equal. Twenty protons gives this element an atomic number = 20.

(b) There are 2 s electrons in each of 4 shells : 8 s electrons total

(c) There are 6 p electrons in each of 2 shells:12 p electrons total

(d) There are 0 d electrons

(e) With its outer electrons in an "s" sublevel--specifically the 4s sublevel, the element is a **metal**-- specifically elemental **Calcium**.

41. The set of quantum numbers which is incorrect is (b). The maximum value of ℓ is (n- 1). With an ℓ =1, this value could not exceed 0.

43. (a) Electron configuration for Nd, Fe, and B

For Nd:

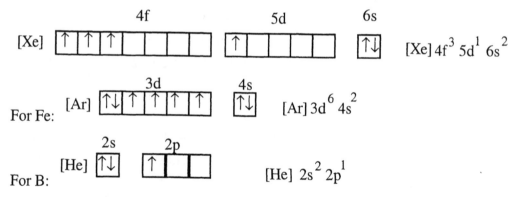

For Fe: $[Ar] \ 3d^6 \ 4s^2$

For B: $[He] \ 2s^2 \ 2p^1$

(b) Paramagnetic or diamagnetic?

All three species have unpaired electrons—therefore all are paramagnetic

(c) For the ions Nd^{3+} and Fe^{3+} their electron configurations:

For the Nd^{3+} ion

For the Fe^{3+} ion:

$[Ar] \ 3d^5$

Both ions have unpaired electrons, and are paramagnetic.

45. Arranged in order of increasing ionization energy for K, Ca, Si, and P.

   As we move across Period 4, we find K, then Ca. In Period 3, we find Si, then P. With IE increasing across a period (which means that K < Ca and Si < P), and decreasing down a group (so elements in Period 4 generally have lower IE than analogues in Period 3), we expect the ionization energy to increase in the order: K < Ca < Si < P.

47. For element A = [Kr] $5s^1$ and B = [Ar] $3d^{10}4s^24p^4$

   (a) Element A is a metal (1 electron in the s sublevel).
   (b) Element B has the greater IE. B's atomic radius would be smaller than that of A.
   (c) Element B has the more positive EA. In general EA for nonmetals is more positive than that of the metals in the same period
   (d) A would have the larger atomic radius—with its outermost electrons in the "5th" shell.

49. Ions not likely to be found include: In^{4+}, Fe^{6+}, and Sn^{5+}

   Indium would form a cation by losing **three** electrons, forming the In^{3+} cation.
   Iron can lose either 2 or 3 electrons, forming Fe^{2+} and Fe^{3+} ions. Loss of more electrons would necessitate loss of d electrons from a half-filled d subshell.
   Tin can lose up to 4 electrons (2 s and 2 p) to form the 4+ cation. Loss of more electrons would require removal of electrons from a filled d subshell.

51. (a) Element with the largest atomic radius: Se has the largest radius—being in period 4, group 6A. S would be smaller. Cl would be smaller than S. Recall the fact that atomic radius decreases across a period.

   (b) Br^- is larger than Br—having more electrons per proton.
   (c) Na would have the largest **difference** between the 1st and 2nd IE. Removing one electron would provide the stable 1+ cation and provide a specie with a filled shell.
   (d) Element with the largest ionization energy: N (IE is inversely proportional to atomic radius).
   (e) Largest radius: N^{3-} would be the largest, since the number of electrons/proton is greatest for the N^{3-} ion and smallest for the F^- ion.

53. For the elements Na, B, Al, and C:
   (a) Largest atomic radius: Na and Al have electrons in level 3, while B and C have valence electrons in level 2. So Na and Al are larger than B and C. With 3 electrons in the outer level, Al is going to be smaller than Na (with only 1 electron), so **Na** is the largest atom.

   (b) Largest electron affinity: Since electron affinity increases across the periodic table, we anticipate that the electron affinity for C will be greater than that of B. Also EAs tend to decrease down a group, so we anticipate that Na and Al will be less than C or B, so **C** has the largest EA.

   (c) Increasing order of ionization energy: With B and C being in period 2, and Na and Al in period 3, we would expect B and C to have greater IEs than that of Na and Al. With Na in group 1, and Al in group 3, we expect Al to have a greater IE than Na, so the order of increasing IE is Na < Al < B < C.

55. (a) The element (containing 27 electrons) is Cobalt
   (b) The sample contains unpaired electrons—so it is paramagnetic
   (c) The 3+ ion of cobalt would be formed by the loss of the two 4s electrons and 1 of the d electrons— leaving four unpaired electrons.

Summary and Conceptual Questions

57. The radius of Li^+ is smaller than that of Li owing to the fact that the lithium ion has electrons in the 1st energy level, while atomic Li has electrons in the 1st and 2nd energy levels. Also note that the Li^+ has 3 protons and only 2 electrons, while atomic Li has 3 protons and 3 electrons. This proton:electron ratio would also "explain" the relatively larger size of the Li atom over its cation. Using similar logic, the radius of F^- is anticipated to be larger than that of F, since F^- has 9 protons and 10 electrons, while atomic F has 9 protons and 9 electrons. The proton:electron ratio would have us anticipate that the atomic F is smaller than the anion.

59. Electron configurations for the two ionizations of the K atom:
Electron configuration for K: $1s^2 2s^2 2p^6 3s^2 3p^6 4s^1$. The loss of one electron (1st ionization) gives a configuration of: $1s^2 2s^2 2p^6 3s^2 3p^6$. The loss of the second electron (2nd ionization) gives a configuration of: $1s^2 2s^2 2p^6 3s^2 3p^5$. Note that the 1st ionization results in a specie that is isoelectronic with a noble gas (Ar), and is not anticipated to be very large. However, the loss of the 2nd electron disrupts the filled 3p sublevel—and we anticipate (and find to be experimentally true) that this 2nd IE will be much greater than the first IE.

61. Explain the trends in size:
   (a) The decrease in atomic size across a period is attributable to the increasing nuclear charge with an increasing number of protons. Given that the electrons are in the same outer energy level, the nuclear attraction for the electrons results in a diminishing atomic radius.

(b) The slight decrease in atomic radius of the transition metals is a result of increased repulsions of (n-1)d electrons for (n)s electrons. Thus repulsion reduces the effects of the increasing nuclear charge across a period.

63. The existence of Mg^{2+} and O^{2-} rather than their monopositive and mononegative counterparts is easily argued from the experimental evidence that members of group 2A (Mg) typically lose 2 electrons while members of group 6A (O) typically gain 2 electrons—both with the aim of becoming isoelectronic with a noble gas. One form of experimental evidence is in measuring the relative melting points of MgO versus a +1 salt (e.g. NaCl). MgO melts about 2850 °C, while NaCl (a +1:-1) salt melts about 800 °C. The greater charges associated with the +2/-2 ions are one factor for the much greater melting point of MgO over that of the +1/-1 NaCl salt.

65. (a) Orbital energies decrease across period 2, owing to the increased effective nuclear charge. With the electrons in "p" sublevels—and hence about the same average distance from the nucleus, the increasing nuclear charge (with increasing numbers of protons in the nucleus) results in a greater attraction for the valence electrons—and a lower orbital energy.

(b) As the orbital energies decrease (become more negative), the amount of energy needed to remove an electron (the Ionization Energy) increases. With this decreasing orbital energy, an atom has a greater affinity to **gain** an electron.

(c) The data show:

Li	-520.0 kJ/mol	(2s)
Be	-899.2 kJ/mol	(2s)
B	-800.8 kJ/mol	(2p)
C	-1029 kJ/mol	(2p)

Li has the lowest ioinization energy, owing to the fact that the loss of the sole 2s electron results in a noble gas configuration for the ion. Be has a higher IE, since removal of an electron disrupts the pairing of electrons (in 2s), The slightly lower IE for B (compared to Be) isn't surprising, since removal of the 2p electron would give B an electron configuration equal to that of Be (with a pair of electrons). Finally the 2nd 2p electron (for C) is not spin paired with the first 2p electron, and has a higher effective nuclear charge than B, so we anticipate that C will have a greater IE than B.

67. A plot of the atomic radii of the elements K—V shows a decrease. With the mass of these elements increasing from K through V (as more protons, neutrons, and electrons are added), the density is expected to increase.

69. (a) Electron configurations for elements 113 and 115

Element 113 is located in period 7, group 3A while 115 is in group 5A.

The anticipated electron configrations are:

113     $[Rn]5f^{14}6d^{10}7s^27p^1$

115     $[Rn]5f^{14}6d^{10}7s^27p^3$

(b) Name an element in the same periodic group as 113 and 115.

Thallium is a group member of 3A –like 113, while Bismuth is a group member of 5A—like 115.

(c) What atom could be used as a projectile to bombard Am to produce the element 113?

With 95 protons in an Americium nucleus, we'd need to add 18 protons to form the 113 nucleus. Hence an element with 18 protons would be called for (Argon).

71. (a) Orbital box notation for sulfur:

1s	2s	2p	3s	3p
↑↓	↑↓	↑↓ ↑↓ ↑↓	↑↓	↑↓ ↑ ↑

(b) Quantum numbers for the "last electron": $n = 3$, $\ell = 1$, $m\ell = +1$ (or 0 or -1), $m_s = -1/2$

(c) Element with the smallest ionization energy : S

Element with the smallest radius: O

(d) S: Negative ions are always larger than the element from which they are derived.

(e) Grams of SCl_2 to make 675 g of $SOCl_2$:

$$\frac{675 \text{ g } SOCl_2}{1} \cdot \frac{1 \text{ mol } SOCl_2}{119.0 \text{ g } SOCl_2} \cdot \frac{1 \text{ mol } SCl_2}{1 \text{ mol } SOCl_2} \cdot \frac{102.97 \text{ g } SCl_2}{1 \text{ mol } SCl_2} = 584 \text{ g } SCl_2$$

(f) The theoretical yield of $SOCl_2$ if 10.0 g of SO_3 and 10.0 g of SCl_2 are used:

$$\text{Moles of } SO_3 = 10.0 \text{ g } SO_3 \cdot \frac{1 \text{ mol } SO_3}{80.06 \text{ g } SO_3} = 0.125 \text{ moles } SO_3$$

$$\text{Moles of } SCl_2 = 10.0 \text{ g } SCl_2 \cdot \frac{1 \text{ mol } SCl_2}{102.97 \text{ g } SCl_2} = 0.0971 \text{ moles } SCl_2$$

$$\text{Moles-available ratio: } \frac{0.125 \text{ mol } SO_3}{0.0971 \text{ mol } SCl_2} = \frac{1.29 \text{ mol } SO_3}{1 \text{ mol } SCl_2}$$

$$\text{Moles-required ratio: } \frac{1 \text{ mol } SO_3}{1 \text{ mol } SCl_2} \quad \text{so } \mathbf{SCl_2 \text{ is the limiting reagent.}}$$

$$0.0971 \text{ moles } SCl_2 \cdot \frac{1 \text{ mol } SOCl_2}{1 \text{ moles } SCl_2} \cdot \frac{119.0 \text{ g } SOCl_2}{1 \text{ mol } SOCl_2} = 11.6 \text{ g } SOCl_2$$

(g) For the reaction:

$$SO_3(g) + SCl_2(g) \rightarrow SOCl_2(g) + SO_2(g) \qquad \Delta H°_{rxn} = -96.0 \text{ kJ/mol } SOCl_2$$

$$\Delta H°_{rxn} = [1 \bullet \Delta H°_f SOCl_2(g) + 1 \bullet \Delta H°_f SO_2(g)]$$

$$- [1 \bullet \Delta H°_f SO_3(g) + 1 \bullet \Delta H°_f SCl_2(g)]$$

-96.0 kJ = [-212.5kJ + -296.84 kJ] - [- 395.77 kJ + 1 • ($\Delta H°_f SCl_2$ (g)])]

-96.0 kJ = [-509.34 kJ] – [-395.77 kJ + $\Delta H°_f SCl_2$(g)]

-96.0 kJ = (395.77 kJ –509.34 kJ) - $\Delta H°_f SCl_2$(g)

17.6 kJ = - $\Delta H°_f SCl_2$(g) and –17.6 kJ = $\Delta H°_f SCl_2$(g)

73. Distances between atoms in molecules:

Molecule	Atomic Distance	Calculated(pm)	Measured (pm)
BF_3	B-F	(83+71) =154	130
PF_3	P-F	(115 + 71) = 186	178
CH_4	C-H	(77 + 37) = 114	109
H_3COH	C-O	(77 + 66) = 143	150

Calculated distances are arrived at by simple addition of atomic radii.

Hence for B (86 pm) and F (71 pm) the calculated distance is the sum, (83+71) or 154 pm. Figure 8.10 of your textbook provides the appropriate radii

The agreement between calculated and measured is good for C-O, C-H, and P-F. The greatest difference is in the B-F distance in BF_3.

Chapter 9:
Bonding and Molecular Structure: Fundamental Concepts

Practicing Skills
Valence Electrons and the Octet Rule

1. | Element | Group Number | Number of Valence Electrons |
|---------|--------------|------------------------------|
| (a) O | 6A | 6 |
| (b) B | 3A | 3 |
| (c) Na | 1A | 1 |
| (d) Mg | 2A | 2 |
| (e) F | 7A | 7 |
| (f) S | 6A | 6 |

3. | Group Number | Number of Bonds |
|--------------|-----------------|
| 3A | 3 |
| 4A | 4 |
| 5A | 3 (or 4 in species such as NH_4^+) |
| 6A | 2 (or 3 as in H_3O^+) |
| 7A | 1 |

Ionic Compounds

5. Compound with the most negative energy of ion pair formation? least negative?

Coulomb's Law tells us that the most negative IP energy will result from (a) increased charges on the ions and (b) decreased distance between the ions. Compiling those data for the ions involved we get:

ion	charge	ionic radius (pm)
Na^+	1	98
Mg^{2+}	2	79
K^+	1	133
Cl^-	1	181
I^-	1	220
S^{2-}	2	184

Ignoring for the moment the charges on the electrons (since they are the same on all electrons, we can calculate terms for the numerators (# of + and - charges) and denominators (sum of ionic radii) for the three compounds:

(a) NaCl $\quad \dfrac{(1 \bullet 1)}{(98 + 181)} = \dfrac{1}{279} = \dfrac{1}{279}$

(b) MgS $\quad \dfrac{(2 \bullet 2)}{(79 + 184)} = \dfrac{4}{263} = \dfrac{1}{65.75}$

(c) KI $\quad \dfrac{(1 \bullet 1)}{(133 + 220)} = \dfrac{1}{353} = \dfrac{1}{353}$

The third column represents the reduction of all numerators to unity. The result is that **MgS would have the largest negative energy of ion pair formation, and KI would have the least negative value.**

7. Arrange lattice energies from least negative to most negative:

Since CaO involves 2+ and 2- ions, the lattice energy for this compound is greater than the other compounds. Given that lattice energy is **inversely** related to the distance between the ions, the lattice energy for LiI is greater(more negative) than that for RbI (Rb$^+$ > Li$^+$). The small diameter of the fluoride ion (compared to iodide) indicates that the lattice energy for LiF would be more negative than for either LiI or RbI. The lattice energies are then:

least   -----   RbI   -----   LiI   -----   LiF   -----   CaO   **most**
negative   **negative**

9. Since melting a solid involves disassembling the crystal lattice of cations and anions, the lesser the distance between cations and anions—the greater the attraction between the cation and anion, and the harder it becomes to disassemble the lattice, and hence the **higher the melting point**.

Lewis Electron Dot Structures

11. (a) NF$_3$: [1(5) + 3(7)] = 26 valence electrons

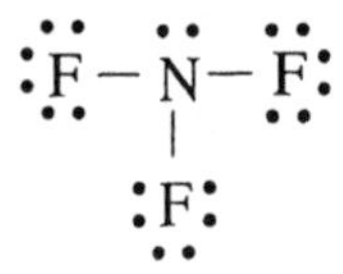

$$:\overset{..}{\underset{..}{F}} - \overset{..}{N} - \overset{..}{\underset{..}{F}}:$$
$$|$$
$$:\overset{}{\underset{..}{F}}:$$

(b) ClO$_3^-$: [1(7) + 3(6) + 1] = 26 valence electrons

$\uparrow$

ion charge

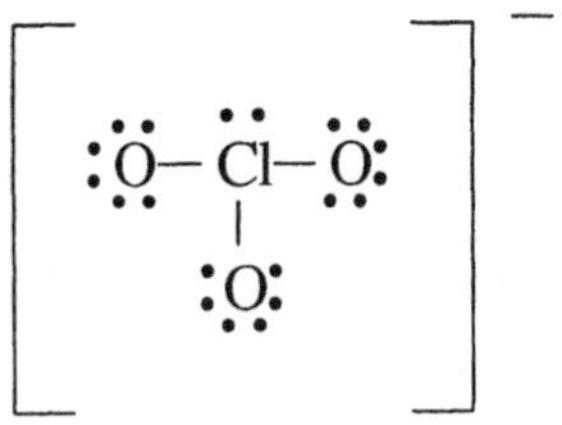

$$\left[:\overset{..}{\underset{..}{O}} - \overset{..}{Cl} - \overset{..}{\underset{..}{O}}: \right]^-$$
$$|$$
$$:\overset{}{\underset{..}{O}}:$$

(c) HOBr: [1(1) + 1(6) + 1(7)] = 14 valence electrons

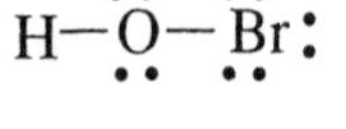

$$H - \overset{..}{\underset{..}{O}} - \overset{..}{\underset{..}{Br}}:$$

(d) SO_3^{2-} : $[1(6) + 3(6) + 2]$ = 26 valence electrons

ion charge

$$\left[\ddot{\underset{..}{O}}-\underset{|}{\overset{..}{S}}-\ddot{\underset{..}{O}}\colon\underset{\underset{:\ddot{O}:}{|}}{}\right]^{2-}$$

13. (a) $CHClF_2$: $[1(4) +1(1) + 1(7) + 2(7)]$ = 26 valence electrons

$$\ddot{\underset{..}{F}}-\underset{\underset{:\overset{..}{Cl}:}{\overset{|}{\underset{}{}}}}{\overset{\overset{H}{|}}{C}}-\ddot{\underset{..}{F}}\colon$$

(b) CH_3CO_2H: $[3(1) + 2(4) + 2(6) + 1(1)]$ = 24 valence electrons

$$H-\underset{\underset{H}{|}}{\overset{\overset{H}{|}}{C}}-\underset{}{\overset{\overset{:\ddot{O}:}{\|}}{C}}-\ddot{\underset{..}{O}}-H$$

(c) H_3CCN: $[3(1) + 2(4) + 1(5)]$ = 16 valence electrons

$$H-\underset{\underset{H}{|}}{\overset{\overset{H}{|}}{C}}-C\equiv N\colon$$

(d) H_2CCCH_2: $[4(1) + 3(4)]$ = 16 valence electrons

$$H-\underset{}{\overset{\overset{H}{|}}{C}}=C=\underset{}{\overset{\overset{H}{|}}{C}}-H$$

15. Resonance structures for:

(a) SO_2:

$$\ddot{O}=\ddot{S}-\ddot{\underset{..}{O}}\colon \longleftrightarrow \colon\ddot{\underset{..}{O}}-\ddot{S}=\ddot{O}$$

(b) NO_2^-:

$$\left[\ddot{\underset{..}{O}}=\ddot{N}-\ddot{\underset{..}{O}}\colon\right]^- \longleftrightarrow \left[\colon\ddot{\underset{..}{O}}-\ddot{N}=\ddot{O}\right]^-$$

(c) SCN^-:

$$\left[\colon\ddot{\underset{..}{N}}-C\equiv S\colon\right]^- \longleftrightarrow \left[\ddot{\underset{..}{N}}=C=\ddot{\underset{..}{S}}\right]^- \longleftrightarrow \left[\colon N\equiv C-\ddot{\underset{..}{S}}\colon\right]^-$$

17. (a) BrF_3 : $[1(7) + 3(7)] = 28$ valence electrons

$$\overset{\cdot\cdot}{\underset{\cdot\cdot}{:F}}-Br-\overset{\cdot\cdot}{\underset{\cdot\cdot}{F:}}$$
$$|$$
$$\overset{}{\underset{\cdot\cdot}{:F:}}$$

(b) I_3^- : $[3(7) + 1] = 22$ valence electrons

$$\overset{\cdot\cdot}{\underset{\cdot\cdot}{:I}}-\overset{\cdot\cdot}{\underset{\cdot\cdot}{I}}-\overset{\cdot\cdot}{\underset{\cdot\cdot}{I:}}$$

(c) XeO_2F_2 : $[1(8) + 2(6) + 2(7)] = 34$ valence electrons

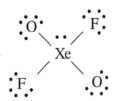

(d) XeF_3^+ : $[1(8) + 3(7) - 1] = 28$ valence electrons

$$\left[\overset{\cdot\cdot}{\underset{\cdot\cdot}{:F}}-Xe-\overset{\cdot\cdot}{\underset{\cdot\cdot}{F:}} \atop | \atop \overset{}{\underset{\cdot\cdot}{:F:}} \right]^+$$

Molecular Geometry

19. Using the Lewis structure describe the Electron-pair and molecular geometry:

(a).

$$H-\overset{\cdot\cdot}{N}-\overset{\cdot\cdot}{\underset{\cdot\cdot}{Cl:}}$$
$$|$$
$$H$$

Electron-pair: tetrahedral
Molecular : trigonal pyramidal

(b)

$$\overset{\cdot\cdot}{\underset{\cdot\cdot}{:Cl}}-\overset{\cdot\cdot}{\underset{\cdot\cdot}{O}}-\overset{\cdot\cdot}{\underset{\cdot\cdot}{Cl:}}$$

Electron-pair: tetrahedral
Molecular: bent or angular

(c)

$$\left[\overset{\cdot\cdot}{\underset{\cdot\cdot}{N}}= C = \overset{\cdot\cdot}{\underset{\cdot\cdot}{S}} \right]^-$$

Electron-pair: linear
Molecular: linear

(d)

$$H-\overset{\cdot\cdot}{\underset{\cdot\cdot}{O}}-\overset{\cdot\cdot}{\underset{\cdot\cdot}{F:}}$$

Electron-pair: tetrahedral
Molecular: bent or angular

21. Using the Lewis structure describe the Electron-pair and molecular geometry:

(a)

$$\ddot{O} = C = \ddot{O}$$

Electron-pair: linear
Molecular : linear

(b)

$$\left[\ddot{O} = \ddot{N} - \ddot{O} : \right]^{-}$$

Electron-pair: trigonal planar
Molecular: bent or angular

(c)

$$: \ddot{O} - \ddot{O} = \ddot{O}$$

Electron-pair: trigonal planar
Molecular: bent or angular

(d)

$$\left[: \ddot{O} - \ddot{Cl} - \ddot{O} : \right]^{-}$$

Electron-pair: tetrahedral
Molecular: bent or angular

For the species shown above, having at least one lone pair on the central atom **changes the molecular geometry from linear to bent.**

23. Using the Lewis structure describe the electron-pair and molecular geometry.

[Lone pairs on F have been omitted for clarity.]

(a)

$$\left[F - \ddot{Cl} - F \right]^{-}$$

Electron-pair: trigonal bipyramidal
Molecular : linear

(b)

$$F - \ddot{Cl} - F$$
$$|$$
$$F$$

Electron-pair: trigonal bipyramidal

Molecular: T-shaped

(c)

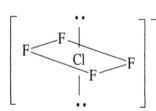

Electron-pair: octahedral
Molecular: square planar

(d)

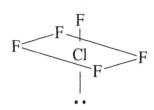

Electron-pair: octahedral
Molecular: square pyramidal

25. (a) O-S-O angle in SO_2 :   Slightly less than 120°; The lone pair of S should reduce the predicted 120° angle slightly.

(b) F-B-F angle in BF_3 :   120°

(c) Cl-C-Cl in Cl_2CO Slightly less than 120°; The two lone pairs of electrons on O will reduce the predicted 120° angle slightly.

(d) (1) H-C-H angle in CH_3CN: 109°

    (2) C-C ≡N angle in CH_3CN: 180°

27. Estimate the values of the angles indicated in the model of phenylalanine below:

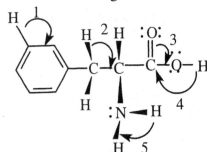

Angle 1:   H-C-C 120°   three groups around the C atom

Angle 2:   H-C-C 109°   four groups around the C atom

Angle 3:   O-C-O 120°   three groups around the C atom

Angle 4:   C-O-H 109°   four groups around the O atom

Angle 5:   H-N-H 109°   four groups around the N atom

The CH_2-$CH(NH_2)$-CO_2H can not be a straight line, since the first two carbons will have bond angles of 109 degrees(with their connecting atoms) and the third C (the C of the CO_2H group) has a 120 bond angle with the C and O on either side.

Formal Charges

29. Formal charge on each atom in the following:

(a) N_2H_4

Atom	Formal Charge
H	1 - 1/2(2) = 0
N	5 - 2 - 1/2(6) = 0

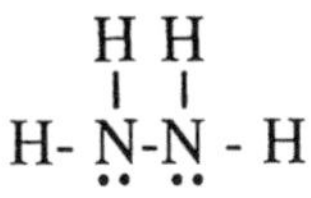

H   H
| |
H- N-N - H
 .. ..

130

(b) PO_4^{3-} Atom   Formal Charge

P $5 - 1/2(8) = +1$
O $6 - 6 - 1/2(2) = -1$
                        Sum = -3 (charge on ion)

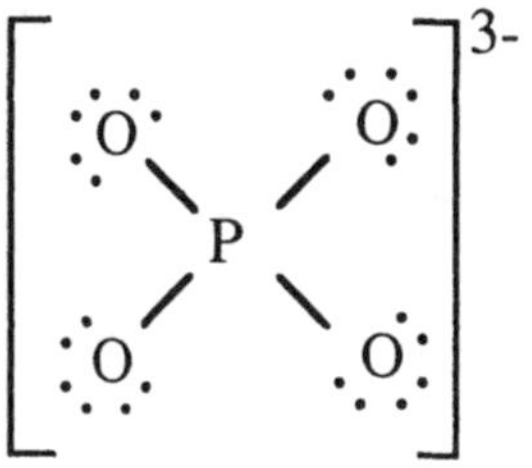

(c) BH_4^- Atom   Formal Charge

B $3 - 1/2(8) = -1$
H $1 - 1/2(2) = 0$
                        Sum = -1 (charge on ion)

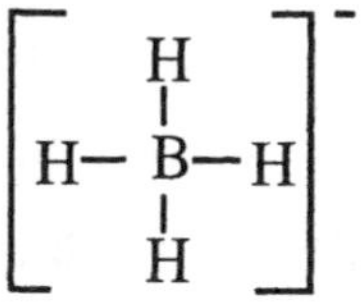

(d) NH_2OH Atom   Formal Charge
                    N $5 - 2 - 1/2(6) = 0$
                    H $1 - 1/2(2) = 0$
                    O $6 - 4 - 1/2(4) = 0$

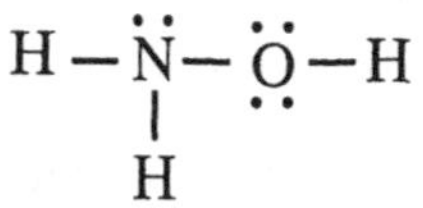

31. Formal charge on each atom in the following:
    (a) NO_2^+ Atom   Formal Charge
                    O $6 - 4 - 1/2(4) = 0$
                    N $5 - 0 - 1/2(8) = +1$

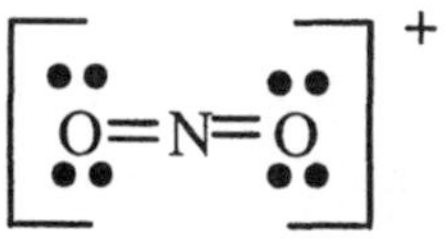

    (b) NO_2^- Atom   Formal Charge
                    O1 $6 - 4 - 1/2(4) = 0$
                    O2 $6 - 6 - 1/2(2) = -1$
                    N $5 - 2 - 1/2(6) = 0$

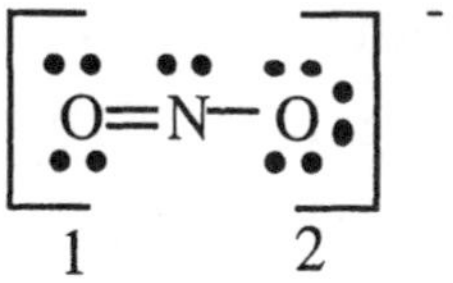

    (c) NF_3 Atom   Formal Charge
                    F $7 - 6 - 1/2(2) = 0$
                    N $5 - 2 - 1/2(6) = 0$

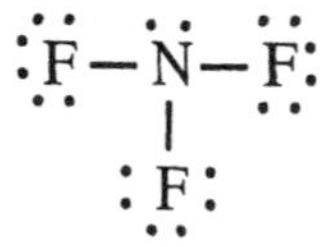

    (d) HNO_3 Atom   Formal Charge
                    O1 $6 - 4 - 1/2(4) = 0$
                    O2 $6 - 6 - 1/2(2) = -1$
                    O3 $6 - 4 - 1/2(4) = 0$
                    N $5 - 0 - 1/2(8) = +1$
                    H $1 - 1/2(2) = 0$
                    [Note: This is only 1 possible structure]

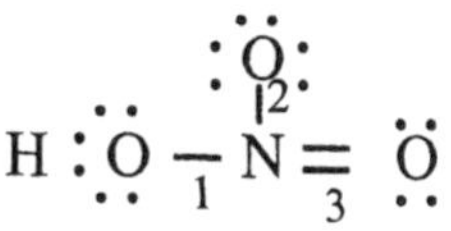

Bond Polarity and Electronegativity

33. Indicate the more polar bond (Arrow points toward the more negative atom in the dipole).

    (a) C-O >   C-N   (b) P-Cl >   P-Br
        →   →   →   →

131

(c) B - O > B - S   (d) B-F > B-I
   $\rightarrow$ $\rightarrow$ $\rightarrow$ $\rightarrow$

35. For the bonds in acrolein the polarities are as follows:

	H-C	C-C	C=O
$\dfrac{\Delta\chi}{\Sigma\chi}$	0.09	0	0.16

(Note that χ represents electronegativity)

(a) The C-C bonds are nonpolar, the C-H bonds are slightly polar, and the C=O bond is polar.

(b) The most polar bond in the molecule is the C=O bond, with the oxygen atom being the negative end of the dipole.

Bond Polarity and Formal Charge

37. Atom(s) on which the negative charge resides in:

(a) OH^- Formal charges: O = -1 ; H = 0   Oxygen is much more electronegative than H so the negative charge resides on the Oxygen atom

(b) BH_4^- Formal charges: B = -1: H = 0   Hydrogen is only slightly more electronegative than B (2.1 compared to 2.0), so the negative charge would reside on the H atoms (although the B-H bonds are **not very polar**).

(c) $CH_3CO_2^-$

Formal charges : C = 0; O1 = 0, O2 = -1   Oxygen is more electronegative than C, so the charge would reside on the oxygens as opposed to the C. The picture shown here is a bit misleading, since **either oxygen** could have the double bond to the C, so **two resonance structures are available** with the negative charge distributed (delocalized) over both C-O bonds.

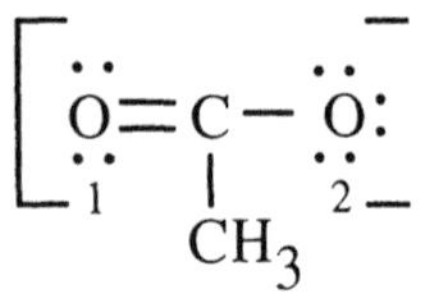

39. (a) Resonance structures of N_2O :

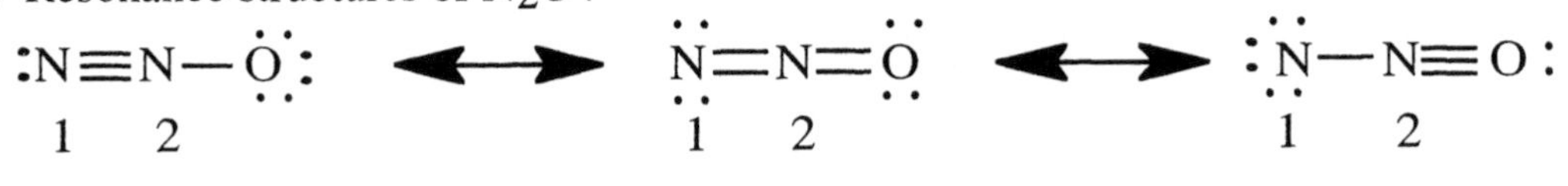

(b) Formal Charges:

N_1 5 - 2 - 1/2(6) = 0	5 - 4 - 1/2(4) = -1	5 - 6 - 1/2(2) = -2
N_2 5 - 0 - 1/2(8) = +1	5 - 0 - 1/2(8) = +1	5 - 0 - 1/2(8) = +1
O 6 - 6 - 1/2(2) = -1	6 - 4 - 1/2(4) = 0	6 - 2 - 1/2(6) = +1

(c) Of these three structures, the first is the most reasonable in that the most electronegative atom, O, bears a formal charge of -1.

41. Resonance structures for NO_2^- :

$$\left[\ddot{O} = N - \ddot{\underset{..}{O}} : \right]^- \longleftrightarrow \left[: \ddot{\underset{..}{O}} - N = \ddot{O} \right]^-$$

         1   2   1   2

Formal charges:

           O_1 $6 - 4 - 1/2(4) = 0$ O_1 $6 - 6 - 1/2(2) = -1$

           N $5 - 2 - 1/2(6) = 0$ N $5 - 2 - 1/2(6) = 0$

           O_2 $6 - 6 - 1/2(2) = -1$ O_2 $6 - 4 - 1/2(4) = 0$

Given that the formal charge on the oxygen atoms is -1, an H^+ ion will attach to the more negative oxygen atom.

Molecular Polarity

43. For the molecules:

          H_2O NH_3 CO_2 ClF CCl_4

(i) Using the electronegativities to determine bond polarity:

$\dfrac{\Delta \chi}{\Sigma \chi}$	$\dfrac{1.4}{5.6}$	$\dfrac{0.9}{5.1}$	$\dfrac{1.0}{6.0}$	$\dfrac{10}{7.0}$	$\dfrac{0.5}{5.5}$

Reducing these fractions to a decimal form indicates that the H-O bonds in water are the most polar of these bonds. [Note: $\Delta \chi$ represents the **difference** in electronegativities between the elements while $\Sigma \chi$ is the **sum** of the electronegativities of the elements.]

(ii) The nonpolar compounds are:

         CO_2 The O-C-O bond angle is $180°$, thereby canceling the C-O dipoles.

         CCl_4 The Cl-C-Cl bond angles are approximately $109°$, with the Cl atoms directed at the corners of a tetrahedron. Such an arrangement results in a net dipole moment of zero.

(iii) The F atom in ClF is more negatively charged.(Electronegativity of F = 4.0, Cl = 3.0)

45. Molecular polarity of the following: (a) $BeCl_2$, (b) HBF_2, (c) CH_3Cl, (d) SO_3

$BeCl_2$ and SO_3 are nonpolar—since the linear geometry (of $BeCl_2$) and the trigonal planar geometry (of SO_3) would give a net dipole moment of zero. For HBF_2, the hydrogen and fluorine atoms are arranged at the corners of a triangle. The "negative end" of the molecule lies on the plane between the fluorine atoms, and the H atom is the "positive end." For CH_3Cl, with the H and Cl atoms arranged at the corners of a tetrahedron, the chlorine atom is the negative end and the H atoms form the positive end.

Bond Order and Bond Length

47.

Specie	Number of bonds	Bond Order : Bonded Atoms
(a) H_2CO	3	1 : CH 2: C = O
(b) SO_3^{2-}	3	1 : SO
(c) NO_2^+	2	2 : NO
(d) NOCl	2	1: N-Cl 2: N=O

Bond order is calculated: (number of shared electron pairs between atoms/number of links)

49. In each case the shorter bond length should be between the atoms with smaller radii--if we assume that the bond orders are equal.

    (a) B-Cl   B is smaller than Ga   (b) C-O   C is smaller than Sn

    (c) P-O   O is smaller than S   (d) C=O   O is smaller than N

51. The bond order for NO_2^+ is 2, for NO_2^- is 3/2 while the bond order for NO_3^- is 4/3. The Lewis dot structure for the NO_2^+ ion indicates that both NO bonds are double, while in the nitrate ion, any resonance structure (there are three) shows one double bond and two single bonds. The nitrite ion has—in either resonance structure (there are two)—one double and one single bond. Hence the **NO bonds in the nitrate ion will be longest** while those in the **NO_2^+ ion will be shortest.**

Bond Energy

53. The CO bond in carbon monoxide is shorter. The CO bond in carbon monoxide is a **triple bond**, thus it requires more energy to break than the CO double bond in H_2CO. (See bond order for HCHO in SQ47).

55.

$$H_3C\text{-}CH_2\text{-}\overset{\overset{H}{|}}{C}=\overset{\overset{H}{|}}{C}\text{-}H \quad + \quad H_2 \quad \longrightarrow \quad H_3C\text{-}CH_2\text{-}\overset{\overset{H}{|}}{\underset{\underset{H}{|}}{C}}\text{-}\overset{\overset{H}{|}}{\underset{\underset{H}{|}}{C}}\text{-}H$$

Energy input:	1 mol C=C =	1 mol • 610 kJ/mol =	610 kJ
	1 mol H-H =	1 mol • 436 kJ/mol =	436 kJ
		Total input =	1046 kJ
Energy release:	1 mol C-C =	1 mol • 346 kJ/mol =	346 kJ
	2 mol C-H =	2 mol • 413 kJ/mol =	826 kJ
		Total released =	1172 kJ
Energy change:	1046 kJ - 1172 kJ =	- 126 kJ	

57. OF_2 (g) + H_2O (g) $\rightarrow$ O_2 (g) + 2 HF (g) ΔH = - 318 kJ

   Energy input :   2 mol O-F = 2 x (where x = O-F bond energy)

                        2 mol O-H = 2 mol • 463 kJ/mol = <u>926 kJ</u>

                                    Total input   = (926 + 2x) kJ

   Energy release: 1 mol O=O = 1 mol • 498 kJ/mol =   498 kJ

                        2 mol H-F = 2 mol • 565 kJ/mol = <u>1130 kJ</u>

                                    Total release = 1628 kJ

        - 318 kJ   = 926 kJ + 2x - 1628 kJ

        384 kJ   = 2x

        192 kJ/mol = O-F bond energy

General Questions on Bonding and Molecular Structure

59. Number of valence electrons for Li, Ti, Zn, Si, and Cl:

Specie	Li	Ti	Zn	Si	Cl
# valence electrons	1	4	2	4	7

The number of valence electrons for any main group(representative) element is easily determined by viewing the GROUP NUMBER in which that element resides. Ti—which is a transition element, has 2d electrons and 2s electrons , accounting for the 4—as the number of valence electrons. Zn also a transition element has a filled d sublevel and 2s electrons—giving rise to "2" valence electrons.

61. Prediction on the bonding in:

   KI   ionic   bonding between metal and nonmetal

   MgS   ionic   bonding between metal and nonmetal

   CS_2 covalent   bonding between two nonmetals

   P_4O_{10} covalent   bonding between two nonmetals

63. $CaCl_4$ is not likely to exist. Calcium (Group 2A) forms a 2+ cation and chlorine (Group 7A) forms a 1- anion, making $CaCl_2$ the likely formula of the compound between these two elements.

65. Which of the following do not have an octet surrounding the central atom:

BF_4^-, SiF_4, SeF_4, BrF_4^-, XeF_4

A simple solution to this problem is gained by asking (a) how many valence electrons does the central atom have and (b) how many electrons are added by all the combined atoms.

BF_4^-

B has 3 electrons, 4 F each contribute 1 electron, -1 charge adds one electron : (3 + 4 + 1) = 8--an octet

SiF_4,

Si has 4 electrons, 4 F each contribute 1 electron (4 + 4) = 8 electrons--an octet.

SeF_4

Se has 6 electrons, 4 F each contribute 1 electron (6 + 4)= 10 electrons—NOT an octet

BrF_4^-

Br has 7 electrons, 4 F each contribute 1 elecron, -1 charge adds 1 electron (7 + 4 + 1) = 12 electrons—NOT an octet.

XeF_4

Xe has 8 electrons, 4 F each contribute 1 electron (8 + 4) = 12 electrons—NOT an octet.

67. Bond order in acetylene and phosgene:

$$H-C\equiv C-H \qquad \overset{\ddot{\ddot{Cl}}}{\underset{}{:\ddot{Cl}-C}} = \ddot{\ddot{O}}$$

The C-H bonds are of bond order =1 (single bond); the C-C bond is of bond order 3.
The Cl-C bonds (in phosgene) are bond order = 1; the C-O bond is of bond order 2.

69. The N-O bond order in the nitrate ion:

The N-O bonds labeled (1) are of bond order 1, the N-O bond labeled (2) is of bonder order 2. This is only one structure that we could draw for the nitrate ion. The others would show a similar electron distribution. We have **4 pairs** of electrons making 3 bonds so the "average" bond order is 4/3 of 1 1/3.

$$\left[:\ddot{O}-\overset{\ddot{O}:}{\underset{1}{N}}\overset{2}{=}\ddot{O} \right]^-$$

71. To estimate the enthalpy change for the formation of water (g) from hydrogen and oxygen one **needs**: O=O (double bond) energy , H-H bond energy, H-O bond energy

$\Delta H_{reaction}$ = Σ E(bonds broken) - Σ E (bonds made)

= [E (O=O) bond + 2 E (H-H)bond] - 4[E (O-H) bond]

= [498 kJ + 2•436 kJ] – [4 • 463 kJ] = - 482 kJ

73. Lewis structure(s) for the following: What are similarities and differences ?

(a) CO_2

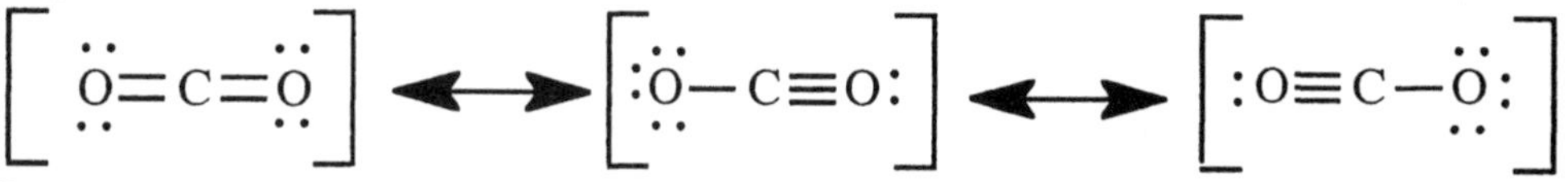

$$\left[\ddot{O} = C = \ddot{O} \right] \longleftrightarrow \left[:\ddot{O} - C \equiv O: \right] \longleftrightarrow \left[:O \equiv C - \ddot{O}: \right]$$

(b) N_3^-

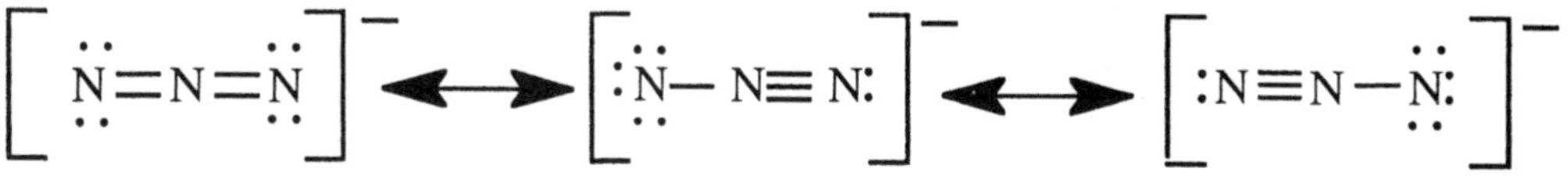

$$\left[\ddot{N} = N = \ddot{N} \right]^- \longleftrightarrow \left[:\ddot{N} - N \equiv N: \right]^- \longleftrightarrow \left[:N \equiv N - \ddot{N}: \right]^-$$

(c) OCN^-

$$\left[\ddot{O} = C = \ddot{N} \right]^- \longleftrightarrow \left[:\ddot{O} - C \equiv N: \right]^- \longleftrightarrow \left[:O \equiv C - \ddot{N}: \right]^-$$

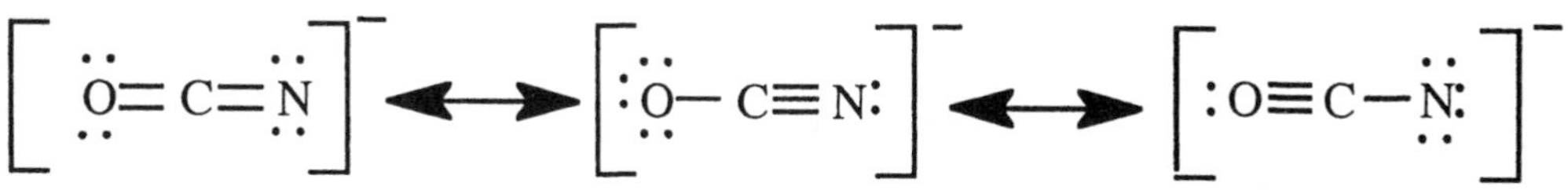

Each of these species has 16 electrons, and each is linear. Carbon dioxide is neutral while the azide ion and isocyanate ion are charged. The isocyanate ion is also polar, while carbon dioxide and azide are not polar.

75. Bond orders in NO_2^- and NO_2^+:

The Lewis dot structure for the NO_2^+ ion (SQ31a) indicates that both NO bonds are double. The nitrite ion(NO_2^-—SQ31b) has—in either resonance structure (there are two)—one double and one single bond, so 110 pm is the N-O bond distance in NO_2^+ and 124 pm is the N-O bond distance in NO_2^-.

77. Compare the F-Cl-F angles in ClF_2^+ and ClF_2^-

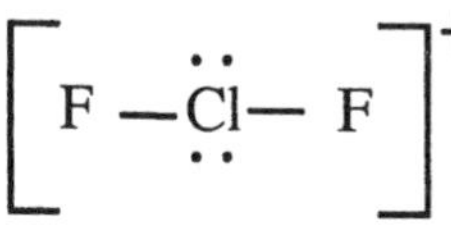

$$\left[F - \ddot{Cl} - F \right]^+ \text{ and } \left[F - \ddot{\ddot{Cl}} - F \right]^-$$

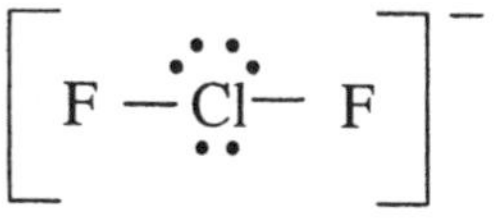

Note that in the cation, there are (7 Cl electrons + 1 electron from each of 2 F + (-) an electron owing to the + charge) a total of 8 electrons—an octet around the Cl atom. In an anion, there are (7 Cl electrons + 1 electron from each of 2 F + 1 electron owing to the - charge) a total of 10 electrons. The cation will have 4 groups around the Cl atom—at a F-Cl-F bond angle of 109°. The anion will have 5 groups around the Cl atom, with a F-Cl-F bond angle of 180°.

79. Draw the electron dot for SO_3^{2-}. Does the H^+ ion attach to the S or the O atom?

The Lewis dot structure for sulfite ion shows single bonds between each O and S atom, and lone pair of electrons on the S atom. The formal charges show: for the S atom $(6 - 2 - 1/2(6) = +1)$. [group # - lone pair electrons $-1/2$(bonding electrons)]. For the O atoms, the formal charges (all are identical) are $6 - 6 - 1/2(2) = -1$. So one would predict that the H^+ would attach itself to the more negative O atoms.

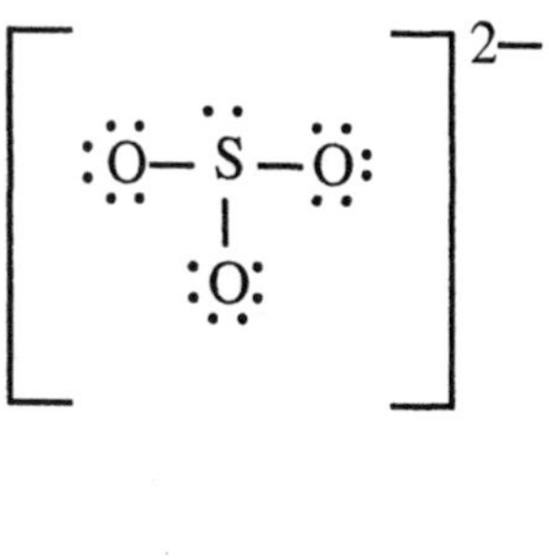

$$\left[:\ddot{O} - \overset{..}{S} - \ddot{O}: \atop \underset{:\ddot{O}:}{|} \right]^{2-}$$

81. (a) Estimate the enthalpy change for the reaction and the ΔHcombusion for 1mole of methanol.

$$2 \; CH_3OH(g) + 3 \; O_2 \; (g) \rightarrow 2 \; CO_2(g) + 4 \; H_2O(g)$$

Energy input:
- 6 mol C-H = 6 mol • 413 kJ/mol = 2478 kJ
- 2 mol C-O = 2 mol • 358 kJ/mol = 716 kJ
- 2 mol O-H = 2 mol • 463 kJ/mol = 926 kJ
- 3 mol O=O = 3 mol • 498 kJ/mol = 1494 kJ
- Total input = 5614 kJ

Energy release:
- 4 mol C=O = 4 mol • 732 kJ/mol = 2928 kJ
- 8 mol H-O = 8 mol • 463 kJ/mol = 3704 kJ
- Total released = 6632 kJ

Energy change: = 5614 – 6632 = -1018 kJ for the reaction

Noting that the equation above indicates the combustion of 2 moles of methanol, the ΔHcombusion for 1mole of methanol is –1018 kJ/2mol = -509 kJ/mol CH_3OH

(b) The calculation using ΔH values:

$\Delta Hrx = [2 \; \Delta H_f \; (CO_2(g)) + 4 \; \Delta H_f \; (H_2O(g))] - [2 \; \Delta H_f \; (CH_3OH(g)) + 3 \; \Delta H_f \; (O_2(g))]$

$\Delta Hrx = [2 \; mol • -393.509 \; kJ/mol + 4mol • -241.83 \; kJ/mol] - [2 \; mol • -201.0 \; kJ/mol + 0]$

$\Delta Hrx = [-787.018 \; kJ + -967.32 \; kJ] - [-402.0 \; kJ] = -1352.3 \; kJ$ and on a per mol CH_3OH

basis the number is -1352.3 kJ/2 = -676.2 kJ/mol CH_3OH.

83. (a) Resonance structures for CNO^- with formal charges.

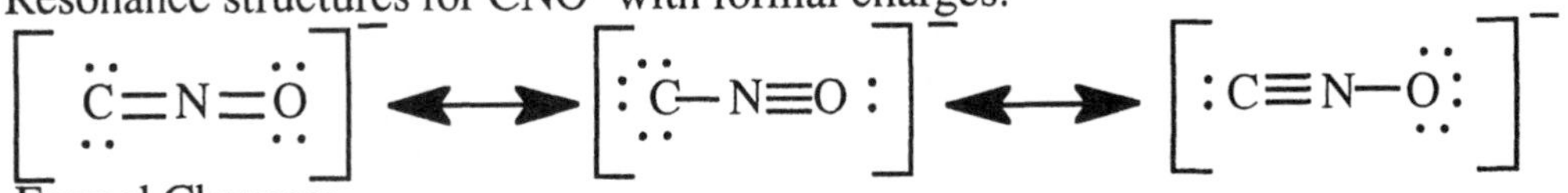

Formal Charges:
-2   +1   0   -3   +1   +1   -1   +1   -1

(b) The most reasonable structure is the one at the far right because it is the one in which the formal charges on the atoms are at a minimum, and oxygen has the negative formal charge.

(c) The instability of the ion could be attributed to the fact that the least electronegative atom in the ion bears a negative charge in all resonance structures.

85. (a) In XeF_2 the bonding pairs occupy the axial positions (of a trigonal bipyramid) with the lone pairs located in the equatorial plane. (See question 17(b) for an isoelectronic specie, (I_3^-).

(b) In ClF_3 two of the three equatorial positions are occupied by the lone pairs of electrons on the Cl atom. (See question 23b)

87. Hydroxyproline has the structure:

(a) Values for the selected angles:

Angle 1: 109° since the N has four groups of electrons around the atom.

Angle 2: 120° since the C atom has three groups of electrons around it.

Angle 3: 109° since the C atom has four groups of electrons around it.

Angle 4: 109° since the O atom has four groups of electrons around it

    (2 bonding and 2 non-bonding pairs)

Angle 5: 109° since the C atom has four groups of bonding pairs of electrons around it.

(b) The most polar bonds in the molecule are the O-H bonds. The electronegativity of O is 3.5, while that of H is 2.1 (a Δ of 1.4). This difference represents the **greatest** differences in electronegativity in the hydroxyproline molecule.

89. Enthalpy change for decomposition of urea:

Break N-C bonds (2)	$2 \cdot 305$ kJ/mol
C=O bond (1)	$1 \cdot 745$ kJ/mol bonds broken: $610 + 745 = 1355$ kJ
Make N-N bond (1)	$1 \cdot 163$ kJ/mol
C≡O bond (1)	$1 \cdot 1046$ kJ/mol bonds made: $163 + 1046 = 1209$ kJ

Change in energy: 1355 kJ - 1209 kJ = 146 kJ

Since there was no change in the number of N-H bonds (in urea) or hydrazine, those bond energies were omitted in the calculation.

91. (a) Bond energies for the conversion of acetone to dihydroxyacetone:

acetone   dihydroxyacetone

Energy input:

6 mol C-H	=	6 mol • 413 kJ/mol	=	2478 kJ
1 mol C=O	=	1 mol • 732 kJ/mol	=	732 kJ
2 mol C-C	=	2 mol • 346 kJ/mol	=	692 kJ
1 mol O=O	=	1 mol • 498 kJ/mol	=	498 kJ
		Total input	=	4400 kJ

Energy release:

1 mol C=O	=	1 mol • 732 kJ/mol	=	732 kJ
2 mol H-O	=	2 mol • 463 kJ/mol	=	926 kJ
2 mol C-C	=	2 mol • 346 kJ/mol	=	692 kJ
2 mol C-O	=	2 mol • 358 kJ/mol	=	716 kJ
4 mol C-H	=	4 mol • 413 kJ/mol	=	1652 kJ
		Total released	=	4718 kJ

Energy change: = 4400 – 4718 = -318 kJ for the reaction, so the reaction is **exothermic**.

(b) The central C atom contains a doubly-bonded O, containing 2 lone pairs of electrons, giving the molecule an overall net polarity—yes, acetone is polar!

(c) Based on the electronegativity of O, the O-H bonds will be very polar, and the H attached to those O will be the most positive.

93. For the synthesis of acrolein from ethylene:

(a) The C=C bond in acrolein is stronger than the C-C bond. Double bonds are stronger than single bonds between the same two atoms.

(b) The C-C bond is longer than a C=C bond. The stronger the bond, the shorter the bond length.

(c) Polarity of ethylene and acrolein:

For the bonds in acrolein the polarities are as follows:

	H-C	C-C	C=O
$\frac{\Delta\chi}{\Sigma\chi}$	0.09	0	0.16

The C-C and C=C bonds are nonpolar, the C-H bonds are very slightly polar, and the C=O bond is polar. [$\Delta\chi$ represents the **difference** in electronegativities between the two bonded atoms while $\Sigma\chi$ represents the **sum** of the electronegativities of the two bonded atoms.] The structure of ethylene (coupled with the relative lack of polarity of the C-H and C=C bonds) results in a molecule that is nonpolar (net dipole moment of 0). Acrolein however substitutes the polar C=O group (essentially inserted into a C-H bond, and gives a **polar** molecule.

(d) Is the reaction endothermic or exothermic?:

Careful examination of the structures reveals that to form acrolein from ethylene, we must break a C-H bond (which reforms in the product), and a C≡O bond (which forms C=O bond in the product). Using bond energies, the energy change is:

Break C≡O bond (1)		1 • 1046 kJ/mol
C-H bond (1)		1 • 413 kJ/mol
Energy input:		1459 kJ
Make C-C bond (1)		1 • 346 kJ/mol
C-H bond (1)		1 • 413 kJ/mol
C=O bond (1)		1 • 745 kJ/mol
Energy released:		1504 kJ

Change in energy: 1459 kJ - 1504kJ = -45 kJ , so the reaction is **exothermic**.

95. For the reaction of acetylene with chlorine:

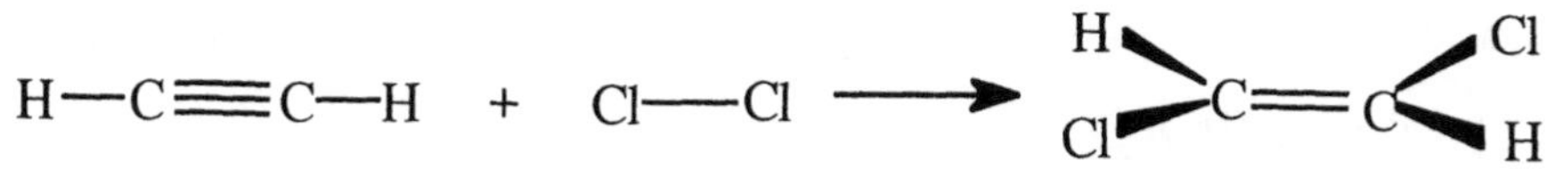

Energy input:	2 mol C- H = 2 mol • 413 kJ/mol = 826 kJ
	1 mol C ≡ C = 1 mol • 835 kJ/mol = 835 kJ
	1 mol Cl-Cl = 1 mol • 242 kJ/mol = 242 kJ
	Total input = 1903 kJ
Energy release:	1 mol C=C = 1 mol • 602 kJ/mol = 610 kJ
	2 mol C- H = 2 mol • 413 kJ/mol = 826 kJ
	2 mol C - Cl = 2 mol • 339 kJ/mol = 678 kJ
	Total released = 2114 kJ

Energy change: = 1903 − 2114 = -211 kJ for the reaction, so the reaction is **exothermic**.

97. For the molecule epinephrine:

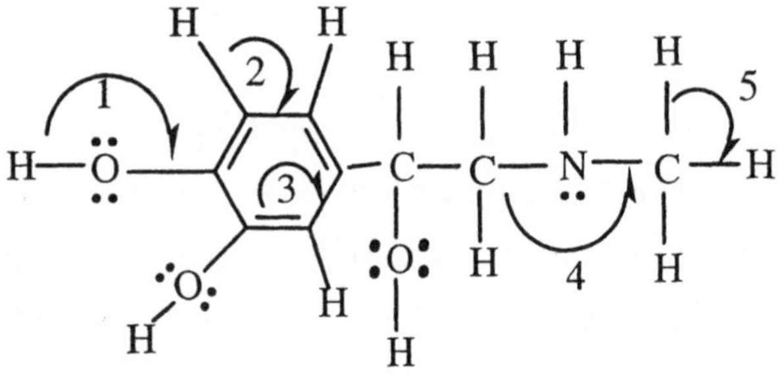

(a) The indicated bond angles are:

       Angle 1: 109°   Angle 2: 120°

       Angle 3: 120°   Angle 4: 109°

       Angle 5: 109°

(b) The most polar bonds in the molecule:

    The O-H bonds are most polar (with a difference in electronegativity of 3.5-2.1)

Summary and Conceptual Questions:

99. Four electron pairs form a **pyramidal molecule if** one of the electron pairs is a non–bonding electron pair. A **bent molecule** is obtained if **two** electron pairs are non-bonding pairs. In either case, the angle between two electron pairs and the central atom is approximately 109°.

101. (a) Odd electron molecules:

HBr	BrO	HOBr	OH
1 + 7 = 8	7 + 6 = 13	1 + 6 + 7 = 14	6 + 1 = 7

(b) Estimate energy reactions of the three reactions:

    $Br_2(g) \rightarrow 2\ Br(g)$ 1 • Br-Br bond energy = 1 • 193 kJ/mol = 193 kJ

    $2\ Br(g) + O_2(g) \rightarrow 2\ BrO(g)$

                Energy input:

                      O=O = 1 • 498 kJ/mol   - 498 kJ

                Energy released:

                      2 mol Br-O = 2 • 201 kJ/mol = 402 kJ

                Energy change: 498 − 402 = 96 kJ

    $BrO(g) + H_2O(g) \rightarrow HOBr(g) + OH(g)$

                Energy input: 1 mol Br-O = 1 • 201 kJ/mol = 201 kJ

                              2 mol H-O = 2 • 463 kJ/mol = <u>926 kJ</u>

                              Total energy input: 1127 kJ

142

Energy released:

$$1 \text{ mol Br-O} = 1 \cdot 201 \text{ kJ/mol} = 201 \text{ kJ}$$
$$2 \text{ mol H-O} = 2 \cdot 463 \text{ kJ/mol} = \underline{926 \text{ kJ}}$$

Total energy released: 1127 kJ

Energy change: 1127 –1127 = 0 kJ

(c) Molar heat of formation of HOBr:

$$Br_2(g) + O_2(g) + H_2(g) \rightarrow 2 \text{ HOBr}(g)$$

Energy input: 1 mol Br-Br = 1 • 193 kJ/mol = 193 kJ

1 mol O=O = 1 • 498 kJ/mol = 498 kJ

1 mol H - H = 1 • 436 kJ/mol = $\underline{436 \text{ kJ}}$

Total energy input: 1127 kJ

Energy released:

2 mol Br-O = 2 • 201 kJ/mol = 402 kJ

2 mol H-O = 2 • 463 kJ/mol = $\underline{926 \text{ kJ}}$

Total energy released: 1328 kJ

Energy change: 1127 – 1328 = -201 kJ, and noting that we form 2 moles of HOBr in this process, we divide to obtain –101 kJ/mol HOBr.

(d) The first two reactions in (b) are **endothermic**, while the 3rd is neither endo- nor exothermic. In (c), the process releases energy, therefore is **exothermic**.

103. The much more negative lattice energy of NaF is attributable to the very small ionic radius of the fluoride ion. The result is a greater attraction between the Na and F ions in the solid lattice.

105. For the molecules in question:

(a) Is BF_3 polar? The trigonal planar geometry around B, results in a net dipole moment of 0 around B, hence the molecule is **nonpolar**.

Replacing F by H would certainly change the net dipole moment around B, and would result in polar species—for HBF_2 and for FBH_2.

(b) Is $BeCl_2$ polar? The linear geometry around Be, results in a net dipole moment of 0 around Be, hence a **nonpolar** molecule. With differing electronegativities , replacing one Cl atom with a Br atom (Br-Be-Cl) would definitely change the net dipole moment around Be, and result in a **polar** molecule.

Chapter 10
Bonding and Molecular Structure: Orbital Hybridization and Molecular Orbitals

Practicing Skills

Valence Bond Theory

1. The Lewis electron dot structure of $CHCl_3$:

   Electron pair geometry = tetrahedral

   Molecular geometry = tetrahedral

   The H-C bonds are a result of the overlap of the
   hydrogen **s** orbital with **sp³** hybrid orbitals on carbon.

   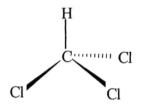

   The Cl-C bonds are formed by the overlap of the **sp³** hybrid orbitals on carbon with the **p** orbitals on chlorine. [Lone pairs on the chlorine atoms have been omitted for the sake of clarity.]

3. Orbital sets used by the underlined atoms:
   - (a) $\underline{B}Br_3$: sp^2 3 groups to be attached to the B
   - (b) $\underline{C}O_2$: sp   2 groups to be attached to the C
   - (c) $\underline{C}H_2Cl_2$: sp^3 4 groups to be attached to the C
   - (d) $\underline{C}O_3{}^{2-}$: sp^2 3 groups to be attached to the C

5. Hybrid orbital sets used by the underlined atoms:
   - (a) the C atoms and the O atom in dimethylether: $\underline{C}$: sp^3 ; $\underline{O}$: sp^3

     In the case of either the carbon OR oxygen atoms in dimethylether, each atom is bound to *four other groups*. This would require **four orbitals**
   - (b) The carbon atoms in propene: $\underline{C}H_3$: sp^3 ; $\underline{C}H$ and $\underline{C}H_2$: sp^2

     The methyl carbon is attached to four groups (three H and 1 C), and needs then four orbitals. The methylene and methine have bonds to four groups (2H and 2C in the case of CH_2 and four groups (1H and 3 C) in the case of CH-.
   - (c) The C atoms and the N atom in glycine: $\underline{N}$: sp^3; $\underline{C}H_2$: sp^3, $\underline{C}=O$: sp^2

     The N atom is attached to 4 groups (3 atoms and 1 lone pair), the CH_2 carbon has 4 groups attached (2 H atoms and 1N and 1 C atom). The carbonyl carbon is attached to only 3 groups (1C and 2 O atoms)

7. Hybrid orbital sets used by the underlined atoms:

    (a) $\underline{S}iF_6{}^{2-}$: sp^3d^2 (b) $\underline{S}eF_4$: sp^3d (c) $\underline{I}Cl_2{}^-$: sp^3d (d) $\underline{X}eF_4$: sp^3d^2

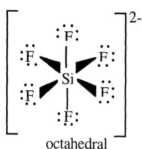 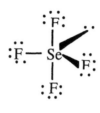 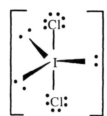 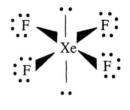

Electron Pair geometry:	octahedral	trigonal bipyramidal	trigonal bipyramidal	octahedral
Molecular geometry:	octahedral	distorted tetrahedron (see-saw)	linear	square planar

9. For the acid HPO_2F_2 and the anion $PO_2F_2{}^-$:

    Structure

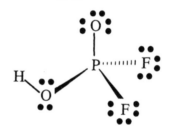

Valence electrons	1P (5)+ 1H(1) + 2 O(6) + 2 F(7) = 32 electrons	1P (5)+ 1charge (1) + 2 O(6) + 2 F(7) = 32 electrons
Molecular geometry	Tetrahedral around the P	Tetrahedral around the P
Hybridization of P	sp^3	sp^3

11. For the molecule $COCl_2$: Hybridization of C = sp^2

    Bonding:1 sigma bond between each chlorine and carbon

        (**sp** hybrid orbitals)

        1 sigma bond between carbon and oxygen

        (**sp** hybrid orbitals)

        1 pi bond between carbon and oxygen

        (**p** orbital)

13. For the following compounds, the other isomer is:

(a)	H_3C—C=C—H / H ... CH_3	H_3C—C=C—CH_3 / H ... H
(b)	Cl—C=C—CH_3 / H ... H	Cl—C=C—H / H ... CH_3

Molecular Orbital Theory

15. Configuration for H_2^+: $(\sigma 1s)^1$

Bond order for H_2^+: 1/2 (no. bonding e^- - no. antibonding e^-) = 1/2

The bond order for molecular hydrogen is <u>one</u> (1), and the H-H bond is <u>stronger</u> in the H_2 molecule than in the H_2^+ ion.

17. The molecular orbital diagram for C_2^{2-}, the acetylide ion:

$\sigma^* 2p$ ___

$\pi^* 2p$ ___   ___

$\sigma 2p$ ↑↓

$\pi 2p$ ↑↓   ↑↓

$\sigma^* 2s$ ↑↓

$\sigma 2s$ ↑↓

There are 2 net pi bonds and 1 net sigma bond in the ion, giving a bond order of 3. On adding two electrons to C_2 (added to $\sigma 2p$) to obtain C_2^{2-}, the bond order increases by one. The ion is **diamagnetic**.

19. (a) The electron configuration (showing only the outer level electrons) for CO is:

$\sigma^* 2p$ ___

$\pi^* 2p$ ___   ___

$\sigma 2p$ ↑↓

$\pi 2p$ ↑↓   ↑↓

$\sigma^* 2s$ ↑↓

$\sigma 2s$ ↑↓

(b) The HOMO is the $\sigma 2p$

(c) There are no unpaired electrons, hence CO is diamagnetic.

(d) There is one net sigma bond, and two net pi bonds for an overall bond order of 3.

General Question on Valence Bond and Molecular Orbital Theory

21. Lewis structure for AlF_4^-:

Electron-pair geometry: tetrahedral (4 groups)

Molecular geometry: tetrahedral

Orbitals on Al and F overlap: on Al: sp^3 and from

F: p orbitals

$$\left[\begin{array}{c} :\ddot{F}: \\ | \\ :\ddot{F} - Al - \ddot{F}: \\ | \\ :\ddot{F}: \end{array} \right]^-$$

23. The O-S-O bond angle and the hybrid orbitals used by sulfur:

    (a) SO_2 $120°$ angle   3 electron-pair groups (1 lp on S) sp^2

    (b) SO_3 $120°$ angle   3 electron-pair groups sp^2

    (c) SO_3^{2-} $109°$ angle   4 electron-pair groups (1 lp on S) sp^3

    (d) SO_4^{2-} $109°$ angle   4 electron-pair groups sp^3

25. Resonance structures for the nitrite ion:

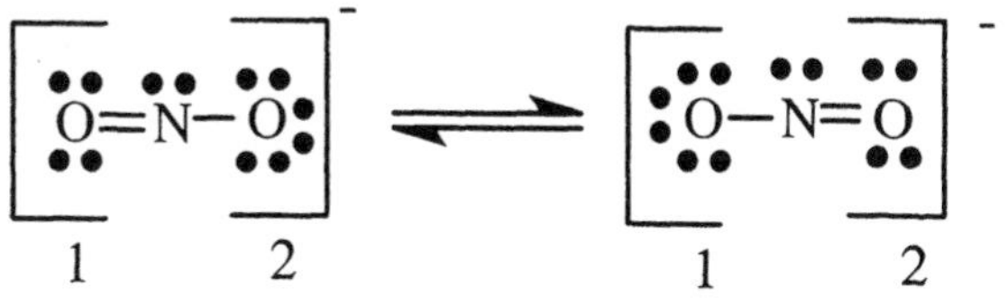

The electron-pair geometry of the ion is *trigonal planar*, and the molecular geometry is *bent* (or angular). With three electron-pairs around the N, the O-N-O bond angle is $120°$. The average bond order is 3/2 (three bonds divided connecting the two O atoms). The hybridization associated with 3 electron-pair groups is sp^2.

27. Resonance structures for N_2O:

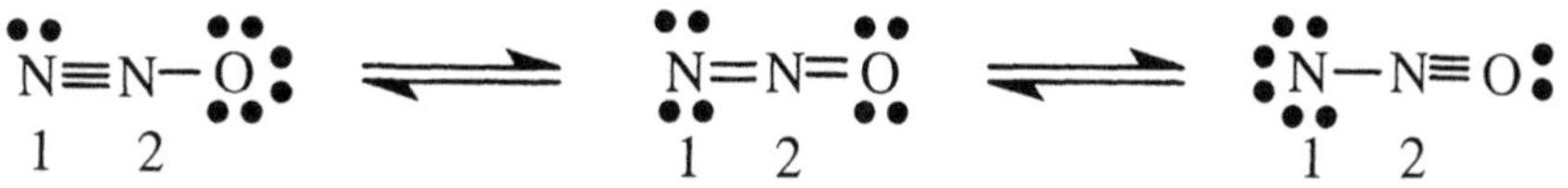

Hybridization	N1:sp **N2:sp**		N1:sp^2 **N2: sp**		N1: sp^3 **N2:sp**
Central N orbitals	The s & p orbitals		The s & p orbitals		The s & p orbitals

29. Regarding ethylene oxide, acetaldehyde, and vinyl alcohol:

	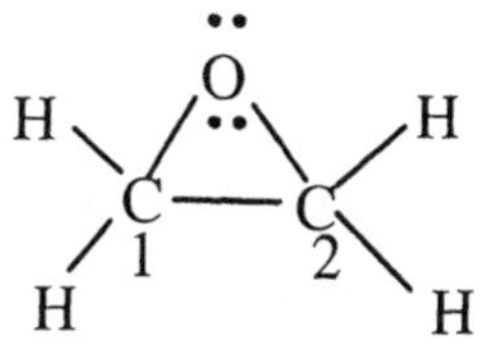	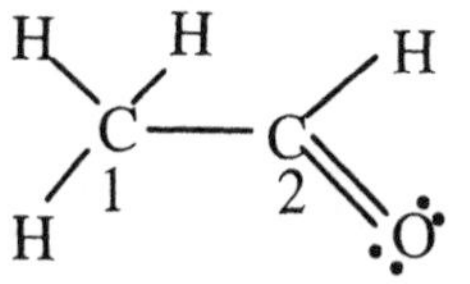	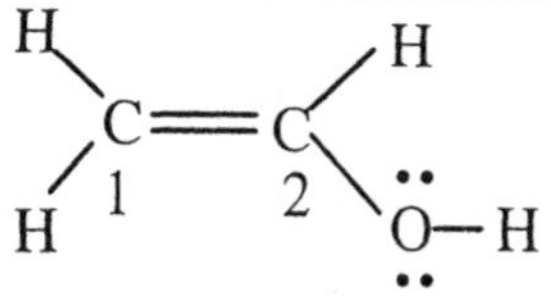
(a)Formula:	C_2H_4O	C_2H_4O	C_2H_4O—these **are** isomers of one another
(b)Hybridization	C1 and C2 - sp^3	C1 sp^3 C2-sp^2	C1 and C2 - sp^2
(c)Bond angles	$109°$ (anticipated) $60°$ due to geometry	C1 $109°$ C2-$120°$	$120°$
(d)Polarity	Polar	Polar	Polar
(e)Strongest bond		C-O bond	C-C bond

31. For the oxime shown:
    (a) Hybridization of the C atoms and the N atom: The leftmost C in the diagram is attached to 3 H and another C, so 4 bonds are necessary—sp^3 hybridization would accomplish this. For the rightmost C atom, (attached to a C,H, and N), 3 bond orbitals are needed for the σ bond—sp^2 hybridization would accomplish this. The N atom needs 3 sigma bonding orbitals (1 lp, bond to C, bond to O), so sp^2 hybridization would accomplish this,
    (b) Approximate C-N-O angle: With a N that is sp^2 hybridized, we would anticipate a bond angle of approximately 120°.

33. For phosphoserine:
    (a) Hybridizations of atoms 1-5:
        Atom 1: 3 major groups attached(3 atoms): sp^2
        Atom 2: 4 major groups attached (2 atoms; 2 lp): sp^3
        Atom 3: 4 major groups attached (4 atoms): sp^3
        Atom 4: 4 major groups attached (4 atoms): sp^3
        Atom 5: 4 major groups attached (4 atoms): sp^3 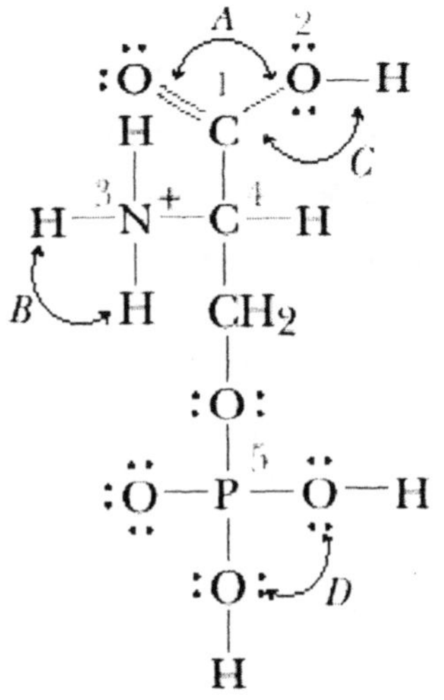
    (b) Approximate bond angles A-D:
        Angle A: For an sp^2 hybridized C atom: 120°
        Angle B: for an sp^3 hybridized N atom: 109°
        Angle C: for an sp^3 hybridized O atom: 109°
        Angle D: for an sp^3 hybridized P atom: 109°
    (c) Most polar bonds in the molecule: Owing to differing electronegativities, we expect the P-O bonds and the O-H bonds to be the most polar.

35. For BF_3 and NH_3BF_3 :
    (a) With 3 groups attached to the B in BF_3 , the geometry is trigonal planar. With 4 groups attached to the B in NH_3BF_3, the geometry is tetrahedral.
    (b) In BF_3 the B is sp^2 hybridized, in NH_3BF_3 the B is sp^3 hybridized.
    (c) The hybridization of B changes.

37. For the molecule cinnamaldehyde:
    (a) The C=O is the most polar bond (electronegativity differences).
    (b) 2π bonds (outside the ring) and 3 π bonds (in the aromatic ring) There are 18 σ bonds.

(c) cis-trans isomerism is possible. The two isomers are shown below.

(d) Since each C is attached to 3 electron pairs, all C are sp^2 hybridized.

(e) Since the angles indicated have an sp^2 hybridized C at the center, all the angles are 120°.

39. (a) Hybridizations: In SbF_5, the Sb atom has five pairs of electrons around it, and the hybridization would be sp^3d. In the SbF_6^- anion, with six pairs of electrons around it, the hybridization would be sp^3d^2.

   (b) The Lewis structure for H_2F^+:

   The electron-pair geometry is *tetrahedral* and the molecular geometry is *bent (or angular)*.

   With 4 electron pairs around the F, the hybridization would be sp^3.

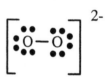

41. (a) A Lewis dot structure of the peroxide ion indicates that the bond order is 1.

   (b) The molecular orbital electron configuration of the peroxide ion is:(Showing only valence electrons)
   $(\sigma_{2s})^2(\sigma^*_{2s})^2(\pi_{2p})^4(\sigma_{2p})^2(\pi^*_{2p})^4$

   Using the MO configuration, we have 8 bonding electrons and 6 antibonding electrons or 2 bonding electrons (Bond order =1)

   (c) Both theories show *no unpaired electrons*, and a bond order of 1.
   The magnetic character of both is the same: diamagnetic.

43. Consider the diatomic molecules of Li_2 through Ne_2

molecule	electron configuration	magnetic property	bond order
Li_2	$(\sigma_{2s})^2$	diamagnetic	1
Be_2	$(\sigma_{2s})^2 (\sigma^*_{2s})^2$	diamagnetic	0
B_2	$(\sigma_{2s})^2(\sigma^*_{2s})^2(\pi_{2p})^2$	paramagnetic	1
C_2	$(\sigma_{2s})^2(\sigma^*_{2s})^2(\pi_{2p})^4$	diamagnetic	2
N_2	$(\sigma_{2s})^2(\sigma^*_{2s})^2(\pi_{2p})^4(\sigma_{2p})^2$	diamagnetic	3

molecule	electron configuration	magnetic property	bond order
O_2	$(\sigma2s)^2(\sigma^*2s)^2(\pi2p)^4(\sigma2p)^2(\pi^*2p)^2$	paramagnetic	2
F_2	$(\sigma2s)^2(\sigma^*2s)^2(\pi2p)^4(\sigma2p)^2(\pi^*2p)^4$	diamagnetic	1
Ne_2	$(\sigma2s)^2(\sigma^*2s)^2(\pi2p)^4(\sigma2p)^2(\pi^*2p)^4(\sigma2p)^2$ diamagnetic		0

45. Using the orbital diagram we predict the electron configuration to be:
(with 4 electrons from C and 5 from N) $(\sigma2s)^2(\sigma^*2s)^2(\pi2p)^4(\sigma2p)^1$

(a) The highest energy MO to which an electron is assigned is the $\sigma2p$.

(b) Bond order = 1/2 (# bonding electrons - # non-bonding electrons)
= 1/2 (7 - 2) or 5/2 or 2.5.

(c) There is a **net of 1/2** sigma bond —$(\sigma2p)^1$, and 2 net π bonds— $(\pi2p)^4$.

(d) The molecule has an unpaired electron so it is paramagnetic

47. For the molecule menthol:

(a) Every carbon is surrounded by 4 electron pairs, so the hybridization is sp^3.

(b) The O atom has 4 electron pairs (two lone pairs and two bonded pairs), so the C-O-H bond
angle is predicted to be the tetrahedral angle of 109°.

(c) With the pendant OH group, the molecule will be slightly polar.

(d) The ring is *not planar*. Each of the carbons is bonded to two other carbons with a 109°
bond angle, and *unlike the benzene ring—which is planar*, this ring will be puckered.

49. The *enol* → *keto* transformation:

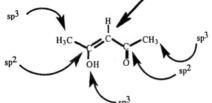

(a) The keto and enol forms are **not resonance hybrids.** Resonance hybrids are in general
structures that differ one from the other by the location of a multiple bond. While the
transformation does result in the "movement" of the C-C double bond, it also has H atoms
that move.

(b) Hybridization in *enol* form: This C atom changes from sp^2 in the *enol* form to sp^3 in the
keto form.

(c) C atom geometries in *enol* and *keto* forms. Geometries around the C atoms are shown in the picture below. Since there are no lone pairs of electrons on the C atoms, the electron-pair and the molecular geometries are identical.

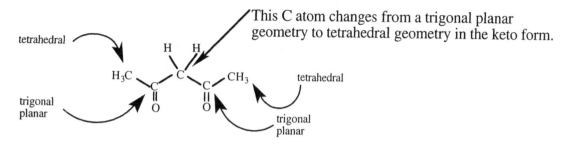

(d) Resonance structures for acac:

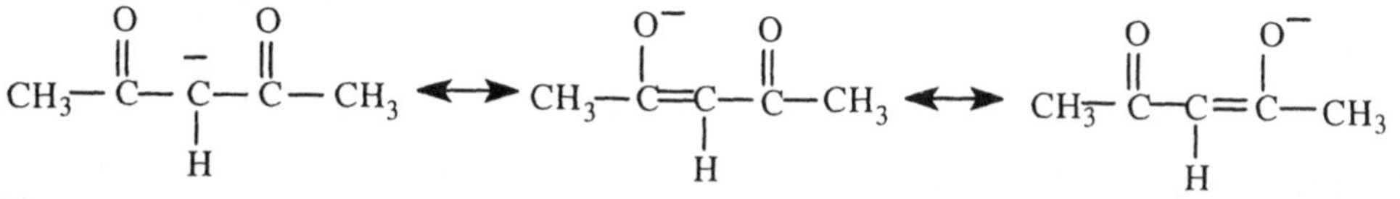

(e) The *enol* form would have possibilities for cis- and trans- forms. Note that in the *keto* form, there are no C-C double bonds around which to form cis- or trans- isomers.

Summary and Conceptual Questions

51. C has four valence orbitals (s and 3p). Hence hybridizing the **minimum** number would provide 2 orbitals (sp), while the **maximum** number of hybrid orbitals would involve all of the orbitals and result in **4 hybrid** orbitals (sp^3).

<u>53</u>. The essential part of the Lewis structure for the peptide linkage is shown here: 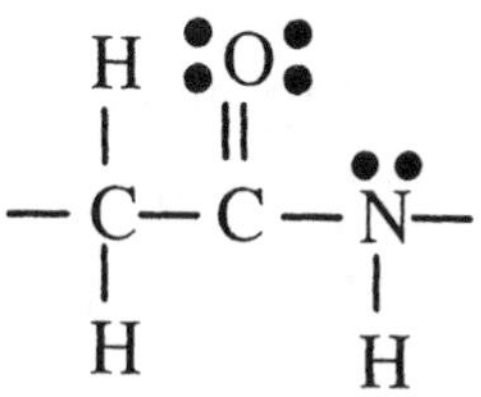
   (a) The carbonyl carbon (C=O) is attached to 3 groups and so has a hybridization of sp^2. The N atom is connected to 4 groups (3 atoms, 1 lone pair) and has a hybridization of sp^3.
   (b) Another structure is feasible: 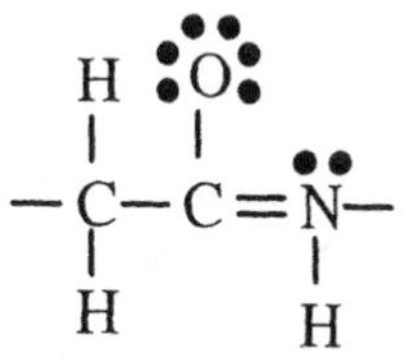
   This structure would have a formal charge on: carbonyl carbon of 0, on the oxygen of -1, and on the N a formal charge +1. Since the "preferred" structure has 0 formal charges on **all three atoms**, the first structure is the more favorable.
   (c) If one views the "less preferred" resonance structure as a contributor, one can see that both the carbonyl carbon and the N are sp^2 hybridized. This leaves one "p" orbital on O,C, and N unhybridized, and capable of side-to-side overlap (also

151

known as π type overlap) between the C and the O and between the C and the N, forming a planar region in the molecule.

55. Valence bond theory rests on an assumption that bonds are formed by pairing of electrons in orbitals on neighboring atoms. This theory has difficulties when one considers odd-electron molecules. VB theory also fails in explaining the magnetic properties of some molecules, e.g. O_2. MO theory, on the other hand, becomes more complex when molecules consist of more than two atoms, and for this reason, we fall back on VB theory in such cases.

57. Regarding hybrid orbitals:
   (a) If n atomic orbitals are combined, n hybrid orbitals will be formed—in other words, orbitals are conserved.
   (b) Hybrid orbitals do not form between p orbitals without involvement of an s orbital.
   (c) Hybrid orbitals are intermediate in energy to the atomic orbitals from which they were formed. An sp hybrid orbital would be lower in energy than the p orbital and higher in energy than the s orbital from which the hybrid orbital was formed.
   (d) The shapes of a sp_x hybrid and an sp_z hybrid are identical, and differ only in their orientation. The sp_x hybrid lies along the p_x axis while the p_z hybrid would lie along the z axis.
   (e) The shapes of the orbital formed from the s,p_x and p_y orbitals are identical to the orbital formed from the s, p_x, p_z. As in (d) above, the only difference is in the orientation of the orbital in space.

59. (a) Explain why allene is not flat: If one examines the sigma bonding in allene, one sees that the C (1 and 3) are sp^2 hybridized, while C(2) is sp hybridized. The picture below indicates that the geometry around C (1) and C(3) is expected to be trigonal planar.

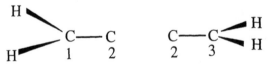

C(2) has **two** unhybridized p orbitals (It uses 1 to form the sp hybrids, yes?). If we assume that the p_x orbital was used to form the sp hybrids, the p_y and p_z orbitals are the remaining unused p orbitals. Recall that the π bonds formed between C(1) and C(2) and between C(2) and C(3) are formed by overlap of the unused p orbitals on C(2) with the unused p orbital on C(1) and C(3). The result is that these mutually perpendicular p orbitals, force the CH_2 group associated with C(1) to be **perpendicular** to the CH_2 group associated with C(3)—and an allene molecule that is **not flat**.

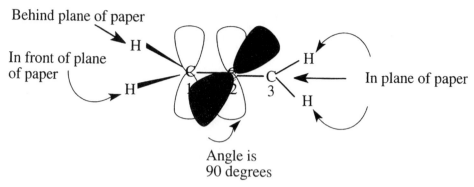

Behind plane of paper

In front of plane
of paper

In plane of paper

Angle is
90 degrees

(b) Using orbital hybridization, explain why benzene is a planar symmetrical molecule

The diagram shows the 6-member C ring which is the basis
for benzene. Each C has a H atom attached. The overall
geometry around each C is trigonal planar around each C
(using sp^2 hybridization).

The **unhybridized p orbitals** on each C sit perpendicular

to the plane of the ring, giving rise to a network of overlapping p orbitals above and below the
plane of the ring. A view from slightly above the plane of the ring can be pictured as below: (H
atoms are omitted for clarity.) The resulting overlap **both above and below** constrict the C-C
ring to be planar.

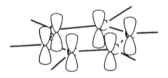

(c) Hybrid orbitals used by the C atoms of allyl alcohol:

sp^2 hybridization

sp^2 hybridization

sp^3 hybridization

Chapter 11
Carbon: More Than Just Another Element

Practicing Skills

Alkanes and Cycloalkanes

1. The straight chain alkane with the formula C_7H_{16} is heptane. The prefix "hept" tells us that there are seven carbons in the chain.

3. Of the formulas given, which represents an alkane? Which a cycloalkane?

    Alkanes have saturated carbon atoms—with 4 bonds. Since bridging C atoms have two bonds to C, then the other 2 bonds are to H atoms. Terminal C atoms on the end of the chains have only 1 bond to the chain, leaving 3 bonds to H. The net result is that alkanes have the general formula: C_nH_{2n+2} (c) has this general formula: $C_{14}H_{30}$. Cycloalkanes "lose 2H" atoms as the ends of the chains are "joined", so cycloalkanes have the general formula: C_nH_{2n} so (b) C_5H_{10} has this general formula. While (a) fits the general formula, it is impossible to make a ring of fewer than 3 C atoms.

5. Systematic name for the alkane: Numbering the longest C chain from one end to the other indicates that the longest C chain has 4 carbons, the root name is therefore *butane*. On the 2nd and 3rd C there is a 1-C chain. Since we truncate the –ane ending and change it to –yl, these are *methyl groups*. Indicating that the methyl groups are on the 2nd and 3rd carbon atoms, we get 2,3-dimethylbutane.

7. Structures for the compounds:

    (a) 2,3-dimethylhexane

    $$CH_3$$
    $$CH_3CHCHCH_2CH_2CH_3$$
    $$CH_3$$

    (b) 2,3-dimethyloctane

    $$CH_3$$
    $$CH_3CHCHCH_2CH_2CH_2CH_3$$
    $$CH_3$$

    (c) 3-ethylheptane

    $$CH_3CH_2CHCH_2CH_2CH_2CH_3$$
    $$CH_2CH_3$$

    (d) 3-ethyl-2-methylhexane

    $$CH_3$$
    $$CH_3CHCHCH_2CH_2CH_3$$
    $$CH_2CH_3$$

9. Structures for all compounds with a seven-carbon chain and one methyl substituent.

$$CH_3$$
$$|$$
$$CH_3CHCH_2CH_2CH_2CH_2CH_3$$
2-methylheptane

$$CH_3$$
$$|$$
$$CH_3CH_2CHCH_2CH_2CH_2CH_3$$
3-methylheptane

$$CH_3$$
$$|$$
$$CH_3CH_2CH_2CHCH_2CH_2CH_3$$
4-methylheptane

Assuming that we deal *only* with alkanes, there are 3 structures. *3-methylheptane has a chiral carbon*, since on C-3, there is a methyl group (CH_3) an ethyl group (CH_3CH_2), a H, and a butyl group ($CH_2CH_2CH_2CH_3$).

11. Structures, and names of the two ethylheptanes:

$$CH_2CH_3$$
$$|$$
$$CH_3CH_2CHCH_2CH_2CH_2CH_3$$
3-ethylheptane

$$CH_2CH_3$$
$$|$$
$$CH_3CH_2CH_2CHCH_2CH_2CH_3$$
4-ethylheptane

Neither of the isomers have a chiral carbon.

13. Physical properties of C_4H_{10}: Assuming that we're talking about butane, the following properties are pertinent: mp = -138.4 °C, bp = –0.5°C (it's a colorless gas at room T). According to the CRC Handbook of Chemistry and Physics, it is slightly soluble in water, but more so in less polar ether, chloroform, and ethanol.

Predicted properties for $C_{12}H_{26}$: One would guess that the material is colorless (no features that would invoke color). Given the much longer chain, one could guess that it might be a liquid (It is! mp = -9.6°C and bp = 216°C. We would also guess that it would be less water soluble than butane--(and it is indeed reported *not* to be water soluble), but soluble in alcohol and ether.

Alkenes and Alkynes

15. Cis- and trans- isomers of 4-methyl-2-hexene:

cis   trans

17. (a) Structures and names for alkenes with formula C_5H_{10}.

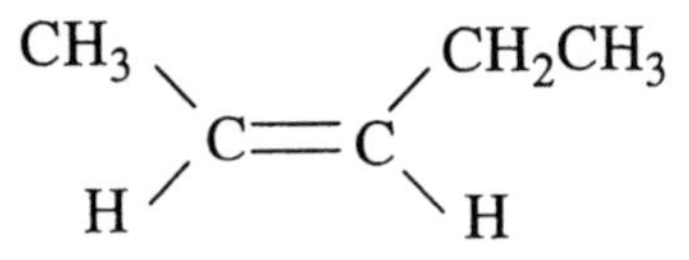

CH₃   CH₂CH₃
   \   /
    C == C
   /   \
  H   H

cis-2-pentene

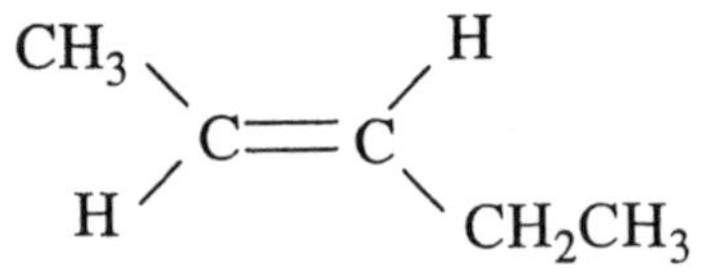

CH₃   H
   \   /
    C == C
   /   \
  H   CH₂CH₃

trans-2-pentene

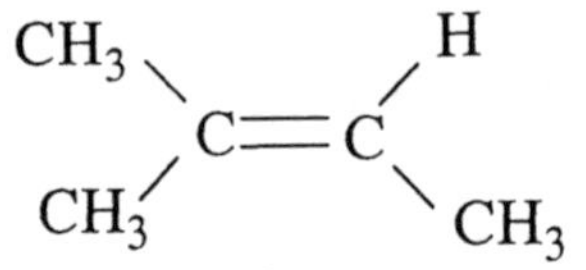

CH₃   H
   \   /
    C == C
   /   \
  CH₃   CH₃

2-methyl-2-butene

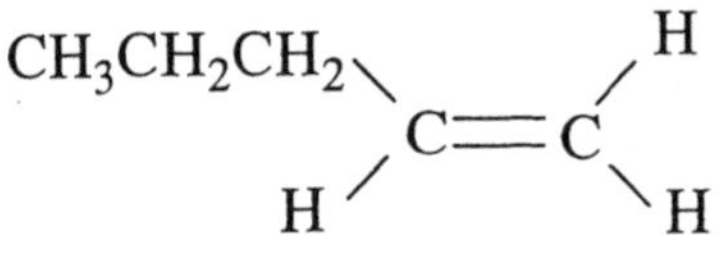

CH₃CH₂CH₂   H
        \   /
         C == C
        /   \
       H   H

1-pentene

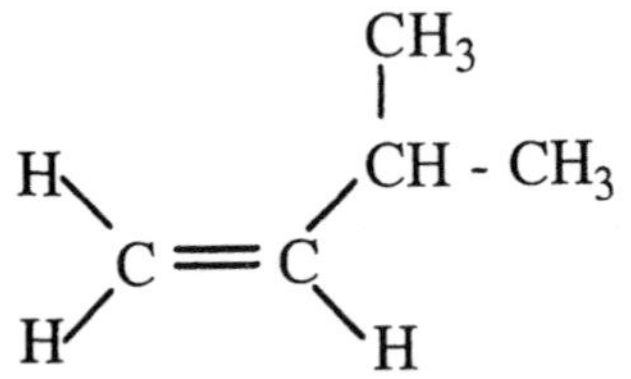

       CH₃
        |
       CH - CH₃
H\   /
  C == C
 /   \
H   H

3-methyl-1-butene

H\   CH₂ - CH₃
  \   /
   C == C
  /   \
 H   CH₃

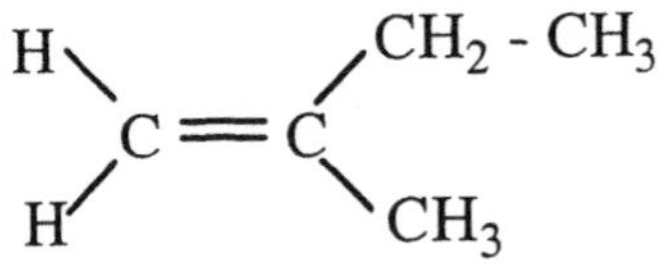

2-methyl-1-butene

(b) The cycloalkane with the formula C_5H_{10}

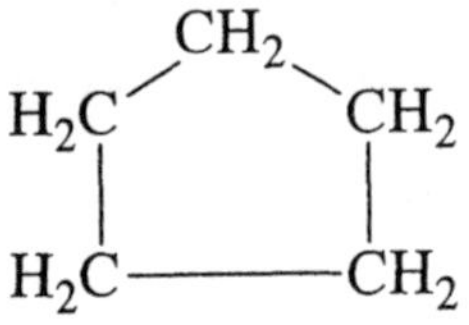

       CH₂
      /   \
   H₂C   CH₂
    | |
   H₂C ───── CH₂

19. Structure and names for products of:

(a) CH3CH=CH2 + Br2 → CH3CHBrCH2Br

1,2-dibromopropane

(b) CH3CH2CH=CHCH3 + H2 → CH3CH2CH2CH2CH3

pentane

21. The alkene which upon addition of HBr yields: CH3CH2CH2BrCH3

The addition of HBr is accomplished by adding (H) on one "side" of the double bond and (Br) on the other. If we can mentally "subtract" an H and Br from the formula above we get:

CH3CH2CH2BrCH3 →   CH3CH2CH=CH2 (1-butene)

23. Alkenes with the formula C3H5Cl:

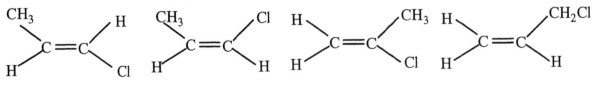

trans-1-chloropropene   cis-1-chloropropene   2-chloropropene   3-chloropropene

Aromatic Compounds

25. Structural formulas for :

(a) m-dichlorobenzene

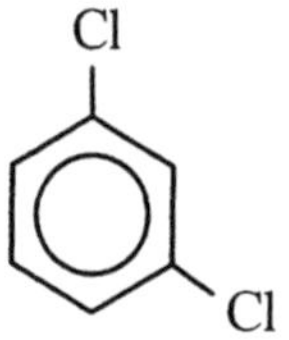

(b) p-bromotoluene

27. Ethylbenzene may be prepared from benzene in the following manner:

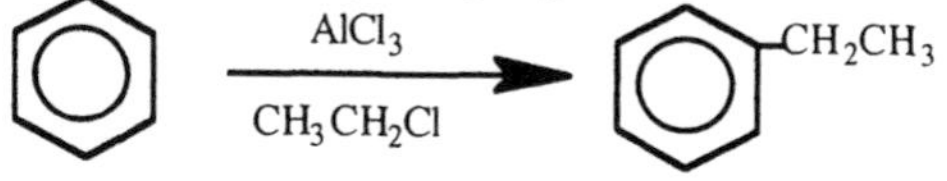

29. The alkylated product of p-xylene:

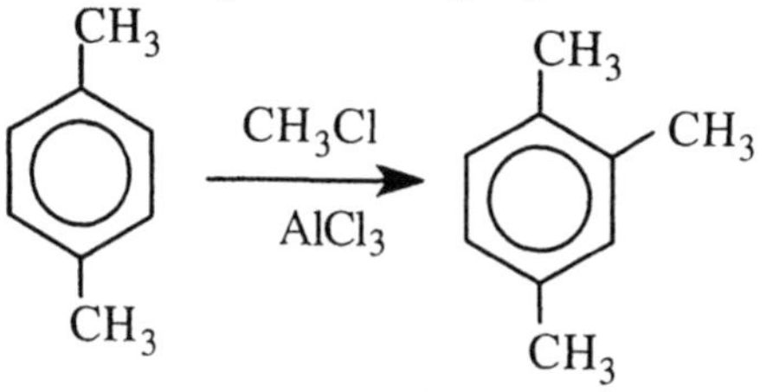

1,2,4-trimethylbenzene

Alcohols, Ethers, and Amines

31. Systematic names of the alcohols:

(a) 1-propanol   primary alcohol (OH group on C connected to 1 other C "group"

(b) 1-butanol   primary alcohol (OH group on C connected to 1 other C "group"

(c) 2-methyl-2-propanol tertiary alcohol (OH group on C connected to 3 other C "groups"

(d) 2-methyl-2-butanol   tertiary alcohol (OH group on C connected to 3 other C "groups"

33. Formulas and structures

(a) ethylamine	(b) dipropylamine
$C_2H_5NH_2$	$(CH_3CH_2CH_2)_2NH$
CH_3CH_2 N H H	$CH_3CH_2CH_2$ N $CH_2CH_2CH_3$ H
(c) butyldimethylamine	(d) triethylamine
$(CH_3CH_2CH_2CH_2)(CH_3)_2N$ 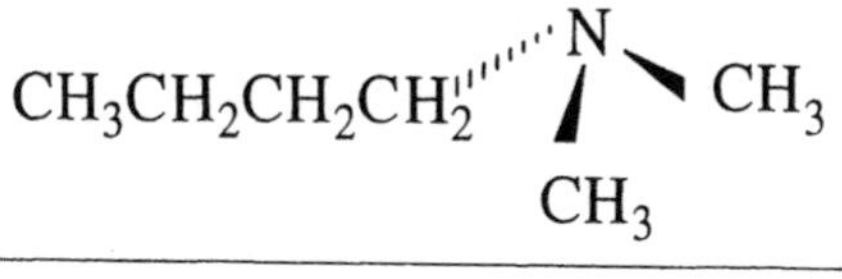	$(CH_3CH_2)_3N$
$CH_3CH_2CH_2CH_2$ N CH_3 CH_3	CH_3CH_2 N CH_2CH_3 CH_2CH_3

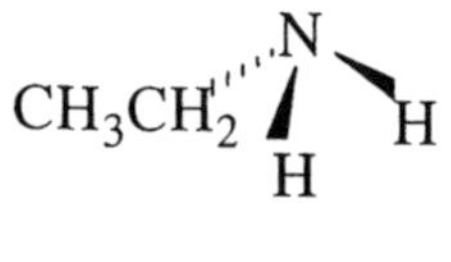

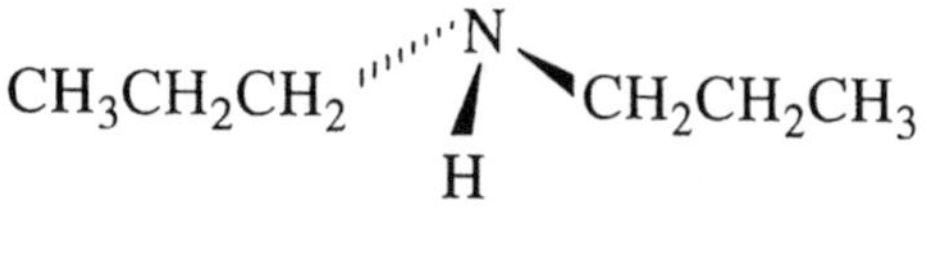

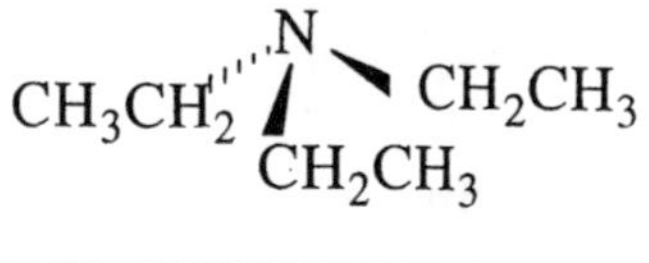

35. Structural formulas for alcohols with the formula C4H10O:

   1-butanol $CH_3CH_2CH_2CH_2OH$

   2-butanol $CH_3CH_2CHCH_3$
                                    |
                                   OH

   2-methyl-1-propanol CH_3CHCH_2OH
                                |
                               CH_3

                                OH
                                |
   2-methyl-2-propanol CH_3CCH_3
                                |
                               CH_3

37. Amines treated with acid:

   (a) $C_6H_5NH_2 + HCl \longrightarrow C_6H_5NH_3^+Cl^-$

   (b) $(CH_3)_3N + H_2SO_4 \longrightarrow (CH_3)_3NH^+ \ HSO_4^-$

Compounds with a Carbonyl Group

39. Structural formulas for :

   (a) 2-pentanone   (b) hexanal

   $CH_3-CH_2-CH_2-C-CH_3$ $CH_3-CH_2-CH_2-CH_2-CH_2-C-H$
                              || ||
                              O   O

   (c) pentanoic acid

   $CH_3CH_2CH_2CH_2CO_2H$ or $CH_3(CH_2)_3CO_2H$

41. Name the following compounds:

   (a) $CH_3CH_2CHCH_2CO_2H$ 3-methylpentanoic acid
                       | a carboxylic acid
                     CH_3

                        O
                        ||
   (b) $CH_3CH_2COCH_3$ methyl propanoate
                                                           an ester

                     O
                     ||
   (c) $CH_3COCH_2CH_2CH_2CH_3$ butyl ethanoate
                                                           an ester

                 CO_2H
                    |
   (d)   p-bromobenzoic acid
                                                        an aromatic carboxylic acid

                 Br

43. (a) The product of oxidation is pentanoic acid:

$$\begin{array}{c} O \\ \parallel \end{array}$$
$CH_3CH_2CH_2CH_2C\text{-}OH$

(b) The reduction product is 1-pentanol:

$CH_3CH_2CH_2CH_2CH_2OH$

(c) The reduction yields 2-octanol:

$CH_3CH_2CH_2CH_2CH_2CH_2CHCH_3$
         |
        OH

(d) The oxidation of 2-octanone with potassium permanganate gives <u>no reaction</u>.

45. The following equations show the preparation of propyl propanoate:

$CH_3CH_2CH_2OH \xrightarrow{\ KMnO_4\ } CH_3CH_2CO_2H$

$CH_3CH_2CO_2H + HOCH_2CH_2CH_3 \xrightarrow{\ H^+\ } CH_3CH_2COCH_2CH_2CH_3$

47. The products of the hydrolysis of the ester are 1-butanol and sodium acetate:

$CH_3CH_2CH_2CH_2OH$ CH_2CO_2Na

49. Regarding phenylalanine:

(a) Carbon 3 has three groups attached, and a trigonal planar geometry.

(b) The O-C-O bond angle would be 120°.

(c) Carbon 2 has four different groups attached, so the molecule is chiral.

(d) The H attached to the carboxylic acid group (Carbon 3) is acidic.

Functional Groups

51. Functional groups present:

(a) the –OH group makes this molecule an *alcohol*.

(b) the carbonyl group (C=O) adjacent to the N-H makes this molecule an *amide*.

(c) the carbonyl group (C=O) adjacent to the O-H makes this molecule a *carboxylic acid*.

(d) the carbonyl group (C=O) with an attached O-C makes this molecule an *ester*.

Polymers

53. (a) An equation for the formation of polyvinyl acetate from vinyl acetate:

n CH$_2$CHOCOCH$_3$ $\longrightarrow$ [-CH$_2$CH(OCOCH$_3$)-] n

(b) The structure for polyvinylacetate:

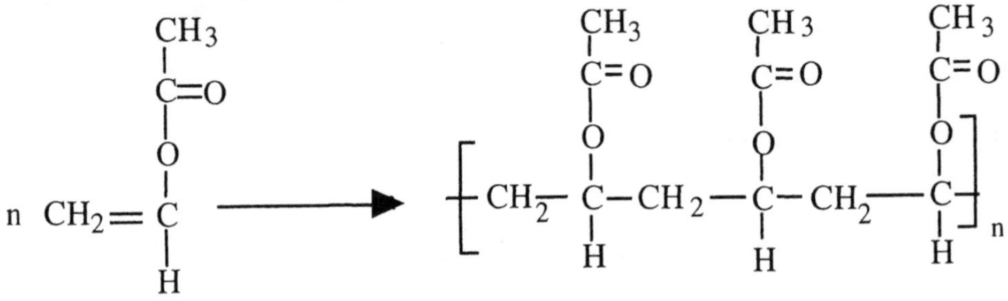

(c) Prepare polyvinyl alcohol from polyvinyl acetate:

Polyvinyl acetate is an ester. Hydrolysis of the ester (structure shown above) with NaOH will produce the sodium salt, NaC$_2$H$_3$O$_2$, and polyvinyl alcohol. Acidification with a strong acid (e.g. HCl) will produce polyvinyl alcohol (structure below).

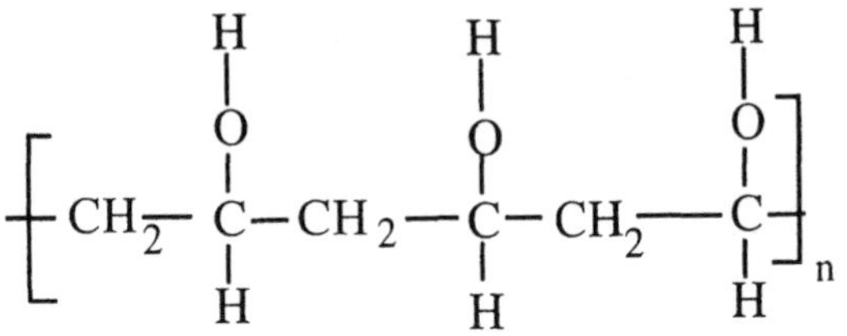

55.

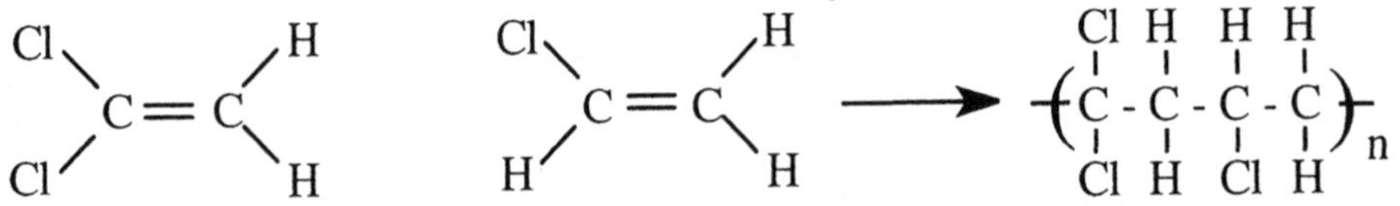

1,1-dichloroethene   chloroethene

The reaction proceeds with the free radical addition of the copolymers:

General Questions on Organic Chemistry

57. (a) geometric isomers of C$_2$H$_2$Cl$_2$:

cis isomer   trans isomer

(b) structural isomer of $C_2H_2Cl_2$:

59. structural isomers of C_6H_{12}:   **H** have been omitted for clarity. In addition, these structures
    have been drawn with "-" instead of "··" to simplify the pictures.

61. Reactions of cis-2-butene:

(c) 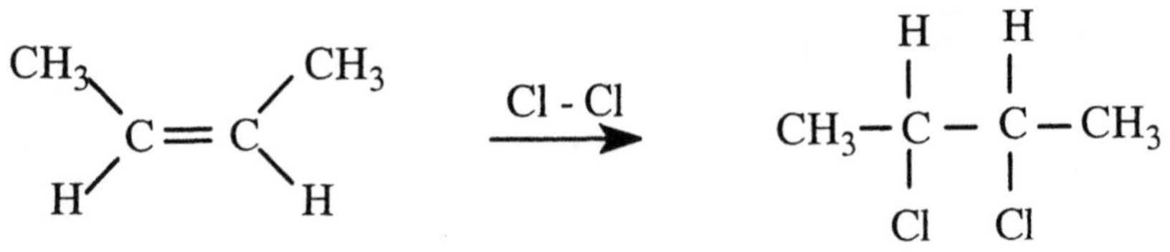
$$CH_3\text{, }H \quad C=C \quad CH_3\text{, }H$$

Cl - Cl →

$$CH_3-\overset{\overset{\displaystyle H}{|}}{\underset{\underset{\displaystyle Cl}{|}}{C}}-\overset{\overset{\displaystyle H}{|}}{\underset{\underset{\displaystyle Cl}{|}}{C}}-CH_3$$

63. In (a) acetic acid reacts with NaOH(a base) to form a salt (an ionic compound). In (b) there is also an acid-base reaction. The amine (base) reacts with HCl (the acid) to form a salt.

(a) 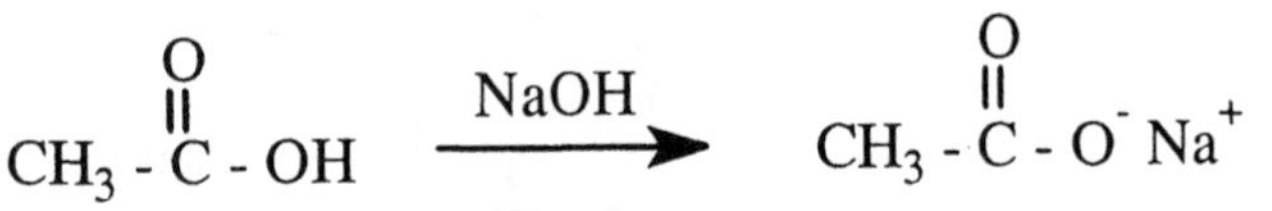 $CH_3 - \overset{\overset{\displaystyle O}{\|}}{C} - OH$ → NaOH → $CH_3 - \overset{\overset{\displaystyle O}{\|}}{C} - O^- \; Na^+$

(b) 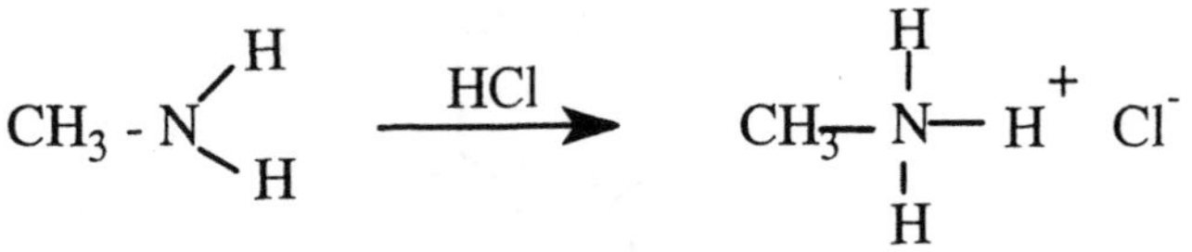 $CH_3 - N \overset{\displaystyle H}{\underset{\displaystyle H}{\big<}}$ → HCl → $CH_3 - \overset{\overset{\displaystyle H}{|}}{\underset{\underset{\displaystyle H}{|}}{N}} - H^+ \; Cl^-$

65. The equation for the formation of:

(a) 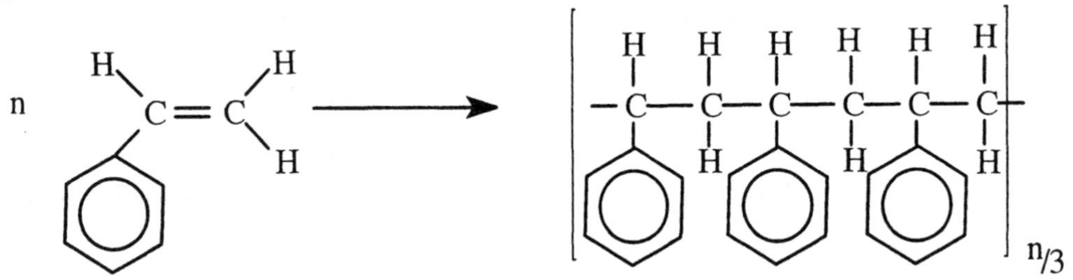 n $\overset{\displaystyle H}{\underset{}{}}$ $C=C$ $\overset{\displaystyle H}{\underset{\displaystyle H}{}}$ (with phenyl group) → polymer chain with $n/3$

(b) 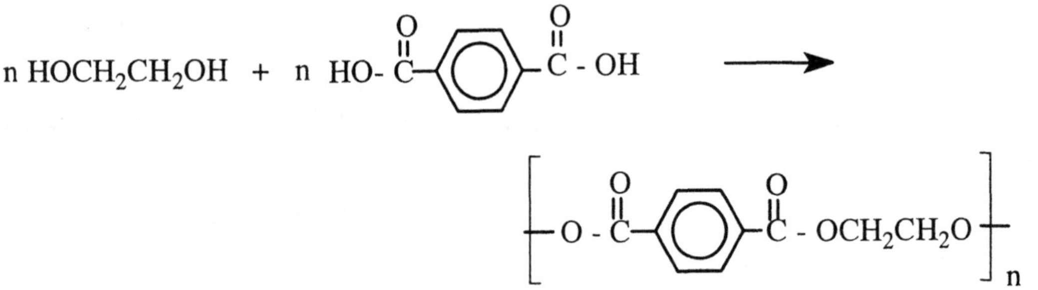 n HOCH$_2$CH$_2$OH + n HO-$\overset{\overset{\displaystyle O}{\|}}{C}$-(ring)-$\overset{\overset{\displaystyle O}{\|}}{C}$ - OH →

$$\left[-O-\overset{\overset{\displaystyle O}{\|}}{C}-(ring)-\overset{\overset{\displaystyle O}{\|}}{C}- OCH_2CH_2O- \right]_n$$

67. Structures for:

(a) 2,2-dimethylpentane

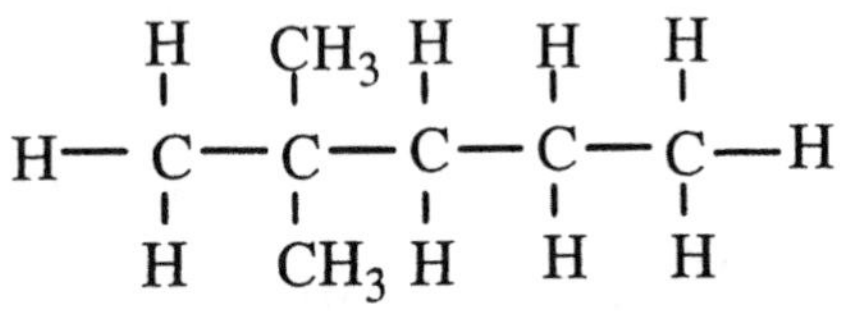

$$H-\overset{\overset{\displaystyle H}{|}}{\underset{\underset{\displaystyle H}{|}}{C}}-\overset{\overset{\displaystyle CH_3}{|}}{\underset{\underset{\displaystyle CH_3}{|}}{C}}-\overset{\overset{\displaystyle H}{|}}{\underset{\underset{\displaystyle H}{|}}{C}}-\overset{\overset{\displaystyle H}{|}}{\underset{\underset{\displaystyle H}{|}}{C}}-\overset{\overset{\displaystyle H}{|}}{\underset{\underset{\displaystyle H}{|}}{C}}-H$$

(b) 3,3-diethylpentane

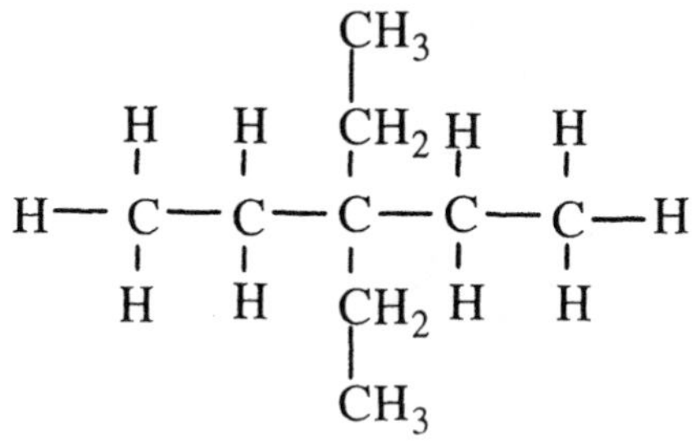

$$
\begin{array}{c}
& & CH_3 & & \\
H & H & CH_2 & H & H \\
| & | & | & | & | \\
H-C-C-C-C-C-H \\
| & | & | & | & | \\
H & H & CH_2 & H & H \\
& & CH_3 & &
\end{array}
$$

(c) 3-ethyl-2-methylpentane

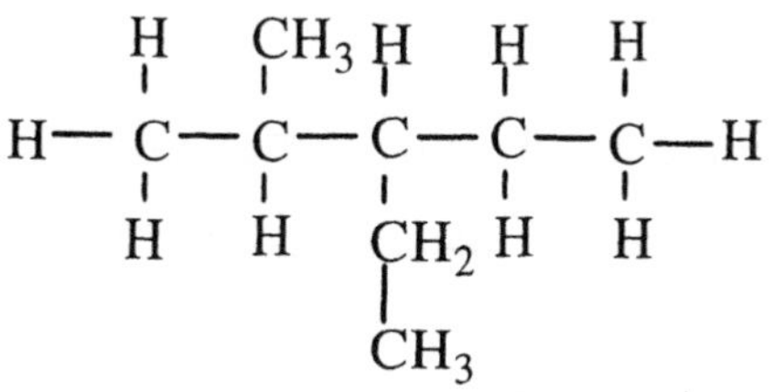

$$
\begin{array}{c}
H & CH_3 & H & H & H \\
| & | & | & | & | \\
H-C-C-C-C-C-H \\
| & | & | & | & | \\
H & H & CH_2 & H & H \\
& & CH_3 & &
\end{array}
$$

(d) 3-ethylhexane

$$
\begin{array}{c}
H & H & H & H & H \\
| & | & | & | & | \\
H-C-C-C-C-C-CH_3 \\
| & | & | & | & | \\
H & H & CH_2 & H & H \\
& & CH_3 & &
\end{array}
$$

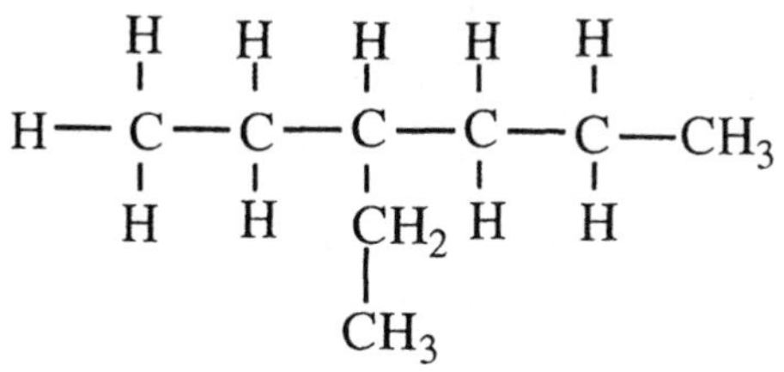

<u>69</u>. Structural isomers for $C_3H_6Cl_2$:

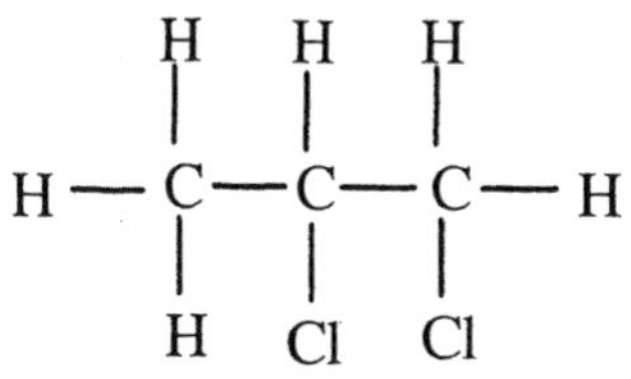

$$
\begin{array}{c}
H & H & H \\
| & | & | \\
H-C-C-C-H \\
| & | & | \\
H & Cl & Cl
\end{array}
$$

1,2-dichloropropane

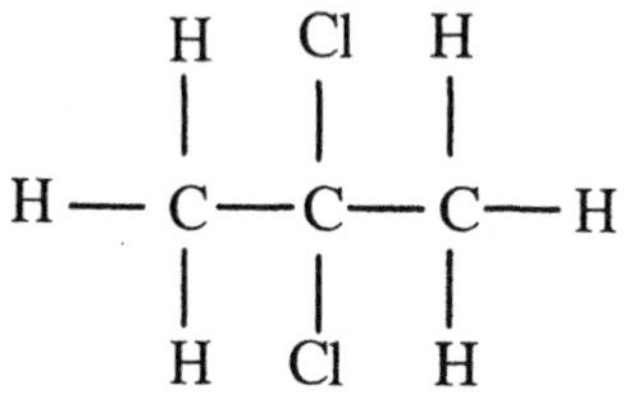

$$
\begin{array}{c}
H & Cl & H \\
| & | & | \\
H-C-C-C-H \\
| & | & | \\
H & Cl & H
\end{array}
$$

2,2-dichloropropane

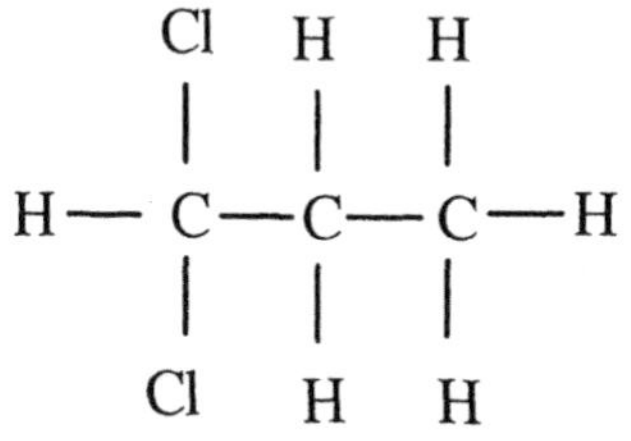

$$
\begin{array}{c}
Cl & H & H \\
| & | & | \\
H-C-C-C-H \\
| & | & | \\
Cl & H & H
\end{array}
$$

1,1-dichloropropane

$$
\begin{array}{c}
H & H & H \\
| & | & | \\
H-C-C-C-H \\
| & | & | \\
Cl & H & Cl
\end{array}
$$

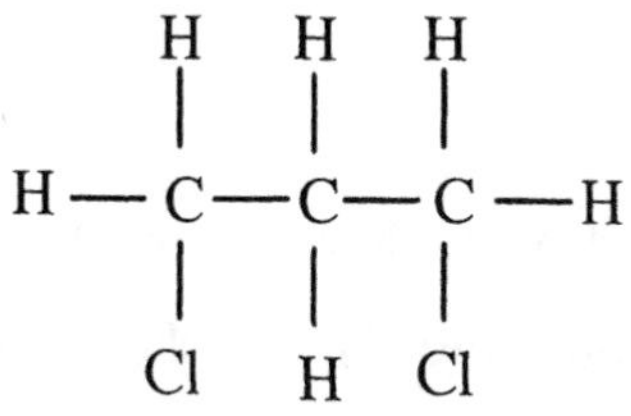

1,3-dichloropropane

71. Structural isomers for trimethylbenzene:

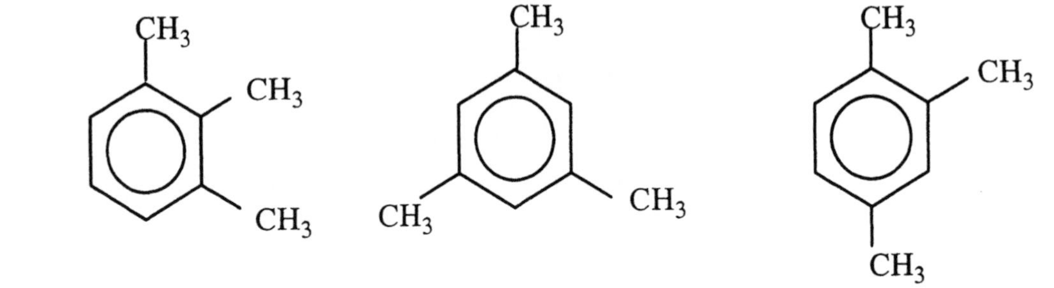

1,2,3-trimethylbenzene   1,3,5-trimethylbenzene   1,2,4-trimethylbenzene

73. The decarboxylase enzyme would remove the COOH functionality-- releasing CO_2 and appending the H in the former location of the COOH group.

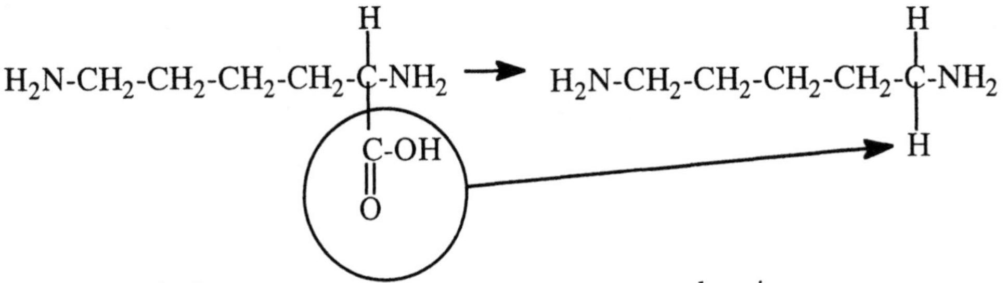

lysine   cadaverine

75. Reaction of cis-2-butene:

(a)

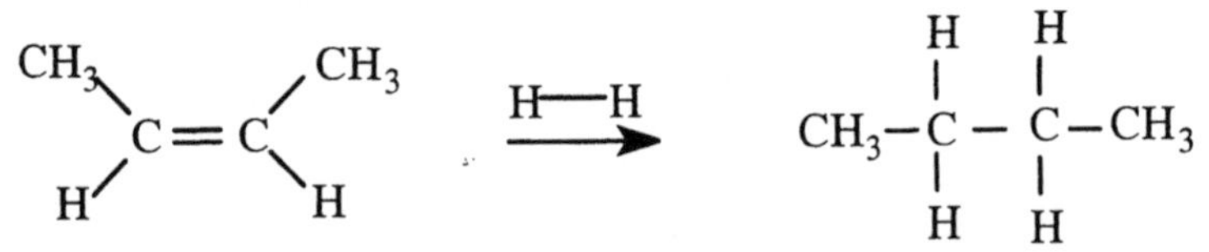

butane, not chiral

(b)   an isomer of butane

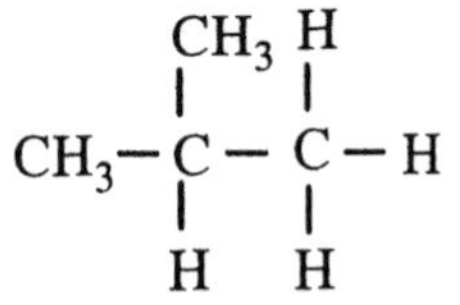

77. Which of the following produce acetic acid when reacted with $KMnO_4$:

Both (b) and (c) produce acetic acid. The alcohol(c) is oxidized to the aldehyde (b) which subsequently is oxidized to acetic acid.

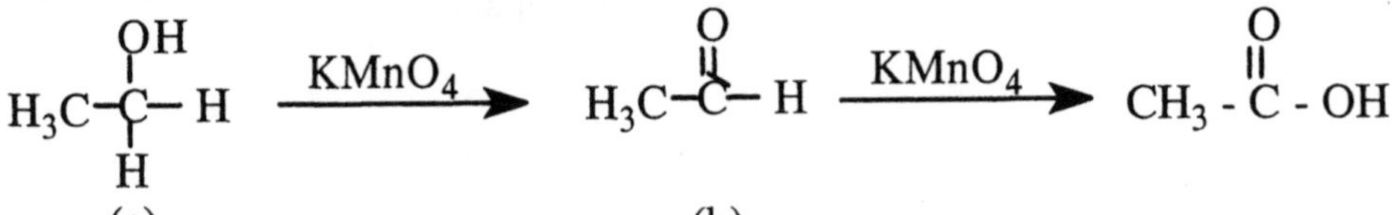

(c)   (b)

79. Repeating unit of the polymer, Kevlar

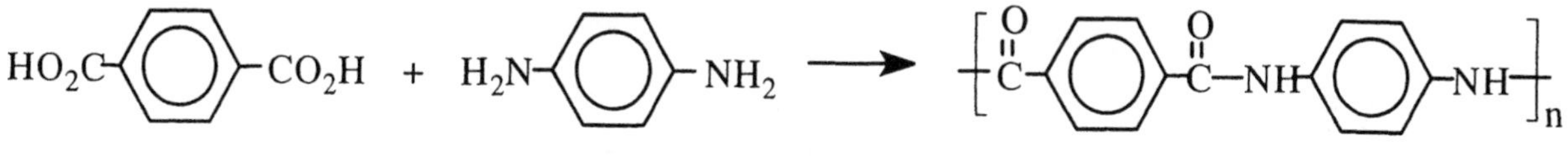

$$HO_2C-\bigcirc-CO_2H \quad + \quad H_2N-\bigcirc-NH_2 \quad \longrightarrow \quad \left[\overset{O}{\overset{\|}{C}}-\bigcirc-\overset{O}{\overset{\|}{C}}-NH-\bigcirc-NH \right]_n$$

81. Structure for glyceryl trilaurate and saponification products:

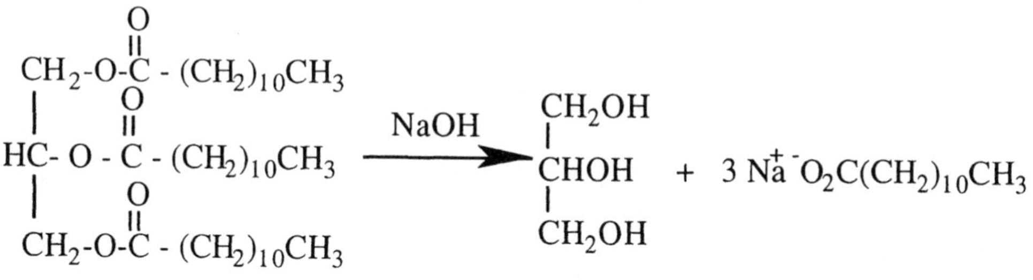

$$
\begin{array}{l}
\qquad\quad\; \overset{O}{\overset{\|}{}} \\
CH_2\text{-}O\text{-}C - (CH_2)_{10}CH_3 \\
\;\;|\qquad\quad \overset{O}{\overset{\|}{}} \\
HC\text{-}\,O\,\text{-}\,C\,\text{-}\,(CH_2)_{10}CH_3 \\
\;\;|\qquad\quad\; \overset{O}{\overset{\|}{}} \\
CH_2\text{-}O\text{-}C - (CH_2)_{10}CH_3
\end{array}
\quad\xrightarrow{\;NaOH\;}\quad
\begin{array}{l}
CH_2OH \\
| \\
CHOH \\
| \\
CH_2OH
\end{array}
\;+\; 3\,Na^+\,{}^-O_2C(CH_2)_{10}CH_3
$$

glyceryl trilaurate   glycerol   sodium laurate

83. The reaction between bromine and cyclohexene:

The double bond in cyclohexene absorbs elemental bromine—adding across the double bond so that the adjacent carbon atoms (originally participating in the double bond) each have one Br atom bound to them.

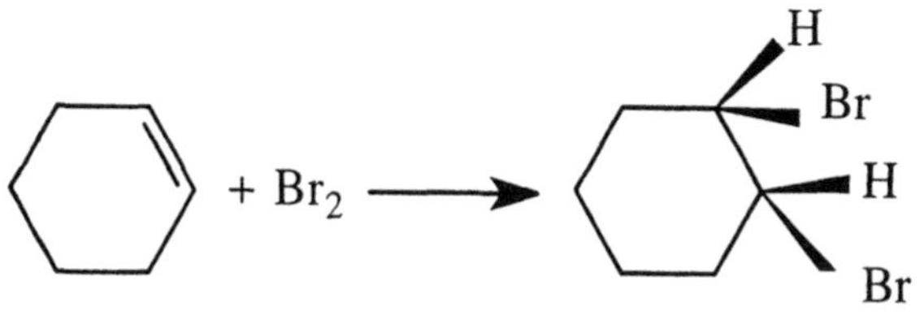

$$\bigcirc \!\!\!\!\!\! \diagdown \quad + \; Br_2 \quad \longrightarrow \quad \text{(dibromocyclohexane with H, Br, H, Br substituents)}$$

85. (a) Two structures with formula C_3H_6O:

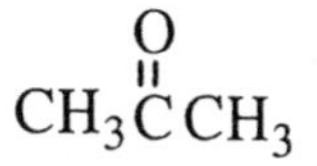

$$CH_3\overset{O}{\overset{\|}{C}}CH_3 \qquad\qquad CH_3CH_2\overset{O}{\overset{\|}{C}}H$$
    propanone (acetone) 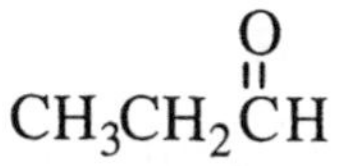   propanal

(b) Oxidation of the compound gives an acidic solution (a carboxylic acid, perhaps?), indicating that the unknown is propanal.

(c) Oxidation of propanal gives propanoic acid (structure below):

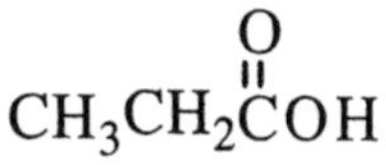

$$CH_3CH_2\overset{O}{\overset{\|}{C}}OH$$

165

87. Addition of water to an alkene , X, gives an alcohol, Y. Oxidation of Y

gives 3,3-dimethyl-2-pentanone. The structure for X and Y:

$$CH_3\overset{O}{\overset{||}{C}}\overset{CH_3}{\overset{|}{C}}CH_2CH_3 \overset{CH_3}{\underset{}{}} \longleftarrow CH_3\overset{HO}{\overset{|}{C}H}\overset{CH_3}{\overset{|}{C}}CH_2CH_3 \overset{CH_3}{\underset{}{}} \longleftarrow CH_2=CH\overset{CH_3}{\overset{|}{C}}CH_2CH_3 \overset{CH_3}{\underset{}{}}$$

                                            Y   X

89. Show products of the following reactions of CH_2CHCH_2OH:

(a) Hydrogenation of the compound
$$CH_2=CHCH_2OH \longrightarrow CH_3CH_2CH_2OH$$

(b) Oxidation of the compound:
$$CH_2=CHCH_2OH \qquad\qquad CH_2=CHCO_2H$$

(c) Addition polymerization:
$$n\ CH_2=CHCH_2OH \longrightarrow [-CH_2-\overset{}{\underset{CH_2OH}{C}H-]_n}$$

(d) Ester formation with acetic acid:
$$CH_2=CHCH_2OH\ +\ CH_3CO_2H \longrightarrow CH_3CO_2CH_2CH=CH_2\ +\ H_2O$$

Summary and Conceptual Questions:

91. Modes by which C can achieve an octet of electrons. Since C has 4 valence electrons (it is in

group 4A). It needs to acquire an octet by gaining 4 electrons. This can be done by forming 4

single bonds to other atoms (e.g. 4H atoms in CH_4). It can also participate in a double bond

and two single bonds (e.g. ethylene), two double bonds (as in allene), OR a triple bond and one

single bond (e.g. ethyne—or acetylene).

$$H-\overset{H}{\underset{H}{\overset{|}{\underset{|}{C}}}}-H \quad H-\overset{H}{\overset{|}{C}}=\overset{H}{\overset{|}{C}}-H \quad H-\overset{H}{\overset{|}{C}}=C=\overset{H}{\overset{|}{C}}-H \quad H-C\equiv C-H$$

93. Properties imparted by the following characteristics:

(a) Cross-linking in polyethylene: Brings structural integrity to the polymeric chain. This also

increases the rigidity of the chain. CLPE (the material of soda bottle caps) is a good

example of a cross-linked polymer.

(b) OH groups in polyvinyl alcohol: These OH groups increase water solubility as hydrogen-

bonding between polyvinyl alcohol and water is possible.

Additionally these OH group provide a locus for cross-linking agents (e.g. borax is a good cross-linking agent in the commercial polymer called Slime™.

(c) Hydrogen bonding in a polyamide: Causes polyamides (e.g. peptides) to form coils and sheets.

95. (a) Combustion of ethane and ethanol:

$$2\ CH_3CH_3(g) + 7\ O_2(g) \rightarrow\ 4\ CO_2(g) + 6\ H_2O(g)$$
$$CH_3CH_2OH(\ell) + 3\ O_2(g) \rightarrow 2\ CO_2(g) + 3\ H_2O(g)$$

$\Delta H°rx\ (C_2H_6)\ = [4 \cdot \Delta H°_f\ (CO_2) + 6 \cdot \Delta H°_f\ (H_2O)] - [2 \cdot \Delta H°_f\ (CH_3CH_3) + 7 \cdot \Delta H°_f\ (O_2)]$

$= [\ (4\ mol \cdot -393.509\ kJ/mol) + (6\ mol \cdot -241.83\ kJ/mol)]$

$- [2\ mol \cdot -83.85kJ/mol) + 7\ mol \cdot 0]$

$= [-1574.036\ kJ +\ -1450.98\ kJ] - [-167.7\ kJ] = -2857.32\ kJ$ for 2 moles

of ethane or -1428.66 kJ/mol

$\Delta H°rx(C_2H_5OH)\ = [2 \cdot \Delta H°_f\ (CO_2) + 3 \cdot \Delta H°_f\ (H_2O)] - [1 \cdot \Delta H°_f(CH_3CH_3) + 3 \cdot \Delta H°_f\ (O_2)]$

$= [(2\ mol \cdot -393.509\ kJ/mol) + (3\ mol \cdot -241.83\ kJ/mol)]$

$- [1\ mol \cdot -277.0\ kJ/mol) + 3\ mol \cdot 0]$

$= [-787.018\ kJ + -725.49\ kJ] - [-277.0\ kJ] = -1235.5\ kJ/mol$ ethanol

On a per gram basis:

-1428.66 kJ/mol • 1 mol/30.0694 g or approximately - 47.51 kJ/g of ethane and

-1235.5 kJ/mol ethanol • 1 mol/46.0688 g or approximately – 26.82 kJ/g ethanol

(b) If ethanol is partially oxidized ethane, the ΔH of reaction is **less negative**. So partially oxidized ethane has less energy to release during the combustion process.

97. Test to distinguish between 2-propanol and methyl ethyl ether: Methyl ethyl ether will not react with an oxidizing agent like $KMnO_4$. The alcohol will react with $KMnO_4$, forming a ketone.

99. (a) Combustion of 0.125 g (125 mg) of maleic acid gives 0.190 g (190. mg) of CO_2 and 0.0388 g(38.8 mg) of H_2O. The empirical formula of maleic acid:

The combustion of maleic acid can be represented: $C_xH_yO_z + O_2 \rightarrow\ CO_2 + H_2O$

It's important to note that while **all** the C in CO_2 originated in the maleic acid and **all** the H in H_2O originated in the maleic acid —**not all** of the oxygen originates in the maleic acid.

So we begin by calculating the masses of C and H , and subtracting those masses from the 125 mg of compound to determine the mass of O present in maleic acid.

$$190. \text{ mg CO}_2 \cdot \frac{12.01 \text{ g C}}{44.02 \text{ g CO}_2} = 51.8 \text{ mg C}$$

$$38.8 \text{ mg H}_2\text{O} \cdot \frac{2.02 \text{ g H}}{18.02 \text{ g H}_2\text{O}} = 4.35 \text{ mg H}$$

Mass O = 125 mg - (51.8 mg C + 4.35 mg H) = 68.8 mg O

Now we can calculate the # of moles of each of these atoms:

$$51.8 \times 10^{-3} \text{ g C} \cdot \frac{1 \text{ mol C}}{12.01 \text{ g C}} \qquad = 4.32 \times 10^{-3} \text{ mol C}$$

$$4.35 \times 10^{-3} \text{ g H} \cdot \frac{1 \text{ mol H}}{1.008 \text{ g H}} \qquad = 4.31 \times 10^{-3} \text{ mol H}$$

$$68.8 \times 10^{-3} \text{ g O} \cdot \frac{1 \text{ mol O}}{16.00 \text{ g O}} \qquad = 4.30 \times 10^{-3} \text{ mol O}$$

Note that the **ratio** of C,H, and O present in maleic acid indicates equal numbers of C, H, and O atoms. The empirical formula is then $C_1H_1O_1$—or CHO.

(b) 0.261 g of maleic acid requires 34.60 mL of 0.130 M NaOH. The molecular formula of maleic acid:

The text indicates that there are 2 carboxylic acid groups per molecule of maleic acid. The titration indicates: $\frac{0.130 \text{ mol NaOH}}{1 \text{ L}} \cdot \frac{0.03460 \text{ L}}{1} = 0.004498 \text{ mol NaOH}$.

Since 1 mol of NaOH reacts with 1 mol of an acidic group (a –COOH group), we know that there are 0.002249 mol of maleic acid present.

The molecular weight of maleic acid is: 0.261 g maleic acid/0.002249 mol maleic acid or 116 g/mol. The empirical formula (CHO) would have a formula weight of (12 + 1 + 16) 29 g. So the molecular formula is : $\frac{116 \text{ g}}{1 \text{ mol}} \cdot \frac{1 \text{ empirical formula}}{29 \text{ g}} = 4 \text{ empirical formulas/mol}$

or a molecular formula of $C_4H_4O_4$

(c) A Lewis structure for maleic acid:

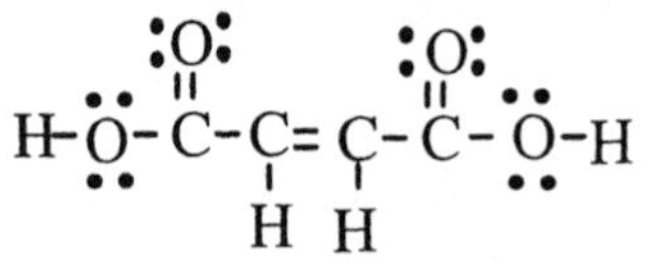

(d) Hybridization used by C atoms: All four C atoms are attached to three other groups (and a double bond), making the hybridization of all sp^2.

(e) Bond angles around each carbon will be 120°.

101. For the reactions in question:

(a) 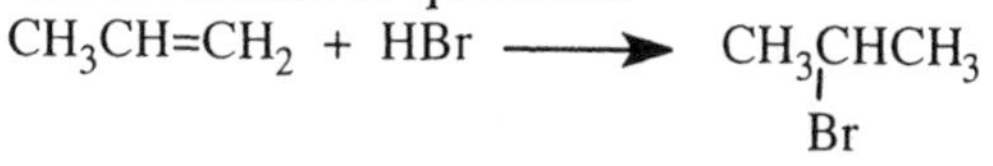 $CH_3CH=CH_2 + HBr \longrightarrow CH_3CHCH_3$
   $\overset{|}{Br}$

2-bromopropane

(b)

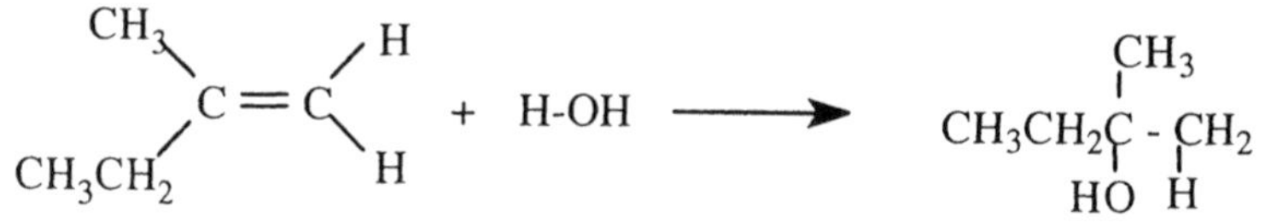

$CH_3CH_2C - CH_2$ with CH_3 above, HO H below

2-methyl-2-butanol

(c)

$CH_3CH - CCH_3$ with CH_3 above, H OH below

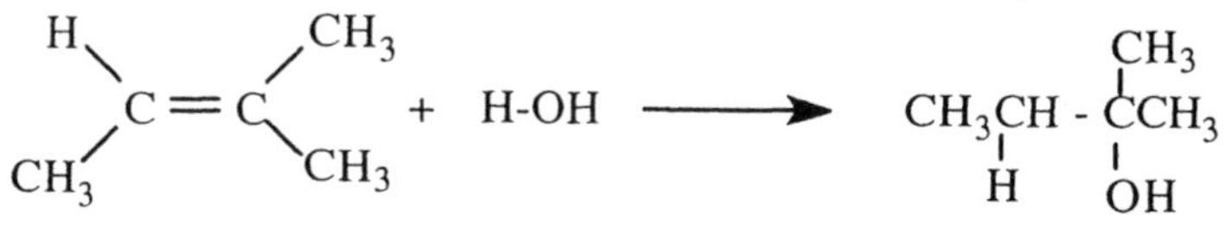

2-methyl-2-butanol

As you can see by the identical names for the product formed in (b) and (c), the products of these two reactions are the same.

103. (a) Difference between substitution and elimination reactions:

In substitution reactions, one atom (say X) is replaced by another atom (say Y). We might envision that as CH_3X being transformed into CH_3Y. In elimination reactions, atoms are lost in a molecule, resulting in a less saturated bond being formed. So we might imagine the molecule ethane, CH_3CH_3, eliminating H_2 and forming ethylene, $CH_2=CH_2$.

(b) The two reactions are similar in that the hydrocarbon is reacting with another small molecule. In an elimination reaction, an **alkene** is being formed (a **product**). In an addition reaction, the **alkene** is the starting material (**reactant**)—and not the product.

105. Addition polymerization:

(a) The primary structural feature to form addition polymers is a C=C double bond.

(b) The chain can be shorter OR longer than 14-Carbons. The necessary feature is a C=C double bond to begin the addition process.

(c) Reaction conditions control the length of the polymer chain formed. If the monomer concentration is too great, for example, a frequent termination of the chain-growth sequence results when two chains—each containing a free radical—combine.

(d) Addition polymerization would most reasonably be classified as an *Addition.*

Chapter 12
Gases and Their Properties

Practicing Skills
Pressure

1. (a) $440 \text{ mm Hg} \cdot \dfrac{1 \text{ atm}}{760 \text{ mm Hg}} = 0.58 \text{ atm}$ (2 significant figures in 440)

   (b) $440 \text{ mm Hg} \cdot \dfrac{1.013 \text{ bar}}{760 \text{ mm Hg}} = 0.59 \text{ bar}$

   (c) $440 \text{ mm Hg} \cdot \dfrac{101.325 \text{ kPa}}{760 \text{ mm Hg}} = 59 \text{ kPa}$

3. The higher pressure in each of the pairs: (using appropriate conversion factors)
   (a) 534 mm Hg or 0.754 bar

   $534 \text{ mm Hg} \cdot \dfrac{1.013 \text{ bar}}{760 \text{ mm Hg}} = 0.712 \text{ bar}$ so 0.754 bar is higher

   (b) 534 mm Hg or 650 kPa

   $534 \text{ mm Hg} \cdot \dfrac{101.325 \text{ kPa}}{760 \text{ mm Hg}} = 71.2 \text{ kPa}$ so 650 kPa is higher

   (c) 1.34 bar or 934 kPa

   $1.34 \text{ bar} \cdot \dfrac{101.325 \text{ kPa}}{1.01325 \text{ bar}} = 134 \text{ kPa}$ so 934 kPa is higher

Boyle's Law and Charles's Law

5. **Boyle's law** states that the pressure a gas exerts is inversely proportional to the volume it occupies, or for a given amount of gas--PV = constant . We can write this as: $P_1V_1 = P_2V_2$

   So $(67.5 \text{ mm Hg})(500. \text{ mL}) = (P_2)(125 \text{ mL})$

   and $\dfrac{(67.5 \text{ mm Hg})(500. \text{ mL})}{125 \text{ mL}} = 270. \text{ mm Hg}$

   Note that the **volume decreased** by a factor of 4, and the **pressure increased** by a factor of 4.

7. **Charles' law** states that $V \propto T$ (in Kelvin) or $\dfrac{V_1}{T_1} = \dfrac{V_2}{T_2}$

   $\dfrac{3.5 \text{ L}}{295 \text{ K}} = \dfrac{V_2}{310 \text{ K}}$ and rearranging to solve for V_2 yields

   $V_2 = \dfrac{(3.5 \text{ L})(310 \text{ K})}{(295 \text{ K})} = 3.7 \text{ L}$

   Note that with the **increase in T** there has been an **increase in volume.**

The General Gas Law

9. Pressure of 3.6 L of H_2 at 380 mm Hg and 25 °C, if transferred to a 5.0 L flask at 0.0 °C:

Substituting into : $\dfrac{P_1V_1}{T_1} = \dfrac{P_2V_2}{T_2}$, will permit us to calculate P_2. Rearranging the equation:

$$P_2 = P_1 \cdot \frac{T_2}{T_1} \cdot \frac{V_1}{V_2} = 380 \text{ mm Hg} \cdot \frac{273 \text{ K}}{298 \text{ K}} \cdot \frac{3.6 \text{ L}}{5.0 \text{ L}} = 250 \text{ mm Hg}$$

11. Using the general gas law we can write: $\dfrac{P_1V_1}{T_1} = \dfrac{P_2V_2}{T_2}$

and for a fixed volume: $\dfrac{P_1}{T_1} = \dfrac{P_2}{T_2}$ or if we rearrange, we obtain $P_2 = P_1 \cdot \dfrac{T_2}{T_1}$

$$P_2 = 360 \text{ mm Hg} \cdot \frac{268.2 \text{ K}}{298.7 \text{ K}} = 320 \text{ mm Hg} \ (2 \text{ significant figures})$$

13. Using the general gas law we can write: $\dfrac{P_1V_1}{T_1} = \dfrac{P_2V_2}{T_2}$

The volume is changing from 400. cm^3 to 50.0 cm^3, and the temperature from 15°C to 77 °C. We can rearrange the equation to solve for the new pressure:

$$P_2 = P_1 \cdot \frac{T_2}{T_1} \cdot \frac{V_1}{V_2} = 1.00 \text{ atm} \cdot \frac{350. \text{ K}}{288 \text{ K}} \cdot \frac{400. \text{ cm}^3}{50.0 \text{ cm}^3} = 9.72 \text{ atm}$$

Avogadro's Hypothesis

15. (a) The balanced equation indicates that 1 O_2 is needed for 2 NO. At the same conditions of T and P, the amount of O_2 is 1/2 the amount of NO, so 75.0 mL of O_2 is required.

(b) The amount of NO_2 produced will have the same volume as the amount of NO (since their coefficients are equal in the balanced equation—150 mL of NO_2 .

Ideal Gas Law

17. The pressure of 1.25 g of gaseous carbon dioxide may be calculated with the ideal gas law:

$$1.25 \text{ g CO}_2 \cdot \frac{1 \text{ mol CO}_2}{44.01 \text{ g CO}_2} = 0.0284 \text{ mol CO}_2$$

Rearranging $PV = nRT$ to solve for P, we obtain:

$$P = \frac{nRT}{V} = \frac{(0.0284 \text{ mol})(0.082057 \frac{\text{L} \cdot \text{atm}}{\text{K} \cdot \text{mol}})(295.7 \text{ K})}{0.750 \text{ L}} = 0.919 \text{ atm}$$

19. The volume of the flask may be calculated by realizing that the gas will expand to fill the flask.

$$2.2 \text{ g CO}_2 \cdot \frac{1 \text{ mol CO}_2}{44.0 \text{ g CO}_2} = 0.050 \text{ mol CO}_2$$

$$P = 318 \text{ mm Hg} \cdot \frac{1 \text{ atm}}{760 \text{ mm Hg}} = 0.418 \text{ atm}$$

$$V = \frac{(0.050 \text{ mol})(0.082057 \frac{\text{L} \cdot \text{atm}}{\text{K} \cdot \text{mol}})(295 \text{ K})}{0.418 \text{ atm}} = 2.9 \text{ L}$$

21. Rearranging $PV = nRT$ to solve for n, we obtain: $n = \frac{PV}{RT}$

Converting 737 mm Hg to atmospheres, we obtain

$$737 \text{ mm Hg} \cdot \frac{1 \text{ atm}}{760 \text{ mm Hg}} = 0.970 \text{ atm}$$

$$n = \frac{(0.970 \text{ atm})(1.2 \times 10^7 \text{ L})}{(0.082057 \frac{\text{L} \cdot \text{atm}}{\text{K} \cdot \text{mol}})(298.2 \text{ K})} = 4.8 \times 10^5 \text{ moles of He}$$

and since each mole of He has a mass of 4.00 g,

$$4.8 \times 10^5 \text{ mol He} \cdot \frac{4.00 \text{ g He}}{1 \text{ mol He}} = 1.9 \times 10^6 \text{ g He}$$

Gas Density

23. Write the ideal gas law as: Molar Mass $= \frac{dRT}{P}$ where d = density in grams per liter.

Solving for d, we obtain: $\frac{(\text{Molar Mass}) \cdot P}{R \cdot T} = d$

The average molar mass for air is approximately 28.96 g/mol.

$$\frac{(28.96 \text{ g/mol})(0.20 \text{ mm Hg} \cdot \frac{1 \text{ atm}}{760 \text{ mm Hg}})}{(0.082057 \frac{\text{L} \cdot \text{atm}}{\text{K} \cdot \text{mol}})(250 \text{ K})} = 3.7 \times 10^{-4} \text{ g/L} = d$$

25. Molar mass $= \dfrac{(0.355 \text{ g/L})(0.082057 \frac{\text{L} \cdot \text{atm}}{\text{K} \cdot \text{mol}})(290. \text{K})}{(189 \text{ mmHg})\left(\frac{1 \text{atm}}{760 \text{ mmHg}}\right)} = 34.0 \text{ g/mol}$

Ideal Gas Laws and Determining Molar Mass

27. Rearranging $PV = nRT$ to solve for n, we obtain: $n = \dfrac{PV}{RT}$

Converting 715 mm Hg to atmospheres, we obtain

$$715\text{mm Hg} \cdot \frac{1\text{ atm}}{760\text{ mm Hg}} = 0.941 \text{ atm}$$

$$n = \frac{(0.941 \text{ atm})(0.452 \text{ L})}{(0.082057\dfrac{L \bullet atm}{K \bullet mol})(296.2 \text{ K})} = 0.0175 \text{ moles of unknown gas.}$$

Since this number of moles of the gas has a mass of 1.007 g, we can calculate the molar mass

$$\frac{1.007 \text{ g of unknown gas}}{0.0175 \text{ moles of unknown gas}} = 57.5 \text{ g/mol}$$

29. To calculate the molar mass:

$$\text{Molar mass} = \frac{(0.0125 \text{ g}/0.125\text{L})(0.082057\dfrac{L \bullet atm}{K \bullet mol})(298.2 \text{ K})}{(24.8 \text{ mmHg} \bullet 1\text{atm}/760\text{mmHg})} = 74.9 \text{ g/mol}$$

B_6H_{10} has a molar mass of 74.94 grams.

See SQ 25 for a similar problem.

Gas Laws and Stoichiometry

31. Determine the amount of H_2 generated when 2.2 g Fe reacts:

$$2.2 \text{ g Fe} \cdot \frac{1 \text{ mol Fe}}{55.85 \text{ g Fe}} \cdot \frac{1 \text{ mol H}_2}{1 \text{ mol Fe}} = 0.0394 \text{ mol H}_2 \text{ (0.039 to 2 sf)}$$

Note that the latter factor is achieved by examining the balanced equation!

The pressure of this amount of H_2 is :

$$P = \frac{nRT}{V} = \frac{(0.039 \text{ mol H}_2)(62.4 \dfrac{L \bullet torr}{K \bullet mol})(298 \text{ K})}{10.0 \text{ L}} = 73 \text{ torr or 73 mm Hg}$$

33. Calculate the moles of N_2 needed:

$$n = \frac{PV}{RT} = \frac{(1.3 \text{ atm})(75.0 \text{ L})}{(0.082057\dfrac{L \bullet atm}{K \bullet mol})(298 \text{ K})} = 3.99 \text{ mol N}_2$$

The mass of NaN_3 needed is obtained from the stoichiometry of the equation:

$$3.99 \text{ mol } N_2 \cdot \frac{2 \text{ mol } NaN_3}{3 \text{ mol } N_2} \cdot \frac{65.0 \text{ g } NaN_3}{1 \text{ mol } NaN_3} = 170 \text{ g } NaN_3 \text{ (to 2 sf)}$$

35. $N_2H_4 (g) + O_2 (g) \rightarrow N_2 (g) + 2 H_2O (g)$

$$1.00 \text{ kg } N_2H_4 \cdot \frac{1.0 \times 10^3 \text{ g } N_2H_4}{1.0 \text{ kg } N_2H_4} \cdot \frac{1 \text{ mol } N_2H_4}{32.0 \text{ g } N_2H_4} \cdot \frac{1 \text{ mol } O_2}{1 \text{ mol } N_2H_4}$$

$$= 3.13 \times 10^1 \text{ mole } O_2$$

$$P(O_2) = \frac{n(O_2) \cdot R \cdot T}{V} = \frac{(3.13 \times 10^1 \text{mol})(0.082057 \text{ L} \cdot \text{atm/K} \cdot \text{mol})(296 \text{ K})}{450 \text{ L}}$$

$$P(O_2) = 1.69 \text{ atm or } 1.7 \text{ atm to 2 sf}$$

Gas Mixtures and Dalton's Law

37. We know that the total pressure will be equal to the sum of the pressure of each gas (also called the *partial pressure* of each gas).

$$1.0 \text{ g } H_2 \cdot \frac{1 \text{ mol } H_2}{2.02 \text{ g } H_2} = 0.50 \text{ mol } H_2 \text{ and } 8.0 \text{ g } Ar \cdot \frac{1 \text{ mol } Ar}{39.9 \text{ g } Ar} = 0.20 \text{ mol } Ar$$

We can calculate the **total pressure** using the ideal gas law:

$$P = \frac{n \cdot R \cdot T}{V} = \frac{(0.70 \text{ mol})(0.082057 \text{ L} \cdot \text{atm/K} \cdot \text{mol})(300 \text{ K})}{3.0 \text{ L}} = 5.7 \text{ atm}$$

The pressure of **each gas** can be calculated by multiplying the total pressure (5.7 atm) by the mole fraction of the gas.

$$\text{Pressure of } H_2 = \frac{0.50 \text{ mol } H_2}{0.70 \text{ mol } H_2 + Ar} \cdot 5.7 = 4.1 \text{ atm and the}$$

$$\text{Pressure of } Ar = \frac{0.20 \text{ mol } Ar}{0.70 \text{ mol } H_2 + Ar} \cdot 5.7 = 1.6 \text{ atm}$$

Note that the total pressure is indeed (4.1+1.6) or 5.7 atm

39. $P_{total} = P_{halothane} + P_{oxygen} = 170 \text{ mm Hg} + 570 \text{ mm Hg} = 740 \text{ mm Hg}$

(a) Since we know that the pressure a gas exerts is **proportional** to the # of moles of gas present we can calculate the ratio of moles by using their partial pressures:

$$\frac{\text{moles of halothane}}{\text{moles of oxygen}} = \frac{170 \text{ mm Hg}}{570 \text{ mm Hg}} = 0.30$$

(b) $160 \text{ g oxygen} \cdot \dfrac{1 \text{ mol oxygen}}{32.0 \text{ g oxygen}} \cdot \dfrac{0.30 \text{ mol halothane}}{1 \text{ mol oxygen}} \cdot \dfrac{197.38 \text{ g halothane}}{1 \text{ mol halothane}} =$

$$3.0 \times 10^2 \text{ g halothane (2 significant figures)}$$

Kinetic-Molecular Theory

41. (a) Kinetic energy depends only on the temperature so the average kinetic *energies of these two gases are equal.*

(b) Since the kinetic energies are equal, we can state:

$$KE(H_2) = KE(CO_2)$$

$$1/2 \; m(H_2) \cdot \overline{V}^2 (H_2) = 1/2 \; m(CO_2) \cdot \overline{V}^2(CO_2)$$

Where m = mass of a molecule and $\overline{V}$ = average velocity of a molecule

So

$$m(H_2) \cdot \overline{V}^2(H_2) = m(CO_2) \cdot \overline{V}^2(CO_2)$$

and

$$\frac{\overline{V}^2 (H_2)}{\overline{V}^2 (CO_2)} = \frac{m(CO_2)}{m(H_2)}$$

Now the molar mass of H_2 = 2.0 g and the molar mass of CO_2 = 44 g

$$\frac{\overline{V}_{H_2}}{\overline{V}_{CO_2}} = \sqrt{\frac{m_{CO_2}}{m_{H_2}}} = \sqrt{\frac{44}{2.0}} = 4.7$$

The hydrogen molecules have an average velocity which is 4.7 times the average velocity of the CO_2 molecules.

(c) Since the volumes are equal for these two gas samples, the pressure is proportional to the amount of gas present.

$$V_A = \frac{n_A R T_A}{P_A} \qquad \text{and} \quad V_B = \frac{n_B R T_B}{P_B} \quad \text{and} \; V_A = V_B \; \text{so}$$

$$\frac{n_A R T_A}{P_A} = \frac{n_B R T_B}{P_B} \text{and rearranging to solve for the ratio of molecules present}$$

$$\frac{P_B T_A}{T_B P_A} = \frac{n_B}{n_A} \quad \text{substituting yields:} \; \frac{2 \; atm \bullet 273 \; K}{298 \; K \bullet 1 \; atm} = \frac{n_B}{n_A} = 1.8$$

There are 1.8 times as many moles (and molecules) of gas in Flask B (CO_2) as there are in Flask A (H_2).

(d) Since Flask B contains 1.8 times as many moles of CO_2 as Flask A contains of H_2, the *ratio* of masses of gas present are:

$$\frac{\text{Mass (Flask B)}}{\text{Mass (Flask A)}} = \frac{(1.8 \; \text{mole} \; CO_2)(44 \; g \; CO_2/\text{mol} \; CO_2)}{(1 \; \text{mol} \; H_2)(2 \; g \; H_2/\text{mol} \; H_2)} = \frac{40}{1}$$

Note that any number of moles of CO_2 and H_2 (in the ratio of 1.8:1) would provide the same answer.

43. Since two gases at the same temperature have the same kinetic energy

$$KE_{O_2} = KE_{CO_2}$$

and since the average $KE = 1/2\, M\bar{U}^2$

where $\bar{U}$ is the average speed of a molecule, we can write.

$$1/2\, M_{O_2}\bar{U}^2_{O_2} = 1/2\, M_{CO_2}\bar{U}^2_{CO_2} \qquad \text{or} \qquad M_{O_2}\bar{U}^2_{O_2} = M_{CO_2}\bar{U}^2_{CO_2}$$

and

$$\frac{M_{O_2}}{M_{CO_2}} = \frac{\bar{U}^2_{CO_2}}{\bar{U}^2_{O_2}}$$

and solving for the average velocity of CO_2 :

$$\bar{U}^2_{CO_2} = \frac{M_{O_2}}{M_{CO_2}} \cdot \bar{U}^2_{O_2}$$

Taking the square root of both sides

$$\bar{U}_{CO_2} = \sqrt{\frac{M_{O_2}}{M_{CO_2}}} \cdot \bar{U}_{O_2} = \sqrt{\frac{32.0\ g\ O_2\ /\ mol\ O_2}{44.0\ g\ CO_2\ /\ mol\ CO_2}} \cdot 4.28 \times 10^4\, cm/s$$

$$= 3.65 \times 10^4\ cm/s$$

45. The species will have average molecular speeds which are inversely proportional to their molar masses.

	Slowest					Fastest	
	CH_2F_2	<	Ar	<	N_2	<	CH_4
	54		40		28		16 (to integral values)

Diffusion and Effusion

47. Relative rates of effusion for the following pairs of gases:

(a) CO_2 or F_2 : *Fluorine* effuses faster, since the molar mass of F_2 is 38 g/mol and that of CO_2 is 44 g/mol

(b) O_2 or N_2: *Nitrogen* effuses faster. (MM N_2 = 28 g/mol; for O_2 = 32 g/mol)

(c) C_2H_4 or C_2H_6 : *Ethylene* effuses faster. (MM C_2H_4 = 28 g/mol; for C_2H_6 = 30 g/mol)

(d) $CFCl_3$ or $C_2Cl_2F_4$: *$CFCl_3$* effuses faster.(MM $CFCl_3$ = 137 g/mol; for $C_2Cl_2F_4$ = 171 g/mol)

49. Determine the molar mass of a gas which effuses at a rate 1/3 that of He:

$$\frac{\text{Rate of effusion of He}}{\text{Rate of effusion of unknown}} = \sqrt{\frac{\text{M of unknown}}{\text{M of He}}}$$

$$\frac{3}{1} = \sqrt{\frac{\text{M of unknown}}{4.0 \text{ g/mol}}}$$

Squaring both sides gives: $9 = \frac{M}{4.0}$ or M = 36 g/mol

Nonideal Gases

51. According to the Ideal Gas Law, the pressure would be:

$$P = \frac{n \cdot R \cdot T}{V} = \frac{(8.00 \text{ mol})(0.082057 \frac{L \cdot atm}{K \cdot mol})(300 \text{ K})}{4.00 \text{ L}} = 49.3 \text{ atm}$$

The van der Waal's equation is: $\left[P + a\left(\frac{n}{V}\right)^2\right][V - bn] = nRT$. Substituting, we get:

$$\left[P + 6.49\frac{atm \cdot L^2}{mol^2}\left(\frac{8.00 \text{ mol}}{4.00 \text{ L}}\right)^2\right]\left[4.00L - 0.0562\frac{L}{mol} \cdot 8.00mol\right] = 8.00mol \cdot 0.082057\frac{atm \cdot L}{K \cdot mol} \cdot \text{?}$$

Simplifying: $[P + 25.96atm][4.00L - 0.45L] = 196.94atm \cdot L$ and

$$P = \frac{196.94 \text{ atm} \cdot L}{3.55L} - 25.96 \text{ atm} = 29.5 \text{ atm}$$

General Questions

53.

	atm	mm Hg	kPa	bar
Standard atmosphere:	1	$1 \text{ atm} \cdot \frac{760. \text{ mm Hg}}{1 \text{ atm}}$ $= 760. \text{ mm Hg}$	$1 \text{ atm} \cdot \frac{101.325 \text{ kPa}}{1 \text{ atm}}$ $= 101.325 \text{ kPa}$	$1 \text{ atm} \cdot \frac{1.013 \text{ bar}}{1 \text{ atm}}$ $= 1.013 \text{ bar}$
Partial pressure of N_2 in the atmosphere	$593 \text{ mm Hg} \cdot \frac{1 \text{ atm}}{760 \text{ mm Hg}}$ $= 0.780 \text{ atm}$	**593 mm Hg**	$0.780 \text{ atm} \cdot \frac{101.3 \text{ kPa}}{1 \text{ atm}}$ $= 79.1 \text{ kPa}$	$0.780 \text{ atm} \cdot \frac{1.013 \text{ bar}}{1 \text{ atm}}$ $= 0.791 \text{ bar}$
Tank of compressed H_2	$133 \text{ bar} \cdot \frac{1 \text{ atm}}{1.013 \text{ bar}}$ $= 131 \text{ atm}$	$131 \text{ atm} \cdot \frac{760. \text{ mm Hg}}{1 \text{ atm}}$ $= 99800 \text{ mm Hg}$	$131 \text{ atm} \cdot \frac{101.3 \text{ kPa}}{1 \text{ atm}}$ $= 13300 \text{ kPa}$	**133 bar**
Atmospheric pressure at top of Mt. Everest	$33.7 \text{ kPa} \cdot \frac{1 \text{ atm}}{101.3 \text{ kPa}}$ $= 0.333 \text{ atm}$	$0.333 \text{ atm} \cdot \frac{760 \text{ mm Hg}}{1 \text{ atm}}$ $= 253 \text{ mm Hg}$	**33.7 kPa**	$0.333 \text{ atm} \cdot \frac{1.013 \text{ bar}}{1 \text{ atm}}$ $= 0.337 \text{ bar}$

55. 1.0 L of a compound gives 2.0 L CO_2, 3.5 L H_2O, and 0.50 L of N_2 at STP.

At STP, 1 mol of a gas occupies 22.4 L so we can calculate moles of CO_2 and N_2.

$$2.0 \text{ L } CO_2 \bullet \frac{1 \text{ mol } CO_2}{22.4 \text{ L}} \bullet \frac{1 \text{ mol C}}{1 \text{ mol } CO_2} = 0.089 \text{ mol C}$$

$$3.5 \text{ L } H_2O \bullet \frac{1 \text{ mol } H_2O}{22.4 \text{ L}} \bullet \frac{2 \text{ mol H}}{1 \text{ mol } H_2O} = 0.31 \text{ mol H}$$

$$0.50 \text{ L } N_2 \bullet \frac{1 \text{ mol } N_2}{22.4 \text{ L}} \bullet \frac{2 \text{ mol N}}{1 \text{ mol } N_2} = 0.045 \text{ mol N}$$

Establish the ratio of C:N:H by dividing the amount of each element by the *smallest amount present* (in this case, 0.045 mol N).

$$\frac{0.089 \text{ mol C}}{0.045 \text{ mol N}} = 2 \qquad \frac{0.31 \text{ mol H}}{0.045 \text{ mol N}} = 7 \quad \text{The empirical formula is then } C_2H_7N.$$

57. To increase the average speed of helium atoms by 10.0%, we must know the average speed initially.

$$\sqrt{u^2} = \sqrt{\frac{3RT}{M}} \quad \text{and substituting for He: } = \sqrt{\frac{3 \bullet (8.314 \text{ J/K} \bullet \text{mol}) \bullet 240K}{4.00 \times 10^{-3} \text{ kg/mol}}} = 1220 \text{ m/s}$$

NOTE: A **J**oule is a kg•m/s, so it's necessary to express the molar mass of helium in units of kg/mol.

The new average speed = 110%(1220 m/s) = 1350 m/s. Substituting this value as u:

$$\frac{Mu^2}{3R} = T; \quad \frac{4.00 \times 10^{-3} \text{ kg/mol} \bullet (1350 \text{ m/s})^2}{3 \bullet 8.314 \text{ J/K} \bullet \text{mol}} = 290 \text{ K} \quad \text{or } (290-273)°C \text{ or } 17°C.$$

59. (a) To calculate the balloon containing the greater number of molecules we need only to calculate the ratio of the amounts of gas present in the balloons. Using the Ideal Gas Law allows us to solve for the number of moles of each gas:

Pick a volume of gas—say 5L for He. Since the

$$\boxed{n = \frac{P \bullet V}{R \bullet T}}$$

hydrogen balloon is twice the size of the He balloon,

we'll use 10 L for the volume of H. We can then establish a ratio of this expression for the

two gases. $\dfrac{n_{He}}{n_{H_2}} = \dfrac{\dfrac{P_{He} \bullet V_{He}}{R \bullet T_{He}}}{\dfrac{P_{H_2} \bullet V_{H_2}}{R \bullet T_{H_2}}} = \dfrac{\dfrac{2atm \bullet 5L}{296K}}{\dfrac{1atm \bullet 10 L}{268K}} = \dfrac{268 \text{ K}}{296 \text{ K}} = \dfrac{0.9 \text{ mol He}}{1 \text{ mol } H_2}$

Note that **R** cancels in the numerator and denominator, simplifying the calculation.

(b) Two calculate which balloon contains the greater mass of gas, we can use the ratio found:

0.9 mol He • 4.0 g/mol He = 3.6 g He; 1 mol H_2 • 2.0 g/mol H_2 = 2 g H_2

The balloon containing the *HELIUM* has the greater mass. Note that the *ratio* of moles of helium and hydrogen are **independent** of the sizes of the balloons.

61. Using the Ideal Gas Law,

$$n = \frac{P \bullet V}{R \bullet T} = \frac{\left(8 \text{ mm Hg} \bullet \dfrac{1 \text{ atm}}{760 \text{ mm Hg}}\right) \bullet \left(10. \text{ m}^3 \bullet \dfrac{1000L}{1 \text{ m}^3}\right)}{0.082057 \dfrac{L \bullet atm}{K \bullet mol} \bullet 300. \text{ K}} = 4 \text{ mol (1sf)}$$

63. This problem has two parts: 1) How many moles of Ni are present?

                                              2) How many moles of CO are present?

1) # moles of Ni present:

$$0.450 \text{ g Ni} \bullet \frac{1 \text{ mol Ni}}{58.693 \text{g Ni}} = 7.67 \times 10^{-3} \text{ mol Ni}$$

and since 1 mol $Ni(CO)_4$ is formed for **each mol of Ni**, one can form

$$7.67 \times 10^{-3} \text{ mol Ni(CO)}_4$$

2) # moles of CO present:

Using the Ideal Gas Law:

$$n = \frac{\dfrac{418}{760} \text{ atm} \bullet (1.50L)}{(0.082057 \dfrac{L \bullet atom}{K \bullet mol})(298K)} = 0.0337 \text{ mol CO which would be capable of}$$

forming $0.0337 \text{ mol CO} \bullet \dfrac{1 \text{ mol Ni(CO)}_4}{4 \text{ mol CO}} = 8.43 \times 10^{-3} \text{ mol Ni(CO)}_4$

Since the amount of *nickel limits the maximum amount of Ni(CO)₄ that can be formed*, the maximum mass of $Ni(CO)_4$ is then:

$$7.67 \times 10^{-3} \text{ mol Ni(CO)}_4 \bullet \frac{170.7 \text{ g Ni(CO)}_4}{1 \text{ mol Ni(CO)}_4} = 1.31 \text{ g Ni(CO)}_4$$

65. For the four samples given:

  (1) 1.0 L of H_2 at STP   (2) 1.0 L of Ar at STP

  (3) 1.0 L of H_2 at 27°C and 760 mm Hg   (4) 1.0 L of He at 0°C and 900 mm Hg

For samples (1) and (2), the calculation is identical for # of particles:

Rearranging the Ideal Gas Law: $n = \dfrac{PV}{RT} = \dfrac{(1 \text{ atm})(1.0 \text{ L})}{(0.082057 \dfrac{L \bullet atm}{K \bullet mol})(273 \text{ K})} = 4.4 \times 10^{-2} \text{ mol}$

Note that we can alternatively use the factor: 22.4 L = 1 mol (since samples 1 and 2 are at

STP) For sample (3) n = $\dfrac{(1\ atm)(1.0\ L)}{(0.082057\dfrac{L\bullet atm}{K\bullet mol})(300\ K)}$ = 4.1x 10^{-2} mol of H_2

For sample (4) n = $\dfrac{(\dfrac{900}{760}\ atm)(1.0\ L)}{(0.082057\dfrac{L\bullet atm}{K\bullet mol})(273\ K)}$ = 5.3 x 10^{-2} mol of He

(a) Which sample has largest number of gas particles?

Sample 4 has the largest number of gas particles.

(b) Which sample has the smallest number of gas particles?

Sample 3 has the smallest number of gas particles.

(c) Which sample represents the largest mass?

Given the number of moles of each gas, we can calculate the masses:
Sample (1): 4.4 x 10^{-2} mol H_2 • 2.0 g H_2/mol H_2 = 8.8 x 10^{-2} g H_2

Sample (2): 4.4 x 10^{-2} mol Ar • 39.95 g Ar /mol Ar = 1.7 g Ar

Sample (3): 4.1 x 10^{-2} mol H_2 • 2.0 g H_2 /mol H_2 = 8.2 x 10^{-2} g H_2

Sample (4): 5.3 x 10^{-2} mol of He • 4.00 g He/mol He = 0.21 g He

Sample 2 has the largest mass.

67. The total pressure of B_2H_6 and O_2 is 228 mm Hg.

Dalton's Law of Partial Pressures: P (B_2H_6) + P (O_2) = 228 mm Hg

Given the stoichiometric ratio of 1 B_2H_6 for each 3 O_2 , we know that the pressure of O_2 will

be 3 times the pressure of B_2H_6. and Dalton's Law can be rewritten :

P(B_2H_6) + 3P(B_2H_6) = 228 mm Hg or 4P(B_2H_6) = 228 mm Hg and P(B_2H_6) = 57 mm Hg

The partial pressure of oxygen will then be 3 x 57 or 171 mm Hg.

Assuming that P and T don't change during the reaction, the pressure of water vapor produced
will be exactly equal to that of O_2.

69. Determine the empirical formula of the S_xF_y compound:

Given the data, let's calculate the # of mol of the compound:

Note that we convert 89 mL to the volume in liters (0.089 L). We also need to express 83.8 mm

Hg in units of atmospheres. With those conversions, we can calculate the # mol of the

compound.

$$n = \frac{(\frac{83.8}{760} \text{ atm})(0.089 \text{ L})}{(0.082057 \frac{\text{L} \cdot \text{atm}}{\text{K} \cdot \text{mol}})(318 \text{ K})} = 3.8 \times 10^{-4} \text{ mol compound (to 2 sf)}$$

The molecular weight is then: 0.0955 g/ 3.8×10^{-4} mol = 253.9 g/mol or 250 (to 2 sf).

The formula is 25.23% S (and therefore 74.77 % F). So we can anticipate that of the 250 g, 25.23% is S (0.2523 • 250) = 64.1 g S. The amount of F is (0.7477 • 250) = 190 g F.

Now we can calculate the # of moles of each element present:

$$\frac{64.1 \text{ g S}}{1} \cdot \frac{1 \text{ mol S}}{32.07 \text{ g S}} = 2.0 \text{ mol (to 2 sf)}$$

and for F: $\frac{190 \text{ g F}}{1} \cdot \frac{1 \text{ mol F}}{19.00 \text{ g}} = 10. \text{ mol (to 2 sf). We note the formula is then } S_2F_{10}.$

71. To determine the theoretical yield of $Fe(CO)_5$, we need to know the # of moles of both Fe and of CO. Let's use the Ideal Gas Law to calculate moles of CO.

$$n = \frac{(\frac{732}{760} \text{ atm})(5.50 \text{ L})}{(0.082057 \frac{\text{L} \cdot \text{atm}}{\text{K} \cdot \text{mol}})(296 \text{ K})} = 0.218 \text{ mol CO}$$

For Fe: $\frac{3.52 \text{ g Fe}}{1} \cdot \frac{1 \text{ mol Fe}}{55.85 \text{ g Fe}} = 0.0630 \text{ mol Fe}$

The equation is: $Fe + 5 CO \rightarrow Fe(CO)_5$, indicating that we need 5 mol of CO per mol of Fe.

Moles available $= \frac{0.218 \text{ mol CO}}{0.0630 \text{ mol Fe}} = 3.46$, indicating that CO is the Limiting Reagent.

The theoretical yield of $Fe(CO)_5$ will then be:

$$\frac{0.218 \text{ mol CO}}{1} \cdot \frac{1 \text{ mol Fe(CO)}_5}{5 \text{ mol CO}} \cdot \frac{195.90 \text{ g Fe(CO)}_5}{1 \text{ mol Fe(CO)}_5} = 8.54 \text{ g Fe(CO)}_5.$$

73. (a) Average molar mass of air at 20 km above the earth's surface:

Converting volume to L, we note the conversion factor, 1 L = 1 x 10^{-3} m^3 so 1 m^3 is 1000 L. Noting also that –63°C will be 210K, we can substitute:

$$n = \frac{(\frac{42}{760} \text{ atm})(1000 \text{ L})}{(0.082057 \frac{\text{L} \cdot \text{atm}}{\text{K} \cdot \text{mol}})(210 \text{ K})} = 3.2 \text{ mol air. Given the density as 92 g/m}^3, \text{ we can}$$

calculate the molar mass: 92 g/3.2 mol = 29 g/mol. (to 2 sf)

Calculations for part (b) will be made using 2 sf. Using the un-rounded molar mass in part (a), the answers in part (b) will be different from those shown here.

(b) If the atmosphere is only O_2 and N_2, what is the mole fraction of each gas?

We know that the mf O_2 + mf N_2 = 1 . We know that the mixture of oxygen and nitrogen has a molar mass of 29 g/mol. We can then calculate the % (or mole fraction) of this weighted average.

(mf O_2) • 32.00 g/mol + (mf N_2) • 28.01g/mol = 29 g/mol.

Using the mf relationship from above, we get:

(mf O_2) • 32.00 g/mol + (1 – mf O_2) • 28.01g/mol = 29 g/mol.

For simplicity let's use x to represent mf O_2.

x • 32.00 g/mol + (1 – x) • 28.01g/mol = 29 g/mol.

32.00x + 28.01 –28.01x = 29 and

3.99 x = (29 - 28.01) and x = (29 - 28.01)/3.99 or 0.25.

The mf O_2 = 0.25 and the mf N_2 = 0.75

75. Pressure of PH_3 gas at a toxic concentration of 7 x 10^{-5} mg/L:

We can use the Ideal Gas Law, if we convert mass of PH_3 to moles of PH_3.

$$\frac{7 \times 10^{-8} \text{ g PH}_3}{1} \bullet \frac{1 \text{ molPH}_3}{34.00 \text{ g PH}_3} = 2.0 \times 10^{-9} \text{ mol PH}_3$$

$$P = \frac{n \bullet R \bullet T}{V} = \frac{(2.0 \times 10^{-9} \text{ mol})\left(0.082057 \frac{L \bullet atm}{K \bullet mol}\right)(298 \text{ K})}{1 \text{ L}} = 5 \times 10^{-8} \text{atm}$$

expressed in mm Hg: 4 x 10^{-5} mm Hg (to 1 sf)

77. Calculate the # of moles of gas present in the flask:

$$n = \frac{(\frac{17.2}{760} \text{ atm})(1.850 \text{ L})}{(0.082057 \frac{L \bullet atm}{K \bullet mol})(294 \text{ K})} = 1.74 \times 10^{-3} \text{ mol ClxOyFz}$$

This amount of the gas has a mass of 0.150 g, so the molar mass is 0.150 g/1.74 x 10^{-3} mol or 86.4 g/mol. If the gas contains Cl,O, and F, the "molar mass" of ClOF = (35.5 + 16 + 19)= 70.5 g/mol. With our calculated value of 86.4, we would hypothesize the formula to be ClO_2F.

79. To calculate percent yield, we need both an actual and a theoretical yield. Calculate the

theoretical yield: $\frac{2.65 \text{ g CaC}_2}{1} \bullet \frac{1 \text{ mol CaC}_2}{64.10 \text{ g CaC}_2} = 4.13 \times 10^{-2} \text{ mol CaC}_2$

We would then anticipate (given the stoichiometric ratio from the balanced equation an equal number of moles of acetylene (our theoretical yield).

To calculate the actual yield, we need to use the Ideal Gas Law:

$$n = \frac{(\frac{735.2}{760}\text{ atm})(0.795 \text{ L})}{(0.082057\frac{\text{L} \bullet \text{atm}}{\text{K} \bullet \text{mol}})(298.35 \text{ K})} = 3.14 \times 10^{-2} \text{ mol CaC}_2$$

The percent yield is : 3.14×10^{-2} mol $CaC_2/4.13 \times 10^{-2}$ mol $CaC_2 \bullet 100 = 76.0 \%$

81. Calculate moles of O_2 to determine the amount of $KClO_3$ originally present.

$$n = \frac{(\frac{735}{760}\text{ atm})(0.327 \text{ L})}{(0.082057\frac{\text{L} \bullet \text{atm}}{\text{K} \bullet \text{mol}})(292 \text{ K})} = 1.32 \times 10^{-2} \text{ mol O}_2$$

The mole ratio from the equation tells us that for each mol of O_2, we had 2/3 mole of $KClO_3$.

The number of moles of $KClO_3$ will be 1.32×10^{-2} mol $O_2 \bullet$ (2 mol $KClO_3$/3 mol O_2)=

8.80×10^{-3} mol of $KClO_3$. The mass to which this corresponds is:

$$\frac{8.80 \times 10^{-3} \text{ mol KClO}_3}{1} \bullet \frac{122.55 \text{ g KClO}_3}{1\text{mol KClO}_3} = 1.08 \text{ g KClO}_3. \text{ The percent of perchlorate in}$$

the original mixture is: 1.08 g $KClO_3$/ 1.56 g mixture $\bullet$ 100 = 69.1 %

83. (a) NO, O_2 and NO_2 in increasing velocity at 298 K. We know that the velocity of gases is
    inversely related to the molar masses. So the NO molecules would be moving fastest, and
    the NO_2 molecules slowest, with O_2 molecules intermediate in velocity.

    (b) Partial pressure of O_2 when mixed in the appropriate ratio with NO: The equation shows
    that 2 molecules of NO are needed per molecule of O_2. With the partial pressure of NO =
    150 mm Hg, the partial pressure of O_2 would be 150/2 or 75 mm Hg.

    (c) The partial pressure of NO_2 after reaction should be equal to that of NO before
    reaction(150 mm Hg), since the partial pressures are proportional to the number of moles, 2
    moles of NO (and 1 mole of O_2) would form 2 moles of NO_2.

85. Calculate the pressure of O_2 in the tank:

    Mass of $O_2 = 0.0870$ g or (0.0870 g $O_2 \bullet$ 1 mol O_2/32.00 g O_2) = 0.00272 mol O_2

$$P = \frac{n \bullet R \bullet T}{V} = \frac{(2.72 \times 10^{-3} \text{ mol})\left(0.082057\frac{\text{L} \bullet \text{atm}}{\text{K} \bullet \text{mol}}\right)(297 \text{ K})}{0.550 \text{ L}} = 0.121 \text{ atm (to 3 sf)}$$

With the total pressure (of O_2, CO, and CO_2) = 1.56 atm, the pressure of CO and CO_2 will be

1.56 atm = P (O_2) + P(CO) + P(CO_2) = 0.121 atm + P(CO) + P(CO_2).

Then 1.44 atm = P(CO) + P(CO_2).

We know that P(CO) + P (O_2) = 1.34 atm. So P(CO) + 0.121 = 1.34 atm and

P(CO) = 1.22 atm.

Since 1.44 atm = P(CO) + P(CO_2), and P(CO) = 1.22 atm, then P(CO_2) = 0.22 atm.

The masses of CO and CO_2 can be found by substitution into the Ideal Gas Law:

$$n = \frac{(1.22 \text{ atm})(0.550 \text{ L})}{(0.082057 \frac{\text{L} \bullet \text{atm}}{\text{K} \bullet \text{mol}})(297 \text{ K})} = 2.75 \times 10^{-2} \text{ mol CO}$$

$$n = \frac{(0.22 \text{ atm})(0.550 \text{ L})}{(0.082057 \frac{\text{L} \bullet \text{atm}}{\text{K} \bullet \text{mol}})(297 \text{ K})} = 4.96 \times 10^{-3} \text{ mol CO}_2$$

The masses would then be (4.96 x 10^{-3} mol CO_2 • 44 g/mol) or 0.22 g of CO_2 and

(2.75 x 10^{-2} mol CO • 28 g/mol) = 0.77 g CO

87. The CO_2 evolved is:

$$n = \frac{(\frac{44.9}{760} \text{ atm})(1.50 \text{ L})}{(0.082057 \frac{\text{L} \bullet \text{atm}}{\text{K} \bullet \text{mol}})(298 \text{ K})} = 3.62 \times 10^{-3} \text{ mol CO}_2$$

Note that the CO in the compound is oxidized to CO_2, so we know the original number of

moles of CO in the compound (3.62 x 10^{-3} mol).

The mass of this amount of CO would be: (3.62 x 10^{-3} mol • 28.0 g CO/mol) = 0.101 g CO.

From 0.142 g sample of the compound, the mass of Fe is (0.142- 0.101) or 4.0 x 10^{-2}

grams Fe, and corresponding to (4.0 x 10^{-2} g Fe • 1 mol Fe/55.85 g Fe) = 7.26 x 10^{-4} mol Fe.

The ratio of Fe: CO would be: 7.26 x 10^{-4} mol Fe: 3.62 x 10^{-3} mol CO or 1:5.

The empirical formula would be Fe(CO)$_5$.

89. Since the total pressure is 120 mm Hg, and we know (according to the balanced equation) that

the SiH_4 and O_2 are present in a 1mol:2mol ratio. This tells us that the pressure of O_2 will be

twice that of SiH_4.

P(total) = P(SiH_4) + P(O_2) = P(SiH_4) + 2P(SiH_4) = 3 P(SiH_4) = 120 mm Hg.

The pressure of silane is then 40 mm Hg and that of oxygen is then 80 mm Hg.

After the reaction, 2 H_2O are formed for each 1 SiH_4 and 2 O_2.

The pressure would then be equal to the pressure of the oxygen (before the reaction) or 2 x the pressure of the silane (before the reaction) or 80 mm Hg.

91. (a) Pressure of B_2H_6 formed:

$$\frac{0.136 \text{ g NaBH}_4}{1} \bullet \frac{1 \text{ mol NaBH}_4}{37.83 \text{ g NaBH}_4} = 3.60 \times 10^{-3} \text{ mol NaBH}_4.$$

The equation shows that we get 1 mol of B_2H_6 for each 2 mol of $NaBH_4$.

We can now calculate the pressure of the B_2H_6.

$$P = \frac{\left(1.80 \times 10^{-3} \text{ mol}\right)\left(0.082057 \dfrac{\text{L} \bullet \text{atm}}{\text{K} \bullet \text{mol}}\right)(298 \text{ K})}{2.75 \text{ L}} = 0.0160 \text{ atm (to 3 sf)}$$

(b) The total pressure is that of B_2H_6 and H_2: The balanced equation shows 2 mol of H_2 gas for each 1 mol of B_2H_6. If the pressure of $B_2H_6 = 0.0160$ atm, the pressure of H_2 will be 2 x that amount or 0.0320 atm and the total pressure = 0.0160 + 0.0320 = 0.0480 atm.

93. The amount of HCl used is: (1.50 mol HCl/L • 0.0120 L) = 0.0180 mol HCl.

Let's use x = mass of Na_2CO_3, and (1.249 – x) = mass of $NaHCO_3$.

We know that each mol of Na_2CO_3 consumes 2 mol HCl and each mol of $NaHCO_3$ consumes 1 mol of HCl---and that the total number of mol of HCL consumed is 0.0180 mol.

Mol HCl (consumed by Na_2CO_3) + mol HCl (consumed by $NaHCO_3$) = 0.0180 mol [1]

Express the mol HCl consumed by each substance:

$$\frac{x}{1} \bullet \frac{1 \text{ mol Na}_2\text{CO}_3}{106 \text{ g Na}_2\text{CO}_3} \bullet \frac{2 \text{ mol HCl}}{1 \text{ mol Na}_2\text{CO}_3} = \text{mol HCl consumed by Na}_2\text{CO}_3 \text{ and}$$

$$\frac{(1.249 - x)}{1} \bullet \frac{1 \text{ mol NaHCO}_3}{84.0 \text{ g NaHCO}_3} \bullet \frac{1 \text{ mol HCl}}{1 \text{ mol NaHCO}_3} = \text{mol HCl consumed by NaHCO}_3$$

Substituting into equation [1]:

$$\frac{x}{1} \bullet \frac{1 \text{ mol Na}_2\text{CO}_3}{106 \text{ g Na}_2\text{CO}_3} \bullet \frac{2 \text{ mol HCl}}{1 \text{ mol Na}_2\text{CO}_3} + \frac{(1.249 - x)}{1} \bullet \frac{1 \text{ mol NaHCO}_3}{84.0 \text{ g NaHCO}_3} \bullet \frac{1 \text{ mol HCl}}{1 \text{ mol NaHCO}_3} = 0.0180 \text{ mol}$$

Simplifying yields:

$$\frac{2x}{106} + \left[\frac{1.249}{84} - \frac{x}{84}\right] = 0.0180 \text{ and}$$

0.01887 x + 0.01487 – 0.01190x = 0.0180 and

0.006965 x = 0.00313 and x = 0.449 g (Na_2CO_3), indicating that mass of $NaHCO_3$

= (1.249 – 0.449) = 0.800 g $NaHCO_3$.

Converting to moles of each substance:

$$\frac{0.449 \text{ g Na}_2\text{CO}_3}{1} \bullet \frac{1 \text{ mol Na}_2\text{CO}_3}{106 \text{ g Na}_2\text{CO}_3} = 0.00424 \text{ mol Na}_2\text{CO}_3.$$

$$\frac{0.800 \text{ g NaHCO}_3}{1} \cdot \frac{1 \text{ mol NaHCO}_3}{84.0 \text{ g NaHCO}_3} = 0.00951 \text{ mol NaHCO}_3$$

The two equations tell us that 1 mol of either substance produces 1 mol of CO_2.

The amount of CO_2 produced is then: 0.00424 mol $Na_2CO_3 + 0.00951$ mol $NaHCO_3$ or

0.0138 mol CO_2. This gas would occupy a volume of:

$$V = \frac{(0.0138 \text{ mol})(0.082057 \frac{L \cdot atm}{K \cdot mol})(298 \text{ K})}{\left(\frac{745}{760} \text{ atm}\right)} = 0.343L \text{ or } 343 \text{ mL}$$

95. Mass of water per Liter of air under the conditions:

(a) At 20 °C and 45% relative humidity

Recalling the definition of Relative Humidity (RH):

$$RH = \frac{P(H_2O) \text{ in air}}{\text{Vapor Pressure(VP) of water}}$$

We can calculate the $P(H_2O)$ in air by multiplying the VP of water by the RH.

At 20 °C, the VP of water is 17.5 torr (Appendix at back of textbook)

$P(H_2O) = RH \cdot VP = 0.45 \cdot 17.5$ torr

Recalling that water vapor is a gas, we can use the Ideal Gas Law to calculate the # of moles—and the mass of water present in 1 L of air. Substituting the $P(H_2O)$ that we

calculated above into "P". Rearranging the Ideal Gas Law: $\frac{P \cdot V}{R \cdot T} = n$, and recalling that

$n = \text{mass(g)}/MW$ we can further rearrange to produce: $\frac{P \cdot V \cdot MW}{R \cdot T} = \text{mass}$.

Given that VP is given in torr, use the value of $R = 62.4 \frac{L \cdot torr}{K \cdot mol}$.

[Alternatively convert 17.5 torr into atm, by dividing by 760, and using the value

of $R = 0.082057 \frac{L \cdot atm}{K \cdot mol}$]

$$\frac{(0.45 \cdot 17.5 \text{ torr}) \cdot 1 \text{ L} \cdot 18.02 \text{ g/mol}}{62.4 \frac{L \cdot torr}{K \cdot mol} \cdot 293 \text{ K}} = \text{mass} = 7.8 \times 10^{-3} \text{ g (2 sf)}$$

(b) At 0 °C and 95% relative humidity: (VP of water at 0°C = 4.6 torr)

$$\frac{(0.95 \cdot 4.6 \text{ torr}) \cdot 1 \text{ L} \cdot 18.02 \text{ g/mol}}{62.4 \frac{L \cdot torr}{K \cdot mol} \cdot 273 \text{ K}} = \text{mass} = 4.6 \times 10^{-3} \text{ g (2 sf)}$$

97. In a 1.0-L flask containing 10.0 g each of O_2 and CO_2 at 25 °C.

(a) The gas with the greater partial pressure:

Partial pressure is a relative measure of the number of moles of gas present. The molar mass of oxygen is approximately 32 g/mol while that of CO_2 is approximately 44 g/mol.

There will be a greater number of moles of oxygen in the flask—hence the **partial pressure of O_2 will be greater**.

(b) The gas with the greater average speed:

The kinetic energy of each gas is given as: $KE = 1/2\ mu^2$, where u is the average speed of the molecules. Since the average KE of both gases are the same (They are at the same T), The lighter of the gases **(O_2)will have a greater average speed**.

(c) The gas with the greater average kinetic energy:

The KE depends upon temperature, and since both gases are at the same T, the **average KE of the two gases is the same**.

99. Two containers with 1.0 kg of CO and 1.0 kg C_2H_2

(a) Cylinder with the greater pressure:

Since P is proportional to the amount of substance present, let's calculate the # of moles of each gas.

Molar Mass of CO = 28 g/mol   Molar Mass of C_2H_2= 26 g/mol

While we could calculate the # of moles, note that since acetylene has a small molar mass, 1.0 kg of acetylene would have more moles of gas (and *a greater pressure*) than 1.0 kg of CO.

(b) Cylinder with the greater number of molecules:

Since we know that the cylinder with acetylene has more moles of gas, the cylinder *with the acetylene* will have the greater number of molecules (since to convert between moles of gas and molecules of gas, we multiply by the constant, Avogadro's number.

101. Which of the following samples is a gas?

(a) Material expands 10% when a sample of gas originally at 100 atm is suddenly allowed to exist at one atmosphere pressure: This sample is **not a gas**, since a gas would expand in volume by 100-fold (to exist at 1 atm).

(b) A 1.0-ml sample of material weighs 8.2 g: This sample is **not a gas** since the density of the sample (at 8.2 g/mL) is too great.

(c) Material is transparent and pale green in color: **Insufficient information** to tell. Liquids could also be pale green and transparent.

(d) One cubic meter of material contains as many molecules as an equal volume of air at the same temperature and pressure: This material **is a gas**, since one cubic meter of a liquid or solid would contain a greater number of molecules than a cubic meter of air.

103. (a) Theoretical yield of sodium azide:

$$65.0 \text{ g Na} \bullet 1 \text{ mol Na}/23.00 \text{ g Na} = 2.83 \text{ mol Na}$$

$$n_{ClF_3} = \frac{PV}{RT} = \frac{2.12 \text{ atm} \bullet 35.0 \text{ L}}{0.082057 \dfrac{L \bullet atm}{K \bullet mol} \bullet 296 \ K} = 3.05 \text{ mol N}_2\text{O}$$

The ratio of required substances is: $\dfrac{4 \text{ mol Na}}{3 \text{ mol N}_2\text{O}} = 1.33$

The ratio that is available is: $\dfrac{2.83 \text{ mol Na}}{3.05 \text{ mol N}_2\text{O}} = 0.928$, so Na is the Limiting Reagent.

$$2.83 \text{ mol Na} \bullet \frac{1 \text{ mol NaN}_3}{4 \text{ mol Na}} \bullet \frac{65.01 \text{g NaN}_3}{1 \text{ mol NaN}_3} = 46.0 \text{ g NaN}_3$$

(b) Lewis structure for the azide ion:

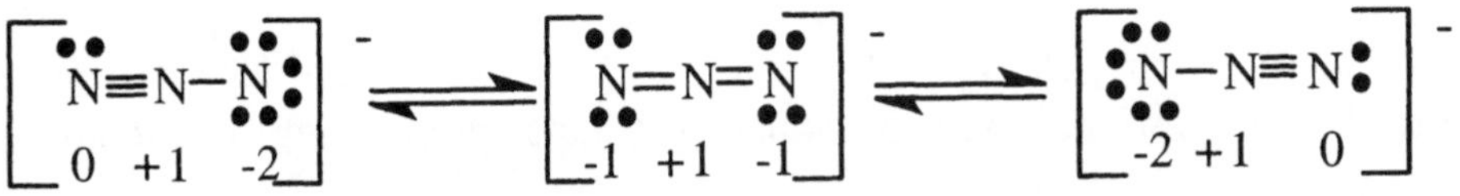

The central structure (having the smallest set of formal charges on each atom) is the most likely.

(c) The central N has two groups of electron pairs around it, making the *linear* geometry the preferred geometry for the azide ion.

105. The change of average speed of a gaseous molecules when T doubles:

Since the speed of a gaseous molecule is related to the square root of the *Absolute* T (Equation 12.9 on page 569 of your text.), a change by a factor of 2 in the Absolute T will lead to a change of $\sqrt{2}$ in the speed.

Chapter 13
Intermolecular Forces, Liquids, and Solids

Practicing Skills

Intermolecular Forces

1. The <u>intermolecular</u> forces one must overcome to:

change	intermolecular force
(a) melt ice	hydrogen bonds and dipole-dipole (molecule with OH bonds)
(b) sublime solid I_2	induced dipole-induced dipole (nonpolar molecule)
(c) convert $NH_3(\ell)$ to $NH_3(g)$	hydrogen bonds and dipole-dipole (molecule with NH bonds)

3. To convert <u>species</u> from a liquid to a gas the <u>intermolecular</u> forces one must overcome:

species	intermolecular force
(a) liquid O_2	induced dipole-induced dipole (nonpolar molecule)
(b) mercury	induced dipole-induced dipole (nonpolar atom)
(c) methyl iodide	dipole-dipole (polar molecule)
(d) ethanol	hydrogen bonding and dipole-dipole (polar molecule with OH bonds)

5. Increasing strength of intermolecular forces:

$$Ne \ < \ CH_4 \ < \ CO \ < \ CCl_4$$

Neon and methane are nonpolar species and possess only induced dipole-induced dipole interactions. Neon has a smaller molar mass than CH_4, and therefore weaker London (dispersion) forces. Carbon monoxide is a polar molecule. Molecules of CO would be attracted to each other by dipole-dipole interactions, but the CO molecule is not a very strong dipole. The CCl_4 molecule is a non-polar molecule, but very heavy (when compared to the other three). Hence the greater London forces that accompany larger molecules would result in the strongest attractions of this set of molecules.

The lower molecular weight molecules with weaker interparticle forces should be gases at 25 °C and 1 atmosphere: Ne, CH_4, CO.

7. Compounds which are capable of forming hydrogen bonds with water are those containing polar O-H bonds and lone pairs of electrons on N,O, or F.

(a) CH_3-O-CH_3 no; no "polar H's" and the C-O bond is not very polar

(b) CH_4 no

(c) HF   yes: lone pairs of electrons on F and a "polar hydrogen".

(d) CH_3COOH yes: lone pairs of electrons on O atoms, and a "polar hydrogen"

attached to one of the oxygen atoms

(e) Br_2 no; nonpolar molecules

(f) CH_3OH yes: "polar H" and lone pairs of electrons on O

9. For each pair, which has the greater heat of hydration (more negative)?

(a) LiCl –since Li^{1+} is a smaller cation than Cs^{1+}.

(b) $Mg(NO_3)_2$; two effects here. Mg^{2+} is a smaller cation than Na^{1+}, and secondly

magnesium has a 2+ charge while Na only a 1+ charge.

(c) $NiCl_2$ – same effects here as in (b) above. Ni^{2+} is a smaller cation than Rb^{1+}, and is

doubly charged.

Liquids

11. Heat required: $125 \text{ mL} \cdot \dfrac{0.7849 g}{1 \text{ mL}} \cdot \dfrac{1 \text{ mol}}{46.07 \text{ g}} \cdot \dfrac{42.32 \text{ kJ}}{1 \text{mol}} = 90.1 \text{ kJ}$

13. Using Figure 13.18:

(a) The equilibrium vapor pressure of water at 60 °C is approximately 150 mm Hg.

Appendix G lists this value as 149.4 mm Hg

(b) Water has a vapor pressure of 600 mm Hg at 93 °C.

(c) At 70 °C the vapor pressure of water is approximately 225 mm Hg while that of ethanol is
approximately 520 mm Hg.

15. The vapor pressure of $(C_2H_5)_2O$ at 30°C is **590 mm Hg.**

Calculate the amount of $(C_2H_5)_2O$ to furnish this vapor pressure at 30 °C (303K).

$$n = \frac{PV}{RT} = \frac{590 \text{ mm} \cdot \dfrac{1 \text{ atm}}{760 \text{ mm}} \cdot 0.100 \text{ L}}{0.082057 \dfrac{L \cdot atm}{K \cdot mol} \cdot 303 \text{ K}} = 3.1 \times 10^{-3} \text{ mol}$$

The total mass of $(C_2H_5)_2O$ [FW = 74.1 g] needed to create this pressure is about 0.23 g.

Since there is adequate ether to provide this pressure, we anticipate the pressure in the flask to

be approximately 590 mm Hg.

As the flask is cooled from 30.°C to 0 °C, **some of the gaseous ether will condense** to form

liquid ether.

17. Member of each pair with the higher boiling point:

    (a) O_2 would have a higher boiling point than N_2 owing to its greater molar mass.

    (b) SO_2 would boil higher since SO_2 is a polar molecule while CO_2 is non-polar.

    (c) HF would boil higher since strong hydrogen bonds exist in HF but not in HI

    (d) GeH_4 would boil higher. While both molecules are non-polar, germane has the greater molar mass and therefore stronger London forces.

19. (a) From the figure, we can read the vapor pressure of CS_2 as approximately 620 mmHg and for nitromethane as approximately 80 mm Hg.

    (b) The principle intermolecular forces for CS_2 (a non-polar molecule) are **induced dipole-induced dipole**; for nitromethane (polar molecule)--**dipole-dipole**.

    (c) The normal boiling point from the figure for CS_2 is 46 °C and for CH_3NO_2, 100 °C.

    (d) The temperature at which the vapor pressure of CS_2 is 600 mm Hg is about 39 °C.

    (e) The vapor pressure of CH_3NO_2 is 60 mm Hg at approximately 34 °C.

21. Regarding benzene:

    (a) Using the data provided, note that the vapor pressure of benzene is 760 mm Hg at 80.1°C (the definition of the normal boiling point).

    (b) The graph of the data:

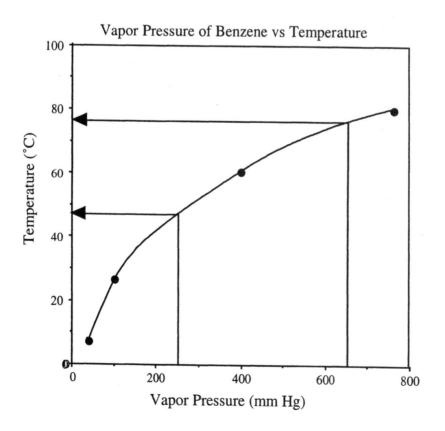

The temperature at which the liquid has a vapor pressure of 250 mm Hg is about 47°C.

The temperature at which the vapor pressure is 650 mm Hg is about 77 °C.

(c) The molar enthalpy of vaporization from the Clausius-Clapeyron equation:

The Clausius-Clapeyron equation is: $\ln\left(\dfrac{P_2}{P_1}\right) = \dfrac{\Delta H}{R}\left(\dfrac{1}{T_1} - \dfrac{1}{T_2}\right)$

From the graph using (arbitrarily) two of the data points:

$T_1 = 47$ °C and $P_1 = 250.$ mm Hg and $T_2 = 77$ °C and $P_2 = 650.$ mm Hg. Recall however

that the units of R include temperature units of K, So convert the two temperatures to the K

scale and substitute. $\ln\left(\dfrac{650.}{250.}\right) = \dfrac{\Delta H}{8.3145\ \dfrac{J}{K\bullet mol}}\left(\dfrac{1}{320\ K} - \dfrac{1}{350\ K}\right)$

Simplifying the equation:

$8.3145\ \dfrac{J}{K\bullet mol} \bullet \ln(2.60) = \Delta H\left(2.180\ \text{x}\ 10^{-4}\right)$ and $\Delta H = 29660$ J/mol or 30.0 kJ/mol

Using the graph, or two other data points than those used here, one may obtain a slightly

different value.

Metallic and Ionic Solids

23. This compound would have the formula AB_8 since each black square (A) has eight

corresponding white squares (B).

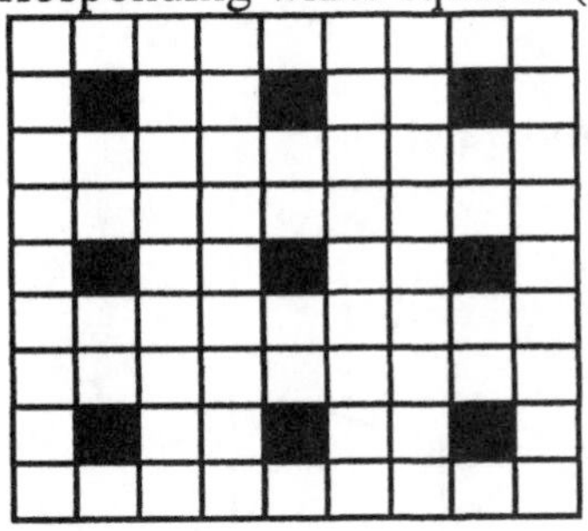

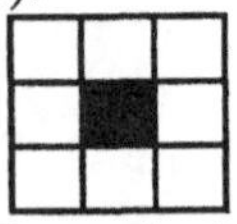

The unit shown above is the smallest
"cookie cutter" that would reproduce the
larger pattern shown at the left. The
formula for such a compound would be
AB_8

25. To determine the perovskite formula, determine the number of each atom belonging **uniquely**

to the unit cell shown. The Ca atom is wholly contained within the unit cell. There are Ti atoms

at each of the eight corners. Since each of these atoms belong to eight unit cells, the portion of

each Ti atom belonging to the pictured unit cell is 1/8 so 8 Ti atoms x 1/8 = 1 Ti atom. The O

atoms on an edge belong to 4 unit cells, so the fraction contained within the pictured cell is 1/4.

There are twelve such O atoms, leading to 12 x 1/4 = 3 O atoms— and a formula of $CaTiO_3$.

27. For cuprite:

   (a) Formula for cuprite: There are 8 oxygen atoms at the corner of the cell (each 1/8 within the cell) and 1 oxygen atom internal to the cell (wholly within the cell).

   The number of oxygen atoms is then [(8 • 1/8) +(1)] 2. There are 4 Cu atoms wholly within the cell, so the ratio was Cu_2O.

   (b) With a formula of Cu_2O, and the oxidation state of O = -2, the oxidation state of copper must be +1.

Other Types of Solids

29. For the unit cell of diamond:

   (a) The unit cell has 8 corner atoms (1/8 in the cell), 6 face atoms (1/2 in the cell), and 4 atoms wholly within the cell, for a total of **8 carbon atoms**.

   (b) Diamond uses a fcc unit cell (The structure shown also has 4 atoms occupying holes in the lattice. The holes are **tetrahedral**.

Physical Properties of Solids

31. The heat evolved when 15.5 g of benzene freezes at 5.5 °C :

$$15.5 \text{ g benzene} \cdot \frac{1 \text{ mol benzene}}{78.1 \text{ g benzene}} \cdot \frac{9.95 \text{ kJ}}{1 \text{ mol benzene}} = -1.97 \text{ kJ}$$

Note once again the negative sign indicates that heat is evolved.

The quantity of heat needed to remelt this 15.5 g sample of benzene would be +1.97 kJ.

Phase Diagrams and Phase Changes

33 (a) The positive slope of the solid/liquid equilibrium line means the liquid CO_2 is **less dense** than solid CO_2.

   (b) At 5 atm and 0 °C, CO_2 is in the **gaseous phase**.

   (c) The phase diagram for CO_2 shows the critical pressure for CO_2 to be 73 atm, and the critical temperature to be +31 °C, so CO_2 cannot be liquefied at 45 °C.

35. The heat required is a summation of three "steps":

   1. heat the liquid(at -50.0 °C) to its boiling point (-33.3 °C)

   2. "boil" the liquid—converting it to a gas and

   3. warm the gas from -33.3 °C to 0.0 °C

   1. To heat the liquid(at -50.0 °C) to its boiling point (-33.3 °C):

$$q_{liquid} = 1.2 \text{ x } 10^4 \text{ g} \cdot 4.7 \frac{J}{g \cdot K} \cdot (239.9 \text{ K} - 223.2 \text{ K}) = 9.4 \text{ x } 10^5 \text{ J}$$

2. To boil the liquid:

$$23.3 \times 10^3 \frac{J}{mol} \cdot \frac{1 \text{ mol NH}_3}{17.03 \text{ g NH}_3} \cdot 1.2 \times 10^4 \text{ g} = 1.6 \times 10^7 \text{ J}$$

3. To heat the gas from -33.3 °C to 0.0 °C:

$$q_{gas} = 1.2 \times 10^4 \text{ g} \cdot 2.2 \frac{J}{g \cdot K} \cdot (273.2 \text{ K} - 239.9 \text{ K}) = 8.8 \times 10^5 \text{ J}$$

The total heat required is:

$$9.4 \times 10^5 \text{ J} + 1.6 \times 10^7 \text{ J} + 8.8 \times 10^5 \text{ J} = 1.8 \times 10^7 \text{ J or } 1.8 \times 10^4 \text{ kJ}$$

37. Critical temperature of chloromethane is 416K and critical pressure is 66.1 atm. So at room temperature (well below the critcal T) chloromethane can be liquefied.

General Questions

39. Increasing strength of intermolecular forces: Ar $< CO_2 < CH_3OH$

Argon and CO_2 are nonpolar species and possess only induced dipole-induced dipole (London) interactions. Ar has a smaller mass than CO_2, so London forces are expected to be weaker for Ar than for CO_2. The polar molecular CH_3OH is capable of forming the stronger hydrogen-bonds(with an O-H bond in the molecule)

41.

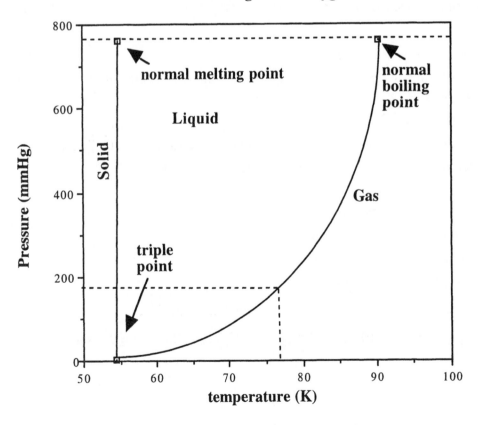

Phase Diagram of Oxygen

The estimated vapor pressure at 77 K (-196 °C) is between 150-200 mm Hg. The very slight positive slope of the solid/liquid equilibrium line indicates the **solid is more dense than the liquid.**

43. Volume of room = 3.0×10^2 cm • 2.5×10^2 cm • 2.5×10^2 cm = 1.9×10^7 cm^3

    Convert volume to L:

    $$1.9 \times 10^7 \text{ cm}^3 \bullet \frac{1 \text{ L}}{1.0 \times 10^3 \text{ cm}^3} = 1.9 \times 10^4 \text{ L}$$

    To produce a pressure of 59 mmHg, calculate the amount of ethanol required.

    $$P = 59 \text{ mm} \bullet \frac{1 \text{ atm}}{760 \text{ mm}} = 7.8 \times 10^{-2} \text{ atm}$$

    $$V = 1.9 \times 10^4 \text{ L} \qquad \text{and } T = 25 + 273 = 298 \text{ K}$$

    $$\frac{PV}{RT} = \frac{7.8 \times 10^{-2} \text{ atm} \bullet 1.9 \times 10^4 \text{ L}}{0.082057 \frac{\text{L} \bullet \text{atm}}{\text{K} \bullet \text{mol}} \bullet 298 \text{ K}} = 6.0 \times 10^1 \text{ mol ethanol}$$

    $$6.0 \times 10^1 \text{ mol ethanol} \bullet \frac{46.1 \text{ g ethanol}}{1 \text{ mol ethanol}} = 2.8 \times 10^3 \text{ g ethanol}$$

    This mass of ethanol would occupy a volume of:

    $$2.8 \times 10^3 \text{ g C}_2\text{H}_5\text{OH} \bullet \frac{1 \text{ cm}^3}{0.785 \text{ g}} = 3.5 \times 10^3 \text{ cm}^3$$

    As only 1 L of C_2H_5OH (1.0×10^3 cm^3) was introduced into the room, **all the ethanol would evaporate**.

45. Which salt has the *more exothermic* Enthalpy of Hydration? Between the two salts, the LITHIUM cation—being the smaller of the two—will have greater interaction with the water—and a resulting greater negative (exothermic) enthalpy of hydration.

47. Using the vapor pressure curves:

    (a) The vapor pressure of ethanol at 60 °C is: 350 mm Hg (to the limits of this readers ability to read the graph)

    (b) The stronger intermolecular forces in the liquid state are those of: Ethanol has a lower vapor pressure than carbon disulfide at every temperature—and hence stronger intermolecular forces. This is quite expected since ethanol has hydrogen bonding as the intermolecular force while CS_2 has only induced dipole forces.

    (c) The temperature at which heptane has a vapor pressure of 500 mm Hg is: 84 °C

(d) The approximate normal boiling points of the three substances are:

Carbon disulfide   bp = 46°C (literature value 46.5 °C)

Ethanol   bp = 78°C (literature value 78.5 °C)

Heptane   bp = 99°C (literature value 98.4 °C)

(e) At a pressure of 400 mm Hg and 70 °C, the state of the three substances is:

Carbon disulfide   state = gas

Ethanol   state = gas

Heptane   state = liquid

49. Silver crystallizes in the face-centered cubic cell, with a side of 409 pm. The radius of a silver atom can be found by examining the geometry of a face of the cell. The diagram shows such a face with an edge of 409 pm. A look at the diagram will reveal that the diagonal is the hypotenuse of a right triangle—two sides of which are 409 pm, and the hypotenuse^2 is then $(409pm)^2 + (409pm)^2$. Let d be the hypotenuse. Then we have: $d^2 = 2 \cdot (409pm)^2$.

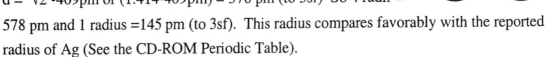

The interest in d lies in the fact that d equals 1 diameter (of the center atom) and two-halves of two other atoms (or 4 radii).

d = √2 •409pm or (1.414•409pm) = 578 pm (to 3sf) So 4 radii = 578 pm and 1 radius =145 pm (to 3sf). This radius compares favorably with the reported radius of Ag (See the CD-ROM Periodic Table).

51. The noncubic unit cell for calcium carbide shows 8 Ca atoms at the corners, and 1 Ca atom wholly within the cell for a total of [(8*1/8) +1] 2 Ca atoms. There are 8 C atoms along edges (each 1/4 within a cell) and 2 C atoms wholly within the cell for a total of [(8 • 1/4) +2] 4 C atoms. The ratio of atoms indicates a formula for calcium carbide of CaC_2.

53. Increasing molar enthalpy of vaporization: CH_3OH, C_2H_6, HCl.

Owing to the H-bonding of CH_3OH, we predict that this substance will have the greatest molar enthalpy of vaporization. Ethane, on the other extreme, being a nonpolar compound, and of the 3 substances here the lowest molecular mass, will have the lowest. HCl—with a intermediate molecular mass and being polar—will have the intermediate molar enthalpy of vaporization.

$$C_2H_6 < HCl < CH_3OH$$

55. D for Iridium = 22.56 g/cm^3 The radius of an iridium atom is:

Ir crystallizes in the fcc lattice. There are 4 atoms in the unit cell (8 •1/8 corner atoms + 6•1/2 face atoms).

The mass/unit cell is: $\dfrac{4 \text{ Ir atoms}}{1 \text{ unit cell}} \bullet \dfrac{192.22 \text{ g Ir}}{6.0221 \times 10^{23} \text{atoms}} = 1.277 \times 10^{-21}$ g/unit cell

The volume of the unit cell is:

$$\frac{1.277 \times 10^{-21} \text{g}}{1 \text{ unit cell}} \bullet \frac{1 \text{ cm}^3}{22.56 \text{ g}} = \frac{5.659 \times 10^{-23} \text{cm}^3}{\text{unit cell}}$$

Since the cell is a cube, we can determine an edge by taking the cube root of this volume:

Edge = 3.839 x 10^{-8} cm

The diagonal of the unit cell (which corresponds to 4•radii) = 1.414 • 3.839 x 10^{-8} cm

So 4•radii = 1.414 • 3.839 x 10^{-8} cm and the radius =

$\dfrac{1.414 \bullet 3.839 \times 10^{-8} \text{ cm}}{4}$ = 1.356 x 10^{-8}cm = or 135.6 x 10^{-12} m or 135.6 pm.

This value compares favorably with the literature value of 136pm (See the CD-ROM Periodic Table).

57. Let's calculate the density for simple cubic, body-centered cubic, and face-centered cubic lattice. The number of vanadium atoms in the unit cell for each lattice is 1 (simple) 1 (body-centered) or 4 (face-centered).

First, let's calculate the volume of a body-centered cell. The body diagonal (d) of such a cell would be d^2 = 3ℓ^2 , and since d would contain 4 radii:

(4 x 132 pm/radius x $\dfrac{1\text{x } 10^2 \text{ cm}}{1 \times 10^{12} \text{ pm}}$)^2 = 3 ℓ^2

2.79 x 10^{-15} cm^2 = 3 ℓ^2

3.05 x 10^{-8} = ℓ The volume would be ℓ^3 or 2.8 x 10^{-23} cm^3

The mass of one Vanadium ion is:

$\dfrac{50.9415 \text{ g V}}{1 \text{ mol V}}$ • $\dfrac{1 \text{ mol V}}{6.022 \times 10^{23} \text{ atoms}}$ = 8.459 x 10^{-23} $\dfrac{\text{g V}}{\text{atom}}$

So the density of a body-centered cell would be $\dfrac{(2 \text{ atom})(8.459 \times 10^{-23} \frac{\text{g V}}{\text{atom}})}{2.8 \times 10^{-23} \text{ cm}^3}$

= 5.97 g/cm^3

For a face-centered lattice the edge (ℓ) wouldbe related to the face diagonal (f)

$$f^2 = \ell^2 + \ell^2$$

Since f contains 4 radii as calculated earlier

$$f^2 = 2.79 \times 10^{-15}\ cm^2 = 2\ell^2$$

$$= 3.73 \times 10^{-8}\ cm = \ell$$

And a corresponding volume of ℓ^3 or $5.20 \times 10^{-23}\ cm^3$

and a density of $\dfrac{(4\ atoms)(8.459 \times 10^{-23}\ \frac{g\ V}{atom})}{5.20 \times 10^{-23}\ cm^3} = 6.50\ g/cm^3$

For a simple cubic lattice, the vanadium atoms at the corners of the unit cell are in contact. The length of the side would then be equal to two radii. The volume of the cube would be:

$$[(2\ radii)(132\tfrac{pm}{radii})(\tfrac{1 \times 10^2\ cm}{1 \times 10^{12}\ pm})]^3 = 1.84 \times 10^{-23}\ cm^3$$

With 1 atom of V per unit cell, the density would be

$$\dfrac{(8\ atoms)(\frac{1\ atom}{8\ unit\ cells})(8.459 \times 10^{-23}\frac{g\ V}{atom})}{1.84 \times 10^{-23}\ cm^3} = 4.60\ g/cm^3$$

So the V atom appears to use the **body-centered cubic lattice**.

59. Given the edge of a CaF_2 unit cell is 5.46295×10^{-8} cm, and the density of the solid is $3.1805\ g/cm^3$, calculate the value of Avogadro's number.

The unit cell contains 4 calcium ions and 8 fluoride ions (or 4 CaF_2 ion pairs). With the density and the length of the unit cell, we can calculate the mass of these 4 ion pairs.

$$\dfrac{\left(5.46295 \times 10^{-8}\ cm\right)^3}{1} \cdot \dfrac{3.1805\ g}{1\ cm^3} = 5.1853 \times 10^{-22}\ g.$$

The mass of one ion pair would be 1/4 of that mass or 1.2963×10^{-22} g. If we add the atomic masses of Ca and 2 F we get a molar mass of 78.077 g. Since this is the mass corresponding to Avogadro's number of formula units (also known as 1 mol), we can calculate:

$$\dfrac{78.077\ g\ CaF_2}{1\ mol\ CaF_2} \cdot \dfrac{1\ CaF_2\ formula\ unit}{1.2963 \times 10^{-22}\ g\ CaF_2} = 6.0230 \times 10^{23}\ formula\ units/\ 1mol\ CaF_2$$

61. For the two unit cells, determine the amount of filled space in each.

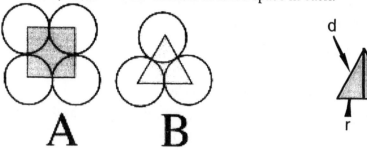

Assign a diameter of each of the atoms = d.

For unit cell A:

the length of the square inscribed is equal to 2 radii or 1 diameter (d). The area associated with the square is then d • d or d^2. Each circle has **one-fourth** of its area covered by the inscribed square. The area of a circle is Πr^2. So the r (the radius) = d/2. So the area of the circle is $\Pi(d/2)^2$ or $(\Pi d^2)/4$. Only one-fourth of each circle is covered by the portion of the inscribed square, so the area covered by the inscribed square is $1/4 \cdot (\Pi d^2)/4$. Noting that there are 4 circles, each of which has the area we've calculated, the **total** area covered by the inscribed square is $\Pi d^2/4$. The area **not covered** is the *difference* between the area of the square (d^2) and the area of the circles inside the square ($\Pi d^2/4$): $d^2 - \Pi d^2/4$. Arbitrarily assign a length to d (say 2 inches). Then the area not covered is $(2)^2 - 3.14(2)^2/4$. Or 4-3.14 = 0.86 sq. in. The amount of coverage is: 3.14/4.00 or 78.5 %.

For unit cell B:

Once again, assign a diameter of **d** to each circle. With the equilateral triangle shown, each interior angle (at the corners of the triangle) is 60 degrees. Since there are 360 degrees in the circle, this corresponds to 1/6 of each of the circles covered by the triangle. With three circles involved, $3 \cdot 1/6 = 1/2$ of the area of the 3 circles. Since the area of a circle (in terms of the diameter) is $(\Pi d^2)/4$, the total area of the 3 circles covered by the inscribed triangle is $1/2 \cdot (\Pi d^2)/4$ or $(\Pi d^2)/8$. The relationship for the area of the triangle is = 1/2 b•h, where b = base and h = height. The base of the triangle is equal to 2r, as shown by the diagram, above right.

The height may be calculated by using the Pythagorean Theorem, since we know the length of the hypotenuse (d) and the base (of the triangle, r)

$d^2 = r^2 + h^2$ and isolating the height, $d^2 - r^2 = h^2$ and noting that the radius is 1/2•d, $d^2 - (d/2)^2 = h^2$ or $3/4(d)^2 = h^2$. Substituting the arbitrary value of 2 for d, gives $3/4(2)^2 = h^2$ or $3 = h^2$ and 1.732 = h.

The area of the triangle is: $1/2 \cdot d \cdot h$, and A= $1/2 \cdot 2 \cdot 1.732 = 1.732$ sq. inches

The area of the circles (covered by the triangle) is $(\Pi d^2)/8$, and substituting the arbitrary value 2 for d gives: $(\Pi \bullet 4)/8$ or $3.14/2 = 1.57$ sq in.

The percent occupied is then $1.57/1.732 \bullet 100 = 90.7\ \%$.

63. Regarding dichlorodimethylsilane, and given the data:

   (a) The normal boiling point is easily determined, since it is *defined* as the temperature at which the vapor pressure of a substance is equal to 760 mm Hg. The normal boiling point is 70.3 °C.

   (b) Plotting the data:

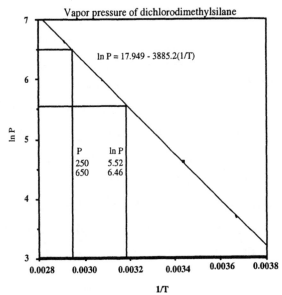

Note that on the graph, we calculate the value for lnP at P values of 250 and 650 mm Hg. Using those **ln P** values, we drop a perpendicular to the bottom axis. Using the equation for the line will give slightly different answers from those shown here.

Converting those 1/T values, we get:

For P = 250 mm Hg, ln P = 5.52 and T = approximately 310K (39.4 °C).

For P = 650 mm Hg, ln P = 6.46 and T = approximately 339K (65.8 °C).

   (c) The Molar Enthalpy of vaporization for dichlorodimethylsilane from the Clausius-Clapeyron equation:

$$\ln\!\left(\frac{P_2}{P_1}\right) = \frac{\Delta H_{vap}}{R}\left[\frac{1}{T_1} - \frac{1}{T_2}\right]$$

Substituting a pair of T,P from the data table

For T = 17.5 °C (290.7 K), P = 100. mm Hg, for T = 51.9 °C (325.1), P = 400. mm Hg:

$$\ln\!\left(\frac{400.\ \text{mm Hg}}{100.\ \text{mm Hg}}\right) = \frac{\Delta H_{vap}}{8.3145 \times 10^{-3}\,\text{kJ}/\text{K} \bullet \text{mol}}\left[\frac{1}{290.7\text{K}} - \frac{1}{325.1\text{K}}\right]$$

$$\ln 4.00 \bullet 8.3145 \times 10^{-3}\,\text{kJ}/\text{K} \bullet \text{mol} = \Delta H_{vap}\left[\frac{1}{290.7\text{K}} - \frac{1}{325.1\text{K}}\right]$$

$$\ln 4.00 \bullet 8.3145 \times 10^{-3} \text{kJ/K} \bullet \text{mol} = \Delta H_{vap} (3.6399 \times 10^{-4} \text{ K}^{-1})$$

$$\frac{1.1526 \times 10^{-2} \text{ kJ/K} \bullet \text{mol}}{3.6399 \times 10^{-4} \text{ K}^{-1}} = 31.7 \text{ kJ/mol}$$

Summary and Conceptual Questions

65. Acetone readily absorbs water owing to *hydrogen bonding* between the C = O oxygen atom and the O—H bonds of water.

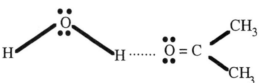

67. The **viscosity of ethylene glycol would be predicted to be greater** than that of ethanol since the glycol possesses two O-H groups per molecule while ethanol possesses one. Two OH groups/molecule would provide more hydrogen bonding!

69. Explain the fact that

(a) Ethanol has a lower boiling point than water. Both molecules are polar, and both can form hydrogen bonds. Water has **two** polar H atoms, and two lone pairs of electrons on the O atom. Ethanol has only **one** polar H atom, with the accompanying two lone pairs on the O atom. Given the increased ability of water to form hydrogen bonds with other water molecules, one would expect that water would have a higher boiling point.

(b) Mixing 50 mL of water with 50 mL of ethanol results in less than 100 mL of solution. The H-bonding that occurs not only between water molecules and other water molecules and ethanol molecules and other molecules also occurs between ethanol and water molecules. The attraction results in the molecules occupying less space than one would anticipate—and a non-additive volume.

71. CaCl$_2$ cannot have the NaCl structure. As shown in Figure 13.27 of your text (page 619), the cubic structure possesses 4 net lattice ions (occupied by anions) per face-centered lattice and 4 octahedral holes (occupied by cations). This is suitable for salts of a 1:1 composition.

73. Evidence that water molecules in the liquid state exert attractive forces on one another:

1) Water has a relatively large specific heat capacity (4.18 J/g •K). This large value is a reflection of the strong forces that hold water molecules together and require a large amount of energy to overcome.

2) Water is a liquid at room temperature—even though it has a molar mass of approximately 18 g/mol. With this molar mass, one would anticipate a boiling point well below 0°C (and a *gaseous* physical state).

75. Referring to Figure 13.13:
    (a) The hydrogen halide with the *largest total intermolecular force* is HI.
    (b) The dispersion forces are greater for HI than for HCl owing to the fact that HI (specifically I) has a larger volume than HCl (or Cl).
    (c) Dipole-Dipole forces are greater for HCl than for HI owing to the more polar H-Cl bond. The electronegativities (H= 2.2; Cl = 3.2; I = 2.7) indicate this difference in polarity (3.2-2.2) versus (2.7-2.2).
    (d) The HI molecule has the largest dispersion forces of the molecules shown in the Figure. This is quite reasonable given the large volume of the Iodine atom (see part b).

77. The can collapses as a result of the condensation of the gas in the can—which has filled the heated can—to a liquid. The distances between the particles of liquid are **much less** than the distances between the particles of gas. The resulting decrease in pressure inside the can causes the greater pressure outside the can to crush it.

79. Regarding dichlorodifluoromethane:
    (a) The normal boiling point for this substance is the temperature *at which the vapor pressure of the substance is equal to atmospheric pressure*. From the diagram we see that this is approximately –27 °C.
    (b) The pressure will be equal to the vapor pressure of the substance at 25 °C. From the diagram, this pressure is approximately 6.5 atmospheres.
    (c) The flow is rapid at first, since the liquid has evaporated to form the gas to the maximum extent possible at that temperature. As the gas leaves the cylinder, the lower energy molecules remain behind, and the gas absorbs heat—resulting in the formation of ice on the exterior of the cylinder.
    (d) Knocking the valve off the top is a **big safety risk**. Gas cylinders in laboratories are required to be anchored to a fixed surface so that they do not overturn and possibly knock the valve off in the process. The safest alternative given is (2). Cooling the container to –78 °C should cause most of the vapor to condense to a liquid. The valve can safely be opened.

81. Figure 13.41 shows the substance (a) on the left—an equilibrium between the liquid and the gaseous states, and (b) on the right--the supercritical fluid

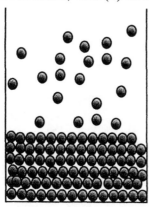 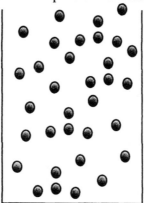

Using Molecular Models to Explore Intermolecular Forces &the Solid State

83. (a) ZnS uses a fcc arrangement, with the Zn^{2+} ions occupying the lattice sites, and S^{2-} ions occupying half the tetrahedral holes.

(b) With 4 Zn^{2+} ions and 4 S^{2-} ions in the unit cell, the formula ZnS follows logically.

85. (a) Calcium fluoride uses the fcc arrangement, with calcium ions occupying the unit cell sites, and fluoride ions filling tetrahedral holes in the unit cell.

(b) The formula for CaF_2 follows logically from the unit cell with four Ca^{2+} ions in the unit cell, and eight F^- ions filling the holes. The ratio 4 Ca^{2+}:8 F^- gives the formula 1Ca:2F or a formula of CaF_2.

(c) Both structures use the face-centered-cubic lattice.

87. Regarding aspartame:

(a) The molecular structure of aspartame:

$$HO_2C-CH_2-CH-C-NH-CH$$

with NH₂, O, CH₂, C=O, OCH₃ groups.

(b) Capable of hydrogen bonding? Sites at which hydrogen bonding can occur?
The molecule is capable of hydrogen bonding. The sites indicated by arrows are possible hydrogen bonding sites. Additionally the 3 carbonyl groups (C=O) will have lone pairs of electrons on the O—providing additional sites.

Chapter 14
Solutions and Their Behavior

Practicing Skills

Concentration

1. For 2.56 g of succinic acid in 500. mL of water:

The molality of the solution:

Molality = #mole solute/kg solvent:

With a density of water of 1.00 g/cm3, 500. mL= 0.500 kg

$$\text{Molality} = \frac{0.0217 \text{ mol}}{0.500 \text{ kg}} = 0.0434 \text{ molal}$$

The mole fraction of succinic acid in the solution:

For mole fraction we need *both* the #moles of solute *and* #moles of solvent.

$$\text{Moles of water} = 500. \text{ g H}_2\text{O} \bullet \frac{1 \text{ mol H}_2\text{O}}{18.02 \text{ g H}_2\text{O}} = 27.7 \text{ mol H}_2\text{O}$$

$$\text{The mf of acid} = \frac{0.0217 \text{ mol}}{(0.0217 \text{ mol} + 27.7 \text{ mol})} = 7.81 \times 10^{-4}$$

The weight percentage of succinic acid in the solution:

The fraction of *total* mass of solute +solvent which is solute:

$$\text{Weight percentage} = \frac{2.56 \text{ g succinic acid}}{502.56 \text{ g acid} + \text{ water}} \bullet 100 = 0.509\% \text{ succinic acid}$$

3. Complete the following transformations for

NaI:

Weight percent:

$$\frac{0.15 \text{ mol NaI}}{1 \text{ kg solvent}} \bullet \frac{149.9 \text{ g NaI}}{1 \text{ mol NaI}} = \frac{22.5 \text{ g NaI}}{1 \text{ kg solvent}}$$

$$\frac{22.5 \text{ g NaI}}{1000 \text{ g solvent} + 22.5 \text{ g NaI}} \bullet 100 = 2.2 \% \text{ NaI}$$

Mole fraction:

1000 g H$_2$O = 55.51 mol H$_2$O

$$X_{\text{NaI}} = \frac{0.15 \text{ mol NaI}}{55.51 \text{ mol H}_2\text{O} + 0.15 \text{ mol NaI}} = 2.7 \times 10^{-3}$$

C$_2$H$_5$OH:

Molality:

$$\frac{5.0 \text{ g C}_2\text{H}_5\text{OH}}{100 \text{ g solution}} \bullet \frac{1 \text{ mol C}_2\text{H}_5\text{OH}}{46.07 \text{ g C}_2\text{H}_5\text{OH}} \bullet \frac{100 \text{ g solution}}{95 \text{ g solvent}} \bullet \frac{1000 \text{ g solvent}}{1 \text{ kg solvent}} = 1.1 \text{ molal}$$

Mole fraction:

$$\frac{5.0 \text{ g C}_2\text{H}_5\text{OH}}{1} \cdot \frac{1 \text{ mol C}_2\text{H}_5\text{OH}}{46.07 \text{ g C}_2\text{H}_5\text{OH}} = 0.11 \text{ mol C}_2\text{H}_5\text{OH}$$

and for water :
$$\frac{95 \text{ g H}_2\text{O}}{1} \cdot \frac{1 \text{ mol H}_2\text{O}}{18.02 \text{ g H}_2\text{O}} = 5.27 \text{ mol H}_2\text{O}$$

$$X_{\text{C}_2\text{H}_5\text{OH}} = \frac{0.11 \text{ mol C}_2\text{H}_5\text{OH}}{5.27 \text{ mol H}_2\text{O} + 0.11 \text{ mol C}_2\text{H}_5\text{OH}} = 0.020$$

$C_{12}H_{22}O_{11}$:

Weight percent:

$$\frac{0.15 \text{ mol C}_{12}\text{H}_{22}\text{O}_{11}}{1 \text{ kg solvent}} \cdot \frac{342.3 \text{ g C}_{12}\text{H}_{22}\text{O}_{11}}{1 \text{ mol C}_{12}\text{H}_{22}\text{O}_{11}} = \frac{51.3 \text{ g C}_{12}\text{H}_{22}\text{O}_{11}}{1 \text{ kg solvent}}$$

$$\frac{51.3 \text{ g C}_{12}\text{H}_{22}\text{O}_{11}}{1000 \text{ g H}_2\text{O} + 51.3 \text{ g C}_{12}\text{H}_{22}\text{O}_{11}} \times 100 = 4.9 \text{ \% C}_{12}\text{H}_{22}\text{O}_{11}$$

Mole fraction:

$$X_{\text{C}_{12}\text{H}_{22}\text{O}_{11}} = \frac{0.15 \text{ mol C}_{12}\text{H}_{22}\text{O}_{11}}{55.51 \text{ mol H}_2\text{O} + 0.15 \text{ mol C}_{12}\text{H}_{22}\text{O}_{11}} = 2.7 \times 10^{-3}$$

5. To prepare a solution that is 0.200 m Na_2CO_3:

$$\frac{0.200 \text{ mol Na}_2\text{CO}_3}{1 \text{ kg H}_2\text{O}} \cdot \frac{0.125 \text{ kg H}_2\text{O}}{1} \cdot \frac{106.0 \text{ g Na}_2\text{CO}_3}{1 \text{ mol Na}_2\text{CO}_3} = 2.65 \text{ g Na}_2\text{CO}_3$$

$$\text{mol Na}_2\text{CO}_3 = \frac{0.200 \text{ mol Na}_2\text{CO}_3}{1 \text{ kg H}_2\text{O}} \cdot \frac{0.125 \text{ kg H}_2\text{O}}{1} = 0.025 \text{ mol}$$

The mole fraction of Na_2CO_3 in the resulting solution:

$$\frac{125. \text{ g H}_2\text{O}}{1} \cdot \frac{1 \text{ mol H}_2\text{O}}{18.02 \text{ g H}_2\text{O}} = 6.94 \text{ mol H}_2\text{O}$$

$$X_{\text{Na}_2\text{CO}_3} = \frac{0.025 \text{ mol Na}_2\text{CO}_3}{0.025 \text{ mol Na}_2\text{CO}_3 + 6.94 \text{ mol H}_2\text{O}} = 3.59 \times 10^{-3}$$

7. To calculate the number of mol of $C_3H_5(OH)_3$:

$$0.093 = \frac{x \text{ mol C}_3\text{H}_5(\text{OH})_3}{x \text{ mol C}_3\text{H}_5(\text{OH})_3 + (425 \text{ g H}_2\text{O} \cdot \frac{1 \text{ mol H}_2\text{O}}{18.02 \text{ g H}_2\text{O}})}$$

$$0.093 = \frac{x \text{ mol C}_3\text{H}_5(\text{OH})_3}{x \text{ mol C}_3\text{H}_5(\text{OH})_3 + 23.58 \text{ mol H}_2\text{O}}$$

$0.093(x + 23.58) = x$ and solving for x we get 2.4 mol $C_3H_5(OH)_3$

Grams of glycerol needed: 2.4 mol $C_3H_5(OH)_3 \cdot \dfrac{92.1 \text{ g}}{1 \text{ mol}} = 220$ g $C_3H_5(OH)_3$

The molality of the solution is (2.4 mol $C_3H_5(OH)_3$, 0.425 kg H_2O)= 5.7 m

9. Concentrated HCl is 12.0M and has a density of 1.18 g/cm^3.

   (a) The molality of the solution:

   Molality is defined as moles HCl/kg solvent, so begin by deciding the mass of 1 L, and the mass of water in that 1L. Since the density = 1.18g/mL, then 1 L (1000 mL) will have a mass of 1180g.

   The mass of HCl present in 12.0 mol HCl =

   $$12.0 \text{ mol HCl} \bullet \frac{36.46 \text{ g HCl}}{1 \text{ mol HCl}} = 437.52 \text{ g HCl}$$

   Since 1 L has a mass of 1180 g and 437.52 g is HCl, the difference (1180-437.52) is solvent. So 1 L has 742.98 g water.

   $$\frac{12.0 \text{ mol HCl}}{1 \text{ L}} \bullet \frac{1 \text{ L}}{742.98 \text{ g H}_2\text{O}} \bullet \frac{1000 \text{ g H}_2\text{O}}{1 \text{ kg H}_2\text{O}} = 16.2 \text{ m}$$

   (b) Weight percentage of HCl:

   12.0 mol HCl has a mass of 437.52 g, and the 1 L of solution has a mass of 1180 g.

   $$\%HCl = \frac{437.52 \text{ g HCl}}{1180 \text{ g solution}} \bullet 100 = 37.1 \%$$

11. The concentration of ppm expressed in grams is:

$$0.18 \text{ ppm} = \frac{0.18 \text{ g solute}}{1.0 \times 10^6 \text{ g solvent}} = \frac{0.18 \text{ g solute}}{1.0 \times 10^3 \text{ kg solvent}} \quad \text{or} \quad \frac{0.00018 \text{ g solute}}{1 \text{ kg water}}$$

$$\frac{0.00018 \text{ g Li}^+}{1 \text{ kg water}} \bullet \frac{1 \text{ mol Li}^+}{6.939 \text{ g Li}^+} = 2.6 \times 10^{-5} \text{ molal Li}^+$$

The Solution Process

13. Pairs of liquids that will be miscible:

   (a) $H_2O/CH_3CH_2CH_2CH_3$

   Will **not** be miscible. Water is a polar substance, while butane is nonpolar.

   (b) C_6H_6/CCl_4

   Will **be** miscible. Both liquids are nonpolar and are expected to be miscible.

   (c) H_2O/CH_3CO_2H

   Will **be** miscible. Both substances can hydrogen bond, and we know that they mix—since a 5% aqueous solution of acetic acid is sold as "vinegar"

15. The enthalpy of solution for LiCl:

The process can be represented as LiCl (s) $\rightarrow$ LiCl (aq)

The $\Delta H_{reaction}$ $= \Sigma \Delta H_f$ (product) - $\Sigma \Delta H_f$ (reactant)

$$= (-445.6 \text{ kJ/mol})(1 \text{mol}) - (-408.7 \text{ kJ/mol})(1 \text{mol}) = -36.9 \text{ kJ}$$

The similar calculation for NaCl is + 3.9 kJ. Note that the enthalpy of solution for NaCl is endothermic while that for LiCl is exothermic. Note the data (-408.7 kJ/mol) is from Table 14.1.

17. Raising the temperature of the solution will increase the solubility of NaCl in water. Hence to increase the amount of dissolved NaCl in solution one must **(c) raise the temperature of the solution and add some NaCl.**

19. More likely to have a more negative heat of hydration:

(a) **LiF** or RbF—See SQ14.15 for similar data. The aquation of the larger Rb ion requires more energy (ΔH_f is less negative) and the resulting $\Delta H_{hydration}$ is less negative than for the smaller lithium cation.

(b) KNO_3 or **Ca(NO$_3$)$_2$**: The greater charge of the Ca^{2+} compared to the K^+ ion will have a stronger attraction to water, and a *more negative* heat of hydration.

(c) CsBr or **CuBr$_2$**: The smaller dipositive Cu ion will cause $CuBr_2$ to have the more negative heat of hydration. Here two factors are in play: (1)the smaller size of the metal cation, and (2) the greater charge of the copper cation over that of the cesium cation.

Henry's Law

21. Solubility of $O_2 = k \cdot P_{O_2}$

$$= (1.66 \times 10^{-6} \frac{M}{\text{mm Hg}}) \cdot 40 \text{ mm Hg} = 6.6 \times 10^{-5} \text{ M } O_2$$

$$\text{and } 6.6 \times 10^{-5} \frac{\text{mol}}{L} \cdot \frac{32.0 \text{ g } O_2}{1 \text{ mol } O_2} = 2 \times 10^{-3} \frac{\text{g } O_2}{L}$$

23. Solubility $= k \cdot P_{CO_2}$

$$0.0506 \text{ M} = (4.48 \times 10^{-5} \frac{M}{\text{mm Hg}}) \cdot P_{CO_2}$$

$1130 \text{ mm Hg} = P_{CO_2}$ or expressed in units of atmospheres:

$$1130 \text{ mm Hg} \cdot \frac{1 \text{ atm}}{760 \text{ mm Hg}} = 1.49 \text{ atm}$$

Raoult's Law

25. Since $P_{water} = X_{water} P°_{water}$, to determine the vapor pressure of the solution (P_{water}), we need the mf of water.

$$35.0 \text{ g glycol} \bullet \frac{1 \text{ mol glycol}}{62.07 \text{ g glycol}} = 0.564 \text{ mol glycol and}$$

$$500.0 \text{ g H}_2\text{O} \bullet \frac{1 \text{ mol H}_2\text{O}}{18.02 \text{ g H}_2\text{O}} = 27.75 \text{ mol H}_2\text{O}. \text{ The mf of water is then:}$$

$$\frac{27.75 \text{ mol H}_2\text{O}}{(27.75 \text{ mol} + 0.564 \text{ mol})} = 0.9801 \text{ and}$$

$P_{water} = X_{water} P°_{water} = 0.9801 \bullet 35.7 \text{ mm Hg} = 35.0 \text{ mm Hg}$

27. Using Raoult's Law, we know that the vapor pressure of pure water ($P°$) multiplied by the mole fraction(X) of the solute gives the vapor pressure of the solvent above the solution (P).

$$P_{water} = X_{water} P°_{water}$$

The vapor pressure of pure water at 90 °C is 525.8 mmHg (from Appendix G).

Since the P_{water} is given as 457 mmHg, the mole fraction of the water is:

$$\frac{457 \text{ mmHg}}{525.8 \text{ mmHg}} = 0.869$$

The 2.00 kg of water correspond to a mf of 0.869. This mass of water corresponds to:

$$2.00 \times 10^3 \text{ g H}_2\text{O} \bullet \frac{1 \text{mol H}_2\text{O}}{18.02 \text{g H}_2\text{O}} = 111 \text{ mol water.}$$

Representing moles of ethylene glycol as x we can write:

$$\frac{\text{mol H}_2\text{O}}{\text{mol H}_2\text{O} + \text{mol C}_2\text{H}_4(\text{OH})_2} = \frac{111}{111 + x} = 0.869$$

$$\frac{111}{0.869} = 111 + x; \quad 16.7 = x \text{ (mol of ethylene glycol)}$$

$$16.7 \text{ mol C}_2\text{H}_4(\text{OH})_2 \bullet \frac{62.07 \text{ g C}_2\text{H}_4(\text{OH})_2}{1 \text{ mol C}_2\text{H}_4(\text{OH})_2} = 1.04 \times 10^3 \text{ g C}_2\text{H}_4(\text{OH})_2$$

Boiling Point Elevation

29. Benzene normally boils at a temperature of 80.10 °C. If the solution boils at a temperature of 84.2 °C, the change in temperature is (84.2 - 80.10 °C) or 4.1 °C.

Calculate the Δt, using the equation $\Delta t = K_{bp} \bullet m_{solute}$:

The molality of the solution is $\dfrac{0.200 \text{ mol}}{0.125 \text{ kg solvent}}$ or 1.60 m

The K_{bp} for benzene is +2.53 °C/m

So $\Delta t = K_{bp} \bullet m_{solute} = +2.53 \text{ °C/m} \bullet 1.60 \text{ m} = +4.1 \text{ °C.}$

31. Calculate the molality of acenaphthene, $C_{12}H_{10}$, in the solution.

$$0.515 \text{ g } C_{12}H_{10} \cdot \frac{1 \text{ mol } C_{12}H_{10}}{154.2 \text{ g } C_{12}H_{10}} = 3.34 \times 10^{-3} \text{ mol } C_{12}H_{10}$$

and the molality is: $\dfrac{3.34 \times 10^{-3} \text{ mol acenaphthene}}{0.0150 \text{ kg } CHCl_3} = 0.223$ molal

the boiling point elevation is: $\quad \Delta t = m \cdot K_{bp} = 0.223 \cdot \dfrac{+3.63 \text{ }^{\circ}C}{\text{molal}} = 0.808 \text{ }^{\circ}C$

and the boiling point will be $61.70 + 0.808 = 62.51 \text{ }^{\circ}C$

33. The change in the temperature of the boiling point is $(80.51 - 80.10)^{\circ}C$ or $0.41 \text{ }^{\circ}C$.

Using the equation $\Delta t = m \cdot K_{bp}$; $0.41 \text{ }^{\circ}C = m \cdot +2.53 \text{ }^{\circ}C/m$, and the molality is:

$$\frac{0.41 \text{ }^{\circ}C}{+2.53 \text{ }^{\circ}C/m} = m = 0.16 \text{ molal}$$

The solution contains 50.0 g of solvent (or 0.0500 kg solvent). We can calculate the # of moles of phenanthrene:

$0.16 \text{ molal} = \dfrac{x \text{ mol } C_{14}H_{10}}{0.0500 \text{ kg}}$ or 8.0×10^{-3} mol $C_{14}H_{10}$, and since 1 mol of $C_{14}H_{10}$ has a

mass of 178 g, $8.0 \times 10^{-3} \text{ mol } C_{14}H_{10} \cdot \dfrac{178 \text{ g } C_{14}H_{10}}{1 \text{ mol } C_{14}H_{10}} = 1.4 \text{ g } C_{14}H_{10}$

Freezing Point Depression

35. The solution freezes $16.0 \text{ }^{\circ}C$ lower than pure water.

(a) We can calculate the molality of the ethanol:

$$\Delta t = mK_{fp}$$

$$-16.0 \text{ }^{\circ}C = m (-1.86 \text{ }^{\circ}C/\text{molal})$$

$$8.60 = \text{ molality of the alcohol}$$

(b) If the molality is 8.60 then there are 8.60 moles of C_2H_5OH (8.60 x 46.07 g/mol= 396 g) in 1000 g of H_2O.

The weight percent of alcohol is $\dfrac{396 \text{ g}}{1396 \text{ g}}$ x 100 = 28.4 % ethanol

37. Freezing point of a solution containing 15.0 g sucrose in 225 g water:

(1) Calculate the molality of sucrose in the solution:

$$15.0 \text{ g } C_{12}H_{22}O_{11} \bullet \frac{1 \text{ mol } C_{12}H_{22}O_{11}}{342.30 \text{ g } C_{12}H_{22}O_{11}} = 0.0438 \text{ mol}$$

$$\frac{0.0438 \text{ mol } C_{12}H_{22}O_{11}}{0.225 \text{ kg } H_2O} = 0.195 \text{ molal}$$

(2) Use the Δt equation to calculate the freezing point change:

$$\Delta t = mK_{fp} = 0.195 \text{ molal} \bullet (-1.86 \text{ °C/molal}) = -0.362 \text{ °C}$$

The solution is expected to begin freezing at -0.362 °C.

Colligative Properties and Molar Mass Determination

39. The change in the temperature of the boiling point is (80.26 - 80.10)°C or 0.16 °C.

Using the equation $\Delta t = m \bullet K_{bp}$; 0.16 °C = m $\bullet$ +2.53 °C/m, and the molality

is: $\frac{0.16 \text{ °C}}{+2.53 \text{ °C/m}} = m = 0.063 \text{ molal}$

The solution contains 11.12 g of solvent (or 0.01112 kg solvent). We can calculate the # of

moles of the orange compound, since we know the molality:

$$0.063 \text{ molal} = \frac{x \text{ mol compound}}{0.01112 \text{ kg solvent}} \quad \text{or } 7.0 \times 10^{-4} \text{ mol compound.}$$

This number of moles of compound has a mass of 0.255 g, so 1 mol of compound is:

$$\frac{0.255 \text{ g compound}}{7.0 \times 10^{-4} \text{ mol}} = 360 \text{ g/mol.}$$

The empirical formula, $C_{10}H_8Fe$, has a mass of 184 g, so the # of "empirical formula units" in

one molecular formula is : $\frac{360 \text{g/mol}}{184 \text{ g/empirical formula}} = 2 \text{ mol/empirical formulas or a molecular}$

formula of $C_{20}H_{16}Fe_2$.

41. The change in the temperature of the boiling point is (61.82 - 61.70)°C or 0.12 °C.

Using the equation $\Delta t = m \bullet K_{bp}$; 0.12 °C = m $\bullet$ +3.63 °C/m, and the molality is:

$$\frac{0.12 \text{ °C}}{+3.63 \text{ °C/m}} = m = 0.033 \text{ molal}$$

The solution contains 25.0 g of solvent (or 0.0250 kg solvent). We can calculate the # of

moles of benzyl acetate:

$$0.033 \text{ molal} = \frac{x \text{ mol compound}}{0.0250 \text{ kg solvent}} \quad \text{or } 8.3 \times 10^{-4} \text{ mol compound.}$$

This number of moles of benzyl acetate has a mass of 0.125 g, so 1 mol of benzyl acetate is:

$$\frac{0.125 \text{ g compound}}{8.3 \times 10^{-4} \text{ mol}} = 150 \text{ g/mol. (to 2 sf)}$$

43. To determine the molar mass, first determine the molality of the solution

$$-0.040\,^\circ C = m \bullet -1.86\,^\circ C/molal = 0.0215 \text{ molal (or 0.022 to 2 sf)}$$

$$\text{and}\quad 0.022 \text{ molal} = \frac{\dfrac{0.180 \text{ g solute}}{MM}}{0.0500 \text{ kg water}}$$

$$MM = 167 \text{ or } 170 \text{ (to 2 sf)}$$

45. The change in the temperature of the freezing point is $(69.40 - 70.03)\,^\circ C$ or $-0.63\,^\circ C$.

Using the equation $\Delta t = m \bullet K_{fp}$; $-0.63\,^\circ C = m \bullet -8.00\,^\circ C/m$, and the molality is:

$$\frac{-0.63\,^\circ C}{-8.00\,^\circ C/m} = m = 0.079 \text{ molal (to 2 sf)}$$

The solution contains 10.0 g of biphenyl (or 0.0100 kg solvent).

We can calculate the # of moles of naphthalene:

$$0.079 \text{ molal} = \frac{x \text{ mol naphthalene}}{0.0100 \text{ kg solvent}} \quad \text{or } 7.9 \times 10^{-4} \text{ mol compound.}$$

This number of moles of naphthalene has a mass of 0.100 g, so 1 mol of naphthalene is:

$$\frac{0.100 \text{ g naphthalene}}{7.9 \times 10^{-4} \text{ mol}} = 130 \text{ g/mol (to 2 significant figures)}$$

Colligative Properties of Ionic Compounds

47. The number of moles of LiF is : $52.5 \text{ g LiF} \bullet \dfrac{1 \text{ mol LiF}}{25.94 \text{ g LiF}} = 2.02 \text{ mol LiF}$

So $\Delta t_{fp} = \dfrac{2.02 \text{ mol LiF}}{0.306 \text{ kg H}_2\text{O}} \bullet -1.86\,^\circ C/molal \bullet 2 = -24.6\,^\circ C$

The anticipated freezing point is then 24.6 °C lower than pure water (0.0°C) or -24.6 °C

49. Solutions given in order of increasing freezing point (lowest freezing point listed first):

The solution with the **greatest number** of particles will have the lowest freezing point.

The total molality of solutions is:

	solution	Particles / formula unit	Identity of particles	Total molality
(a)	0.1 m sugar	1	covalently bonded molecules	0.1 m • 1 = 0.1 m
(b)	0.1 m NaCl	2	Na^+ , Cl^-	0.2 m • 1 = 0.2 m
(c)	0.08 m CaCl$_2$	3	Ca^{2+}, 2 Cl^-	0.08 m • 3 = 0.24 m
(d)	0.04 m Na$_2$SO$_4$	3	2 Na^+, SO_4^{2-}	0.04 m • 3 = 0.12 m

The freezing points would increase in the order: $CaCl_2 < NaCl < Na_2SO_4 < \text{sugar}$

Osmosis

51. Assume we have 100 g of this solution, the number of moles of phenylalanine is

$$3.00 \text{ g phenylalanine} \cdot \frac{1 \text{ mol phenylalanine}}{165.2 \text{ g phenylalanine}} = 0.0182 \text{ mol phenylalanine}$$

The molality of the solution is: $\frac{0.0182 \text{ mol phenylalanine}}{0.09700 \text{ kg water}} = 0.187 \text{ molal}$

(a) The freezing point :

$\Delta t = 0.187 \text{ molal} \cdot -1.86 \,°\text{C/molal} = -0.348 \,°\text{C}$

The new freezing point is $0.0 - 0.348 \,°\text{C} = -0.348 \,°\text{C}$.

(b) The boiling point of the solution

$\Delta t = m \, K_{bp} = 0.187 \text{ molal} \cdot + 0.5121 °\text{C/molal} = +0.0959 \,°\text{C}$

The new boiling point is $100.000 + 0.0959 = +100.0959 \,°\text{C}$

(c) The osmotic pressure of the solution:

If we assume that the **Molarity** of the solution is equal to the **molality**, then

the osmotic pressure should be

$$\Pi = (0.187 \text{ mol/L})(0.0821 \frac{\text{L} \cdot \text{atm}}{\text{K} \cdot \text{mol}})(298 \text{ K}) = 4.58 \text{ atm}$$

The osmotic pressure will be most easily measured, since the magnitudes of osmotic pressures (large values) result in decreased experimental error.

53. The molar mass of bovine insulin with a solution having an osmotic pressure of 3.1 mm Hg:

$$3.1 \text{ mm Hg} \cdot \frac{1 \text{ atm}}{760 \text{ mm Hg}} = (M)(0.08205 \frac{\text{L} \cdot \text{atm}}{\text{K} \cdot \text{mol}})(298 \text{ K})$$

$$1.67 \times 10^{-4} = \text{Molarity or } 1.7 \times 10^{-4} \text{ (to 2 sf)}$$

The definition of molarity is #mol/L. Substituting into the definition we obtain:

$$1.7 \times 10^{-4} \frac{\text{mol bovine insulin}}{\text{L}} = \frac{\frac{1.00 \text{ g bovine insulin}}{\text{MM}}}{1 \text{ L}}$$

Solving for MM $= 6.0 \times 10^3$ g/mol

Colloids

55. (a) $BaCl_2(aq) + Na_2SO_4(aq) \rightarrow BaSO_4(s) + 2 \, NaCl(aq)$

(b) The $BaSO_4$ initially formed is of a colloidal size — not large enough to precipitate fully.

(c) The particles of $BaSO_4$ grow with time, owing to a gradual loss of charge and become large enough to have gravity affect them —and settle to the bottom.

General Questions

57. Li_2SO_4 is expected to have the more exothermic (negative) heat of solution. See SQ14.19 and 14.77 for additional information on this concept.

59. Arranged the solutions in order of (i) increasing vapor pressure of water and (ii) increasing boiling points:

(i) The solution with the highest water vapor pressure would have the **lowest particle concentration**, since according to Raoult's Law, the vapor pressure of the water in the solution is directly proportional to the mole fraction of the water. The lower the number of particles, the greater the mf of water, and the greater the vapor pressure. Hence the order of *increasing* vapor pressure is:

$$Na_2SO_4 < sugar < KBr < glycol$$

(See part (ii) for particle concentrations—(m•i))

(ii). Recall that $\Delta t = m \cdot Kfp \cdot i$. The difference in these four solutions will be in the product (m • i). The products for these solutions are:

$$glycol = 0.35 \cdot 1 = 0.35$$
$$sugar = 0.50 \cdot 1 = 0.50$$
$$KBr = 0.20 \cdot 2 = 0.40$$
$$Na_2SO_4 = 0.20 \cdot 3 = 0.60$$

Arranged in *increasing* boiling points: $glycol < KBr < sugar < Na_2SO_4$

61. For DMG, $(CH_3CNOH)_2$, the MM is 116.1 g/mol

So 53.0 g is: $53.0 \text{ g} \cdot \dfrac{1 \text{ mol DMG}}{116.1 \text{ g DMG}} = 0.456 \text{ mol DMG}$

525. g of C_2H_5OH is : $525. \text{ g} \cdot \dfrac{1 \text{ mol } C_2H_5OH}{46.07 \text{ g } C_2H_5OH} = 11.4 \text{ mol } C_2H_5OH$

(a) the mole fraction of DMG: $\dfrac{0.456 \text{ mol}}{(11.4 + 0.456) \text{ mol}} = 0.0385 \text{ mf DMG}$

(b) The molality of the solution: $\dfrac{0.456 \text{ mol DMG}}{0.525 \text{ kg}} = 0.869 \text{ molal DMG}$

(c) $P_{alcohol} = P°_{alcohol} \cdot X_{alcohol}$
$$= (760. \text{ mm Hg})(1 - 0.0385) = 730.7 \text{ mm Hg}$$

(d) The boiling point of the solution:

$$\Delta t = m \cdot K_{bp} \cdot i = (0.870)(+1.22 \,°C/molal)(1)$$

$$= 1.06 \,°C$$

The new boiling point is $78.4 \,°C + 1.06 \,°C = 79.46 \,°C$ or $79.5 \,°C$

63. Concentrated NH_3 is 14.8 M and has a density of $0.90 \, g/cm^3$.

(1) The molality of the solution:

Molality is defined as moles NH_3/kg solvent, so begin by deciding the mass of 1 L, and the mass of water in that 1L. Since the density = 0.90g/mL, then 1 L (1000 mL) will have a mass of 900g.

The mass of NH_3 present in 14.8 mol NH_3=

$$14.8 \text{ mol } NH_3 \cdot \frac{17.03 \text{ g } NH_3}{1 \text{ mol } NH_3} = 252 \text{ g } NH_3$$

Since 1 L has a mass of 900 g and 252 g is NH_3, the difference (900-252) is solvent. So 1 L has 648 g water.

$$\frac{14.8 \text{ mol } NH_3}{1 \text{ L}} \cdot \frac{1 \text{ L}}{648 \text{ g } H_2O} \cdot \frac{1000 \text{ g } H_2O}{1 \text{ kg } H_2O} = 22.8 \text{ m or } 23 \text{ m (to 2sf)}$$

(2) The mole fraction of ammonia is:

Calculate the # of moles of water present:

$$648 \text{ g } H_2O \cdot \frac{1 \text{ mol } H_2O}{18.02 \text{ g } H_2O} = 35.96 \text{ mol } H_2O \text{ (retaining 1 extra sf)}$$

The mf NH_3 is : $\dfrac{14.8 \text{ mol } NH_3}{(14.8 \text{ mol } + 35.96 \text{ mol})} = 0.29$

(3) Weight percentage of NH_3:

14.8 mol NH_3 has a mass of 252.0 g, and 1 L of the solution has a mass of 900 g.

$$\% \, NH_3 = \frac{252.0 \text{ g } NH_3}{900 \text{ g solution}} NH_3 \cdot 100 = 28 \,\% \text{ (to 2sf)}$$

65. To make a 0.100 m solution, we need a ratio of #moles of ions/kg solvent that is 0.100.

$$0.100 \text{ m} = \frac{\text{\# mol ions}}{0.125 \text{ kg solvent}} \text{ and solving for \# mol ions:} \quad 0.0125 \text{ mol ions}$$

The salt will dissociate into 3 ions per formula unit (2 Na^+ and 1 SO_4^{2-}).

The amount of Na_2SO_4 is:

$$0.0125 \text{ mol ions} \cdot \frac{1 \text{ mol } Na_2SO_4}{3 \text{ mol ions}} \cdot \frac{142.04 \text{ g } Na_2SO_4}{1 \text{ mol } Na_2SO_4} = 0.592 \text{ g } Na_2SO_4$$

67. Solution properties:

(a) The solution with the higher boiling point:

Recall that $\Delta t = m \cdot Kfp \cdot i$. The difference in these solutions will be in the product $(m \cdot i)$. The products for these solutions are:

sugar $= 0.30 \cdot 1 = 0.30$ (the sugar molecule remains as one unit)

KBr $= 0.20 \cdot 2 = 0.40$ (KBr dissociates into K^+ and Br^- ions)

KBr will provide the larger Δt.

(b) The solution with the lower freezing point:

Using the same logic as in part (a), NH_4NO_3 provides 2 ions/formula unit with Na_2CO_3 provides 3. The product , $(m \cdot i)$, is larger for Na_2CO_3, so **Na_2CO_3** gives the greater Δt and the lower freezing point.

69. The change in temperature of the freezing point is: $\Delta t = m \cdot Kfp \cdot i$

Calculate the molality:

$$35.0 \text{ g CaCl}_2 \cdot \frac{1 \text{ mol CaCl}_2}{111.0 \text{ g CaCl}_2} = 0.315 \text{ mol CaCl}_2 \text{ in } 0.150 \text{ kg water.}$$

$$m = \frac{0.315 \text{ mol CaCl}_2}{0.150 \text{ kg}} = 2.10 \text{ molal CaCl}_2$$

$\Delta t = m \cdot Kfp \cdot i = (2.10 \text{ molal} \cdot -1.86 \text{ °C/molal} \cdot 2.7) = -10.6 \text{ °C.}(-11 \text{ to 2sf})$

The freezing point of the solution is $0.0°C - 11 °C = -11 °C$

71. The molar mass of hexachlorophene if 0.640 g of the compound in 25.0 g of $CHCl_3$ boils at 61.93 °C:

Recalling the Δt equation: $\Delta t = m \cdot Kbp = m \cdot \dfrac{+3.63 \text{ °C}}{\text{molal}} = (61.93 - 61.70) \text{ °C}$

Solving for m: $\dfrac{0.23 \text{ °C}}{3.63 \text{ °C/m}} = 0.0634 \text{ m}$

Substitute into the definition for molality: $m = \#mol/kg$ solvent

$0.0634 \text{ molal} = \dfrac{\dfrac{0.640 \text{ g hexachloraphene}}{MM}}{0.025 \text{ kg}}$ and solving for MM; $4.0 \times 10^2 \text{ g/mol} = MM$

73. Solubility of $N_2 = k \cdot P_{N_2}$

$$= (8.42 \times 10^{-7} \frac{M}{\text{mmHg}}) \cdot 585 \text{ mm Hg} = 4.93 \times 10^{-4} \text{ M } N_2$$

$$\text{and } 4.93 \times 10^{-4} \frac{\text{mol}}{L} \cdot \frac{28.01 \text{ g } N_2}{1 \text{mol } N_2} = 1.38 \times 10^{-2} \frac{\text{g } N_2}{L}$$

75. (a) *Average* MM of starch if 10.0 g starch/L has an osmotic pressure = 3.8 mm Hg at 25 °C.

$$3.8 \text{ mm Hg} \cdot \frac{1 \text{ atm}}{760 \text{ mm Hg}} = (M)(0.08205 \frac{L \cdot atm}{K \cdot mol})(298 \text{ K})$$

$$2.045 \times 10^{-4} = \text{Molarity or } 2.0 \times 10^{-4} \text{ (to 2 sf)}$$

The definition of molarity is #mol/L. Substituting into the definition we obtain:

$$2.0 \times 10^{-4} \frac{\text{mol bovine insulin}}{L} = \frac{\frac{10.0 \text{ g starch}}{MM}}{1L}$$

Solving for MM = 4.9×10^4 g/mol

(b) Freezing point of the solution:

$$\Delta t = m \cdot K_{fp} \cdot i \quad \text{(assume that i=1 and that Molarity = molality)}$$

$$\Delta t = m \cdot (-1.86 \text{ °C/molal})$$

and the M = $\dfrac{\dfrac{10.0 \text{ g starch}}{4.9 \times 10^4 \text{ g/mol}}}{1 \text{ L}} = 2.0 \times 10^{-4}$

so the $\Delta t = 2.0 \times 10^{-4} \cdot (-1.86 \text{ °C/molal}) = -3.8 \times 10^{-4}$ °C. In essence the starch will boil at the temperature of pure water. From this data we can assume that *it will NOT be easy* to measure the molecular weight of starch using this technique.

77. The apparent molecular weight of acetic acid in benzene, determined by the depression of benzene's freezing point.

$$\Delta t = m \cdot K_{fp} \cdot i; \quad (3.37 \text{ °C} - 5.50 \text{ °C}) = m(-5.12 \text{ °C/molal}) i$$

and $\dfrac{-2.13 \text{ °C}}{-5.12 \text{ °C/molal}} = m \cdot i$ so 0.416 molal $= m \cdot i$ (assume i =1)

and the apparent molecular weight is:

$0.416 \text{ molal} = \dfrac{\frac{5.00 \text{ g acetic acid}}{MM}}{0.100 \text{ kg}}$ and solving for MM; 120 g/mol = MM

The apparent molecular weight of acetic acid in water

$$\Delta t = m \cdot K_{fp} \cdot i; \quad (-1.49 \text{ °C} - 0.00 \text{ °C}) = m(-1.86 \text{ °C/molal}) i$$

and $\dfrac{-1.49 \text{ °C}}{-1.86 \text{ °C/molal}} = m \cdot i$ so 0.801 molal $= m \cdot i$

(once again, momentarily i = 1) and the apparent molecular weight is:

$0.801 \text{ molal} = \dfrac{\frac{5.00 \text{ g acetic acid}}{MM}}{0.100 \text{ kg}}$ and solving for MM; 62.4 g/mol = MM

The accepted value for acetic acid's molecular weight is approximately 60 g/mol. Hence the value for i isn't much larger than 1, indicating that the degree of dissociation of acetic acid molecules in water is not great—a finding consistent with the designation of acetic acid as a weak acid. The apparently doubled molecular weight of acetic acid in benzene indicates that the acid must exist primarily as a dimer.

$$CH_3C-O-H$$
$$:O:\qquad :O:$$
$$H-O-CCH_3$$

79. The enthalpies of solution for Li_2SO_4 and K_2SO_4:
The process is MX (s) → MX (aq)

Using the data for **Li_2SO_4**:

$\Delta H_{solution} = \Delta H_f$ (aq) - ΔH_f (s) = (-1464.4 kJ/mol) - (-1436.4 kJ/mol) = - 28.0 kJ/mol

Using the data for **K_2SO_4**:

$\Delta H_{solution} = \Delta H_f$ (aq) - ΔH_f (s) = (-1414.0 kJ/mol) - (-1437.7 kJ/mol) = 23.7 kJ/mol

Note that for Li_2SO_4 *the process is* **exothermic** *while for* K_2SO_4 *the process is endothermic.*

Similar data for LiCl and KCl:

For LiCl: ΔH_f (aq) - ΔH_f (s) = (-445.6 kJ/mol) - (-408.6 kJ/mol) = - 37.0 kJ/mol and

for KCl: ΔH_f (aq) - ΔH_f (s) = (-419.5 kJ/mol) - (-436.7 kJ/mol) = 17.2 kJ/mol

Note the similarities of the chloride salts, with the lithium salt being **exothermic** *while the potassium salt is* **endothermic***.*

81. Graham's law says that the pressure of a mixture of gases (benzene and toluene) is the sum of the partial pressures. So, using Raoult's Law $P_{benzene} = mf_{benzene} \cdot P^{\circ}_{benzene}$ and similarly for toluene. The total pressure is then:

P_{total} = $P_{benzene} + P_{toluene}$

= ($\dfrac{2\ mol\ benzene}{3\ mol}$ • 75 mm Hg) + ($\dfrac{1\ mol\ toluene}{3\ mol}$ • 22 mm Hg)= 57mm Hg

What is the mole fraction of each component in the liquid and in the vapor?

The **mf of the components in the liquid** are: benzene: 2/3 and toluene: 1/3

The **mf of the components in the vapor** are proportional to their pressures in the vapor state.
The mf of benzene is: $\dfrac{50\ mm\ Hg}{57\ mm\ Hg}$ = 0.87; the mf of toluene would be (1-0.87) or 0.13.

83. Using the freezing point depression and boiling point elevation equations, calculate the term (m•i). Since we have no quantitative information about the quantity of benzoic acid dissolved in the benzene, the term (m•i) will be the best metric by which we can judge the degree of dissociation of benzoic acid at the freezing point and boiling point of benzene.

At the freezing point: $\Delta t = m \cdot i \cdot K_{fp}$; (3.1 °C - 5.12 °C) = m•i(-5.12 °C/molal) i

and $\dfrac{-2.02\ °C}{-5.12\ °C/molal}$ = m • i   so 0.395 molal = m • i

At the boiling point: $\Delta t = m \cdot i \cdot K_{bp}$; (82.6 °C – 80.1 °C) = m•i(+2.53 °C/molal) i

and $\dfrac{+2.5\ °C}{+2.53\ °C/molal}$ = m • i   so 0.988 molal = m • i

If we assume the amount of benzoic acid dissolved in benzene is constant over the temperature range described, the conclusion one reaches is that *i* has a greater value at higher temperatures than at lower ones. Another way of expressing this is that *at higher temperatures*, the *degree of association* between benzoic acid molecules *decreases*,

85. A 2.0 % aqueous solution of novocainium chloride(NC) is also 98.0 % in water. Assume that we begin with 100 g of solution.   The molality of the solution is then:

$$\dfrac{2.0\ g \cdot \dfrac{1\ mol\ NC}{272.8\ g\ NC}}{0.0980\ kg\ water} = 0.075\ m$$

Using the "delta T" equation:

$\Delta t = m \cdot K_{fp} \cdot i$, we can solve for i: $\dfrac{\Delta t}{m \cdot K_{fp}} = i$

$$\dfrac{-0.237\ °C}{0.075m \cdot -1.86\ °C/m} = 1.7$$

So approximately **2 moles of ions are present per mole of compound**.

87. (a) We can calculate the freezing point of sea water if we calculate the molality of the solution. Let's imagine that we have 1,000,000 (or 10^6) g of sea water. The amounts of the ions are then equal to the concentration (in ppm). Calculating their concentrations we get:

Cl⁻ $1.95 \times 10^4\ g\ Cl^- \cdot \dfrac{1\ mol\ Cl^-}{35.45\ g\ Cl^-}$ = 550. mol Cl⁻

Na⁺ $1.08 \times 10^4\ g\ Cl^- \cdot \dfrac{1\ mol\ Na^+}{22.99\ g\ Na^+}$ = 470. mol Na⁺

Mg⁺² $1.29 \times 10^3\ g\ Mg^{+2} \cdot \dfrac{1\ mol\ Mg^{+2}}{24.31\ g\ Mg^{+2}}$ = 53.1 mol Mg⁺²

SO₄⁻² $9.05 \times 10^2\ g\ SO_4^{-2} \cdot \dfrac{1\ mol\ SO_4^{-2}}{96.06\ g\ SO_4^{-2}}$ = 9.42 mol SO₄⁻²

Ca^{+2} 4.12×10^2 g $Ca^{+2} \cdot \dfrac{1 \text{mol } Ca^{+2}}{40.08 \text{ g } Ca^{+2}}$ $= 10.3$ mol Ca^{+2}

K^+ 3.80×10^2 g $K^+ \cdot \dfrac{1 \text{mol } K^+}{39.10 \text{ g } K^+}$ $= 9.72$ mol K^+

Br^- 67 g $Br^- \cdot \dfrac{1 \text{mol } Br^-}{79.90 \text{ g } Br^-}$ $= 0.84$ mol Br^-

For a total of: 1103 mol ions

The concentration per gram is: $\dfrac{1103 \text{ mol ions}}{10^6 \text{ g } H_2O}$

The *change* in the freezing point of the sea water is:

$\Delta t = m \cdot K_{fp} = \dfrac{1103 \text{ mol ions}}{10^6 \text{ g } H_2O} \cdot \dfrac{1000 \text{ g } H_2O}{1 \text{ kg } H_2O} \cdot -1.86 \text{ °C/molal} = -2.05 \text{ °C}$

So we expect this sea water to begin freezing at -2.05 °C.

(b) The osmotic pressure (in atmospheres) can be calculated if *we assume the density of sea water is 1.00 g/mL.*

$\Pi = MRT = \dfrac{1.103 \text{ mol}}{1 \text{ L}} \cdot 0.082057 \dfrac{L \cdot atm}{K \cdot mol} \cdot 298 \text{ K} = 27.0 \text{ atm}$

The pressure needed to purify sea water by reverse osmosis would then be a pressure greater than 27.0 atm.

89. A 2.00 % aqueous solution of sulfuric acid is also 98.00 % in water.

Assume that we begin with 100 g of solution.

(a) We can calculate the van't Hoff factor by first calculating the molality of the solution:

$$\dfrac{2.00 \text{ g} \cdot \dfrac{1 \text{mol } H_2SO_4}{98.06 \text{ g } H_2SO_4}}{0.09800 \text{ kg water}} = 0.208 \text{ m}$$

Using the "delta T" equation:

$\Delta t = m \cdot K_{fp} \cdot i$, we can solve for i: $\dfrac{\Delta t}{m \cdot K_{fp}} = i$

$\dfrac{-0.796 \text{ °C}}{0.208 m \cdot -1.86 \text{ °C/m}} = 2.06 = i$

(b) Given the van't Hoff factor of 2 (above), the best representation of a dilute solution of sulfuric acid in water has to be: $H^+ + HSO_4^-$.

91. The vapor pressure data should permit us to calculate the molar mass of the boron compound.
$P_{benzene} = X_{benzene} \cdot P°_{benzene}$

94.16 mm Hg = $X_{benzene} \cdot 95.26$ mm Hg , and rearranging: $X_{benzene} = \dfrac{94.16 \text{ mm Hg}}{95.26 \text{ mm Hg}}$

$X_{benzene} = 0.9885$

Now we need to know the # of moles of the boron compound, so let's use the mf of benzene to

find that: $10.0 \text{ g benzene} \cdot \dfrac{1 \text{ mol benzene}}{78.11 \text{ g benzene}} = 0.128 \text{ mol benzene}$

$$0.9885 = \dfrac{0.128 \text{ mol benzene}}{0.128 \text{ mol benzene} + x \text{ mol } B_xF_y}$$

$0.9885(0.128 + x) = 0.128 \quad x = 0.00147 \text{ mol } B_xF_y$

Knowing that this # of moles of compound has a mass of 0.146 g, we can calculate the molar mass:

$$\dfrac{0.146 \text{ g}}{0.00147 \text{ mol } B_xF_y} = 99.3 \text{ g/mol}$$

We can calculate the empirical formula, since we know that the compound is 22.1% boron and 77.9 % fluorine.

In 100 g of the compound there are $22.1 \text{ g B} \cdot \dfrac{1 \text{ mol B}}{10.81 \text{ g B}} = 2.11 \text{ mol B}$ and

$$77.9 \text{ g F} \cdot \dfrac{1 \text{ mol F}}{19.00 \text{ g F}} = 4.10 \text{ mol F}$$

The empirical formula then is BF_2, which would have a formula weight of 48.8

Dividing the molar mass (found from the vapor pressure experiment) by the mass of the empirical

formula, we get: $\dfrac{99.3}{48.8} = 2.0$

(a) The molecular formula is then B_2F_4.

(b) A Lewis structure for the molecule:

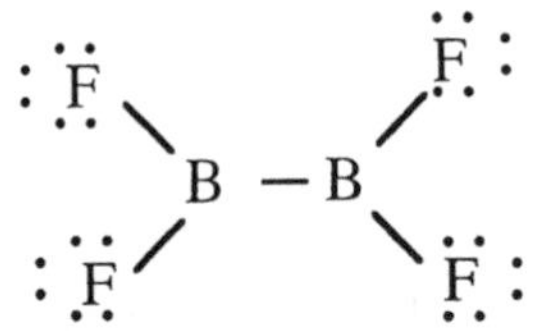

We know that the molecule is nonpolar (does not have a dipole moment), hence the proposed structure. The F-B-F bond angles are 120°, as are the F-B-B bond angles. Hence the molecule is planar (flat). The hybridization of the boron atoms is then sp^2.

Summary and Conceptual Questions

93. Equimolar amounts of $CaCl_2$ and NaCl lower freezing points differently. The formulas tell us that $CaCl_2$ provides 3 particles per formula unit while NaCl provides only two. Hence we expect—given van't Hoff factors of 3 and 2 respectively that $CaCl_2$ should have a freezing point depression that is about 50% greater than that of NaCl.

95. The solution is saturated with NaCl. The *dynamic* equilibrium that exists can be described as:

$$NaCl\ (s) \rightleftharpoons Na^+\ (aq) + Cl^-\ (aq).$$

This indicates that some of the solid NaCl (on the bottom of the container) is dissolving—forming ions—at the same rate that ions (in solution) are precipitating and settling to the bottom of the container. The net result is, of course, no **net** change at the macroscopic level—equilibrium exists.

97. Solutes likely to dissolve in water; and solutes likely to dissolve in benzene:
Substances likely to dissolve in water are polar (ionic compounds) and those polar substances capable of hydrogen-bonding. Substances likely to dissolve in benzene are non-polar substances.

Likely to dissolve in water: (a) $NaNO_3$-ionic; (d) NH_4Cl

Likely to dissolve in benzene: (b) $CH_3CH_2OCH_2CH_3$- only slightly polar, with large fraction of the molecule being non-polar (C-C, and C-H bonds); (c) $C_{10}H_8$--nonpolar

99. Since hydrophilic colloids are those that "love water", we would expect starch to form a hydrophilic colloid since it contains the OH bonds that can hydrogen bond to water. Hydrocarbons on the other hand have non-polar bonds that should have little-to-no attraction to water molecules, and form a hydrophobic colloid.

101. Semipermeable membrane dividing container into two parts; one side containing 5.85 g NaCl in 100 mL solution, and the other side containing 8.88 g KNO_3 in 100 mL solution. Calculate the osmotic pressure of both solutions.: $\Pi = MRT$.

Note that we don't actually have to calculate the osmotic pressures, only to note that the solution with the **greater molarity** will have the greater osmotic pressure.

M for NaCl: $\dfrac{5.85\ g\ NaCl}{1} \bullet \dfrac{1\ mol\ NaCl}{58.5\ g\ NaCl} \bullet \dfrac{1}{0.100L} = 1.00\ M$

M for KNO_3: $\dfrac{8.88\ g\ KNO_3}{1} \bullet \dfrac{1\ mol\ KNO_3}{101.1\ g\ KNO_3} \bullet \dfrac{1}{0.100L} = 0.878\ M$

So the osmotic pressure for NaCl will be greater than that for KNO_3, and the solvent should flow from the KNO_3 to the NaCl, reducing the osmotic pressure for NaCl.

Chapter 15
Principles of Reactivity: Chemical Kinetics

Practicing Skills

Reaction Rates

1. (a) $2 O_3 (g) \rightarrow 3 O_2 (g)$

$$\text{Reaction Rate} = -\frac{1}{2} \bullet \frac{\Delta[O_3]}{\Delta t} = +\frac{1}{3} \bullet \frac{\Delta[O_2]}{\Delta t}$$

(b) $2 HOF (g) \rightarrow 2 HF (g) + O_2 (g)$

$$\text{Reaction Rate} = -\frac{1}{2} \bullet \frac{\Delta[HOF]}{\Delta t} = +\frac{1}{2} \bullet \frac{\Delta[HF]}{\Delta t} = +\frac{\Delta[O_2]}{\Delta t}$$

3. For the reaction, $2 O_3 (g) \rightarrow 3 O_2 (g)$, the rate of formation of O_2 is 1.5×10^{-3} M/L•s. SQ15.1(a) offers a clear assist, indicating that O_2 forms at a rate 1.5 times the rate that ozone decomposes. (2 ozones produce 3 oxygens!) Hence the rate of decomposition of O_3 is -1.0×10^{-3} M/L•s.

5. Plot the data for the hypothetical reaction $A \rightarrow 2 B$

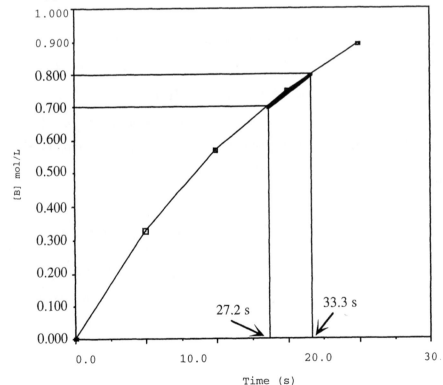

(a) Rate = $\dfrac{\Delta[B]}{\Delta t}$ = $\dfrac{(0.326 - 0.000)}{10.0 - 0.00}$ = $+\dfrac{0.326}{10.0}$ = $+0.0326\ \dfrac{mol}{L\bullet s}$

$= \dfrac{(0.572 - 0.326)}{20.0 - 10.00}$ = $+\dfrac{0.246}{10.0}$ = $+0.0246\ \dfrac{mol}{L\bullet s}$

$= \dfrac{(0.750 - 0.572)}{30.0 - 20.00}$ = $+\dfrac{0.178}{10.0}$ = $+0.0178\ \dfrac{mol}{L\bullet s}$

$= \dfrac{(0.890 - 0.750)}{40.0 - 30.00}$ = $+\dfrac{0.140}{10.0}$ = $+0.0140\ \dfrac{mol}{L\bullet s}$

The rate of change decreases from one time interval to the next *due to a continuing decrease* in the amount of reacting material (A).

(b) Since each A molecule forms 2 molecules of B, the concentration of A will decrease at a rate that is **half** of the rate at which B appears. The negative signs here indicate a **decrease in [A]--not a negative concentration of A!**

T	[B]	$[A] = 1/2([B]_0 - [B])$
10.0 s	0.326	$-1/2(0.326) = -0.163$
20.0 s	0.572	$-1/2(0.572) = -0.286$

Rate at which A changes = $\dfrac{\Delta[A]}{\Delta t}$ = $\dfrac{(-0.286 - -0.163)}{20.0 - 10.00}$ = $\dfrac{-0.123}{10.0}$ or $-0.0123\ \dfrac{mol}{L\bullet s}$

Note that the **negative sign** indicates a <u>reduction in the concentration of A</u> as the reaction proceeds. Compare this change with the change in [B] for the same interval above (+ 0.0246 $\dfrac{mol}{L\bullet s}$). The disappearance of A is half that of the appearance of B.

(c) The instantaneous rate when [B] = 0.750 mol/L:

The instantaneous rate can be calculated by noting the tangent to the line at the point , [B] = 0.750 mol/L. Taking points equidistant ([B] = 0.700 mol/L and [B] = 0.800 mol/L) and determining the times associated with those concentrations, we can calculate the instantaneous rate.

$\dfrac{\Delta[B]}{\Delta t}$ = $\dfrac{\left(0.800\dfrac{mol}{L} - 0.700\dfrac{mol}{L}\right)}{(33.3\ s - 27.2\ s)}$ = $\dfrac{0.100\ \dfrac{mol}{L}}{6.1\ s}$ = $0.0163\ \dfrac{mol}{L\bullet s}$

Concentration and Rate Equations

7. For the rate equation Rate = $k[A]^2[B]$, the reaction is 2nd order in A (superscript 2 with A), 1st order in B (implied superscript of 1 with B), and (2+1) or 3rd order overall.

9. (a) The rate equation :   Rate = $k[NO_2][O_3]$

    (b) Since k is constant, if $[O_3]$ is held constant, the rate would be tripled if the concentration of NO_2 is tripled. Let **C** represent the concentration of NO_2. Substituting into the rate equation:

$$Rate_1 = k[\mathbf{C}][O_3]$$
$$Rate_2 = k[3\mathbf{C}][O_3] = 3 \cdot k[\mathbf{C}][O_3] \text{ or } 3 \cdot Rate_1$$

    (c) Halving the concentration of O_3—assuming $[NO_2]$ is constant, would halve the rate.

$$Rate_1 = k[NO_2][\mathbf{C}]$$
$$Rate_2 = k[NO_2][1/2 \ \mathbf{C}] = 1/2[NO_2][\mathbf{C}] \text{ or } 1/2 \cdot Rate_1$$

11. (a) If we designate the three experiments (data sets in the table as i, ii, and iii respectively,

Experiment	[NO]	[O2]	$-\dfrac{\Delta[NO]}{\Delta t} \ \left(\dfrac{mol}{L \cdot s}\right)$
i	0.010	0.010	2.5×10^{-5}
ii	0.020	0.010	1.0×10^{-4}
iii	0.010	0.020	5.0×10^{-5}

Note that experiment ii proceeds at a rate four times that of experiment i.

$$\frac{\text{experiment ii rate}}{\text{experiment i rate}} = \frac{1.0 \times 10^{-4} \ \frac{mol}{L \cdot s}}{2.5 \times 10^{-5} \ \frac{mol}{L \cdot s}} = 4$$

This rate change was the result of doubling the concentration of NO. The *order of dependence of NO must be second order.* Comparing experiments i and iii, we see that changing the concentration of O_2 by a factor of two, also affects the rate by a factor of two. The *order of dependence of O2 must be first order.*

    (b) Using the results above we can write the rate equation: Rate = $k[NO]^2[O_2]^1$

    (c) To calculate the rate constant we have to have a rate. Note the data provided gives the rate of disappearance of NO. The relation of this concentration to the rate is:

$$Rate = -1/2 \cdot \frac{\Delta[NO]}{\Delta t} \ ; \text{Using experiment ii, the rate is } 5.0 \times 10^{-5} \frac{mol}{L \cdot s}$$

Substituting into the rate law

$$5.0 \times 10^{-5} \frac{mol}{L \cdot s} = k[0.020 \frac{mol}{L}]^2[0.010 \frac{mol}{L}] \text{ gives } 12 \frac{L^2}{mol^2 \cdot s} = k$$

(d) Rate when [NO] = 0.015 M and [O$_2$] = 0.0050 M

$$Rate = k[NO]^2[O_2]$$

$$= 12.5 \, \frac{L^2}{mol^2 \cdot s} \, (0.015 \, \frac{mol}{L})^2 \, (0.0050 \, \frac{mol}{L})$$

$$= -1.4 \times 10^{-5} \, \frac{mol}{L \cdot s}$$

(e) The relation between reaction rate and concentration changes:

$$Rate = -\frac{1}{2} \cdot \frac{\Delta[NO]}{\Delta t} = -\frac{\Delta[O_2]}{\Delta t} = +\frac{1}{2} \cdot \frac{\Delta[NO_2]}{\Delta t}$$

So when NO is reacting at $1.0 \times 10^{-4} \, \frac{mol}{L \cdot s}$ then O$_2$ will be reacting at

$5.0 \times 10^{-5} \, \frac{mol}{L \cdot s}$ and NO$_2$ will be forming at $1.0 \times 10^{-4} \, \frac{mol}{L \cdot s}$

13. For the reaction 2 NO(g) + O$_2$ (g) → 2 NO$_2$ (g):

(a) The rate law can be determined by examining the effect on the rate by changing the concentration of *either* NO *or* O$_2$.

In Data sets 1 and 2, the [O$_2$] doubles, and the rate doubles—a first-order dependence.

In Data sets 2 and 3, the [NO] is halved, and the rate is quartered —a second-order dependence.
The rate law will be: Rate = k[O$_2$][NO]^2

(b) The rate constant is: $3.4 \times 10^{-8} = k[5.2 \times 10^{-3}][3.6 \times 10^{-4}]^2$

Solving for k: k = 50.45 (or 50. L^2/mol^2 •h to 2sf).

Note that I selected the data from Experiment 1. Any of the data sets, (1, 2, or 3) would have provided the same value of k.

(c) The initial rate for Experiment 4 is determined by substitution into the rate law (with the value of k determined in (b):

Rate = $50[5.2 \times 10^{-3}][1.8 \times 10^{-4}]^2 = 8.4 \times 10^{-9}$ mol/L • h.

15. For the reaction: 2 CO(g) + O$_2$ (g) → 2 CO$_2$ (g):

(a) Determine m and n for the expression: Rate = k[CO]^n[O$_2$]^m

Let's call the 3 experiments i, ii, and iii. Note that the Rate of experiment iii is 4 •Rate in experiment i. $\frac{1.47 \times 10^{-4}}{3.68 \times 10^{-5}} = 4$. This change in rate was caused by a change in [CO] by a factor of **2**. The order of dependence on CO must be second-order—or (change)^{order}.

Compare experiments i and iii: The ratio of the rates is $\frac{7.36 \times 10^{-5}}{3.68 \times 10^{-5}} = 2$. This doubling of the rate is occasioned by a change in [O$_2$] of 2. The order of dependence on O$_2$ must be first-order. So **n= 2** and **m = 1**.

(b) The order with *respect to CO is second order* (m=2, yes?), and *first order with respect to* O_2 so the reaction is (2+1) *third order overall.*

(c) The rate constant is determined by substituting into the rate expression:

Rate = $k[CO]^2[O_2]^1$ so, using data from experiment i: $3.68 \times 10^{-5} = k(0.02)^2(0.02)^1$

$$\frac{3.68 \times 10^{-5}}{(0.02)^3} = 4.6 \frac{L^2}{mol^2 \bullet min} \text{ or } 5 \frac{L^2}{mol^2 \bullet min} \text{ (with 1sf)}$$

Concentration-Time Equations

17. Note that the reaction is first order. We can write the rate expression:

$$\ln \left(\frac{[C_{12}H_{22}O_{11}]}{[C_{12}H_{22}O_{11}]_0}\right) = -kt$$

Substitute the concentrations of sucrose at $t = 0$ and $t = 2.57$ hr into the equation:

$$\ln\left(\frac{[0.0132 \text{ mol/L}]}{[0.0146 \text{ mol/L}]_0}\right) = -k(2.57 \text{ hr}) \text{ and solve for k to obtain: } k = 0.0392 \text{ hr}^{-1}$$

19. Since the reaction is first order, we can write:

$$\ln\left(\frac{[SO_2Cl_2]}{[SO_2Cl_2]_0}\right) = -kt.$$ Given the rate constant, 2.8×10^{-3} min^{-1}, we can calculate the time

required for the concentration to fall from 1.24×10^{-3} M to 0.31×10^{-3}M

$$\ln\left(\frac{[0.31 \times 10^{-3}M]}{[1.24 \times 10^{-3} M]_0}\right) = -(2.8 \times 10^{-3} \text{ min}^{-1})t.$$

$$\frac{\ln(0.25)}{(-2.8 \times 10^{-3} \text{ min}^{-1})} = t = 495 \text{ min or } 5.0 \times 10^2 \text{ min (to 2 significant figures)}$$

21. The reaction is second order (the exponent for ammonium cyanate in the rate expression is 2). So we use the integrated form of the second-order rate law:

$$\frac{1}{[NH_4NCO]} - \frac{1}{[NH_4NCO]_0} = kt; \text{ and } \left|\frac{1}{[0.180M]}\right| - \left|\frac{1}{[0.229M]}\right| = (0.0113 \frac{L}{mol \bullet min})t.$$

$$\frac{1.189}{(0.0113 \frac{L}{mol \bullet min})} = t \text{ ; Solving for t gives 105 min.}$$

23. (a) Since the reaction is first order, we can write:

$$\ln\left(\frac{[H_2O_2]}{[H_2O_2]_0}\right) = -kt.$$ Given the rate constant, 1.06×10^{-3} min^{-1}, we can calculate the time

required for the concentration to fall from the original concentration to 85% of that value. Note that the concentrations *per se* are not that critical.

Let's assume the initial concentration is 100.M and after the passage of t time the concentration is 85.0M (that's 15% decomposed, yes?)

$$\frac{\ln\left(\frac{85.0}{100}\right)}{1.06 \times 10^{-3}\,\text{min}^{-1}} = \text{-t} \quad \text{and solving for the fraction:} \quad \frac{-1.897}{1.06 \times 10^{-3}\,\text{min}^{-1}} = \text{-t}$$

and t = 153 min (to 3sf).

(b) For 85% of the sample to decompose, we repeat the process, substituting 15.0 for the $[H_2O_2]$ remaining:

$$\frac{\ln\left(\frac{15.0}{100}\right)}{1.06 \times 10^{-3}\,\text{min}^{-1}} = \text{-t} \quad \text{and solving for t = 1790 min (to 3sf).}$$

Half-Life

25. Given that the reaction is first order we can use the integrated form of the rate law:

$$\ln\left(\frac{[N_2O_5]}{[N_2O_5]_0}\right) = \text{-kt.}$$

(a) Since the **definition of half-life** is "the time required for half of a substance to react", the fraction on the left side = 1/2, and $\ln(0.50) = -0.693$

Given the rate constant $5.0 \times 10^{-4}\,\text{s}^{-1}$ we can solve for t:

$$-0.693 = -(5.0 \times 10^{-4}\,\text{s}^{-1})t \quad \text{and t = } 1.4 \times 10^3 \text{ seconds}$$

(b) Time required for the concentration to drop to 1/10 of the original value:

Substitute the ratio 1/10 for the concentration of N_2O_5:

$$\ln(0.10) = -(5.0 \times 10^{-4}\,\text{s}^{-1})t \quad \text{and t = } 4.6 \times 10^3 \text{ seconds}$$

27. Since the decomposition is first order: $\ln\dfrac{[\text{azomethane}]}{[\text{azomethane}]_0} = \text{- kt}$

Converting 2.00 g of azomethane to **moles**, and substituting into the equation:

$$2.00 \text{ g azomethane} \cdot \frac{1 \text{ mol azomethane}}{58.08 \text{ g azomethane}} = 0.0344 \text{ mol}$$

$$\ln\frac{[\text{azomethane}]}{[0.0344 \text{ mol}]_0} = -(40.8 \text{ min}^{-1})(0.0500 \text{ min})$$

$$\ln\frac{[\text{azomethane}]}{[0.0344 \text{ mol}]_0} = -2.04$$

$$\frac{[\text{azomethane}]}{[0.0344 \text{ mol}]_0} = e^{-2.04} = 0.130$$

$$\text{azomethane} = 4.48 \times 10^{-3} \text{ mol}$$

Since 1 mol N_2 is produced when 1 mol of azomethane decomposes, the amount of N_2

formed is: $(0.0344 \text{ mol} - 0.00448 \text{ mol}) = 0.0300 \text{ mol } N_2$ formed

29. Since this is a first-order process, $\ln \dfrac{[Cu^{2+}]}{[Cu^{2+}]_0} = -kt$ and $k = -\dfrac{0.693}{12.70 \text{ hr}}$

What fraction of the copper remains after time, $t = 64$ hr?

Radioactive decay is a first-order process so we use the equation:

$$\ln \frac{[Cu^{2+}]}{[Cu^{2+}]_0} = -\frac{0.693}{12.70 \text{ hr}} \cdot 64 \text{ hr}$$

$\ln \dfrac{[Cu^{2+}]}{[Cu^{2+}]_0} = -3.49$ and $\dfrac{[Cu^{2+}]}{[Cu^{2+}]_0} = e^{-3.49}$ or 0.030 so 3.0 % remains (to 2 sf)

31. For first-order kinetics, we know that $\ln\left(\dfrac{[HCO_2H]}{[HCO_2H]_0}\right) = -kt$.

Substituting into the equation, we solve for **k**.

$\ln\left(\dfrac{[25]}{[100]_0}\right) = -k(72 \text{ s})$ and rearranging to solve for k gives $\dfrac{-1.386}{-72 \text{ s}} = k = 0.01925 \text{ s}^{-1}$

and since we're pursing the $t_{1/2}$, $t_{1/2} = 0.693/k$ so $t_{1/2} = 0.693/0.01925 \text{ s}^{-1} = 36$ s.

A **much simpler** route to this answer is to recognize that 1half-life would consume 50% of the original sample, and the 2nd half-life would consume half of the remaining amount (25%). So two half-lives would result in the consumption of 75% of the original sample—or 1/2(72 s)!

Graphical Analysis:Rate Equations and k

33. (a) Plot of ln[sucrose] and $\dfrac{1}{[\text{sucrose}]}$ versus time.

Since the plot of ln[sucrose] vs time gives a straight line, the reaction is first order.

(b) Since the reaction is first order with respect to sucrose(the plot of ln[sucrose] vs t is linear), the rate expression may be written: Rate = k [sucrose]. The rate constant can be calculated using two data points:

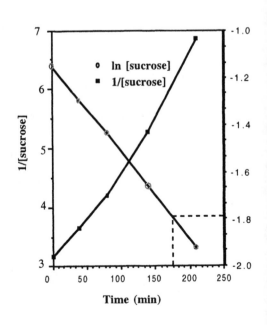

Using the first two points yields:

$$\ln\left(\frac{[A]}{[A]_0}\right) = -kt \quad \text{and substituting :} \quad \ln\left(\frac{0.274}{0.316}\right) = -k(39 \text{ min})$$

$$\frac{\ln(0.867)}{39 \text{ min}} = -k \quad \text{and } 3.7 \times 10^{-3} \text{ min}^{-1} = k$$

(c) Using the graph of ln[sucrose] vs time, an estimate at 175 minutes yields:

ln[sucrose] = -1.8 corresponding to [sucrose] = 0.167 M

35. For the decomposition of N_2O:

Since **ln[N$_2$O] vs t** gives a **straight line**, we know that the reaction is first order with respect to N_2O, and the line has a slope = $-k$.

Taking the natural log (ln) of the concentrations at t=120 min and t =15.0 min gives ln(0.0220) = -3.8167; ln(0.0835) = -2.4829.

$$\text{slope} = -k = \frac{(-3.8167) - (-2.4829)}{(120.0 - 15.0)\text{min}} =$$

$$\frac{1.3338}{105.0 \text{ min}} = 0.0127 \text{ min}^{-1}$$

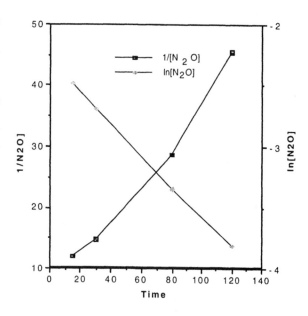

The rate equation is: Rate $= k[N_2O]$.

The rate of decomposition when

$[N_2O] = 0.035$ mol/L : Rate $= (0.0127 \text{ min}^{-1})(0.035 \frac{\text{mol}}{\text{L}}) = 4.4 \times 10^{-4} \frac{\text{mol}}{\text{L} \cdot \text{min}}$

37. Since the graph of reciprocal concentration gives a straight line, we know the reaction is **second-order** with respect to NO_2. Equation 15.2 indicates that the **slope** of the line is k, so k = 1.1 L/mol •s. The rate law is Rate = k•[NO_2]^2

39. The straight line obtained when the reciprocal concentration of C_2F_4 is plotted vs t indicates that the reaction is second-order in C_2F_4.

The rate expression is: Rate $= \frac{-\Delta[C_2F_4]}{\Delta t} = 0.04 \frac{L}{\text{mol} \cdot s}[C_2F_4]^2$

Kinetics and Energy

41. The E^* for the reaction $N_2O_5 (g) \rightarrow 2 NO_2 (g) + 1/2 \; O_2 (g)$

Given k at 25 °C = $3.46 \times 10^{-5} \; s^{-1}$ and k at 55 °C = $1.5 \times 10^{-3} \; s^{-1}$

The rearrangement of the Arrhenius equation (in your text as Equation 15.7) is helpful here.

$$\ln \frac{k_2}{k_1} = - \frac{E^*}{R}\left(\frac{1}{T_2} - \frac{1}{T_1}\right) \; ; \quad \ln \frac{1.5 \times 10^{-3} \; s^{-1}}{3.46 \times 10^{-5} \; s^{-1}} = - \frac{E^*}{8.31 \times 10^{-3} \; kJ/mol \cdot K} \left(\frac{1}{328} - \frac{1}{298}\right)$$

and solving for E^* yields a value of 102 kJ/mol for E^*.

43. Using the Arrhenius equation: $\ln \dfrac{k_2}{k_1} = - \dfrac{E^*}{R} \left(\dfrac{1}{T_2} - \dfrac{1}{T_1}\right)$, $T_1 = 800K$, and $T_2 = 850 \; K$

Given $E^* = 260 \; kJ/mol$ and $k_1 = 0.0315 \; s^{-1}$, we can calculate k_1.

$$\ln \frac{k_2}{0.0315 \; s^{-1}} = - \frac{260 \; kJ/mol}{8.3145 \times 10^{-3} \; kJ/mol \cdot K} \left(\frac{1}{850 \; K} - \frac{1}{800K}\right)$$

$$\ln \frac{k_2}{0.0315 \; s^{-1}} = 2.30 = \ln k_2 - \ln(0.0315 \; s^{-1})$$

$$2.30 + \ln(0.0315 \; s^{-1}) = \ln k_2 = 2.30 - 3.458 = -1.16$$

$$k_2 = e^{-1.16} = 0.3 \; s^{-1} \text{ (1 sf owing to a temperature (800K) with 1sf)}$$

45. Energy progress diagram:

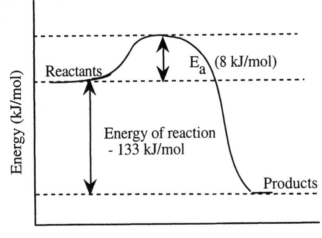

Mechanisms

47. <u>Elementary Step</u>   <u>Rate law</u>

    (a) $NO (g) + NO_3 (g) \rightarrow 2 NO_2 (g)$ Rate = $k[NO][NO_3]$

    Reaction is bimolecular

    (b) $Cl (g) + H_2 (g) \rightarrow HCl (g) + H (g)$ Rate = $k[Cl][H_2]$

    Reaction is bimolecular

    (c) $(CH_3)_3CBr (aq) \rightarrow (CH_3)_3C^+ (aq) + Br^- (aq)$ Rate = $k[(CH_3)_3CBr]$

    Reaction is unimolecular

49. For the reaction reflecting the decomposition of ozone:

    (a) The second step is the slow step, and therefore rate-determining.

    (b) The rate equation involves *only* these substances that affect the rate, (since they participate in the rate determining step), Rate = $k[O_3][O]$.

51. (a) Add the elementary steps:

$$H_2O_2 (aq) \quad + I^- \quad \longrightarrow \quad H_2O (l) + \cancel{OI^-} (aq)$$

$$H^+ (aq) + \cancel{OI^-} (aq) \quad \longrightarrow \quad \cancel{HOI} (aq)$$

$$\underline{\cancel{HOI} (aq) + H^+ (aq) + I^- (aq) \rightarrow I_2 (aq) \quad + \quad H_2O (l)}$$

$$H_2O_2 (aq) + 2 I^- + 2H^+ (aq) \rightarrow I_2 (aq) \quad + \quad 2 H_2O (l)$$

Note that (on the left side of the equations) the I^- and H^+ ions "add". On the right side of the equations, H_2O "adds". HOI and OI^- "cancel", giving the overall stoichiometric equation.

(b) Molecularity: 1st eqn: bimolecular; 2nd eqn: bimolecular; 3rd eqn: termolecular. In elementary steps, each species represents a molecule, so with two molecules as reactants—the equation is bimolecular, 3 molecules—termolecular, etc.

(c) To be consistent with kinetic data, the experimental rate equation should be:

Rate = $k[H_2O_2][I^-]$—first order in both peroxide and iodide.(the **slow** step)

(d) Intermediates are species which are produced in one step and consumed in a subsequent step. In this mechanism, HOI and OI^- are intermediates.

53. For the reaction of NO_2 and CO:

Slow NO_2 + ~~NO_2~~ → NO   + ~~NO_3~~

Fast ~~NO_3~~ + CO → ~~NO_2~~ + CO_2

Net NO_2 + CO   → NO + CO_2

Note that when the two steps are added, the desired overall equation results.

(a) Classify the species:

$NO_2(g)$	Reactant (step 1); Product (step2)
CO (g)	Reactant (step 1)
NO_3 (g)	Intermediate (produced & consumed subsequently)
CO_2 (g)	Product
NO (g)	Product

(b) A reaction coordinate diagram

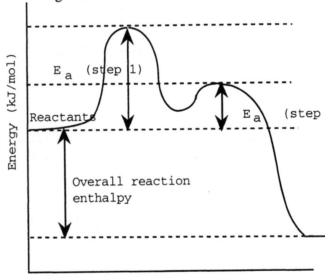

General Questions

55. What happens to the reaction rate for a reaction with the rate equation: Rate = $k[A]^2[B]$

$Rate_1 = k[A]^2[B]$. Double the concentration of A—call it 2A, and halve the concentration of

B—call it B/2.

$Rate_2 = k[2A]^2[B/2]$. Reducing the concentrations gives $Rate_2 = (2)^2 \cdot (1/2) \, k[A]^2[B]$

$Rate_2 = 2 \cdot k \, [A]^2[B]$. So $Rate_2$ is **two times** that of $Rate_1$.

57. To determine second-order dependence, after acquisition of the pH vs time data, plot $1/[OH^-]$ versus time. The reaction is second-order in OH^- if a straight line is obtained,

59. Decomposition of NH_3 is 1st order in NH_3.

   (a) Rate equation: Rate = $k[NH_3]$

   (b) Rate constant : For 1st order processes, the integrated rate equation is:

   $\ln(\frac{[A]}{[A]_0}) = -kt$ and substituting : $\ln(\frac{0.26}{0.67}) = -k(19\ sec)$

   $\frac{\ln 0.388}{19\ sec} = -k$; $\frac{-0.9465}{19\ sec} = -k$ and $5.0 \times 10^{-2}\ sec^{-1} = k$

   (c) Half-life: Since this is a first order process, we use the equation: $t_{\frac{1}{2}} = \frac{0.693}{k}$

   $t_{\frac{1}{2}} = \frac{0.693}{0.050\ sec^{-1}} = 14\ sec$ (to 2sf)

61. The decomposition of CO_2 is first order with respect to CO_2:

   (a) Rate equation: Rate = $k[CO_2]$

   (b) Rate constant : For 1st order processes, the integrated rate equation is:

   $\ln(\frac{[A]}{[A]_0}) = -kt$ and substituting : $\ln(\frac{0.27}{0.38}) = -k(12\ sec)$

   $\frac{\ln 0.711}{12\ sec} = -k$; $\frac{-0.3417}{12\ sec} = -k$ and $2.8 \times 10^{-2}\ sec^{-1} = k$

   (c) Half-life: Since this is a first order process, we use the equation: $t_{\frac{1}{2}} = \frac{0.693}{k}$

   $t_{\frac{1}{2}} = \frac{0.693}{0.028\ sec^{-1}} = 24\ sec$ (to 2sf)

63. For the dimerization to form octafluorocyclobutane the following plot is obtained:

   (a) The plot of reciprocal concentration vs time gives a straight line. Such behavior is indicative of **a second-order process.**

   The rate law is: Rate = $k[C_2F_4]^2$

   (b) The rate constant is equal to the slope of the line: Using a graphical package (I used Cricket Graph™) the equation for the line is:

   y=9.9758 + 0.04488x. So k = 0.045 L/mol•s

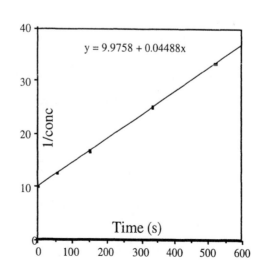

233

(c) The concentration after 600 s is found by using the integrated rate equation for second-order processes:

$$\frac{1}{[C_2F_4]_t} - \frac{1}{[(0.100)]_0} = 0.045\frac{L}{mol \bullet s}(600\ s).\ \text{Rearranging gives}$$

$$\frac{1}{[C_2F_4]_t} = 0.045\frac{L}{mol \bullet s}(600\ s) + \frac{1}{0.100}\ \text{and}\ [C_2F_4]_{600} = 0.027\ \text{(or 0.03 M to 1sf)}$$

(d) Time required for 90% completion: Using the same equation as in part (c), and substituting 10% of the initial concentration as our "concentration at time t", we can solve for t.

$$\frac{1}{[(0.010)]_t} - \frac{1}{[(0.100)]_0} = 0.045\frac{L}{mol \bullet s}(t)$$

$$(100 - 10) = 0.045\frac{L}{mol \bullet s}(t)\ \text{and}\ t = 2000\ s$$

65. For the formation of urea from ammonium cyanate:

(a) Plot the data to determine the order.

The graph shown here has a plot of the $1/[NH_4NCO]$ vs time, and $\ln[NH_4NCO]$ versus time. Note that the plot of reciprocal concentration gives a straight line, indicating the reaction is *second-order in ammonium cyanate*.

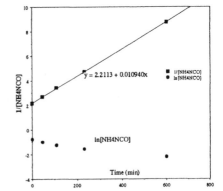

(b) k is the slope of the line, which is 0.0109 L/mol•min

(c) The half-life can be calculated using the integrated rate equation

$$\frac{1}{[0.229]_t} - \frac{1}{[0.458]_0} = 0.0109\frac{L}{mol \bullet min}(t)$$

The concentration at time t is 1/2 that of the original concentration of NH_4NCO (the definition of half-life).

$$4.367 - 2.183 = 0.0109\frac{L}{mol \bullet min}(t)\ \text{and solving for}\ t = 200.\ \text{minutes.}$$

(d) The concentration of ammonium cyanate after 12.0 hours (720. min) is found by using the integrated rate equation. Since we know k and t, we can solve for "concentration at time t =(12.0 hours) "

$$\frac{1}{[NH_4NCO]_t} - \frac{1}{[0.458]_0} = 0.0109\frac{L}{mol \bullet min}(t)$$

$$\frac{1}{[NH_4NCO]_t} = 0.0109\frac{L}{mol \bullet min}(720.\ min) + \frac{1}{[0.458]_0}$$

Solving for $[NH_4NCO]$ we obtain $[NH_4NCO] = 0.0997\ M$

67. The reaction between carbon monoxide and nitrogen dioxide has a rate equation that is second-order in NO_2 This means that the *slowest step* in the mechanism involves **2** molecules of nitrogen dioxide. Mechanism 2 has a SLOW step that fulfills this requirement. Note that Mechanisms 1 and 3 are only 1st order in nitrogen dioxide.

69. The decomposition of dinitrogen pentaoxide has a first-order rate equation. Determine the rate constant and the half-life by substitution into the integrated rate equation for first-order reactions .

$$\ln\left(\frac{[N_2O_5]_t}{[N_2O_5]_0}\right) = - \text{k} \bullet \text{t}$$ The decomposition if 20% complete in 6.0 hours.

The amount of N2O5 remaining after 6.0 h is 80% of the original concentration. The left hand term is then 80/100.

$$\ln\left(\frac{80}{100}\right) = - \text{k} \bullet 6.0 \text{ h} \text{ and k} = 0.037 \text{ h}^{-1}$$

Calculation of the half-life is accomplished by noting that the left-hand side of the equation has a value of 50% --and $\ln(0.50) = - 0.693$. Substituting the value of k we calculated above:

$$\frac{-0.693}{-0.037 \text{ hr}^{-1}} = 19 \text{ h}$$

71. For the decomposition of dimethyl ether:

(a) The mass of dimethyl ether remaining after 125 min and after 145 min:

The half-life is 25.0 min. A period of 125 minutes is 5 half-lives. The fraction remaining after n half-lives is $\left(\frac{1}{2}\right)^n$ and with n = 5, the fraction remaining is 0.03125 (1/32 of the original amount). The *mass remaining is (0.03125)(8.00 g) = 0.251 g dimethyl ether.* Note that 145 minutes is *almost* 6 half-lives. So you should be able to "guess" at a value for the amount remaining. The mass should be slightly greater than 1/64 of the original amount. The exact amount can be found by substitution into the first-order rate equation to solve for the rate constant:

$$\frac{- 0.693}{25.0 \text{ min}} = - \text{k} \text{ and k} = 0.0277 \text{ min}^{-1}$$

Note that the "ln term" is simplified by remembering that a "half-life" is a time for which 50% decomposes. Then $\ln(0.50) = - 0.693$ {A HANDY THING TO REMEMBER}

Substituting our value of k into the equation and solving for the fraction remaining:

$$\ln\left(\frac{[\text{dimethyl ether}]_t}{[\text{dimethyl ether}]_0}\right) = - 0.0277 \text{ min}^{-1} \bullet 145 \text{ min} = - 4.020$$

For simplicity, let's represent the fraction remaining as x. Solving for the ratio of concentrations gives $\ln(x) = -4.020$, for which $x = 0.0179$

The mass of dimethyl ether remaining is $(0.0179)(8.00 \text{ g}) = 0.144$ g

(b) The time required for 7.60 ng of ether to be reduced to 2.25 ng:

Substitute into the first order equation (as we've done above). Now we know the value for the "left side" and we know k; we can solve for t

$$\frac{\ln\left(\frac{[2.25 \text{ ng}]_t}{[7.60 \text{ ng}]_0}\right)}{-0.0277 \text{ min}^{-1}} = t = \frac{\ln(0.296)}{-0.0277 \text{ min}^{-1}} = \frac{-1.217}{-0.0277 \text{ min}^{-1}} = 43.9 \text{ min}$$

(c) The fraction remaining after 150 minutes is easily calculated by noting that 150 minutes is *exactly* 6 half-lives. The fraction remaining is 1/64 or 0.016 (to 2sf)

73. For the reaction of NO with Br_2 to give BrNO:

(a) The balanced equation is: $2 \text{ NO (g)} + Br_2 \text{ (g)} \rightarrow 2 \text{ BrNO (g)}$

(b) The molecularity for each step in each mechanism:

Mechanism 1	Molecularity
	Termolecular (3 reactant species)
Mechanism 2	
Step 1	Bimolecular (2 reactant species)
Step 2	Bimolecular (2 reactant species)
Mechanism 3	
Step 1	Bimolecular (2 reactant species)
Step 2	Bimolecular (2 reactant species)

(c) Intermediates formed in Mechanisms 2 and 3:

In mechanism 2, Br_2NO is produced in step 1 and consumed in step 2—not appearing as reactant or product (the definition of intermediate). In mechanism 3, N_2O_2 fits that description.

(d) Comparison of rate laws from the three mechanisms:

For mechanisms 2 and 3, there are two steps and no indication of the "slow" step. For purposes of our discussion, assume that the first step in each mechanism is the "slow" step.

Mechanism 1	2nd order in NO and 1st order in Br_2
Mechanism 2	1st order in NO and 1st order in Br_2
Mechanism 3	2nd order in NO and 0 order in Br_2

75. Show consistency of mechanism with the rate law: Rate = $k[O_3]^2/[O_2]$:

The rate law for the **slow** step is Rate=$k[O_3][O]$, but the concentration of O is affected by the preceding equilibrium step. Solve for the [O] in that equilibrium step:

The equilibrium constant expression for the fast step is:

$$K = \frac{[O_2][O]}{O_3}; \text{solving for [O]} = \frac{K \bullet [O_3]}{[O_2]}$$

Substitute this concentration into the Rate law for the slow step above:

$$\text{Rate=}k[O_3][O], \; ; \; \text{Rate} = \frac{K \bullet k[O_3][O_3]}{[O_2]}$$

$$\text{Combining } K \bullet k \text{ as } k', \text{ Rate} = \frac{k'[O_3]^2}{[O_2]}$$

77. We can calculate Ea if we know values of rate constants at two temperatures:

$$\ln\frac{k2}{k1} = -\frac{Ea}{R}\left(\frac{1}{T_2} - \frac{1}{T_1}\right)$$

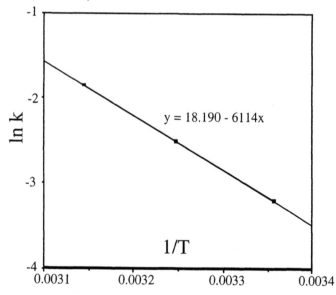

The slope was determined to be –6114 . Since slope = -Ea/R,

$$-6114 = -\frac{Ea}{8.314 \times 10^{-3}\dfrac{kJ}{K \bullet mol}} = 51 \text{ kJ/mol}$$

79. Since the time required to prepare the egg will be inversely proportional to the rate constant, we can calculate the ratio of the time, by calculating the ratio of the rate constants. Using the

Arrhenius equation, $\ln\dfrac{k_2}{k_1} = -\dfrac{E_a}{R}\left|\dfrac{1}{T_2} - \dfrac{1}{T_1}\right|$

and recalling the reciprocal relationship between k and t, we write:

$$\ln\frac{t_{90}}{t_{100}} = -\frac{E_a}{R}\left[\frac{1}{T_2} - \frac{1}{T_1}\right] \text{ or } \ln\frac{t_{90}}{t_{100}} = -\frac{52.0\ \frac{kJ}{mol}}{8.3145 \times 10^{-3}\ \frac{kJ}{K\bullet mol}}\left[\frac{1}{373K} - \frac{1}{363K}\right]$$

$$\ln\frac{t_{90}}{t_{100}} = -6254\ K\left(-7.385 \times 10^{-5}K^{-1}\right) = 0.4619$$

$$\frac{t_{90}}{t_{100}} = 1.59 \text{ or } t_{90} = t_{100}\bullet 1.59 \text{ so } t_{90} = (3\ minutes)\bullet 1.59 = 4.76 \text{ minutes or}$$

$$5 \text{ minutes (to 1 sf)}$$

81. The decomposition of HOF proceeds in a first-order reaction with a half-life of 30 minutes. The equation is: HOF(g) → HF(g) + 1/2 O$_2$ (g)

If the initial pressure of HOF is 1.00×10^2 mm Hg at 25°C, what is the total pressure in the flask and the partial pressure of HOF after 30 minutes?

The rate constant is calculated: $k = \dfrac{0.693}{t_{1/2}} = \dfrac{0.693}{30\ min} = 0.0231\ min^{-1}$

To calculate the concentration (pressure in the case of gases) at **any** time, the integrated first-order rate law can be used:

$$\ln\left(\frac{[HOF]_t}{[HOF]_0}\right) = -k\bullet t$$

For the 30 minute time-frame, the process is simplified. The concentration fraction is 1/2 (since 30 minutes is equal to a half-life). So P$_{HOF}$ = 1/2(100. mm Hg) or 50.0 mm Hg.

The stoichiometry indicates that we get 1 HF and 1/2 O$_2$ for each HOF. Then the P$_{HF}$ = 50.0 mm Hg, and P$_{O2}$ = 25.0 mm Hg. The **total** pressure is then (50.0 + 50.0 + 25.0) or 125.0 mm Hg.

The corresponding values after 45 minutes?

$$\ln\left(\frac{[HOF]_t}{[HOF]_0}\right) = -k\bullet t \text{ becomes } \ln\left(\frac{[HOF]_t}{[100.\ mm\ Hg]_0}\right) = -0.0231 min^{-1}\bullet 45\,min$$

$$\frac{[HOF]_t}{100.\ mmHg} = e^{-(0.0231min^{-1}\bullet 45\,min)} \text{ and } [HOF]_t = 0.3536 \bullet 100.\ mmHg = 35.4\ mm\ Hg$$

P$_{HOF}$ = 35.4 mm Hg; P$_{HF}$ = (100.0-35.4) mm Hg or 65 mm Hg ; P$_{O2}$ = 32 mm Hg

The **total** presssure is then (35.4 mm Hg + 65 mm Hg + 32 mm Hg) or 132 mm Hg.

83. (a) The molecularity of each elementary step:

   Slow step:unimolecular (one molecule: $Ni(CO)_4$)

   Fast step: bimolecular (two molecules: $Ni(CO)_3 + L$)

(b) The doubling of the rate with the doubling of the concentration of $Ni(CO)_4$ indicates a first-order dependence on $Ni(CO)_4$ while the lack of affect of [L] indicates a zero-order dependence on L. This is consistent with the SLOW step in the proposed mechanism shown.

(c) First order kinetics obey the equation: $\ln\dfrac{[Ni(CO)_4]}{[Ni(CO)_4]_0}$ $= - kt$ with $k = 9.3 \times 10^{-3}$ s^{-1} and t =

   3.0×10^2s (Remember that units of time must cancel, so convert 5.0 minutes to seconds.)

   $\ln\dfrac{[Ni(CO)_4]}{[Ni(CO)_4]_0}$ $= - (9.3 \times 10^{-3}$ s^{-1} $)(3.0 \times 10^2$s$) = -2.79$

Taking the inverse natural logarithm of both sides we get:

   $\dfrac{[Ni(CO)_4]}{[Ni(CO)_4]_0}$ $= 0.0614$.

Since the initial concentration of $Ni(CO)_4$ is 0.025M we can solve for the concentration after 5.0 minutes: $[Ni(CO)_4] = 0.0614 \cdot 0.025 = 1.5 \times 10^{-3}$ M

Subtracting this concentration from the *initial* concentration gives the concentration of the product: 0.025 M $- 1.5 \times 10^{-3}$ M $= 0.023$ M.

Summary and Conceptual Questions

85. Finely divided rhodium has a larger surface area than a block of the metal with the same mass. Since hydrogenation reactions depend on adsorption of H_2 on the catalyst surface, the greater the surface area, the greater the locations for such adsorption to occur.

87. For the reaction: H_2 (g) $+ I_2$ (g) $\rightarrow$ 2HI(g), the rate law is Rate $= k[H_2][I_2]$, which of the following statements is false—and why it is incorrect.
   (a) The reaction **must** occur in a single step. **False**. Only an elaboration of the mechanism can clarify this fact. While it might occur in a single step, it *need not occur in a single step*.
   (b) This is a 2nd order reaction overall. **True**. Addition of the order of H_2 and I_2 (1 +1) gives an overall order of 2.
   (c) Raising the temperature will cause the value of k to decrease. **False**. Increases in T lead to increases in the value of k.
   (d) Raising the temperature lowers the activation energy for this reaction. **False**. Temperature does not affect activation energy.

(e) If the concentrations of both reactants are doubled the rate will double. **False**. Doubling the concentration of *either reactant* wil double the rate. Doubling the concentration of both reactants will increase the rate by a factor of 4.

(f) Adding a catalyst in the reaction will cause the initial rate to increase.**True**. A catalyst should increase the rate at which the reaction occurs.

89. (a) True—the rate-determining step in a mechanism **is the slowest step**.

(b) True—the rate constant is a proportionality constant that relates the concentration and rate *at a given temperature* and varies with temperature.

(c) False—As a reaction proceeds the concentration of reactants diminshes and the rate of the reaction slows as fewer collisions occur.

(d) False—If the slow (single) step involves a termolecular process, then only one step is necessary. However termolecular collisions are **not highly likely**.

91. For the reaction: cyclopropane → propene, describe the **change** on the following quantities.

(a) [cyclopropane] – Concentrations of reactants *decrease* as reactions proceed.

(b) [propene] – Concentrations of products *increase* as reactions proceed.

(c) [catalyst]—As catalysts are not consumed during a process, the concentration of the catalyst *will not change*.

(d) rate constant—The rate constant *will not change* as the reaction proceeds.

(e) The order of the reaction—*will not change* as the reaction proceeds.

(f) the half-life of cyclopropane—Since the half-life of cyclopropane is a function of the rate constant (which doesn't change—see (d) above), the half-life *does not change*.

93. For the reaction coorinate diagram shown:

(a) Number of steps: **3**- Note that each "peak" indicates a transition between a "reactant" and a "product".

(b) Is the reaction exo- or endothermic? Noting that the energy of the products is *lower* than that of the reactants, the reaction is **exothermic**.

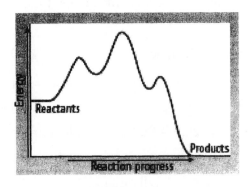

95. (a) The difference between an instantaneous rate and an average rate is similar to the situation you encounter when making a trip. The instantaneous rate is the reading of your speedometer as you glance down at the meter. The average rate is obtained by deciding what distance you traveled over several hours. With chemical reactions, the situation is

identical. The instantaneous rate begins (with the inception of the reaction) at its maximum value, and continues to decrease as reactants are consumed. The average rate is a measure of how long the reaction requires to consume a certain amount of the reactant(s).

(b) The steepness of the plot is an indication of the *instantaneous rate,* with that rate decreasing as the reaction proceeds.

(c) As noted above, as the concentration of dye decreases, the reaction rate slows. The rate of a reaction is proportional to the concentration of reactants!

97. For reactions to occur (as seen on Screen 15.9 of the *General Chemistry Now CD-ROM,*

  (a) Conditions necessary for two molecules to react:

   1. molecules must collide

   2. collisions must be of sufficient energy and

   3. collisions must be with particles in specific orientation with respect to one another

  (b) For the two animations, in animation 2, the molecules are moving faster—so the temperature here must be higher than in animation 1. Recall that molecular velocity is a function of T:

  $1/2 \, m\mathbf{V}^2 = 3/2 \, k\mathbf{T}$.

  (c) The reaction between ozone and N_2 will be *less* sensitive to orientation than that between ozone and NO, since with N_2, any atom contacted will be a "N" atom. With NO, only 1/2 of the atoms contacted are "N" atoms—the other of course is "O".

99. For the reaction showing the iodide ion-catalyzed decomposition of H_2O_2 on Screen 15.14:

  (a) Is the iodide-ion a catalyst in the mechanism described?

   Step 1   (slow rate-determining step)
              $H_2O_2(aq) + I^-(aq) \longrightarrow IO^-(aq) + H_2O(\ell)$

   Step 2 $H_2O_2(aq) + IO^-(aq) \longrightarrow I^-(aq) + H_2O(\ell) + O_2(g)$
   _____
   Overall $2\,H_2O_2(aq) \longrightarrow 2\,H_2O(\ell) + O_2(g)$

   (Mechanism copied from Screen 15.14 with permission)

  Note that the I^- ion is consumed in the first step, but is produced in the second step. The net effect is that I^- is **not** consumed—i.e. it functions as a catalyst.

(b) How does the reaction coordinate diagram show that the catalyzed reaction is expected to be

faster than the uncatalyzed reaction?
(Diagram copied from Screen 15.14
with permission)

Note that the catalyzed reaction has a **lower** Ea than that of the uncatalyzed reaction—with the concommitant expectation of a greater rate.

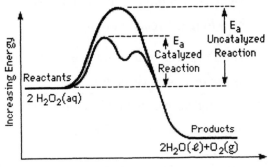

Chapter 16:
Principles of Reactivity : Chemical Equilibria

Practicing Skills
Writing Equilibrium Constant Expressions

1. Equilibrium constant expressions:

(a) $K = \dfrac{[H_2O]^2[O_2]}{[H_2O_2]^2}$ (b) $K = \dfrac{[CO_2]}{[CO][O_2]^{1/2}}$ (c) $K = \dfrac{[CO]^2}{[CO_2]}$ (d) $K = \dfrac{[CO_2]}{[CO]}$

Note that **solids** in the equations are *omitted* in the equilibrium constant expressions.

The Equilibrium Constant and Reaction Quotient

3. The equilibrium expression for the reaction: $I_2 (g) \Leftrightarrow 2 I (g)$ has a $K = 5.6 \times 10^{-12}$

Substituting the molar concentrations into the equilibrium expression:

$$Q = \frac{[I]^2}{[I_2]} = \frac{(2.00 \times 10^{-8})^2}{(2.00 \times 10^{-2})} = 2.0 \times 10^{-14}$$

Since $Q < K$, the system **is not at equilibrium** and will move to **the right to make more product (and reach equilibrium)**.

5. Is the system at equilibrium?

$[SO_2]$	5.0×10^{-3} M
$[O_2]$	1.9×10^{-3} M
$[SO_3]$	6.9×10^{-3} M

Substituting into the equilibrium expression: $\dfrac{[SO_3]^2}{[SO_2]^2[O_2]} = \dfrac{[6.9 \times 10^{-3}]^2}{[5.0 \times 10^{-3}]^2[1.9 \times 10^{-3}]} = 1000$

Since $Q > K$, the system is **not** at equilibrium, and will move "to the left" to make more reactants to reach equilibrium.

Calculating an Equilibrium Constant

7. For the equilibrium: $PCl_5 (g) \Leftrightarrow PCl_3 (g) + Cl_2 (g)$, calculate K

The equilibrium concentrations are: $[PCl_5] = 4.2 \times 10^{-5}$ M

$[PCl_3] = 1.3 \times 10^{-2}$ M

$[Cl_2] = 3.9 \times 10^{-3}$ M

The equilibrium expression is : $\dfrac{[PCl_3][Cl_2]}{[PCl_5]} = \dfrac{[1.3 \times 10^{-2}][3.9 \times 10^{-3}]}{[4.2 \times 10^{-5}]} = 1.2$

9. The equilibrium expression is : $K = \dfrac{[CO]^2}{[CO_2]}$

The quantities given are **moles**, so first calculate the **concentrations at equilibrium**.

$[CO] = \dfrac{0.10 \text{ moles}}{2.0 \text{ L}} = 0.050 \text{ M}$ $[CO_2] = \dfrac{0.20 \text{ moles}}{2.0 \text{ L}} = 0.10 \text{ M}$

(a) $K = \dfrac{[CO]^2}{[CO_2]} = \dfrac{[0.050]^2}{[0.10]} = 2.5 \times 10^{-2}$

(b & c) The only change here is in the amount of carbon. Since C does not appear in the equilibrium expression, K would be the same as in (a).

11. To keep track of concentrations, construct a table:

The reaction is $CO\ (g) + Cl_2\ (g) \Leftrightarrow COCl_2(g)$

	CO	Cl$_2$	COCl$_2$
Initial	0.0102 M	0.00609 M	0 M
Change	- 0.00308 M	- 0.00308 M	+ 0.00308 M
Equilibrium	0.00712 M	0.00301M	+ 0.00308 M

(a) Equilibrium concentrations of CO and COCl$_2$ is found by noting that 0.00308 M Cl$_2$ is consumed, an equimolar amount of CO is consumed and an equal amount of COCl$_2$ is produced.

(b) $K = \dfrac{[+0.00308]}{[0.00712][0.00301]} = 144$

Using Equilibrium Constants

13. For the system butane $\Leftrightarrow$ isobutane K = 2.5

Equilibrium concentrations may be found using a table. First note that the amount of butane must be converted to molar concentrations. So 0.017mol butane/0.50 L = 0.034 M.

	butane	isobutane
Initial	0.034	0
Change	- x	+ x
Equilibrium	0.034 - x	x

Substituting these equilibrium concentrations into the equilibrium expression:

$$K = \dfrac{x}{0.034 - x} = 2.5 \text{ and } x = 2.4 \times 10^{-2}$$

and [isobutane] $= 2.4 \times 10^{-2}$ M, [butane] $= 0.034 - x = 1.0 \times 10^{-2}$ M

15. For the equilibrium, the equilibrium constant is $\frac{[I]^2}{[I_2]}$ = 3.76 x 10^{-3}

The initial concentration of I$_2$ is $\frac{0.105 mol}{12.3 L}$ or 8.54 x 10^{-3}M

The equation indicates that 2 mol of I form for each mol of I$_2$ that reacts.

If some amount, say x M, of I$_2$ reacts, then the amount of I that forms is 2x, and the amount of

I$_2$ remaining at equilibrium is then (8.54 x 10^{-3} - x)

The equilibrium concentrations can then be substituted into the equilibrium expression.

$$\frac{(2x)^2}{8.54 \times 10^{-3} - x} = 3.76 \times 10^{-3}$$

or $4x^2$ = 3.76 x 10^{-3}(8.54 x 10^{-3} - x) rearranging we get:

$4x^2$ + 3.76 x 10^{-3} x - 3.21 x 10^{-5} = 0.

Solve this using the quadratic equation. Using the positive solution to that equation we find

that x = 2.40 x 10^{-3}.

So at equilibrium:

[I$_2$] = 8.54 x 10^{-3} - 2.40 x 10^{-3} = 6.14 x 10^{-3} M

[I] = 2(2.40 x 10^{-3}) = 4.79 x 10^{-3} M

17. Given at equilibrium : $K = \frac{[CO][Br_2]}{[COBr_2]}$ = 0.190 at 73°C

First, calculate concentrations: [COBr2] = $\frac{0.500 \ mol}{2.00 \ L}$ = 0.250 M

Substituting the[COBr2] into the equilibrium expression we get: $\frac{[CO][Br_2]}{(0.250)}$ = 0.190

Note that the stoichiometry of the equation tells us that for **each CO, we obtain 1 Br$_2$**.

	COBr$_2$	CO	Br2
Initial	0.250 M	0 M	0 M
Change	-x	+ x	+ x
Equilibrium	0.250 -x	+ x	+ x

We can rewrite the expression to read: $\frac{[CO][Br_2]}{0.250 - x}$ = $\frac{[x]^2}{0.250 - x}$ = 0.190

Solve this using the quadratic equation. $x^2 + 0.190x - 0.0475 = 0$

Using the positive solution to that equation, x = 0.143 M = [CO] and [COBr2] =

(0.250-0.143) = 0.107M.

The percentage of $COBr_2$ that has decomposed is: $\dfrac{0.143 \text{ M}}{0.250 \text{ M}} \bullet 100 = 57.1\%$

Manipulating Equilibrium Constant Expressions

19. To compare the two equilibrium constants, write the equilibrium expressions for the two.

$$\dfrac{[C]^2}{[A][B]} = K_1 \quad \text{and} \quad \dfrac{[C]^4}{[A]^2[B]^2} = K_2$$

Note that the first expression is equal to the second expression *squared*. So **(b) $K_2 = K_1{}^2$**

21. Comparing the two equilibria:

    (1) SO_2 (g) + $\frac{1}{2}$ O_2 (g) $\Leftrightarrow$ SO_3 (g) K_1

    (2)   2 SO_3 (g) $\Leftrightarrow$ 2 SO_2 (g) + O_2 (g) K_2

The expression that relates K_1 and K_2 is (e): $K_2 = \dfrac{1}{K_1{}^2}$

Reversing equation 1 gives: SO_3 (g) $\Leftrightarrow$ SO_2 (g) + $\frac{1}{2}$ O_2 (g) with the eq. constant $\dfrac{1}{K_1}$

Multiplying the modified equation 1 x 2 gives: 2 SO_3 (g) $\Leftrightarrow$ 2 SO_2 (g) + O_2 (g)   and the modified eq. constant $(\dfrac{1}{K_1})^2$

23. Calculate K for the reaction:
    SnO_2 (s) + 2 CO (g) $\Leftrightarrow$ Sn (s) + 2 CO_2 (g) given:

    (1) SnO_2 (s) + 2 H_2 (g) $\Leftrightarrow$ Sn (s) + 2 H_2O (g)   K = 8.12

    (2) H_2 (g) + CO_2 (g) $\Leftrightarrow$ H_2O (g) + CO (g)   K = 0.771

Take equation 2 and reverse it, then multiply by 2 to give:

    $2 H_2O$ (g) + 2 CO (g) $\Leftrightarrow$ 2 H_2 (g) + 2 CO_2 (g) $K = (\dfrac{1}{0.771})^2 = 1.68$

add equation 1 : SnO_2 (s) + 2 H_2 (g) $\Leftrightarrow$ Sn (s) + 2 H_2O (g)   K = 8.12

    SnO_2 (s) + 2 CO (g) $\Leftrightarrow$ Sn (s) + 2 CO_2 (g) $K_{net} = 8.12 \bullet 1.68 = 13.7$

Disturbing a Chemical Equilibrium

25. The equilibrium may be represented : N_2O_3 (g) + heat $\Leftrightarrow$ NO_2 (g) + NO (g)

Since the process is **endothermic** ($\Delta H = +$), heat is absorbed in the "left to right" reaction.

The effect of:

(a) Adding more N_2O_3 (g): An increase in the Pressure of N_2O_3 (adding more N_2O_3) will shift the equilibrium to the **right**, producing more NO_2 and NO.

(b) Adding more NO_2 (g): An increase in the Pressure of NO_2 (adding more NO_2) will shift the equilibrium to the **left**, producing more N_2O_3 .

(c) Increasing the volume of the reaction flask: If the volume of the flask is increased, the pressure will drop. (Remember that P for gases is inversely related to volume.) A drop in pressure will favor that side of the equilibrium with the "larger total number of moles of gas"—so this equilibrium will shift to the **right**, producing more NO_2 and NO.

(d) Lowering the temperature: You should note that a change in T (up or down) will result in a **change in the equilibrium constant**. (None of the three changes mentioned above change K!) However, the same principle applies. The removal of heat (a decrease in T) favors the exothermic process (shifts the equilibrium to the **left**) producing more N_2O_3

27. K for butane $\Leftrightarrow$ iso-butane is 2.5.

(a) Equilibrium concentration if 0.50 mol/L of iso-butane is added:

	butane	iso-butane
Original concentration	1.0	2.5
Change immediately after addition	1.0	2.5 + 0.50
Change (going to equilibrium)	+ x	- x
Equilibrium concentration	1.0 + x	3.0 - x

Substituting into the equilibrium expression: $K = \dfrac{[\text{iso-butane}]}{[\text{butane}]} = \dfrac{3.0 - x}{1.0 + x} = 2.5$

$$3.0 - x = 2.5\,(1.0 + x) \quad \text{and } 0.14 = x$$

The equilibrium concentrations are:

[butane] = 1.0 + x = 1.1 M and [iso-butane] = 3.0 - x = 2.9 M

(b) Equilibrium concentrations if 0.50 mol /L of butane is added:

	butane	iso-butane
Original concentration	1.0	2.5
Change immediately after addition	1.0 + 0.50	2.5
Change (going to equilibrium)	- x	+ x
Equilibrium concentration	1.5 - x	2.5 + x

$K = \dfrac{[\text{iso-butane}]}{[\text{butane}]} = \dfrac{2.5 + x}{1.5 - x} = 2.5$ and solving: $2.5 + x = 2.5\,(1.5 - x)$ and $x = 0.36$

The eq. concentrations are:[butane] = 1.5 - x = 1.1 M and [iso-butane] = 2.5 + x = 2.9 M

General Questions

29. For the equilibrium: Br_2 (g) $\Leftrightarrow$ $2Br$ (g), we must first calculate the concentration of Br_2

$$[Br_2] = \frac{0.086 \text{ mol}}{1.26 \text{ L}} = 0.068 \text{ M}$$

Then we can complete an equilibrium table:

	Br_2	Br
Initial	0.068 M	0
Change	- (0.037)(0.068) M	+2 (0.037)(0.068) M
Equilibrium	0.066 M	+0.0051 M

Substituting into the K expression: $K = \frac{[Br]^2}{[Br_2]} = \frac{(0.0051)^2}{(0.066)} = 3.9 \times 10^{-4}$

31. K_p is 6.5×10^{11} at 25°C for $CO(g) + Cl_2$ (g) $\Leftrightarrow$ $COCl_2$ (g)

What is the value of K_p for $COCl_2$ (g) $\Leftrightarrow$ $CO(g) + Cl_2$ (g)

Writing the equilibrium expressions for these two processes will show that one is the *reciprocal* of the other. Then K_p for the two will be mathematically related by the reciprocal relationship. So K_p for the second reaction is $1/K_p$ for the first reaction

or $1/6.5 \times 10^{11}$ or 1.5×10^{-12}.

33. Calculate K for the system CS_2 (g) + 3 Cl_2 $\Leftrightarrow$ S_2Cl_2 (g) + CCl_4 (g)

An equilibrium table will help keep track of amounts:

Note that this occurs in a 1.00 L flask. The **molar concentrations** and the **number of moles** will be numerically identical.

	CS_2	Cl_2	S_2Cl_2	CCl_4
Initial	1.2	3.6	0	0
Change	- 0.90	- 3 • 0.90	+0.90	+0.90
Equilibrium	0.3	0.9	0.90	0.90

Equilibrium concentrations are easily determined by noting the stoichiometry of the reaction. If 0.90 mol of CCl_4 are formed (to reach equilibrium), an equal amount of S_2Cl_2 is formed and an equal amount of CS_2 reacted. The amount of Cl_2 that reacts is found by noting the 1:3 ratio of CS_2 to Cl_2 . If 0.90 mol of CS_2 react then 3 x 0.90 mol of Cl_2 react.

Substituting into the equilibrium expression gives: $\dfrac{[S_2Cl_2][CCl_4]}{[CS_2][Cl_2]^3} = \dfrac{[0.90][0.90]}{[0.3][0.9]^3} = 4$ (1 sf)

35. K for butane $\Leftrightarrow$ iso-butane is 2.5. Is system at equilibrium if 1.75 mol of butane and 1.25 mol of iso-butane are mixed.

Substituting into the equilibrium expression: $K = \dfrac{[\text{iso-butane}]}{[\text{butane}]} = \dfrac{1.25 \text{ mol}}{1.75 \text{ mol}} = 0.714$.

The system is not at equilibrium (since Q is not equal to K). Since Q is less than K, the system will need to shift to the right (producing more isobutane) to reach equilibrium.

	butane	iso-butane
Original concentration	1.75	1.25
Change (going to equilibrium)	- x	+ x
Equilibrium concentration	1.75 - x	1.25 + x

$K = \dfrac{[\text{iso-butane}]}{[\text{butane}]} = \dfrac{1.25 + x}{1.75 - x} = 2.5$ and solving for x: $2.5(1.75 - x) = (1.25 + x)$

Expanding: $4.375 - 2.5x = 1.25 + x$; $(4.375-1.25) = 3.5x$; $3.125 = 3.5 x$; $x = 0.89$

The equilibrium concentrations are: $[\text{butane}] = 1.75 - x = 0.86$ M and $[\text{iso-butane}] = 1.25 + x = 2.14$ M

37. For the two equilibria shown, which correctly relates the two equilibrium constants?
K_1 $NOCl(g) \Leftrightarrow NO (g) + 1/2 Cl_2 (g)$
K_2 $2 NO (g) + Cl_2 (g) \Leftrightarrow 2 NOCl(g)$

One way to answer the question is to write the equilbrium expressions for the two equilibria:

$K_1 = \dfrac{[NO][Cl_2]^{1/2}}{[NOCl]}$ and $K_2 = \dfrac{[NOCl]^2}{[NO]^2[Cl_2]}$.

Note two differences between the K expressions: (1) they are reciprocals of each other—in one case with NOCl in the denominator, and in the other with NOCl in the numerator. Mentally invert K_2. Note the 2nd difference. K_2 is $(K_1)^2$. So the relationship between the two is:

(c) $K_2 = \dfrac{1}{K_1^2}$

39. For the equilibrium : $BaCO_3 (s) \Leftrightarrow CO_2 (g) + BaO (s)$, the effect of:

(a) adding $BaCO_3$ (s): (i) no effect. Since the solid does not appear in the equilibrium expression, the addition of more solid has no effect on the position of equilibrium.

(b) adding CO_2 : Increasing the concentration of CO_2 would shift the equilbrium (ii) to the left—consuming CO_2 until the system returned to equilibrium.

(c) Adding BaO: As in (a) above, since the solid, BaO, does not appear in the equilibrium expression, adding BaO would have (i) no effect.

(d) Raising the temperature: Raising T would have the effect of accelerating the decomposition of the carbonate, producing more CO_2.—shifting the equilibrium (iii) to the right

(e) Increasing the volume of the flask: This decreases the concentration of CO_2. and (iii) shifts the equilibrium to the right—just as in (b) increasing the CO_2 concentration shifted the equilibrium to the left.

41. For the equilibrium: PCl_5 (g) $\Leftrightarrow$ PCl_3 (g) + Cl_2 (g), calculate K

	$[PCl_5]$	$[PCl_3]$	$[Cl_2]$
Equilibrium (expressed as mol)	3.120g/208.24 = 0.01498 mol	3.845/137.33 = 0.02800 mol	1.787/70.91 = 0.02520 mol
Concentration immediately after addition	0.01498 mol	0.02800 mol	0.02520 mol + 0.02000 mol (1.418g/70.91g/mol)
Change (going to equilibrium).	+x	-x	-x
New equilibrium	0.01498 + x	0.02800 - x	0.04520 - x

(a) The addition of the chlorine will shift the equilibrium to the *left*

(b) The new equilibrium concentrations will be found using the values from the table, but we must know the value of K, which we can get by using the equilibrium concentrations given in the problem.

$$K = \frac{[PCl_3][Cl_2]}{[PCl_5]} = \frac{[0.02800][0.02520]}{[0.01498]} = 0.0470$$

Substituting the values from our table above: $\frac{[0.02800 - x][0.04520 - x]}{[0.01498 + x]} = 0.0470$

The resulting quadratic equation can be solved for x. The "sensible" root for x= 0.00485 M.

The resulting equilibrium concentrations are:

$[PCl_5]$ = 0.01498 + x = 0.01498 + 0.00485 = 0.0199 M

$[PCl_3]$ = 0.02800 – x = 0.02800 - 0.00485 = 0.0231 M

$[Cl_2]$ = 0.04520 – x = 0.04520 - 0.00485 = 0.0403 M

43. For the system: $NH_4I(s) \Leftrightarrow NH_3(g) + HI(g)$, what is K_p ?

Given that the total pressure is attributable **only** to $NH_3 + HI$, **and** that the stoichiometry of the equation tells us that equal amounts of the two substances are formed, we can easily determine the pressure of **each** of the gases. Since we desire K_p in atmospheres, we must first convert 705 mm Hg to atmospheres of pressure:

$$705 \text{ mm Hg} \cdot \frac{1 \text{ atm}}{760 \text{ mm Hg}} = 0.928 \text{ atm}$$

$P_{total} = P(NH_3) + P(HI) = 0.928$ atm so the pressure of each gas is 1/2(0.928) or 0.464 atm.
$K_p = P(NH_3) \cdot P(HI) = (0.464 \text{ atm})^2 = 0.215$

45. The equilibrium constant K_p for: $N_2O_4(g) \Leftrightarrow 2 NO_2(g)$ is 0.15 at 25 °C. If the pressure of N_2O_4 is 0.85 atm at equilibrium what the the **total** pressure?

$$K_p = \frac{[NO_2]^2}{[N_2O_4]} = 0.15,$$ since we know the value for N_2O_4 concentration, we can calculate the

value for NO_2 concentration. $0.15 \cdot 0.85 = [NO_2]^2$ and 0.36 atm = $[NO_2]$.

The **total pressure is** $P(NO_2) + P(N_2O_4) = 0.36$ atm + 0.85 atm = 1.21 atm

47. For the decomposition of ammonia at 450 °C, K = 6.3.

What are equilibrium concentrations of NH_3, N_2, and H_2?, and

What is the **total** pressure in the flask?

3.6 mol NH_3 in a 2.0 L vessel is 1.8M NH_3. Let's establish a reaction table:

	$[NH_3]$	$[N_2]$	$[H_2]$
Initial concentrations (mol/L)	1.8	0	0
Change (going to equilibrium).	- x	+ x/2	+3x/2
New equilibrium	1.8 - x	x/2	3x/2

Substituting into the K expression:

$$6.3 = \frac{\left[\frac{x}{2}\right]\left[\frac{3x}{2}\right]^3}{[1.8-x]^2} \text{ and simplifying gives: } 6.3 = \frac{\frac{27x^4}{16}}{[1.8-x]^2}$$

Simplifying further: $6.3 = \dfrac{1.6875x^4}{[1.8-x]^2}$, taking the square root of both sides will simplify

the math: $2.51 = \dfrac{1.299x^2}{[1.8-x]}$ and $2.51[1.8-x] = 1.299x^2$.

Expanding gives: $4.4519 - 2.51x = 1.299x^2$, and using the quadratic equation to solve gives $x = 1.13$. Using the table above, we get:

$[NH_3] = 1.8 - x = 0.67$ M; $[N_2] = x/2 = 0.565$ M; $[H_2] = 3x/2 = 1.70$ M

Total pressure in flask: $P = nRT/V$. and since $M = n/M$, we can use $P = MRT$

$P = (0.67 + 0.565 + 1.70) \bullet 0.082057 \dfrac{L \bullet atm}{K \bullet mol} \bullet (723K) = 174$ atm

49. K_c for the decomposition of NH_4HS into ammonia and hydrogen sulfide gases is 1.8×10^{-4}

(a) When pure salt decomposes, the equilibrium concentrations of $[NH_3]$ and $[H_2S]$:

$Kp = [NH_3][H_2S] = 1.8 \times 10^{-4}$. Noting the stoichiometry of the reaction, the two

concentrations will be equal. $[NH_3]^2 = [H_2S]^2 = 1.8 \times 10^{-4}$ and

$[NH_3] = [H_2S] = 1.3 \times 10^{-4}$ M.

(b) If NH_4HS is placed into a flask containing $[NH_3] = 0.020$ M, When system achieves

equilibrium, what are equilibrium concentrations?
The equilibrium table:

	[NH3]	[H2S]
Initial concentration	0.020	0
Change (going to equilibrium).	+ x	+ x
New equilibrium	0.020 + x	+ x

$Kp = [NH_3][H_2S] = 1.8 \times 10^{-4}$ and $[0.020 + x][+ x] = 1.8 \times 10^{-4}$

Simplifying gives: $0.020x + x^2 = 1.8 \times 10^{-4}$. This will require the quadratic equation to

resolve the equation: $x^2 + 0.020x - 1.8 \times 10^{-4} = 0$. Solving gives $x = 6.7 \times 10^{-3}$ M.

The equilibrium concentrations are then $[NH_3] = 0.020 + 6.7 \times 10^{-3}$ or 0.0267 M

(or 0.027 M to 2 sf) and for $[H_2S] = 6.7 \times 10^{-3}$ M

51. Given Kp for $N_2O_4(g) \Leftrightarrow 2 NO_2(g)$ is 0.148 at 25 °C.

(a) If total P is 1.50 atm, what fraction of N_2O_4 has dissociated?

$$P_{total} = P_{NO_2} + P_{N_2O_4} = 1.50 \text{ atm} \; ; \; P_{N_2O_4} = 1.50 - P_{NO_2}$$

$$K_p = \frac{P^2_{NO_2}}{P_{N_2O_4}} = 0.148$$

$$K_p = \frac{P^2_{NO_2}}{1.50 - P_{NO_2}} = 0.148 \text{ or } 0.148(1.50 - P_{NO_2}) = P^2_{NO_2}$$

and rearranging $\quad 0.222 - 0.148 \, P_{NO_2} = P^2_{NO_2}$

and solving for P_{NO_2} (with the quadratic equation) yields:

$P_{NO_2} = 0.403$ atm and $P_{N_2O_4} = 1.10$ atm

To determine the fraction of N_2O_4 that has dissociated, we need to know the amount of N_2O_4 that was originally present. The stoichiometry tells us that we get 2 NO_2 for each N_2O_4 that dissociates. The 0.403 atm of NO_2 that exist indicate that (0.403/2) atm of N_2O_4 dissociated. The **original pressure** of N_2O_4 will then be 1.10 atm + 0.20atm or 1.30 atm. The fraction that has dissociated is then: (1.30-1.10)/1.30 or 0.15.

(b) The fraction dissociated if the total equilibrium pressure falls to 1.00 atm:

$$P_{total} = P_{NO_2} + P_{N_2O_4} = 1.00 \text{ atm} \; ; \; P_{N_2O_4} = 1.00 - P_{NO_2}$$

$$K_p = \frac{P^2_{NO_2}}{P_{N_2O_4}} = 0.148 = \frac{P^2_{NO_2}}{1.00 - P_{NO_2}} \; ; \; 0.148(1.00 - P_{NO_2}) = P^2_{NO_2}$$

Solving via the quadratic equation: $P_{NO_2} = 0.318$ atm and $P_{N_2O_4} = 0.682$ atm

The equilibrium pressure of 0.682 atm for N_2O_4 tells us that the **original pressure of N_2O_4 was 0.682 + 1/2(0.318) or 0.841 atm**. [Recall that 0.318 atm of NO_2 represents 0.159 atm of N_2O_4 decomposing].

The fraction dissociated is: 0.159 atm/0.841 atm = 0.189 or approximately 19%.

<u>53</u>. For the equilibrium : $(NH_3)[B(CH_3)_3] \Leftrightarrow B(CH_3)_3 + NH_3 \qquad Kp = 4.62$

Substituting $(CH_3)_3P$ for NH_3 gives an equilibrium with Kp =0.128 while substituting $(CH_3)_3 N$ gives an equilibrium with Kp = 0.472.

(a) To determine which of the three systems would provide the largest concentration of $B(CH_3)_3$, one needs to ask "What does K tell me?"

One answer to that question is "the extent of reaction—that is the larger the value of K, the more products are formed. Hence **the system with the largest K_p would give the largest concentration of $B(CH_3)_3$ at equilibrium.**

(b) Given $(NH_3)[B(CH_3)_3] \Leftrightarrow B(CH_3)_3 + NH_3$ $K_p = 4.62$

Since we have a K_p, we need to express "concentrations" as Pressures:

$$P = \frac{0.010 \text{ mol} \bullet 0.082057 \dfrac{L \bullet atm}{K \bullet mol} \bullet 373K}{0.100L} = 3.1 \text{ atm}$$

	$(NH_3)[B(CH_3)_3]$	$B(CH_3)_3$	NH_3
Initial:	3.1 atm	0 M	0 M
Change	- x	+ x	+ x
Equilibrium	3.1 - x	+x	+ x

Substituting into the equilibrium expression we have: $K_p = \dfrac{(x)(x)}{3.1 - x} = 4.62$

and simplifying gives : $x^2 = 4.62(3.1 - x)$ and $x^2 + 4.62x - 14.1 = 0$

Solving via the quadratic formula gives $x = 2,1$ arm

So the **concentrations at equilibrium** are:

$[B(CH_3)_3] = [NH_3] = 2.1$ atm and $[(NH_3)[B(CH_3)_3]] = 1.0$ atm

The percent dissociation of $(NH_3)[B(CH_3)_3]$ is:

$$\frac{\text{amount changed}}{\text{original amount}} = \frac{2.1 \text{ atm}}{3.1 \text{ atm}} \times 100 = 69 \text{ \% dissociated}$$

55. For the reaction of hemoglobin (Hb) with CO we can write: $K = \dfrac{[HbCO][O_2]}{[HbO_2][CO]} = 2.0 \times 10^2$

If $\dfrac{[HbCO]}{[HbO_2]} = 1$, substitution into the K expression shows that $\dfrac{[O_2]}{[CO]} = 2.0 \times 10^2$

If $[O_2] = 0.20$ atm then $\dfrac{0.20 \text{ atm}}{[CO]} = 2.0 \times 10^2$ and solving for $[CO] = 1 \times 10^{-3}$ atm.

So a partial pressure of $[CO] = 1 \times 10^{-3}$ atm would likely be fatal.

57. How many O atoms are present if 1.0 mol of O_2 is placed in a 10.L vessel at 1800 K?

Given K_p for $O2 (g) \Leftrightarrow 2 O (g) = 1.2 \times 10^{-10}$

Since we need to express the # of O atoms in a 10. L vessel, let's convert K_p into K_c:

Since $K_p = K_c(RN)^{\Delta n}$ then $K_c = K_p/(RN)^{\Delta n} = 1.2 \times 10^{-10}/(0.082057 \dfrac{L \bullet atm}{K \bullet mol} \bullet 1800K)^1$

So $K_c = 8.12 \times 10^{-13}$.

Substituting into the Kc expression, we get: $\dfrac{[O]^2}{[O_2]} = 8.12 \times 10^{-13}$

If we let a mol/L of O_2 dissociate into atoms, we get: $\dfrac{[2a]^2}{[0.10-a]} = 8.12 \times 10^{-13}$

So $4a^2 = 8.12 \times 10^{-13}(0.10 - a)$ and solving for a: $a = 1.4 \times 10^{-7}$. Note here that with the small size of K, the denominator could be simplified to 0.10, and an approximate value of a calculated. The [O] is then 2a or 2.8×10^{-7}M. To calculate the # of O atoms:

10. L • 2.8×10^{-7} mol/L • 6.02×10^{23} atoms/mol = 1.7×10^{18} O atoms.

59. For the equilibrium with boric acid (BA) + glycerin(gly), K = 0.90

    $B(OH)_3$ (aq) + glycerin(aq) $\Leftrightarrow$ $B(OH)_3$•glycerin(aq)

	BA	gly	BA•gly
Initial	0.10 M	?	0
Change	- 0.60*0.10 M		+ 0.60*0.10M
Equilibrium	0.040M	x	0.060 M

Substituting into the equilibrium expression yields:

$\dfrac{[BA \bullet gly]}{[BA][gly]} = 0.90$, and substituting from our table gives: $\dfrac{[0.060]}{[0.040][gly]} = 0.90$

Rearranging:

$$\dfrac{[0.060]}{[0.040][0.90]} = [gly] = 1.67 \text{ M (or 1.7 to 2 sf)}$$

Solving for [gly] gives the *equilibrium* amount of glycerin.

Recall, however that the equilibrium amount of glycerin represents the amount of glycerin remaining uncomplexed from the original amount. Recall that *some* glycerin is consumed in making the complex (in this case 0.060M). So the initial amount of glycerin present would be 1.67 + 0.060 or 1.73M—which is 1.7 M (to 2 sf).

61. For the reaction of N_2O_4 decomposing into NO_2, a sample of N_2O_4 at a pressure of 1.00atm reaches equilibrium with 20.0% of the N_2O_4 having been converted into NO_2.

(a) What is Kp?

	N_2O_4	NO_2
Initial	1.00 atm	0 atm
Change	- 0.200•1.00 atm	+ 2• 0.200•1.00 atm
Equilibrium	0.80 atm	0.40 atm

Substituting into the equilibrium expression gives:

$$K_p = \frac{P^2(NO_2)}{P(N_2O_4)} = \frac{(0.40)^2}{(0.80)} = 0.20$$

(b) If the initial pressure is 0.10 atm, the percent dissociation:

Now we know that $Kp = 0.20$. Substituting into the equilibrium expression:

$$K_p = \frac{P^2(NO_2)}{P(N_2O_4)} = \frac{(2x)^2}{(0.10 - x)} = 0.20 \quad \text{and rearranging: } 4x^2 = (0.10\text{-x})(0.20)$$

or $4x^2 = 0.020 - 0.20x$, and solving for the quadratic equation, $x = 0.050$ atm.

This represents half (or 50%) of the original N_2O_4.

Does this agree with LcChatelier's Principle: YES. We began with a lower concentration (pressure), so we expect that a larger percentage of the N_2O_4 will dissociate (50% compared to 20%), and it does.

Summary and Conceptual Questions

63. Decide upon the truth of each of the statements:
   (a) The magnitude of the equilibrium constant is always independept of T. **false**- K is always a function of the temperature.
   (b) The equilibrium constant for the net equation is the produce of the equilibrium constants of the summed equations—**true**.
   (c) The equilibrium constant for a reactions has the same value as K for the reverse reaction. **False**. The reverse reaction would have an equilibrium constant that is 1/K(orig).—that is the reciprocal of the original K.
   (d) For an equilibrium with only one non-solid reactant or product, only that concentration (in this case, CO_2) appears in he equilibrium constant expression—**true**.
   (e) For the equilibrium involving the decomposition of $CaCO_3$, the value of K is independent of the expression of the amount of CO_2, --**false; Kp = Kc(RT)$^{\Delta n}$**. with Δn equal to a non-zero number, Kp and Kc will differ by the factor $(RT)^{\Delta n}$

65. Characterize each of the following as product- or reactant-favored:
   (a) with $Kp = 1.2 \times 10^{45}$; The large magnitude of this K indicates the equilibrium would be **product-favored**.
   (b) with $Kp = 9.1 \times 10^{-41}$; The small magnitude of this K indicates the equilibrium would "lie to the left"—that is would be **reactant-favored**.

(c) $Kp = 6.5 \times 10^{11}$; equlibrium would "lie to the right"—that is would be **product-favored.**

<u>67</u>. (a) For the reaction of hydrogen and iodine to give HI, if Kc = 56 at 435 °C, what is Kp ?

The relation between the equilibrium constants is: $Kp = Kc(RT)^{\Delta n}$. The equation is:

$$H_2(g) + I_2(g) \Leftrightarrow 2 \; HI(g)$$

Note that the **total** number of moles of gas is 2 on both sides of the equilibrium, so that $\Delta n = 0$. So Kp = 56 .

(b) Mix 0.45 mol of each gas in a 10.0 L flask at 435 °C, what is the total pressure of the mixture before and after equilibrium?

Before equilibrium: P = nRT/V or

$$\frac{0.90 \; mol \; \bullet \; 0.082057 \frac{L \bullet atm}{K \bullet mol} \; \bullet \; 708 \; K}{10.0 \; L} = 5.2 \; atm$$ (Note that we can simply add the amount

of each gas (obviously the P of either gas will be 2.6 atm).

After equilibrium:

Substituting into the equilibrium expression gives:

$$K_p = \frac{P^2(HI)}{P(H_2) \bullet P(I_2)} = \frac{(2x)^2}{(2.6 - x)^2} = 56$$ Simplify the expression by taking the square root of

both sides to give: $\dfrac{(2x)}{(2.6 - x)} = 7.48$. Solving for x gives x = 2.06 atm, so

$P_{HI} = 2 \bullet 2.06$ or 4.1 atm (to 2 sf), $P(H_2) = P(I_2) = (2.6 - 2.06) = 0.54$ atm

The total pressure is : 4.1 atm + 0.54 atm + 0.54 atm = 5.2 atm

(c) The partial pressures as noted above are: $P_{HI} = 4.1$ atm , $P(H_2) = P(I_2) = 0.54$ atm

69. Since the addition of heat shifts an equilibrium to favor the "endothermic process", Since the $CoCl_4{}^{2-}$ ion is blue, and the tube on the left –in hot water--is blue (indicating the preponderance of the $CoCl_4{}^{2-}$ ion),   the formation of this ion is **endothermic.**

71.The principle of "microscopic reversibility" simply means that in an equilibrium step, the reaction can proceed in "either direction". So if we mix aqueous solution of Pb^{2+} and Cl^- ions, solid $PbCl_2$ will form. Alternatively, if we put solid $PbCl_2$ into water, Pb^{2+} and Cl^- ions will form.

Chapter 17
Principles of Reactivity: Chemistry of Acids and Bases

Practicing Skills

The Bronsted Concept

1. Conjugate Base of: Formula   Name
   (a) HCN CN^- cyanide ion

   (b) HSO_4^- SO_4^{2-} sulfate ion

   (c) HF F^- fluoride ion

3. Products of acid-base reactions:

   (a) HNO_3 (aq) + $H_2O(\ell)$ $\rightarrow$ H_3O^+(aq) + NO_3^-(aq)

         acid   base   conjugate acid   conjugate base

   (b) HSO_4^- (aq) + $H_2O(\ell)$ $\rightarrow$ H_3O^+(aq) + SO_4^{2-}(aq)

         acid   base   conjugate acid   conjugate base

   (c) H_3O^+(aq) + F^-(aq) $\rightarrow$ HF(aq) + $H_2O(\ell)$

         acid   base   conjugate acid   conjugate base

5. Hydrogen oxalate acting as Bronsted acid and Bronsted base:

   Bronsted acid: $HC_2O_4^-$ + OH^- $\rightarrow$ $C_2O_4^{2-}$ + H_2O

   Bronsted base: $HC_2O_4^-$ + HNO_3 $\rightarrow$ $H_2C_2O_4$ + NO_3^-

   The action of the hydrogen oxalate ion is determined by the substance in solution. If that

   substance is a *stronger acid than hydrogen oxalate*(HNO_3)' then hydrogen oxalate acts *as a*

   base. If the substance is a *weaker acid (or stronger base, e.g. hydroxide ion)*, then hydrogen

   oxalate acts *as an acid.*

7. (a) HCO_2H (aq) + H_2O (ℓ) $\rightarrow$ HCO_2^- (aq) + H_3O^+ (aq)

         acid   base   conjugate   conjugate
                                     of HCO_2H of H_2O

   (b) NH_3 (aq) + H_2S (aq) $\Leftrightarrow$ NH_4^+ (aq) + HS^- (aq)

         base   acid   conjugate   conjugate
                                     of NH_3 of H_2S

(c) HSO_4^- (aq) + OH^- (aq) $\Leftrightarrow$ SO_4^{2-} (aq) + H_2O (ℓ)

    acid   base   conjugate   conjugate

                                     of HSO_4^- of OH^-

pH Calculations

9. Since pH = 3.75, $[H_3O^+]$ = 10^{-pH} or $10^{3.75}$ or 1.8×10^{-4} M

Since the $[H_3O^+]$ is greater than 1×10^{-7} (pH< 7), the solution is acidic.

11. pH of a solution of 0.0075 M HCl:

Since HCl is considered a strong acid, a solution of 0.0075 M HCl has

$[H_3O^+]$ = 0.0075 or 7.5×10^{-3} ; pH = $- \log[H_3O^+]$ = $- \log[7.5 \times 10^{-3}]$ = 2.12

The hydroxide ion concentration is readily determined since $[H_3O^+] \cdot [OH^-]$ = 1.0×10^{-14}

$$[OH^-] = \frac{1.0 \times 10^{-14}}{7.5 \times 10^{-3}} = 1.3 \times 10^{-12} \text{ M}$$

13. pH of a solution of 0.0015 M $Ba(OH)_2$:

Soluble metal hydroxides are strong bases. To the extent that it dissolves ($Ba(OH)_2$ is not that

soluble), $Ba(OH)_2$ gives two OH^- for each formula unit of $Ba(OH)_2$.   0.0015 M $Ba(OH)_2$

would provide 0.0030 M OH^-. Since $[H_3O^+] \cdot [OH^-]$ = 1.0×10^{-14}

$[H_3O^+]$ would then be:

$$[H_3O^+] = \frac{1.0 \times 10^{-14}}{3.0 \times 10^{-3}} = 3.3 \times 10^{-12} \text{ M} \text{ and pH} = - \log[3.3 \times 10^{-12}] = 11.48.$$

Equilibrium Constants for Acids and Bases

15. Concerning the following acids:

Phenol		Formic acid		Hydrogen oxalate ion	
C_6H_5OH	1.3×10^{-10}	HCO_2H	1.8×10^{-4}	$HC_2O_4^-$	6.4×10^{-5}

(a) The strongest acid is formic. The weakest acid is phenol. Acid strength is proportional to the magnitude of K_a.

(b) Recall the relationship between acids and their conjugate base: $K_a \cdot K_b = K_w$

Since K_w is a constant, The greater the magnitude of K_a, the smaller the value of the K_b for the conjugate base. The strongest acid (formic) has the weakest conjugate base.

(c) The weakest acid (phenol) has the strongest conjugate base.

17. The substance which has the smallest value for K_a will have the strongest conjugate base. One can prove this quantitatively with the relationship: $K_a \cdot K_b = K_W$. An examination of Appendix H or Table 17.3 shows that--of these three substances--HClO has the smallest K_a, and ClO$^-$ will be the strongest conjugate base.

19. The equation for potassium carbonate dissolving in water:

$$K_2CO_3 \text{ (aq)} \rightarrow 2\,K^+ \text{ (aq)} + CO_3^{2-} \text{ (aq)}$$

Soluble salts--like K_2CO_3--dissociate in water. The carbonate ion formed in this process is a base, and reacts with the acid, water.

$$CO_3^{2-} \text{ (aq)} + H_2O \text{ (}\ell\text{)} \Leftrightarrow HCO_3^- \text{ (aq)} + OH^- \text{ (aq)}$$

The production of the hydroxide ion, a strong base, in this second step is responsible for the basic nature of solutions of this carbonate salt.

21. Most of the salts shown are sodium salts. Since Na^+ does not hydrolyze, we can estimate the acidity (or basicity) of such solutions by looking at the extent of reaction of the anions with water (hydrolysis).

The Al^{3+} ion is acidic, as is the $H_2PO_4^-$ ion. From Table 17.3 we see that the K_a for the hydrated aluminum ion is greater than that for the $H_2PO_4^-$ ion, making the Al^{3+} solution more acidic —lower pH than that of $H_2PO_4^-$. All the other salts will produce basic solutions and since the S^{2-} ion has the largest K_b, we will anticipate that the Na_2S solution will be most basic—i.e. have the highest pH.

pKa: A Logarithmic Scale of Acid Strength

23. The pK_a for an acid with a K_a of 6.5×10^{-5}. $pK_a = -\log(6.5 \times 10^{-5}) = 4.19$

25. K_a for epinephrine, whose $pK_a = 9.53$. $K_a = 10^{-pK_a}$ so $K_a = 10^{-9.53}$ or 3.0×10^{-10}

From Table 17.3, we see that epinephrine belongs between:

Hexaaquairon(II) ion $Fe(H_2O)_6^{2+}$ 3.2×10^{-10}

Hydrogen carbonate ion HCO_3^{1-} 4.8×10^{-11}

27. *2-chlorobenzoic* acid has a smaller pK_a than benzoic acid, so it *is the stronger acid.*

Ionization Constants for Weak Acids and Their Conjugate Bases

29. The K_b for the chloroacetate ion:

Recall the relationship between acids and their conjugate base: $K_a \cdot K_b = K_w$

K_b for the chloroacetate ion will be $\dfrac{1.0 \times 10^{-14}}{1.36 \times 10^{-3}} = 7.4 \times 10^{-12}$

31. The K_a for $(CH_3)_3NH^+$ is 10^{-pK_a} or $10^{-9.80} = 1.6 \times 10^{-10}$

Then $K_b = \dfrac{1.0 \times 10^{-14}}{1.6 \times 10^{-10}} = 6.3 \times 10^{-5}$

Predicting the Direction of Acid-Base Reactions

33. $CH_3CO_2H \ (aq) + HCO_3^- \ (aq) \Leftrightarrow CH_3CO_2^- \ (aq) + H_2CO_3 \ (aq)$

Since acetic acid is a stronger acid than carbonic acid, the equilibrium lies predominantly to the right.

35. Predict whether the equilibrium lies predominantly to the left or to the right:

(a) $NH_4^+ \ (aq) + Br^- \ (aq) \Leftrightarrow NH_3 \ (aq) + HBr \ (aq)$

HBr is a stronger acid than NH_4^+; equilibrium lies to left.

(b) $HPO_4^{2-} \ (aq) + CH_3CO_2^- \ (aq) \Leftrightarrow PO_4^{3-} \ (aq) + CH_3CO_2H \ (aq)$

CH_3CO_2H is a stronger acid than HPO_4^{2-}; equilibrium lies to left,

(c) $Fe(H_2O)_6^{3+} \ (aq) + HCO_3^- \ (aq) \Leftrightarrow Fe(H_2O)_5(OH)^{2+} \ (aq) + H_2CO_3 \ (aq)$

$Fe(H_2O)_6^{3+}$ is a stronger acid than H_2CO_3; equilibrium lies to right.

Types of Acid-Base Reactions

37. (a) The net ionic equation for the reaction of NaOH with Na_2HPO_4:

$$OH^-(aq) + HPO_4^{2-}(aq) \Leftrightarrow H_2O(\ell) + PO_4^{3-}(aq)$$

(b) The equilibrium lies to the right (since HPO_4^{2-} is a stronger acid than H_2O and OH^- is a stronger base than PO_4^{3-}), but phosphate and hydroxide ions aren't too different in basic strength, so the position of equilibrium does not lie very far to the right. The result is that remaining OH^- will result in a basic solution.

39. (a) The net ionic equation for the reaction of CH_3CO_2H with Na_2HPO_4:

$$CH_3CO_2H \text{ (aq)} + HPO_4^{2-}\text{(aq)} \Leftrightarrow CH_3CO_2^-\text{(aq)} + H_2PO_4^-\text{(aq)}$$

(b) The equilibrium lies to the right (since CH_3CO_2H ⁻is a stronger acid than $H_2PO_4^-$ and

HPO_4^{2-} is a stronger base than $CH_3CO_2^-$, The result is that the weak acetic acid and the

$H_2PO_4^-$ ion will result in a weakly acidic solution.

Using pH to Calculate Ionization Constants

41. For a 0.015 M HOCN, pH = 2.67:

(a) Since pH = 2.67, $[H_3O^+] = 10^{-pH}$ or $10^{-2.67}$ or 2.1 x 10 $^{-3}$ M

(b) The Ka for the acid is in general: $K_a = \dfrac{[H^+][A^-]}{[HA]}$

The acid is monoprotic, meaning that for each H^+ ion, one also gets a OCN^-

The two numerator terms are then equal. The concentration of the molecular acid is the

(original concentration – concentration that dissociates).

$$K_a = \frac{[2.1 \times 10^{-3}][2.1 \times 10^{-3}]}{[0.015 - 2.1 \times 10^{-3}]} = 3.5 \times 10^{-4}$$

43. With a pH = 9.11, the solution has a pOH of 4.89 and $[OH^-] = 10^{-4.89} = 1.3 \times 10^{-5}$ M.
The equation for the base in water can be written:

$$H_2NOH \text{ (aq)} + H_2O \text{ (}\ell\text{)} \Leftrightarrow H_3NOH^+ \text{ (aq)} + OH^- \text{ (aq)}$$

At equilibrium, $[H_2NOH] = [H_2NOH] - [OH^-] = (0.025 - 1.3 \times 10^{-5})$ or approximately
0.025 M.

$$K_b = \frac{[H_3NOH^+][OH^-]}{[H_2NOH]} = \frac{(1.3 \times 10^{-5})^2}{0.025} = 6.6 \times 10^{-9}$$

45. (a) With a pH = 3.80 the solution has a $[H_3O^+] = 10^{-3.80}$ or 1.6 x 10^{-4} M

(b) Writing the equation for the unknown acid, HA, in water we obtain:

$$HA \text{ (aq)} + H_2O \text{ (}\ell\text{)} \Leftrightarrow H_3O^+ \text{ (aq)} + A^- \text{ (aq)}$$

$[H_3O^+] = 1.6 \times 10^{-4}$ implying that $[A^-]$ is also 1.6 x 10^{-4}. Therefore the equilibrium

concentration of acid, HA, is (2.5 x 10^{-3} - 1.6 x 10^{-4}) or $\approx$ 2.3 x 10^{-3}.

$$K_a = \frac{[H_3O^+][A]}{[HA]} = \frac{(1.6 \times 10^{-4})^2}{2.3 \times 10^{-3}} = 1.1 \times 10^{-5}$$

We would classify this acid as a moderately weak acid.

Using Ionization Constants

47. For the equilibrium system: CH_3CO_2H (aq) + $H_2O(\ell) \Leftrightarrow CH_3CO_2^- + H_3O^+$(aq)

Initial	0.20 M			
Change	- x		+ x	+ x
Equilibrium	0.20 - x		+ x	+ x

Substituting into the K_a expression:

$$K_a = \frac{[H_3O^+][A]}{[HA]} = \frac{(x)^2}{0.20 - x} = 1.8 \times 10^{-5} \text{ and solving for } x = 1.9 \times 10^{-3}$$

So $[CH_3CO_2H] \cong 0.20$ M; $[CH_3CO_2^-] = [H_3O^+] = 1.9 \times 10^{-3}$ M

49. Using the same logic as in question 47 above, we can write:

$$HCN \text{ (aq)} + H_2O \text{ (ℓ)} \Leftrightarrow H_3O^+ \text{ (aq)} + CN^- \text{ (aq)}$$

Initial	0.025M			
Change	- x		+ x	+ x
Equilibrium	0.025 - x		+ x	+ x

Substituting these values into the K_a expression for HCN:

$$K_a = \frac{[CN^-][H_3O^+]}{[HCN]} = \frac{(x)^2}{(0.025 - x)} = 4.0 \times 10^{-10}$$

Assuming that the denominator may be approximated as 0.025 M, we obtain:

$$\frac{x^2}{0.025} = 4.0 \times 10^{-10} \text{ and } x = 3.2 \times 10^{-6}.$$

The equilibrium concentrations of $[H_3O^+] = [CN^-] = 3.2 \times 10^{-6}$ M.

The equilibrium concentration of $[HCN] = (0.025 - 3.2 \times 10^{-6})$ or 0.025 M.

Since x represents $[H_3O^+]$ the pH $= -\log(3.2 \times 10^{-6})$ or 5.50.

51. The equilibrium of ammonia in water can be written:

$$NH_3 + H_2O \Leftrightarrow NH_4^+ + OH^- \qquad K_b = 1.8 \times 10^{-5}$$

Initial	0.15 M		0	0
Change	- x		+ x	+ x
Equilibrium	0.15 - x		+ x	+ x

Substituting into the K_b expression:

$$K_b = \frac{[NH_4^+][OH^-]}{[NH_3]} = \frac{(x)^2}{(0.15 - x)} = 1.8 \times 10^{-5}$$

With K_b small, the extent to which ammonia reacts with water is slight. We can approximate the

denominator (0.15 - x) as 0.15 M, and solve the equation:

$$\frac{(x)^2}{0.15} = 1.8 \times 10^{-5} \text{ and } x = 1.6 \times 10^{-3}$$

So [NH_4^+] = [OH^-] = 1.6 x 10^{-3} M and [NH_3] = (0.15 -1.6 x 10^{-3}) ≈ 0.15 M

The pH of the solution is then:

$$pOH = - \log(1.6 \times 10^{-3}) = 2.80 \text{ and the pH} = (14.00-2.80) = 11.21$$

53. For CH_3NH_2 (aq) + $H_2O(\ell)$ ⇔ $CH_3NH_3^+$(aq) + OH^-(aq) $K_b = 4.2 \times 10^{-4}$

Using the approach of problem 51, the equilibrium expression can be written:

$$K_b = \frac{x^2}{0.25 - x} = 4.2 \times 10^{-4} \approx \frac{x^2}{0.25}$$

Solving the expression **with approximations** gives x = 1.0 x 10^{-2}; [OH^-] = 1.0 x 10^{-2} M
pOH would then be 1.99 and pH 12.01 (to 2 sf)

55. pH of 1.0 x 10^{-3} M HF; $K_a = 7.2 \times 10^{-4}$
Using the same method as in question 49, we can write the expression:

$$K_a = \frac{x^2}{(1.0 \times 10^{-3} - x)} = 7.2 \times 10^{-4}$$

The concentration of the HF and K_a preclude the use of our usual approximation

(1.0 x 10^{-3} - x ≈ 1.0 x 10^{-3})

So we multiply both sides of the equation by the denominator to get:

$$x^2 = 7.2 \times 10^{-4}(1.0 \times 10^{-3} - x)$$

$$x^2 = 7.2 \times 10^{-7} - 7.2 \times 10^{-4} x$$

Using the quadratic equation we solve for x:

and x = 5.6 x 10^{-4} M = [F^-] = [H_3O^+] and pH = 3.25

Acid-Base Properties of Salts

57. Hydrolysis of the NH_4^+ produces H_3O^+ according to the equilibrium:

$$NH_4^+ \text{ (aq)} + H_2O \text{ (l)} \Leftrightarrow H_3O^+ \text{ (aq)} + NH_3 \text{ (aq)}$$

With the ammonium ion acting as an acid, to donate a proton, we can write the K_a expression:

$$K_a = \frac{[NH_3][H_3O^+]}{[NH_4^+]} = \frac{K_w}{K_b} = \frac{1.0 \times 10^{-14}}{1.8 \times 10^{-5}} = 5.6 \times 10^{-10}$$

The concentrations of both terms in the numerator are equal, and the concentration of ammonium ion is 0.20 M. (Note the approximation for the **equilibrium** concentration of NH_4^+ to be equal to the **initial** concentration.) Substituting and rearranging we get

$$[H_3O^+] = \sqrt{0.20 \cdot 5.6 \times 10^{-10}} = 1.1 \times 10^{-5} \text{ and the pH} = 4.98.$$

59. The hydrolysis of CN^- produces OH^- according to the equilibrium:

$$CN^- \text{ (aq)} + H_2O \text{ (l)} \Leftrightarrow HCN \text{ (aq)} + OH^- \text{ (aq)}$$

Calculating the concentrations of Na^+ and CN^- :

$$[Na^+]_i = [CN^-]_i = \frac{10.8 \text{ g NaCN}}{0.500 \text{ L}} \cdot \frac{1 \text{ mol NaCN}}{49.01 \text{ g NaCN}} = 4.41 \times 10^{-1} \text{ M}$$

$$\text{Then } K_b = \frac{K_w}{K_a} = \frac{1.0 \times 10^{-14}}{4.0 \times 10^{-10}} = 2.5 \times 10^{-5} = \frac{[HCN][OH^-]}{[CN^-]}$$

Substituting the $[CN^-]$ concentration into the K_b expression and noting that:

$[OH^-]_e = [HCN]_e$ we may write

$[OH^-]_e = [(2.5 \times 10^{-5})(4.41 \times 10^{-1})]^{\frac{1}{2}} = 3.3 \times 10^{-3} \text{ M}$

$$[H_3O^+] = \frac{1.0 \times 10^{-14}}{3.3 \times 10^{-3}} = 3.0 \times 10^{-12} \text{ M}$$

pH After an Acid-Base Reaction

61. The net reaction is:

$$CH_3CO_2H \text{ (aq)} + NaOH \text{ (aq)} \Leftrightarrow CH_3CO_2^- \text{(aq)} + Na^+ \text{(aq)} + H_2O \text{(l)}$$

The addition of 22.0 mL of 0.15 M NaOH (3.3 mmol NaOH) to 22.0 mL of 0.15 M CH_3CO_2H (3.3 mmol CH_3CO_2H) produces water and the soluble salt, sodium acetate (3.3 mmol $CH_3CO_2^- Na^+$). The acetate ion is the anion of a weak acid and reacts with water according to the equation: $CH_3CO_2^- \text{ (aq)} + H_2O \text{ (l)} \Leftrightarrow CH_3CO_2H \text{ (aq)} + OH^- \text{ (aq)}$

The equilibrium constant expression is: $K_b = \dfrac{[CH_3CO_2H][OH^-]}{[CH_3COO^-]} = 5.6 \times 10^{-10}$

The concentration of acetate ion is: $\dfrac{3.3 \text{ mmol}}{(22.0+ 22.0) \text{ ml}} = 0.075 \text{ M}$

	$CH_3CO_2^-$	CH_3CO_2H	OH^-
Initial concentration	0.075		
Change	-x	+x	+x
Equilibrium	0.075 - x	+x	+x

$K_b = \dfrac{[CH_3CO_2H][OH^-]}{[CH_3CO_2^-]} = \dfrac{x^2}{0.075 - x} = 5.6 \times 10^{-10}$

Simplifying $(100 \cdot K_b \ll 0.075)$ we get $\dfrac{x^2}{0.075} = 5.6 \times 10^{-10}$; $x = 6.5 \times 10^{-6} = [OH^-]$

The hydrogen ion concentration is related to the hydroxyl ion concentration by the equation:

$K_w = [H_3O^+][OH^-] = 1.0 \times 10^{-14}$

$[H_3O^+] = \dfrac{1.0 \times 10^{-14}}{[OH^-]} = \dfrac{1.0 \times 10^{-14}}{6.5 \times 10^{-6}} = 1.5 \times 10^{-9}$ and pH = 8.81

The pH is greater than 7, as we expect for a salt of a strong base and weak acid.

63. Equal numbers of moles of acid and base are added in each case, leaving only the salt of the acid and base. The reaction (if any) of that salt with water (hydrolysis) will affect the pH.

pH of solution	Reacting Species	Reaction controlling pH
(a) >7	CH3CO2H/KOH	Hydrolysis of $CH_3CO_2^-$
(b) <7	HCl/NH3	Hydrolysis of NH_4^+
(c) = 7	HNO3/NaOH	No hydrolysis

Polyprotic Acids and Bases

65. (a) pH of 0.45 M H2SO3: The equilibria for the diprotic acid are:

$K_{a1} = \dfrac{[HSO_3^-][H_3O^+]}{[H_2SO_3]} = 1.2 \times 10^{-2}$ and $K_{a2} = \dfrac{[SO_3^{2-}][H_3O^+]}{[HSO_3^-]} = 6.2 \times 10^{-8}$

For the first step of dissociation:

	H2SO3	HSO3-	H3O+
Initial concentration	0.45 M		
Change	-x	+x	+x
Equilibrium	0.45 - x	+x	+x

Substituting in to the K_{a1} expression: $K_{a1} = \dfrac{(x)(x)}{(0.45-x)} = 1.2 \times 10^{-2}$

We must solve this expression with the quadratic equation since $(0.45 < 100 \cdot K_{a1})$.

The equilibrium concentrations for HSO_3^- and H_3O^+ ions are found to be 0.0677 M.

The further dissociation is indicated by K_{a2}.

Using the equilibrium concentrations from the first step, substitute into the K_{a2} expression.

	HSO_3^-	SO_3^{2-}	H_3O^+
Initial concentration	0.0677	0	0.0677
Change	-x	+x	+x
Equilibrium	0.0677 - x	+x	0.0677 + x

$K_{a2} = \dfrac{[SO_3^{2-}][H_3O^+]}{[HSO_3^-]} = \dfrac{(+x)(0.0677 + x)}{(0.0677 - x)} = 6.2 \times 10^{-8}$

We note that x will be small in comparison to 0.0677, and we simplify the expression:

$K_{a2} = \dfrac{(+x)(0.0677)}{(0.0677)} = 6.2 \times 10^{-8}$

In summary, the concentrations of HSO_3^- and H_3O^+ ions have been virtually unaffected

by the second dissociation.So $[H_3O^+] = 0.0677$ M and pH = 1.17

(b) The equilibrium concentration of SO_3^{2-} :

From the K_{a2} expression above: $[SO_3^{2-}] = 6.2 \times 10^{-8}$ M

67. (a) Concentrations of OH^-, $N_2H_5^+$, and $N_2H_6^{2+}$ in 0.010 M N_2H_4:

The K_{b1} equilibrium allows us to calculate $N_2H_5^+$ and OH^- formed by the reaction of

N_2H_4 with H_2O. $K_{b1} = \dfrac{[N_2H_5^+][OH^-]}{[N_2H_4]} = 8.5 \times 10^{-7}$

	N_2H_4	$N_2H_5^+$	OH^-
Initial concentration	0.010	0	
Change	-x	+x	+x
Equilibrium	0.010 - x	+x	0.0677 + x

Substituting into the K_{b1} expression : $\dfrac{(x)(x)}{0.010 - x} = 8.5 \times 10^{-7}$

We can simplify the denominator $(0.010 > 100 \cdot K_{b1})$.

$\dfrac{(x)(x)}{0.010} = 8.5 \times 10^{-7}$ and $x = 9.2 \times 10^{-5}$ M $= [N_2H_5^+] = [OH^-]$

The second equilibrium (K_{b2}) indicates further reaction of the $N_2H_5^+$ ion with water. The step should consume some $N_2H_5^+$ and produce more OH^-. The magnitude of K_{b2} indicates that the equilibrium "lies to the left" and we anticipate that not much $N_2H_6^{2+}$ (or additional OH^-) will be formed by this interaction.

	$N_2H_5^+$	$N_2H_6^{2+}$	OH^-
Initial concentration	9.2×10^{-5}	0	9.2×10^{-5}
Change	$-x$	$+x$	$+x$
Equilibrium	$9.2 \times 10^{-5} - x$	$+x$	$9.2 \times 10^{-5} + x$

$$K_{b2} = \frac{[N_2H_6^{2+}][OH^-]}{[N_2H_5^+]} = 8.9 \times 10^{-16} = \frac{x \cdot (9.2 \times 10^{-5} + x)}{(9.2 \times 10^{-5} - x)}$$

Simplifying yields $\dfrac{x \cdot (9.2 \times 10^{-5})}{(9.2 \times 10^{-5})} = 8.9 \times 10^{-16}$; $x = 8.9 \times 10^{-16}$ M $= [N_2H_6^{2+}]$

In summary, the second stage produces a negligible amount of OH^- and consumes very little $N_2H_5^+$ ion. The equilibrium concentrations are:

$[N_2H_5^+]$: 9.2×10^{-5} M; $[N_2H_6^{2+}]$: 8.9×10^{-16} M; $[OH^-]$: 9.2×10^{-5} M

(b) The pH of the 0.010 M solution: $[OH^-] = 9.2 \times 10^{-5}$ M so pOH = 4.04
and pH = 14.0 - 4.04 = 9.96

Lewis Acid and Bases

69. (a) H_2NOH electron rich (accepts H^+)   Lewis base

    (b) Fe^{2+} electron poor   Lewis acid

    (c) CH_3NH_2 electron rich (accepts H^+)   Lewis base

71. CO is a Lewis base (donates electron pairs) in complexes with nickel and iron.

Molecular Structure, Bonding, and Acid-Base Behavior

73. HOCN will be the stronger acid. In HOCN the proton is attached to the very electronegative O atom. This great electronegativity will provide a very polar bond, weakening the O-H bond, and making the H more acidic than in HCN.

75. Benzenesulfonic acid is a Bronsted acid owing to the inductive effect of three oxygen atoms attached to the S (and through the S to the benzene ring).These very electronegative O atoms will—through the inductive effect—remove electron density between the O and the H—weakening the OH bond, and making the H acidic.

General Questions on Acids and Bases

77. For the equilibrium: $HC_9H_7O_4$ (aq) + H_2O (ℓ) ⇔ $C_9H_7O_4^-$ (aq) + H_3O^+ (aq)

we can write the K_a expression:

$$K_a = \frac{[C_9H_7O_4^-]\,[H_3O^+]}{[HC_9H_7O_4]} = 3.27 \times 10^{-4}$$

The initial concentration of aspirin is:

$$2 \text{ tablets} \cdot \frac{0.325 \text{ g}}{1 \text{ tablet}} \cdot \frac{1 \text{ mol } HC_9H_7O_4}{180.2 \text{ g } HC_9H_7O_4} \cdot \frac{1}{0.225 \text{ L}} = 1.60 \times 10^{-2} \text{ M } HC_9H_7O_4$$

Substituting into the K_a expression:

$$K_a = \frac{[H_3O^+]^2}{1.60 \times 10^{-2} - x} = 3.27 \times 10^{-4} = \frac{x^2}{1.60 \times 10^{-2} - x}$$

Since $100 \cdot K_a \approx$ [aspirin], the quadratic equation will provide a "good" value.

Using the quadratic equation, $[H_3O^+] = 2.13 \times 10^{-3}$ M and pH = 2.671.

79. 4-chlorobenzoic acid has a larger K_a than benzoic acid, so it is the stronger acid. A 0.010 M solution of *benzoic acid* would produce a lower concentration of hydronium ions than would a similar solution of 4-chlorobenzoic acid, and would be therefore less acidic—*have a higher pH*.

81. The reaction between H_2S and $NaCH_3CO_2$:

$$H_2S \text{ (aq)} + CH_3CO_2^- \text{ (aq)} \Leftrightarrow HS^- \text{ (aq)} + CH_3CO_2H \text{ (aq)}$$

An examination of Table 17.3 reveals that CH_3CO_2H is a stronger acid than H_2S, so the equilibrium will lie *to the left (reactants)*.

83. Monoprotic acid has $K_a = 1.3 \times 10^{-3}$. Equilibrium concentrations of HX, H_3O^+ and pH for 0.010 M solution of HX:

	HX	H^+	X^-
Initial concentration	0.010	0	0
Change	-x	+x	+x
Equilibrium	0.010 - x	+x	+x

$$K_a = \frac{[H_3O^+][A]}{[HA]} = \frac{(x)^2}{0.010-x} = 1.3 \times 10^{-3}$$

Since Ka and the concentration are of the same order of magnitude, the quadratic equation

will help: $x^2 + 1.3 \times 10^{-3}x - 1.3 \times 10^{-5} = 0$. The "reasonable solution" is $x = 3.0 \times 10^{-3}$

Concentrations are then: $[HX] = 0.010 - x = 0.010-0.0030 = 7.0 \times 10^{-3}$ M

$$[H^+] = +x = 3.0 \times 10^{-3} \text{ M} \quad [X^-] = +x = 3.0 \times 10^{-3}M$$

$$pH = -\log(3.0 \times 10^{-3}) = 2.52$$

85. The pKa of m-nitrophenol:

The Ka expression for this weak acid is: $K_a = \dfrac{[H^+][A^-]}{[HA]}$

The pH of a 0.010 M solution of nitrophenol is 3.44, so we know: $[H^+] = 3.63 \times 10^{-4}$

Since we get one "A-"for each "H+"the concentrations of the two are equal.

The $[HA] = (0.010- 3.63 \times 10^{-4})$. Substituting those values into the K_a expression yields:

$$K_a = \frac{[H^+][A^-]}{[HA]} = \frac{[3.63 \times 10^{-4}][3.63 \times 10^{-4}]}{[0.010 - 3.63 \times 10^{-4}]} = 1.4 \times 10^{-5} \text{ (to 2sf)}$$

$$pK_a = -\log(1.4 \times 10^{-5}) = 4.86$$

87. For Novocain, the $pK_a = 8.85$. The pH of a 0.0015M solution is:

First calculate K_a: $K_a = 10^{-8.85}$ so $K_a = 1.41 \times 10^{-9}$

Now treat Novocain as any weak acid, with the appropriate K_a expression:

$$K_a = \frac{[H^+][A^-]}{[HA]} = \frac{[x][x]}{[0.0015 - x]} = 1.4 \times 10^{-9}$$

Since Ka is so small, we can assume that the denominator is approximated by 0.0015 M.

$$\frac{[x]^2}{0.0015} = 1.4 \times 10^{-9} \text{ and } x^2 = 1.45 \times 10^{-6}$$

Since $[H^+] = 1.45 \times 10^{-6}$; pH = $-\log (1.45 \times 10^{-6}) = 5.84$

89. Regarding ethylamine and ethanolamine:

   (a) Since ethylamine has the larger K_b, ethylamine is the stronger base.

(b) The pH of 0.10 M solution of ethylamine:

$$C_2H_5NH_2 + H_2O \Leftrightarrow C_2H_5NH_3^+ + OH^- \qquad K_b = 4.3 \times 10^{-4}$$

Initial	0.10 M		0	0
Change	- x		+ x	+ x
Equilibrium	0.10 - x		+ x	+ x

Substituting into the K_b expression:

$$K_b = \frac{[NH_4^+][OH^-]}{[NH_3]} = \frac{x^2}{(0.10-x)} = 4.3 \times 10^{-4}$$

So x = 6.56×10^{-3} so pOH = $-\log(6.56 \times 10^{-3})$ = 2.18 and pH = 14.00-2.18 = 11.82

91. K_b for pyridine = 1.5×10^{-9} (from Appendix I) ; $K_a = 1.0 \times 10^{-14}/1.5 \times 10^{-9} = 6.7 \times 10^{-6}$

Treating the hydrochloride as a weak acid ($K_a = 6.7 \times 10^{-6}$), we can write:

$$C_5H_5NH^+ + H_2O \Leftrightarrow C_5H_5N + H_3O^+$$

And the equilibrium concentrations are:

	$C_5H_5NH^+$	H^+	C_5H_5N
Initial concentration	0.025	0	0
Change	-x	+x	+x
Equilibrium	0.025 - x	+x	+ x

So $K_a = \dfrac{[C_5H_5N][H^+]}{[C_5H_5NH^+]} = \dfrac{[x][x]}{[0.025-x]} = 6.7 \times 10^{-6}$

Solving for x = 4.05×10^{-4} and pH = $-\log(4.05 \times 10^{-4})$ = 3.39

93. pH of aqueous solutions of

	reaction	pH
(a) $NaHSO_4$	hydrolysis of HSO_4^- produces H_3O^+	< 7
(b) NH_4Br	hydrolysis of HSO_4^- produces H_3O^+	< 7
(c) $KClO_4$	no hydrolysis occurs	= 7
(d) Na_2CO_3	hydrolysis of CO_3^{2-} produces OH^-	> 7
(e) $(NH_4)_2S$	hydrolysis of S^{2-} produces OH^-	> 7
(f) $NaNO_3$	no hydrolysis occurs	= 7
(g) Na_2HPO_4	hydrolysis of HPO_4^{2-} produces OH^-	> 7

	reaction	pH
(h) LiBr	no hydrolysis occurs	= 7
(i) FeCl3	hydrolysis of Fe^{3+} produces H_3O^+	< 7

95. For oxalic acid $K_{a1} = 5.9 \times 10^{-2}$ and $K_{a2} = 6.4 \times 10^{-5}$

Representing oxalic as H_2A. The first step can be written as:

$$H_2A + H_2O \Leftrightarrow HA^- + H_3O^+ \qquad K_{a1} = 5.9 \times 10^{-2}$$

We can write the second step:

$$HA^- + H_2O \Leftrightarrow A^{2-} + H_3O^+ \qquad K_{a2} = 6.4 \times 10^{-5}$$

If we add the two equations we get:

$$H_2A + 2 H_2O \Leftrightarrow A^{2-} + 2 H_3O^+ \qquad \overline{\qquad\qquad\qquad}$$

$$K_{net} = (5.9 \times 10^{-2})(6.4 \times 10^{-5})$$

$$K_{net} = 3.8 \times 10^{-6}$$

97. Confirm the fact that 1.8×10^{10} is the value of the equilibrium constant for the reaction of formic acid and sodium hydroxide :

To do this, we need two equations that have a **net** equation that corresponds to:

$$HCO_2H(aq) + OH^-(aq) \Leftrightarrow HCO_2^-(aq) + H_2O (\ell)$$

Begin with equilibria associated with the weak acid in water, and the Kw for water:

$$HCO_2H(aq) + H_2O (\ell) \Leftrightarrow HCO_2^-(aq) + H_3O^+(aq) \qquad K_a = 1.8 \times 10^{-4}$$

$$H_3O^+(aq) + OH^-(aq) \Leftrightarrow 2 H_2O (\ell) \qquad 1/K_w = 1.0 \times 10^{14}$$

The net equation is: $HCO_2H(aq) + OH^-(aq) \Leftrightarrow HCO_2^-(aq) + H_2O (\ell)$ and

the net K is: $K_a \cdot 1/K_w = (1.8 \times 10^{-4})(1.0 \times 10^{14}) = 1.8 \times 10^{10}$

99. Volume to which 1.00×10^2 mL of a 0.20 M solution of a weak acid, HA, should be diluted to result in a doubling of the percent ionization:

Begin by writing the Ka expression: $K = \dfrac{[H^+][A^-]}{[HA]}$ (1)

The fraction of dissociation (which we want to double) is $\dfrac{[A^-]}{[HA]}$ or equivalently: $\dfrac{[H^+]}{[HA]}$ (2)

Knowing that for a monoprotic acid, $[H^+] = [A^-]$, we can rearrange equation (1) to yield:

$[H^+]^2 = Ka \cdot [HA-H^+]$, where $[HA-H^+]$ represents the concentration of molecular acid.

Expanding gives: $[H^+]^2 = Ka \cdot [HA] - Ka \cdot [H^+]$, and setting this up as a quadratic equation gives: $[H^+]^2 + Ka \cdot [H^+] - Ka \cdot [HA] = 0$.

For the sake of argument, let's pick a value for Ka, say 1.0×10^{-5}

Given the original concentration of acid is 0.2 M, let's substitute the values into the equation.

$1 \cdot [H^+]^2 + (1.0 \times 10^{-5}) \cdot [H^+] - (1.0 \times 10^{-5}) \cdot [0.2] = 0$, and solve the equation--call it (3).

This gives $[H^+] = 1.41 \times 10^{-3}$ M. The fraction of dissociation is (from equation (2) above)

$\dfrac{[H^+]}{[HA]}$ or 0.00705. Now we want this **fraction to double**. So substitute differing

concentrations of acid (increasingly dilute) into equation (3), solve the quadratic equation that results, and calculate the fraction dissociated. The table below shows such a series of calculations:

Acid conc	fraction	Relative dissociation
0.20	0.00704611	1.0
0.15	0.0081317	1.2
0.10	0.00995012	1.4
0.05	0.01404249	2.0

Note that when the acid concentration falls to 1/4 of the original value, the relative dissociation has doubled. So 100 mL of the acid would need to be diluted to 400 mL to double the fraction (or percent) of dissociation.

101. The pH of a solution of 1.25 g sodium sulfanilate in 125 mL: Sulfanilic acid is a weak acid, so its conjugate base should provide a slightly basic solution.

The $pK_a = 3.23$ so $K_a = 10^{-3.23} = 5.9 \times 10^{-4}$.

The formula for sulfanilic acid is complex ($H_2NC_6H_4SO_3H$), so let's use HSA to represent the acid, and SA to represent the anion.

The concentration of the sodium salt (represented as NaSA) will be necessary:

$$\dfrac{1.25 \text{ g}}{0.125 \text{ L}} \bullet \dfrac{1 \text{ mol NaSA}}{195.15 \text{ g NaSA}} = 0.0512 \text{ M NaSA}$$

In water, the NaSA salt will exist predominantly as sodium cations and SA anions, so we can ignore the sodium cations.

The reaction that occurs with the conjugate base and water is: $SA + H_2O \Leftrightarrow HSA + OH^-$

(Note the production of hydroxyl ion –the reason we expect the solution to be basic.)

As the anion is acting as a base, we'll need to calculate an appropriate K for the anion.

$$K_b \text{ (conjugate)} = K_w/K_a \text{ so } \frac{1.0 \times 10^{-14}}{5.9 \times 10^{-4}} = 1.7 \times 10^{-11}$$

The K_b expression is: $\dfrac{[HSA][OH^-]}{[SA]} = \dfrac{[x][x]}{0.0512 - x} = 1.7 \times 10^{-11}$

The size of K ($100 \bullet K < 0.0512$) tells us that we can safely approximate the denominator of the fraction as 0.0512. The resulting equation is: $x^2 = (0.0512)(1.7 \times 10^{-11})$ and

$x = 9.33 \times 10^{-7}$ Since x represents the concentration of OH^- ion,

the pOH = $-\log(9.33 \times 10^{-7}) = 6.03$, and the pH = 7.97.

103. Arrange the following 0.1M solution in order of increasing pH:

Examine the nature of each solute.

(a) NaCl—no hydrolysis of cation or anion

(b) NH_4Cl—hydrolysis of NH_4^+-- to give H_3O^+

(c) HCl—a strong acid

(d) $NaCH_3CO_2$—hydrolysis of the acetate anion—to produce OH^-

(e) KOH—a strong base

From low pH (acidic) to high pH (basic): $HCl < NH_4Cl < NaCl < NaCH_3CO_2 < KOH$

<u>105</u>. Given that $[H_3O^+] = \sqrt{K_1 \bullet K_2}$ for amphiprotic ions, we can calculate the pH of a 0.050 M

solution of KHP (potassium hydrogen phthalate).

$[H_3O^+] = \sqrt{(1.12 \times 10^{-3}) \bullet (3.91 \times 10^{-6})} = 6.62 \times 10^{-5}$ and pH = $-\log(6.62 \times 10^{-5}) = 4.180$

Summary and Conceptual Questions:

107. Water can be both a Bronsted base and a Lewis base. The Bronsted system requires that a

base be able to accept a H^+ ion. Water does that in the formation of the H_3O^+ ion. The

Lewis system defines a base as an electron pair donor. The two lone pairs of electrons on the

O atom provide a source for those electrons. Examine the reaction of H_2O with H^+. The

hydrogen ion accepts the electron pairs from the O atom of the water molecule—fulfilling

water's role as a Lewis base and the H^+'s role as a Lewis acid.

A Bronsted acid furnishes H^+ to another specie. The autoionization of water provides just one example of water acting as a Bronsted acid. Water however cannot function as a Lewis acid, as it has no capacity to accept an electron pair.

109. To determine the relative basic strengths of the substances, one can measure the pH of aqueous solutions of known concentration (e.g. 0.50M) of each solute. Remember the equilibrium associated with bases in water:

$$B + H_2O \Leftrightarrow BH^+ + OH^- \qquad K_b = y.y \times 10^{-z}$$

The strongest base will have an equilibrium lying farther "to the right" than the weaker bases-resulting in a solution with the *greatest* concentration of hydroxyl ions (and the greatest pH). The weakest base will obviously result in a solution with the *lowest* concentration of hydroxyl ions (and the lowest pH).

<u>111</u>. What species exist, and in what concentrations from the preparation of a 0.10M solution of HCN?

The equilibrium $HCN(aq) + H_2O(\ell) \Leftrightarrow CN^-(aq) + H_3O^+(aq)$ for which $Ka = 4.0 \times 10^{-10}$.

and $\dfrac{[CN^-][H_3O^+]}{[HCN]} = 4.0 \times 10^{-10}$. Substituting appropriate values: $\dfrac{[x][x]}{[0.10-x]} = 4.0 \times 10^{-10}$

Noting that the K for this acid is particularly small we can approximate the denominator as 0.10 M. Solving then gives $x^2 = 4.0 \times 10^{-11}$ and $x = 6.3 \times 10^{-6}$ M

The $[OH^-] = 1.0 \times 10^{-14}$ M$/ 6.3 \times 10^{-6}$ M $= 1.6 \times 10^{-9}$.

Summarizing: $[H_2O]$ = approximately 55 M.

$[H_3O^+] = 6.3 \times 10^{-6}$ M (also true of $[CN^-]$).

$[OH^-] = 1.6 \times 10^{-9}$ M

$[HCN] = 0.10$ M

Listed in order of **decreasing concentrations**: $H_2O > HCN > H_3O^+ = CN^- > OH^-$

<u>113</u>. The reaction is:

$$CH_3CO_2H\ (aq) + NaOH\ (aq) \Leftrightarrow CH_3CO_2^-(aq) + Na^+(aq) + H_2O(\ell)$$

The addition of 30.0 mL of 0.15 M NaOH (4.5 mmol NaOH) to 30.0 mL of 0.15 M CH_3CO_2H (4.5 mmol CH_3CO_2H) produces water and the soluble salt, sodium acetate (4.5 mmol $CH_3CO_2^-$ Na^+).

The acetate ion is the anion of a weak acid and reacts with water according to the equation:

$$CH_3CO_2^- \text{ (aq)} + H_2O \text{ (\ell)} \Leftrightarrow CH_3CO_2H \text{ (aq)} + OH^- \text{ (aq)}$$

The equilibrium constant expression is: $K_b = \dfrac{[CH_3CO_2H][OH^-]}{[CH_3COO^-]} = 5.6 \times 10^{-10}$

The concentration of acetate ion is: $\dfrac{4.5 \text{ mmol}}{(30.0 + 30.0)\text{mL}} = 0.075 \text{ M}$

	CH3CO2⁻	CH3CO2H	OH⁻
Initial concentration	0.075		
Change	-x	+x	+x
Equilibrium	0.075 - x	+x	+x

$$K_b = \frac{[CH_3CO_2H][OH^-]}{[CH_3CO_2^-]} = \frac{x^2}{0.075 - x} = 5.6 \times 10^{-10}$$

Simplifying $(100 \cdot K_b \ll 0.075)$ we get $\dfrac{x^2}{0.075} = 5.6 \times 10^{-10}$; $x = 6.5 \times 10^{-6} = [OH^-]$

The hydrogen ion concentration is related to the hydroxyl ion concentration by the equation:

$$K_W = [H_3O^+][OH^-] = 1.0 \times 10^{-14}$$

$$[H_3O^+] = \frac{1.0 \times 10^{-14}}{[OH^-]} = \frac{1.0 \times 10^{-14}}{6.5 \times 10^{-6}} = 1.5 \times 10^{-9}$$

$$pH = 8.81$$

The pH is greater than 7, as we expect for a salt of a strong base and weak acid.

Now the species present are: H_2O, H_3O^+, OH^-, $CH_3CO_2^-$, CH_3CO_2H

In order of **decreasing concentration:**
$$H_2O < Na^+ \text{ and } CH_3CO_2^- < OH^- \text{ and } CH_3CO_2H < H_3O^+$$

115. (a) The reaction of perchloric acid with sulfuric acid:
$$HClO_4 + H_2SO_4 \Leftrightarrow ClO_4^- + H_3SO_4^+$$

(b) Lewis dot structure for sulfuric acid:

Sulfuric acid has two oxygen atoms capable of "donating" an electron pair, e.g. to a proton, H^+, thereby acting as a base.

117. This puzzle is a great way to test your chemical knowledge. Organize what we know.

Cations	Anions
Na^+	Cl^-
NH_4^+	OH^-
H^+	

Experimentally we observe:

$B + Y \rightarrow$ acidic solution

$B + Z \rightarrow$ basic solution

$A + Z \rightarrow$ neutral solution

If A + Z give a neutral solution, then A + Z must be acid and base (in some order).

Since B + Z gives a basic solution, B + Z cannot be acid and base (as A + Z), otherwise B + Z would be neutral (as A + Z). Z must be basic (OH^-), meaning Y must be neutral (Cl^-).

Since B + Y gives an acid solution, then B must contain NH_4^+. If B contains ammonium, then A must contain H^+, and C must contain Na^+.

In summary then: $A = H^+$; $B = NH_4^+$; $C = Na^+$; $Y = Cl^-$; $Z = OH^-$.

Pairing these cations with chloride and potassium ions, we have:

$A = HCl$; $B = NH_4Cl$; $C = NaCl$; $Y = KCl$; $Z = KOH$ .The pH = 14.00-6.03 or 7.97.

119. (a) Determine the form for the equilibrium constant for the reaction:

NH_4^+ (aq) + CN^-(aq) $\Leftrightarrow$ NH_3(aq) + HCN (aq)

Ka for HCN: HCN(aq) + H_2O (ℓ) $\Leftrightarrow$ CN^-(aq) + H_3O^+(aq).

Noting that the HCN and its conjugate are on the opposite sides for our desired equation, we "swap sides" for this equation: CN^-(aq) + H_3O^+(aq) $\Leftrightarrow$ HCN(aq) + H_2O (ℓ)   and K = 1/Ka

Similarly we can write the Kb expression for NH_3, but we note that NH_3 is on the "right side " of our desired expression, so let's write the "reversed equation", for which K would be 1/Kb: NH_4^+(aq) + OH^-(aq) $\Leftrightarrow$ NH_3(aq) + H_2O(ℓ).

Finally note that the components of water, $H_2O(\ell)$ and $OH^-(aq)$ and $H_3O^+(aq)$, are missing from our desired equation: Recall that equilibrium may be written:

$$2\ H_2O(\ell) \Leftrightarrow H_3O^+(aq) + OH^-(aq)\ \text{with the } K = Kw.$$

Summarizing:

$$CN^-(aq) + H_3O^+(aq) \Leftrightarrow HCN(aq) + H_2O\ (\ell) \qquad K = 1/Ka$$

$$NH_4^+(aq) + OH^-(aq) \Leftrightarrow NH_3(aq) + H_2O(\ell) \qquad K = 1/Kb$$

$$2\ H_2O(\ell) \Leftrightarrow H_3O^+(aq) + OH^-(aq) \qquad K = Kw.$$

$$NH_4^+\ (aq) + CN^-(aq) \Leftrightarrow NH_3(aq) + HCN\ (aq) \qquad K_{net} = Kw \cdot 1/Kb \cdot 1/Ka$$

(b) Derive an expression for the $[H_3O^+]$:

Note that the sources of H_3O^+ ion are from the ionization of HCN and the hydrolysis of $NH_4^+(aq)$.

For HCN, $Ka_{(HCN)} = \dfrac{[H^+][CN^-]}{[HCN]}$ or $\dfrac{Ka_{(HCN)} \cdot [HCN]}{[CN^-]} = [H^+]$ (1)

For NH_4^+, $Ka_{(NH_4^+)} = \dfrac{[H^+][NH_3]}{[NH_4^+]}$ or $\dfrac{Ka_{(NH_4^+)} \cdot [NH_4^+]}{[NH_3]} = [H^+]$ (2)

Recall that $Kw = Ka \cdot Kb$ so for the Ka of ammonium ion, we can write:

$Ka_{(NH_4^+)} = \dfrac{Kw}{Kb_{(NH_3)}}$ and substituting into the equation (2) above, we get

$$\dfrac{\dfrac{Kw}{Kb_{(NH_3)}} \cdot [NH_4^+]}{[NH_3]} = [H^+] \quad (3)$$

Noting the square root relationship of our desired equation, let's multiply the $[H^+]$ in equation (1) by the $[H^+]$ in equation (3)--- to obtain a $[H^+]^2$ term.

$$[H^+]^2 = \dfrac{Ka_{(HCN)} \cdot [HCN]}{[CN^-]} \cdot \dfrac{\dfrac{Kw}{Kb_{(NH_3)}} \cdot [NH_4^+]}{[NH_3]}$$

Recalling that the NH$_4$CN salt dissolves to give equal concentrations of the ammonium and cyanide ions, we can simplify:

$$[H^+]^2 = \frac{Ka_{(HCN)} \cdot [HCN]}{1} \cdot \frac{\frac{Kw}{Kb_{(NH_3)}} \cdot 1}{[NH_3]} \quad \text{and similarly } [NH_3] = [HCN] \text{ to produce:}$$

$$[H^+]^2 = \frac{Ka_{(HCN)}}{1} \cdot \frac{\frac{Kw}{Kb_{(NH_3)}} \cdot 1}{1} \quad \text{or} \quad \frac{Ka \cdot Kw}{Kb} \quad \text{and} \quad [H^+] = \sqrt{\frac{Ka \cdot Kw}{Kb}}$$

(c) The pH of 0.15M NH$_4$CN:

Substituting into the expression we derived in part (b) gives:

$$[H^+] = \sqrt{\frac{Ka \cdot Kw}{Kb}} = \sqrt{\frac{(4 \times 10^{-10})(1 \times 10^{-14})}{1.8 \times 10^{-5}}} = 4.71 \times 10^{-10} \quad \text{and pH} = -\log(4.71 \times 10^{-10})$$

or 9.33

Chapter 18
Principles of Reactivity: Other Aspects of Aqueous Equilibria

Practicing Skills

Common Ion Effect

1. To determine how pH is expected to change, examine the equilibria in each case:

(a) NH_3 (aq) + H_2O (ℓ) $\Leftrightarrow$ NH_4^+ (aq) + OH^- (aq)

As the added NH_4Cl dissolves, ammonium ions are liberated—increasing the ammonium ion concentration and shifting the position of equilibrium to the left—reducing OH^-, and decreasing the pH.

(b) CH_3CO_2H (aq) + H_2O(ℓ) $\Leftrightarrow$ $CH_3CO_2^-$ (aq) + H_3O^+ (aq)

As sodium acetate dissolves, the additional acetate ion will shift the position of equilibrium to the left—reducing H_3O^+, and increasing the pH.

(c) $NaOH$ (aq) $\rightarrow$ Na^+ (aq) + OH^- (aq)

NaOH is a strong base, and as such is totally dissociated. Since the added NaCl does not hydrolyze to any appreciable extent—no change in pH occurs.

3. The pH of the buffer solution is:

$$K_b = \frac{[NH_4^+][OH^-]}{[NH_3]} = \frac{(0.20)[OH^-]}{(0.20)} = 1.8 \times 10^{-5}$$

solving for hydroxyl ion yields: $[OH^-] = 1.8 \times 10^{-5}$ M ; pOH = 4.75   pH = 9.25

5. pH of solution when 30.0 mL of 0.015 M KOH is added to 50.0 mL of 0.015 M benzoic acid:

The equilibrium affected is that of benzoic acid in water:

$C_6H_5CO_2H$ + H_2O $\Leftrightarrow$ $C_6H_5CO_2^-$ + H_3O^+ $K_a = 6.3 \times 10^{-5}$

$$K_a = \frac{[C_6H_5CO_2^-][H_3O^+]}{[C_6H_5CO_2H]} = 6.3 \times 10^{-5}$$

The KOH will *consume benzoic acid* and *produce* the conjugate *benzoate anion*.,

The base and benzoic acid react: $KOH + C_6H_5CO_2H \rightarrow C_6H_5CO_2^-K^+ + H_2O$

The pH of the solution will depend on the ratio of the conjugate pairs (acid and anion).

Determine the concentrations of the two:

Amount of benzoic acid present: 50 mL • 0.015 mol/L = 0.75 mmol of benzoic acid.

Amount of KOH added: 30 mL • 0.015 mol/L = 0.45 mmol of KOH

Amount of benzoic acid remaining: (0.75 mmol -0.45 mmol of KOH)= 0.30 mmol benzoic acid

Amount of benzoate anion produced: *0.45 mmol of benzoate anion.*

We can rearrange the K_a expression to solve for H_3O^+:

$$\left[H_3O^+\right] = 6.3 \times 10^{-5} \cdot \frac{\left[C_6H_5CO_2H\right]}{\left[C_6H_5CO_2^-\right]} = 6.3 \times 10^{-5} \cdot \frac{0.30 \text{ mmol}}{0.45 \text{ mmol}} = 4.2 \times 10^{-5}$$

$pH = -\log(4.2 \times 10^{-5}) = 4.38$

Buffer Solutions

7. The original pH of the 0.12M NH_3 solution will be:

$$\frac{[NH_4^+][OH^-]}{[NH_3]} = 1.8 \times 10^{-5} \quad \text{and} \quad \frac{x^2}{0.12 - x} = 1.8 \times 10^{-5}$$

Assuming $(0.12 - x \approx 0.12)$, $x = 1.5 \times 10^{-3}$

$[OH^-] = 1.5 \times 10^{-3}$ and pOH $= 2.83$ with pH $= 11.17$

Adding 2.2 g of NH_4Cl (0.041 mol) to 250 mL will produce an immediate increase of 0.16 M

NH_4^+ (0.041 mol/0.250 L).

Substituting into the equilibrium expression as we did earlier, we get:

$$\frac{(x + 0.16)(x)}{0.12 - x} = 1.8 \times 10^{-5}$$

Assuming $(x + 0.16 \approx 0.16)$ and $(0.12 - x \approx 0.12)$

$$\frac{0.16(x)}{0.12} = 1.8 \times 10^{-5} \quad x = 1.35 \times 10^{-5} = [OH^-]$$

Note the hundred-fold decrease in $[OH^-]$ over the initial ammonia solution, as predicted by
LeChatelier's principle. So pOH $= 4.88$ and pH $= 9.11$ (lower than original)

9. Mass of sodium acetate needed to change 1.00 L solution of 0.10 M CH_3CO_2H to pH $= 4.5$:

The equilibrium affected is that of acetic acid in water:

$$CH_3CO_2H + H_2O \Leftrightarrow CH_3CO_2^- + H_3O^+ \quad K_a = 1.8 \times 10^{-5}$$

The equilibrium expression is: $K_a = \dfrac{[CH_3CO_2^-][H_3O^+]}{[CH_3CO_2H]} = 1.8 \times 10^{-5}$

We know the concentration of acetic acid (0.10M), and we know the desired $[H_3O^+]$:

pH $= 4.5$ so $[H_3O^+] = 10^{-4.5} = 3.2 \times 10^{-5}$

Substituting these values into the equilibrium expression gives:

$$\frac{[CH_3CO_2^-][H_3O^+]}{[CH_3CO_2H]} = 1.8 \times 10^{-5} = \frac{[CH_3CO_2^-][3.2 \times 10^{-5}]}{[0.10]} \, .$$

We can solve for $[CH_3CO_2^-] = 0.057$ M

What mass of $NaCH_3CO_2$ would give this concentration of $CH_3CO_2^-$?

$$\frac{0.057 \text{ mol}}{1 \text{ L}} \cdot \frac{1 \text{L}}{1} \cdot \frac{82.0 \text{ g } NaCH_3CO_2}{1 \text{ mol } NaCH_3CO_2} = 4.7 \text{ g } NaCH_3CO_2$$

Using the Henderson-Hasselbalch Equation

11. The pH of a solution with 0.050 M acetic acid and 0.075 M sodium acetate:

$$pH = pK_a + \log \frac{[\text{conjugate base}]}{[\text{acid}]}$$

The pK_a for acetic acid is $= -\log(K_a) = -\log(1.8 \times 10^{-5})$ or 4.74

$$pH = 4.74 + \log \frac{[0.075]}{[0.050]} = 4.74 + 0.176 = 4.92$$

13. (a) The pK_a for formic acid is $= -\log(K_a) = -\log(1.8 \times 10^{-4})$ or 3.74

$$pH = pK_a + \log \frac{[\text{conjugate base}]}{[\text{acid}]}$$

$$= 3.74 + \log \frac{0.035}{0.050} = 3.74 - 0.15 \text{ or } 3.59$$

(b) Ratio of conjugate pairs to increase pH by 0.5 (to pH= 4.09)

Substituting into the Henderson-Hasselbalch equation

$$4.09 = 3.74 + \log \frac{[\text{conjugate base}]}{[\text{acid}]}$$

$$+0.345 = \log \frac{[\text{conjugate base}]}{[\text{acid}]} \quad \text{so} \quad -0.345 = \log \frac{[\text{acid}]}{[\text{conjugate base}]}$$

$$0.45 = \frac{[\text{acid}]}{[\text{conjugate base}]} \quad \text{(to 2sf)}$$

Preparing a Buffer Solution

15. The best combination to provide a buffer solution of pH 9 is (b) the NH_3/NH_4^+ system. Note that K_a (for NH_4^+) is approximately 10^{-10}. Buffer systems are good when the desired pH is ±1 unit from pK_a (10 in this case).

The HCl and NaCl don't form a buffer. The acetic acid/sodium acetate system would form an acidic buffer ($pK_a \approx 5$) in the pH range 4 - 6.

17. Preparation of a buffer of NaH_2PO_4 and Na_2HPO_4 for a pH = 7.5

The principle species present (other than Na^+ ions) will be $H_2PO_4^-$ and HPO_4^{2-}

So let's use the Ka2 expression for H_3PO_4.

$$\frac{[HPO_4^{2-}][H_3O^+]}{[H_2PO_4^-]} = 6.2 \times 10^{-8}$$

For a pH = 7.5, $[H_3O^+] = 3.2 \times 10^{-8}$. So the ratio of the conjugate pairs is:

$$\frac{[HPO_4^{2-}]}{[H_2PO_4^-]} = \frac{6.2 \times 10^{-8}}{3.2 \times 10^{-8}} = 1.94.$$ The reciprocal of this value is $1/1.94 = 0.51$. So mixing 0.51

moles of NaH_2PO_4 with 1 mol Na_2HPO_4 would provide the desired pH.

Adding an Acid or Base to a Buffer Solution

19. (a) Initial pH

Need to know the concentrations of the conjugate pairs:

The equilibrium expression shows the *ratio of the conjugate pairs*, we can calculate **moles** of the conjugate pairs, and know that the ratio of the # of moles of the species will have the same value as the ratio of their concentrations!

$$CH_3CO_2H = 0.250\,L \cdot 0.150\,M = 0.0375\,mol$$

$$NaCH_3CO_2 = 4.95\,g \cdot \frac{1\,mol}{82.07\,g} = 0.0603\,mol$$

Substituting into the K_a expression:

$$\frac{[CH_3CO_2^-][H_3O^+]}{[CH_3CO_2H]} = 1.8 \times 10^{-5} = \frac{[0.0603][H_3O^+]}{[0.0375]}$$

and solving for $[H_3O^+] = 1.1 \times 10^{-5}\,M$; pH = 4.95

(b) pH after 82. mg NaOH is added to 100. mL of the buffer. The amount of the conjugate pairs in 100/250 of the buffer is $(100/250)(0.0375\,mol) = 0.0150\,mol\ CH_3CO_2H$

and $(100/250)(0.0603\,mol) = 0.0241\,mol\ CH_3CO_2^-$

$$82\,mg\,NaOH \cdot \frac{1\,mmol\,NaOH}{40.0\,mg\,NaOH} = 2.05\,mmol\,NaOH\ or\ 0.00205\,mol\,NaOH$$

or 2.1 mmol NaOH (to 2 significant figures)

This base would consume an equivalent amount of CH_3CO_2H and produce an equivalent amount of $CH_3CO_2^-$.

After that process: (0.0150- 0.0021) or 0.0129 mol CH_3CO_2H and (0.0241 + 0.0021) or 0.0262 mol $CH_3CO_2^-$ are present. Substituting into the K_a expression as in part (a)

$$\frac{(0.0262)[H_3O^+]}{(0.0129)} = 1.8 \times 10^{-5}$$ so $[H_3O^+] = 8.9 \times 10^{-6}$ and pH = 5.05

21. (a) The pH of the buffer solution is:

$$K_b = \frac{[NH_4^+][OH^-]}{[NH_3]} = \frac{(0.250)[OH^-]}{(0.500)} = 1.8 \times 10^{-5}$$

Note : Here the data presented are given as moles (in the case of ammonium chloride) and molar concentration (in the case of ammonia). In SQ18.19 we substituted the # moles of the conjugate pairs into the K expression. Here we must first *decide* whether to substitute # moles or molar *concentrations* into the K_b expression. Either would work!

What is critical to remember is that we have to have both species expression in one **or** the other form—not a mix of the two. Here I chose to convert moles of NH_4Cl into molar concentrations, and substitute.

Solving for hydroxyl ion in the K_b expression above yields: $[OH^-] = 3.6 \times 10^{-5}$ M and pOH = 4.45 so pH = 9.55

(b) pH after addition of 0.0100 mol HCl:

The basic component of the buffer (NH_3) will react with the HCl, producing more ammonium ion. The composition of the solution is then:

	NH_3	NH_4Cl
Moles present (before HCl added)	0.250	0.125
Change (reaction)	- 0.0100	+ 0.0100
Following reaction	0.240	+ 0.135

The amounts of NH_3 and NH_4Cl following the reaction with HCl are only slightly different from those amounts prior to reaction. Converting these numbers into molar concentrations (Volume is 500. mL) and substituting the concentrations into the K_b expression yields:

$$K_b = \frac{[NH_4^+][OH^-]}{[NH_3]} = \frac{(0.270)[OH^-]}{(0.480)} = 1.8 \times 10^{-5}$$

$[OH^-] = 3.2 \times 10^{-5}$; pOH = 4.50, and the new pH = 9.50.

More About Acid-Base Reactions: Titrations

23. We can calculate the amount of phenol present by converting mass to moles:

$$0.515 \text{ g } C_6H_5OH \cdot \frac{1 \text{ mol } C_6H_5OH}{94.11 \text{ g } C_6H_5OH} = 5.47 \times 10^{-3} \text{ mol phenol}$$

(a) The pH of the solution containing 5.47×10^{-3} mol phenol in 125 mL water:

With a $K_a = 1.3 \times 10^{-10}$, the K_a expression is: $K_a = \frac{[C_6H_5O^-][H_3O^+]}{[C_6H_5OH]} = 1.3 \times 10^{-10}$

The stoichiometry of the compound indicates one phenoxide ion:one hydronium ion.

The initial concentration is 5.47×10^{-3} mol/ 0.125 L $= 0.0438$ M

The K_a expression becomes: $\dfrac{[x][x]}{[0.0438 - x]} = 1.3 \times 10^{-10}$

Given the magnitude of the K_a, we can safely approximate the denominator as 0.0438M.

$x^2 = 1.3 \times 10^{-10} \bullet 0.0438$ and $x = 2.39 \times 10^{-6}$ M and pH $= -\log(2.39 \times 10^{-6}) = 5.62$.

(b) At the equivalence point 5.47×10^{-3} mol of NaOH will have been added. Phenol is a monoprotic acid, so one mol of phenol reacts with one mol of sodium hydroxide. The volume of 0.123 NaOH needed to provide this amount of base is:

$$\text{moles} = \text{M} \times \text{V}$$

$$5.47 \times 10^{-3} \text{ mol NaOH} = \frac{0.123 \text{ mol NaOH}}{\text{L}} \times \text{V}$$

or 44.5 mL of the NaOH solution.

The total volume would be (125 + 44.5) or 170. mL solution.

Sodium phenoxide is a soluble salt hence the initial concentration of both sodium and phenoxide ions will be equal to:

$$\frac{5.47 \times 10^{-3} \text{ mol}}{0.170 \text{ L}} = 3.23 \times 10^{-2} \text{ M}$$

The phenoxide ion however is the conjugate ion of a weak acid and undergoes hydrolysis.

$$C_6H_5O^- \text{ (aq)} + H_2O \text{ (}\ell\text{)} \Leftrightarrow C_6H_5OH \text{ (aq)} + OH^- \text{ (aq)}$$

	$C_6H_5O^-$	C_6H_5OH	OH^-
Initial	3.23×10^{-2} M		
Change	-x	+x	+x
Equilibrium	3.23×10^{-2} M - x	x	x

$$K_b = \frac{[C_6H_5OH][OH^-]}{[C_6H_5O^-]} = \frac{1.0 \times 10^{-14}}{1.3 \times 10^{-10}} = \frac{(x)(x)}{3.23 \times 10^{-2} - x} = 7.7 \times 10^{-5}$$

Since $7.7 \times 10^{-5} \bullet 100 < 3.23 \times 10^{-2}$ we simplify: $\dfrac{(x)(x)}{3.23 \times 10^{-2}} = 7.7 \times 10^{-5}$

and x $= 1.5 \times 10^{-3}$ M.

At the equivalence point: $[OH^-] = 1.5 \times 10^{-3}$ M

and $[H_3O^+] = \dfrac{1.0 \times 10^{-14}}{1.5 \times 10^{-3}} = 6.5 \times 10^{-12}$ and pH $= 11.19$

While the phenoxide ion reacts with water (hydrolyzes) to some extent, the **Na$^+$** is a "spectator ion" and its concentration remains unchanged at 3.23×10^{-2} M.

The concentration of phenoxide **is** reduced (albeit slightly) and at equilibrium is

$(3.23 \times 10^{-2}$ M - x) or 3.08×10^{-2} M.

25. (a) At the equivalence point the moles of acid = moles of base.

$(0.03678$ L$)$ $(0.0105$ M HCl$)$ $=$ 3.86×10^{-4} mol HCl

If this amount of base were contained in 25.0 mL of solution, the concentration of NH_3 in

the original solution was 0.0154 M.

(b) At the equivalence point NH_4Cl will hydrolyze according to the equation:

$$NH_4^+ \text{ (aq)} + H_2O \text{ (}\ell\text{)} \Leftrightarrow NH_3 \text{ (aq)} + H_3O^+ \text{ (aq)}$$

$$K_a = \frac{[NH_3][H_3O^+]}{[NH_4^+]} = \frac{1.0 \times 10^{-14}}{1.8 \times 10^{-5}} = 5.6 \times 10^{-10}$$

The salt $(3.86 \times 10^{-4}$ mol$)$ is contained in $(25.0 + 36.78)$ 61.78 mL. Its concentration will

be $\dfrac{3.69 \times 10^{-4} \text{ mol}}{0.06178 \text{ L}}$ or 6.25×10^{-3} M. Substituting into the K_a expression :

$$\frac{[H_3O^+]^2}{6.25 \times 10^{-3}} = 5.6 \times 10^{-10} \quad \text{and } [H_3O^+] = 1.9 \times 10^{-6} \text{ and pH} = 5.73.$$

Since $[H_3O^+][OH^-] = 1.0 \times 10^{-14}$ then $[OH^-] = \dfrac{1.0 \times 10^{-14}}{1.9 \times 10^{-6}} = 5.3 \times 10^{-9}$ M

and the $[NH_4^+] = 6.25 \times 10^{-3}$ M

Titration Curves and Indicators

27. The titration of 0.10 M NaOH with 0.10 M HCl (a strong base vs a strong acid)

The initial pH of a 0.10 M NaOH would be

pOH = - log[0.10] so pOH = 1.00

and pH = 13.00

When 15.0 mL of 0.10 M HCl have

been added, one-half of the NaOH

initially present will be consumed, leaving

0.5 $(0.030$ L $\bullet$ 0.10 mol/L$)$ or 1.50×10^{-3}

mol NaOH in 45.0 mL—therefore a

concentration of 0.0333 M NaOH.

pOH = 1.48 and pH = 12.52.

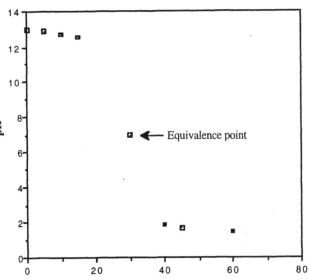

At the equivalence point (30.0 mL of the 0.10 M acid are added) there is only NaCl present. Since this salt does not hydrolyze, the pH at that point is exactly 7.0. The total volume present at this point is 60.0 mL.

Once a total of 60.0 mL of acid are added, there is an excess of 3.0×10^{-3} mol of HCl. Contained in a total volume of 90.0 mL of solution, the [HCl] = 0.0333 and the pH =1.5.

29. (a) pH of 25.0 mL of 0.11 M NH_3:

For the weak base, NH_3, the equilibrium in water is represented as:

$$NH_3 \ (aq) + H_2O \ (\ell) \Leftrightarrow NH_4^+ \ (aq) + OH^- \ (aq)$$

The slight dissociation of NH_3 would form equimolar amounts of NH_4^+ and OH^- ions.

$$K_b \ = \ \frac{[NH_4^+][OH^-]}{[NH_3]} \ = \ \frac{(x)(x)}{0.11 - x} \ = \ 1.8 \times 10^{-5}$$

Simplifying, we get : $\ \frac{x^2}{0.10} \ = \ 1.8 \times 10^{-5}; x \ = \ 1.4 \times 10^{-3} \ M \ = \ [OH^-]$

$$pOH \ = \ 2.87 \quad \text{and} \quad pH \ = \ 11.12$$

(b) Addition of HCl will consume NH_3 and produce NH_4^+ (the conjugate) according to the net equation: $NH_3 \ (aq) + H^+ \ (\ell) \Leftrightarrow NH_4^+ \ (aq)$

The strong acid will drive this equilibrium to the right so we will assume this reaction to be complete. Calculate the moles of NH_3 initially present:

$$(0.0250 \ L) \ (0.10 \ \frac{mol \ NH_3}{L} \ ) \ = \ 0.00250 \ mol \ NH_3$$

Reaction with the HCl will produce the conjugate acid, NH_4^+. The task is two-fold. First calculate the amounts of the conjugate pair present. Second substitute the concentrations into the K_b expression. [One time-saving hint: The ratio of concentrations and the ratio of the amounts (moles) will have the same numerical value. One can substitute the amounts of the conjugate pair into the K_b expression.]

$$K_b = \ \frac{[NH_4^+][OH^-]}{[NH_3]} \ = \ 1.8 \times 10^{-5}$$

When 25.0 mL of the 0.10 M HCl has been added (total solution volume = 50.0 mL), the reaction is at the equivalence point. All the NH_3 will be consumed, leaving the salt, NH_4Cl.

The NH_4Cl (2.50 millimol) has a concentration of 5.0×10^{-2} M. This salt, being formed from a weak base and strong acid, undergoes hydrolysis.

$$NH_4^+(aq) + H_2O\ (\ell) \Leftrightarrow NH_3\ (aq) + H_3O^+\ (aq)$$

	NH_4^+	NH_3	H_3O^+
Initial concentration	$5.0 \times 10^{-2}M$		
Change	$-x$	$+x$	$+x$
Equilibrium	$5.0 \times 10^{-2} - x$	$+x$	$+x$

$$K_a = \frac{[NH_3][H_3O^+]}{[NH_4^+]} = 5.6 \times 10^{-10}$$

$$= \frac{x^2}{5.0 \times 10^{-2} - x} \approx \frac{x^2}{5.0 \times 10^{-2}} = 5.6 \times 10^{-10}$$

and $x = 5.3 \times 10^{-6} = [H_3O^+]$ and pH = 5.28 (equivalence point)

(c) The halfway point of the titration occurs when 12.50 mL of the acid have been added. At that point the amount of base and salt present are equal. An examination of the K_b expression will show that under these conditions the $[OH^-] = K_b$. So pOH of 4.75 and pH = 9.25.

(d) From the table of indicators in your text, one indicator to use is Methyl Red. This indicator would be yellow prior to the equivalence point and red past that point. Bromcresol green would also be suitable, being blue prior to the equivalence point and yellow-green after the equivalence point.

(e)

mL of 0.10 M HCl added	millimol HCl added	millimol NH_3 after reaction	millimol NH_4^+ after reaction	$[OH^-]$ after reaction	pH
5.00	0.50	2.0	0.50	7.2×10^{-5}	9.85
15.0	1.5	1.0	1.5	1.2×10^{-5}	9.08
20.0	2.0	0.50	2.0	4.5×10^{-6}	8.65
22.0	2.2	0.30	2.2	2.5×10^{-6}	8.39

For the pH after 30.0 mL have been added: Addition of acid in excess of 25.00 mL will result in a solution which is essentially a strong acid. After the addition of 30.0 mL, substances present are: millimol HCl added: 3.00

millimol NH_3 present: 2.50

excess HCl present :   0.50 millimol

This HCl is present in a total volume of 55.0 mL of solution, hence the calculation for a strong acid proceeds as follows: $[H_3O^+] = 0.50$ mmol HCl/55.0 mL $= 9.1 \times 10^{-3}$ M and pH = 2.04.

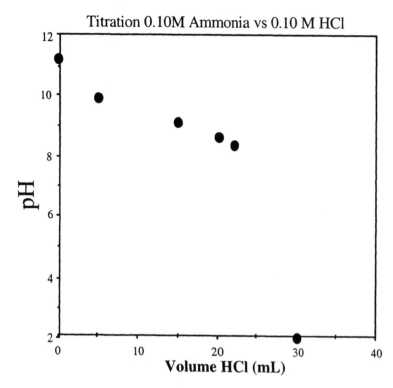

A Summary

mL acid	pH
0.00	11.15
5.00	9.85
15.0	9.08
20.0	8.65
22.0	8.39
30.0	2.04

31. Suitable indicators for titrations:

(a) HCl with pyridine: A solution of pyridinium chloride would have a pH of approximately 3. A suitable indicator would be thymol blue.

(b) NaOH with formic acid: The salt formed at the equivalence point is sodium formate. Hydrolysis of the formate ion would give rise to a basic solution (pH $\approx$ 8.5). Phenolphthalein would be a suitable indicator.

(c) Ethylenediamine and HCl: The base will have two endpoints—as it has two K's ($Kb_1 = 8.5 \times 10^{-5}$ and $Kb_2 = 2.7 \times 10^{-8}$). The first endpoint would contain, as the predominant specie, $H_2N(CH_2)_2NH_3^+$. This ion would hydrolyze with water. The Ka of this conjugate acid would be $1.0 \times 10^{-14} / 8.5 \times 10^{-5}$ or 1.2×10^{-10}. Assume that we have a 0.1 M soln of the anion. The pH can be determined by the expression:

$Ka = \dfrac{[H_3O^+]^2}{0.1} = 1.2 \times 10^{-10}$. [Recall that the concentration of the anion and the hydronium ion would be identical.][H_3O^+]= 3.5×10^{-6} and pH = $-\log(3.5 \times 10^{-6})$ or 5.5. Since the indicator would need to change colors +/- 1 on either side of the endpoint, methyl red would be a suitable indicator for this endpoint. The 2nd endpoint would be reached when the monoprotonated base (let's call it BH^+) has accepted a 2nd proton (to form BH_2^+).

The material formed would hydrolyze: $BH_2^+ + H_2O \Leftrightarrow BH^+ + H_3O^+$, with a

$Ka = 1.0 \times 10^{-14}/ 2.7 \times 10^{-8}$ or 3.7×10^{-7}. Substituting into the Ka expression:

$$Ka = \frac{[H_3O^+]^2}{0.1} = 3.7 \times 10^{-7}.$$ So $[H_3O^+] = 1.9 \times 10^{-4}$ and $pH = 3.7$. A suitable indicator

would be bromphenol blue.

Solubility Guidelines

33. Two insoluble salts of

 (a) Cl^- $AgCl$ and $PbCl_2$

 (b) Zn^{2+} ZnS and $ZnCO_3$

 (c) Fe^{2+} $Fe\,CO_3$ and FeC_2O_4

35. Using the table of solubility guidelines, predict water solubility for the following:

 (a) $(NH_4)_2CO_3$ Ammonium salts are **soluble**.

 (b) $ZnSO_4$ Sulfates are generally **soluble**.

 (c) NiS Sulfides are generally **insoluble**.

 (d) $BaSO_4$ Sr^{2+}, Ba^{2+}, and Pb^{2+} form **insoluble** sulfates.

Writing Solubility Product Constant Expressions

37. Salt dissolving	Ksp expression	Ksp values
(a) $AgCN(s) \Leftrightarrow Ag^+(aq) + CN^-(aq)$	$K_{sp} = [Ag^+][CN^-]$	6.0×10^{-17}
(b) $NiCO_3(s) \Leftrightarrow Ni^{2+}(aq) + CO_3^{2-}(aq)$	$K_{sp} = [Ni^{2+}][CO_3^{2-}]$	1.4×10^{-7}
(c) $AuBr_3(s) \Leftrightarrow Au^{3+}(aq) + 3\,Br^-(aq)$	$K_{sp} = [Au^{3+}][Br^-]^3$	4.0×10^{-36}

Calculating Ksp

39. Here we need only to substitute the equilibrium concentrations into the K_{sp} expression:

 $K_{sp} = [Tl^+][Br^-] = (1.9 \times 10^{-3})(1.9 \times 10^{-3}) = 3.6 \times 10^{-6}$

41. What is K_{sp} for SrF_2?

The temptation is to calculate the number of moles of the solid that are added to water, but one

must recall (a) not all that solid dissolves and (b) the concentration of the solid does not appear

in the K_{sp} expression.

Note that the equilibrium concentration of $[Sr^{2+}] = 1.03 \times 10^{-3}$ M. The stoichiometry of the solid dissolving indicates two fluoride ions accompany the formation of one strontium ion. This tells us that $[F^-] = 2.06 \times 10^{-3}$ M. Substitution into the K_{sp} expression:

$$K_{sp} = [Sr^{+2}][F^-]^2 = (1.03 \times 10^{-3})(2.06 \times 10^{-3})^2 = 4.37 \times 10^{-9}.$$

43. For lead(II) hydroxide, the K_{sp} expression is $K_{sp} = [Pb^{+2}][OH^-]^2$.

Since we know the pH, we can calculate the $[OH^-]$.

pH = 9.15 and pOH = 14.00- 9.15 = 4.85 so $[OH^-]= 1.4 \times 10^{-5}$.

For each mole of $Pb(OH)_2$ that dissolves, we get one mol of Pb^{+2} and two mol of OH^- .

Since $[OH^-]$ at equilibrium = 1.4×10^{-5} M, then $[Pb^{+2}] = 1/2 \cdot 1.4 \times 10^{-5}$ M

$$K_{sp} = [Pb^{+2}][OH^-]^2 = K_{sp} = [7.0 \times 10^{-6}][1.4 \times 10^{-5}]^2 = 1.4 \times 10^{-15}$$

Estimating Salt Solubility from K_{sp}

45. The solubility of AgI in water at 25°C in (a) mol/L and (b) g/L

(a) The K_{sp} for AgI (from Appendix J) is 8.5×10^{-17}. The K_{sp} expression is:

$$K_{sp} = [Ag^+][I^-] = 8.5 \times 10^{-17}$$

Since the solid dissolves to give one silver ion/one iodide ion, the concentrations of Ag+ and I$^-$ will be equal. We can write the K_{sp} as: $[Ag^+]^2 = [I^-]^2 = 8.5 \times 10^{-17}$.

Taking the square root of both sides we obtain: $[Ag^+] = [I^-] = 9.2 \times 10^{-9}$ and recognizing that each mole of AgI that dissolves per liter gives one mol of silver ion, the solubility of AgI will be 9.2×10^{-9} mol/L.

(b) The solubility in g/L: $9.2 \times 10^{-9} \dfrac{mol\ AgI}{L} \cdot \dfrac{234.77\ g\ AgI}{1\ mol\ AgI} = 2.2 \times 10^{-6}$ g AgI/L

47. The K_{sp} for $CaF_2 = 5.3 \times 10^{-11}$
$K_{sp} = [Ca^{2+}][F^-]^2 = 5.3 \times 10^{-11}$
if a mol/L of CaF_2 dissolve, $[Ca^{2+}] = a$ and $[F^-] = 2a$
$K_{sp} = (a)(2a)^2 = 4a^3 = 5.3 \times 10^{-11}$ and $a = 2.36 \times 10^{-4}$

(a) The molar solubility is then 2.4×10^{-4} M (to 2 sf)

(b) Solubility in g/L

$$2.4 \times 10^{-4} \dfrac{mol\ CaF_2}{L} \cdot \dfrac{78.07\ g\ CaF_2}{1\ mol\ CaF_2} = 1.8 \times 10^{-2}\ g\ CaF_2\ /L$$

49. $K_{sp} = [Ra^{2+}][SO_4^{2-}] = 4.2 \times 10^{-11}$ so $[Ra^{2+}] = [SO_4^{2-}] = 6.5 \times 10^{-6}$ M

$RaSO_4$ will dissolve to the extent of 6.5×10^{-6} mol/L

Express this as grams in 100. mL (or 0.1 L)

$$\frac{6.5 \times 10^{-6} \text{ mol } RaSO_4}{1 \text{ L}} \cdot \frac{0.1 \text{ L}}{1} \cdot \frac{322 \text{ g } RaSO_4}{1 \text{ mol } RaSO_4} = 2.1 \times 10^{-4} \text{ g}$$

Expressed as milligrams: 0.21 mg (to 2 sf) of $RaSO_4$ will dissolve!

51. Which solute in each of the following pairs is more soluble.

	Compound	Ksp
(a)	**PbCl$_2$**	1.7×10^{-5}
	PbBr$_2$	6.6×10^{-6}
(b)	HgS (red)	4×10^{-54}
	FeS	6×10^{-19}
(c)	**Fe(OH)$_2$**	4.9×10^{-17}
	Zn(OH)$_2$	3×10^{-17}

To compare relative solubilities of two compounds of the same general formula, one can examine the Ksp value. The larger the K_{sp}, the more soluble the compound.

Common Ion Effect and Salt Solubility

53. The equilibrium for AgSCN dissolving is: $AgSCN$ (s) $\Leftrightarrow$ Ag^+ (aq) + SCN^- (aq).

As x mol/L of AgSCN dissolve in pure water, x mol/L of Ag^+ and x mol/L of SCN^- are produced.

The espression would be: $K_{sp} = [Ag^+][SCN^-] = x^2 = 1.0 \times 10^{-12}$ and $x = 1.0 \times 10^{-6}$ M

So 1.0×10^{-6} mol AgSCN/L dissolve in pure water.

The equilibrium *for AgSCN dissolving in NaSCN* (0.010 M) is like that above. Equimolar amounts of Ag^+ and SCN^- ions are produced as the solid dissolves. However the $[SCN^-]$ is augmented by the soluble NaSCN.

$$K_{sp} = [Ag^+][SCN^-] = (x)(x + 0.010) = 1.0 \times 10^{-12}$$

We can simplify the expression by *assuming* that $x + 0.010 \approx 0.010$. Note that the value of x above (1.0×10^{-6}) lends credibility to this assumption.

$$(x)(0.010) = 1.0 \times 10^{-12} \quad \text{and} \quad x = 1.0 \times 10^{-10} \text{ M}$$

The solubility of AgSCN in 0.010 M NaSCN is 1.0×10^{-10} M -- reduced by four orders of magnitude from its solubility in pure water. LeChatelier strikes again!

55. Solubility in mg/mL of AgI in (a) pure water and (b) water that is 0.020 M in Ag^+ (aq).

$$K_{sp} = [Ag^+][I^-] = 8.5 \times 10^{-17}$$

(a) In SQ 18.45 we found the solubility of AgI to be 9.2×10^{-9} mol/L or 2.2×10^{-6} g/L

Expressing this in mg/mL = 2.2×10^{-6} mg/mL

(b) The difference in solubility arises because of the presence of the 0.020 M in Ag^+.
Recall that in SQ18.45 we knew that the silver and iodide ion concentrations were equal.
With the addition of the 0.020 M silver solution, this is no longer a valid assumption. If we
let x mol/L of the AgI dissolve, the concentrations of Ag^+ and I^- (from the salt dissolving)
will be x mol/L. The $[Ag^+]$ will be amended by 0.020 M, so we write:

$Ksp = [x + 0.020][x] = 8.5 \times 10^{-17}$. This could be solved with the quadratic equation,
but a bit of thought will simplify the process. In the (a) part we discovered that the $[Ag^+]$
was approximately 10^{-9} M. This value is small compared to 0.020. Let's use that
approximation to convert the Ksp expression to:

$[0.020][x] = 8.5 \times 10^{-17}$ and x = 4.25×10^{-15} M.

This molar solubility translates into:

$$4.25 \times 10^{-15} \frac{\text{mol AgI}}{\text{L}} \bullet \frac{234.77 \text{ g AgI}}{1 \text{ mol AgI}} = 9.98 \times 10^{-13} \text{g AgI/L}$$

or 1.0×10^{-12} mg/mL (to 2 sf).

Effect of Basic Anions on Salt Solubility

57. The salt that should be more soluble in nitric acid than in pure water from the pairs:

(a) $PbCl_2$ or PbS: PbS will be more soluble, since the S^{2-} ion will react with the nitric
acid, reducing the $[S^{2-}]$, and increasing the amount of PbS that dissolves.

(b) Ag_2CO_3 or AgI: Ag_2CO_3 will be more soluble. The CO_3^{2-} ion will react with the nitric
acid, and produce HCO_3^- and H_2CO_3 which will decompose to CO_2 and H_2O. The
removal of carbonate shifts the equilibrium to the right, increasing the amount that
dissolves.

(c) $Al(OH)_3$ or AgCl: $Al(OH)_3$ will be more soluble. As in (b) above, the OH^- will react with
the H^+ to form water. The reduction in $[OH^-]$ will increase the amount of the salt that
dissolves.

Precipitation Reactions

59. Given the equation for $PbCl_2$ dissolving in water: $PbCl_2$ (s) $\Leftrightarrow$ Pb^{+2} (aq) + 2 Cl^-

we can write the Ksp expression : $K_{sp} = [Pb^{+2}][Cl^-]^2 = 1.7 \times 10^{-5}$

Substituting the ion concentrations into the Ksp expression we get:

$Q = [Pb^{+2}][Cl^-]^2 = (0.0012)(0.010)^2 = 1.2 \times 10^{-7}$

Since Q is less than K_{sp} , no $PbCl_2$ precipitates.

61. If $Zn(OH)_2$ is to precipitate, the reaction quotient (Q) must exceed the K_{sp} for the salt.

4.0 mg of NaOH in 10. mL corresponds to a concentration of:

$$[OH^-] = \frac{4.0 \times 10^{-3} \text{ g NaOH}}{0.0100 \text{ L}} \cdot \frac{1 \text{ mol NaOH}}{40.0 \text{ g NaOH}} = 0.01 \text{ M}$$

The value of Q is : $[Zn^{2+}][OH^-]^2 = (1.6 \times 10^{-4})(1.0 \times 10^{-2})^2 = 1.6 \times 10^{-8}$

The value of Q is greater than the K_{sp} for the salt (4.5×10^{-17}) , so $Zn(OH)_2$ precipitates.

63. The molar concentration of Mg^{2+} is:

$$\frac{1350 \text{ mg Mg}^{2+}}{1 \text{ L}} \cdot \frac{1 \text{ g Mg}^{2+}}{1000 \text{ mg Mg}^{2+}} \cdot \frac{1 \text{ mol Mg}^{2+}}{24.3050 \text{ g Mg}^{2+}} = 5.55 \times 10^{-2} M$$

For $Mg(OH)_2$ to precipitate, Q must be greater than the K_{sp} for the salt (5.6×10^{-12}).

$Q = [Mg^{2+}][OH^-]^2 = (5.55 \times 10^{-2})(OH^-)^2 = 5.6 \times 10^{-12}$ so

$(OH^-)^2 = 5.6 \times 10^{-12}/5.55 \times 10^{-2}$ or 1.01×10^{-10} , and $[OH^-] = 1.0 \times 10^{-5}$ M.

Solubility and Complex Ions

65. Show that the equation: $AuCl(s) + 2 CN^- (aq) \Leftrightarrow Au(CN)_2^- (aq) + Cl^- (aq)$ is the sum of:

AuCl dissolving: $AuCl(s) \Leftrightarrow Au^+ (aq) + Cl^- (aq)$ $K_{sp} = 2.0 \times 10^{-13}$

$Au(CN)_2^-$ formation: $Au^+ (aq) + 2 CN^- (aq) \Leftrightarrow Au(CN)_2^- (aq)$ $K_f = 2.0 \times 10^{38}$

The net equation $AuCl(s) + 2 CN^- (aq) \Leftrightarrow Au(CN)_2^- (aq) + Cl^-$ $K_{net} = K_{sp} \cdot K_f$

$K_{net} = 2.0 \times 10^{-13} \cdot 2.0 \times 10^{38} = 4.0 \times 10^{25}$

Separations

67. Separate the following pairs of ions:

(a) Ba^{2+} and Na^+ : Since most sodium salts are soluble, it is simple to find a barium salt which is not soluble--e.g. the sulfate. Addition of dilute sulfuric acid should provide a source of SO_4^{2-} ions in sufficient quantity to precipitate the barium ions, but not the sodium ions.

[Ksp $(BaSO_4) = 1.1 \times 10^{-10}$ Ksp (Na_2SO_4) = not listed, owing to the large solubility of sodium sulfate]

(b) Ni^{2+} and Pb^{2+}: The carbonate ion will serve as an effective reagent for selective precipitation of the two ions. A solution of Na_2CO_3 can be added, and the less soluble $PbCO_3$ will begin precipitating first. The 7 orders of magnitude difference in their Ksp's should provide satisfactory separation.

$$[K_{sp} \ (PbCO_3) = 7.4 \times 10^{-14} \quad K_{sp} \ (NiCO_3) = 1.4 \times 10^{-7}]$$

General Questions

69. Solution producing precipitate:

For the solutes listed, note that in water each would exist as the individual cations and anions.

For example, NaBr (aq) more accurately may be represented as: Na^+(aq) and Br^-(aq). The net equations that result are shown:

(a) Ag^+ (aq) + Br^- (aq) $\rightarrow$ AgBr (s)

(b) Pb^{2+} (aq) + 2 Cl^- (aq) $\rightarrow$ $PbCl_2$ (s)

One can make these decisions in one of two ways: (1) Recalling the solubility tables—probably learned earlier or (2) Reviewing Appendix J (which contains compounds that we normally classify as insoluble—and precipitate from a solution containing the appropriate pairs of cations and anions (Ag^+ and Br^- for example).

71. Will $BaSO_4$ precipitate?

Calculate the concentrations of barium and sulfate ions (after the solutions are mixed).

$$\frac{48 \ mL}{72 \ mL} \bullet 0.0012 \ M \ Ba^{2+} = 0.0008 \ M \ Ba^{2+} \ and$$

$$\frac{24 \ mL}{72 \ mL} \bullet 1.0 \times 10^{-6} \ M \ SO_4^{2-} = 3.3 \times 10^{-7} \ M \ SO_4^{2-}$$

Substituting in the K_{sp} expression: $Q = [Ba^{+2}][SO_4^{2-}] = (8.0 \times 10^{-4})(3.3 \times 10^{-7})$

$Q = 2.7 \times 10^{-10}$ Q is larger than the Ksp for the solid(1.1×10^{-10}) so $BaSO_4$ precipitates.

Note that we are assuming that the sulfuric acid totally dissociates into protons and sulfate ions!

73. The pH and $[H_3O^+]$ when 50.0 mL of 0.40 M NH_3 is mixed with 25.0 mL of 0.20 M HCl:

The number of moles of each reactant:

(0.20 mol HCl/L)(0.025 L) = 0.0050 mol HCl and (0.40 mol NH_3/L)(0.050 L) = 0.020 mol

NH_3. So 0.005 mol of NH_3 are consumed—and 0.005 mol NH_4^+ are produced.

	NH_3	H_2O	NH_4^+
Initial	0.020 mol NH_3		
Change	-0.0050 (consumed by HCl)		+0.0050 mol
Equilibrium	0.015 mol		0.0050 mol

So we have a buffer with a conjugate pair (NH_3 and NH_4^+) present. The K_b expression for

NH_3 can be written:

$$\frac{[NH_4^+][OH^-]}{[NH_3]} = 1.8 \times 10^{-5}$$ Substituting the # of mol into the K_b expression yields:

$$\frac{[0.005\ mol][OH^-]}{[0.015\ mol]} = 1.8 \times 10^{-5} \quad or\ [OH^-] = 5.4 \times 10^{-5}.$$

So pOH = 4.27 and pH = 9.73, and a concommitant $[H_3O^+] = 10^{-9.73}$ or 1.9×10^{-10} M.

75. Compounds in order of increasing solubility in H_2O:

Compound	K_{sp}
$BaCO_3$	2.6×10^{-9}
Ag_2CO_3	8.5×10^{-12}
Na_2CO_3	not listed: Na_2CO_3 is very soluble in water and is the most soluble salt of the 3.

To determine the relative solubilities, find the molar solubilities.

For $BaCO_3$: $K_{sp} = [Ba^{2+}][CO_3^{2-}] = (x)(x) = 2.6 \times 10^{-9}$ and $x = 5.1 \times 10^{-5}$

The molar solubility of $BaCO_3$ is 5.1×10^{-5} M.

and for Ag_2CO_3: $K_{sp} = [Ag^+]^2[CO_3^{2-}] = (2x)^2(x) = 8.5 \times 10^{-12}$

$$4x^3 = 8.5 \times 10^{-12} \text{ and } x = 1.3 \times 10^{-4}$$

The molar solubility of Ag_2CO_3 is 1.3×10^{-4} M.

In order of increasing solubility: $BaCO_3 < Ag_2CO_3 < Na_2CO_3$

77. pH of a solution that contains 5.15 g NH_4NO_3 and 0.10 L of 0.15 M NH_3:

The concentration of the ammonium ion (from the nitrate salt) is:

$$5.15\ g\ NH_4NO_3 \bullet \frac{1\ mol\ NH_4NO_3}{80.04\ g\ NH_4NO_3} \bullet \frac{1}{0.10\ L} = 0.64\ M$$

We can calculate the pH using the K_b expression for ammonia:

$$K_b = \frac{[NH_4^+][OH^-]}{[NH_3]} = \frac{(0.64)(OH^-)}{(0.15)} = 1.8 \times 10^{-5} \text{ and solving for the hydroxide ion}$$

concentration, $x = \dfrac{(1.8 \times 10^{-5})(0.15)}{(0.64)} = 4.2 \times 10^{-6}$ M and pOH = 5.38 ; so pH = 8.62

Diluting the solution does not change the pH of the solution, since the dilution affects the concentration of both members of the conjugate pair. Since the pH of the buffer is a function of the *ratio* of the conjugate pair, the pH does not change.

79. For the titration of aniline hydrochloride with NaOH:

(a) initial pH: Recall that the hydrochloride (represented in this example as HA) is a weak acid:

($K_a = 2.4 \times 10^{-5}$) Use a K_a expression to solve for H_3O^+:

$$K_a = \frac{[A^-][H_3O^+]}{[HA]} = 2.4 \times 10^{-5} \bullet 0.10 = x^2 = 2.4 \times 10^{-6}$$

$x = 1.5 \times 10^{-3}$M = $[H_3O^+]$ and pH = 2.81

(b) pH at the equivalence point:

At the equivalence point we know that the number of moles of acid - # moles base (by definition) Since we have 0.00500 mol of HA (M $\bullet$ V), we must ask the question, "How much 0.185 M NaOH contains 0.00500 mol NaOH?"

That volume is 0.00500 mol OH^- $\bullet$ 1L/ 0.185 mol NaOH = 27.0 mL of the NaOH solution. The total volume is then 50 + 27 mL = 0.0770 L. The [HA] = 0.00500 mol/0.0770 L = 0.0649 M

We can calculate the pH of the solution using the K_b expression for A:

$$K_b = \frac{[HA][OH^-]}{[A]} = \frac{1.0 \times 10^{-14}}{2.4 \times 10^{-5}} \text{ and knowing that the } [HA] = [OH^-]$$

$x^2 = 4.2 \times 10^{-12} \bullet 0.0649$ and $x = 5.2 \times 10^{-6}$ M; pOH = 5.28 and pH = 8.72

(c) pH at the midpont:

This portion is easy to solve if you examine the Ka expression or the Henderson-Hasselbalch expression for this system:

From the K_b expression above, note that the conjugate pairs are found in the numerator and denominator of the right-side of the K_b term (the same applies to Ka expressions).

At the mid-point you have reacted half the acid, (using 13.5 mL—see part (b))forming its conjugate base. The result is that the concentrations of the conjugate pairs are equal and the $pOH = pK_b$ and $pH = pK_a$ So $pH = -log(2.4 \times 10^{-5}) = 4.62$.

(d) With a pH= 8.72 at the equivalence point, o-cresolphthalein or phenolphthalein would serve as an adequate indicator.

(e) The pH after the addition of 10.0, 20.0 and 30.0 mL base:

The volumes of base correspond to

10 mL (0.185 mol/L • 0.0100 L)	0.00185 mol NaOH
20.0 mL (0.185 mol/L • 0.0200 L)	0.00370 mol NaOH
30.0 mL(0.185 mol/L • 0.0300L)	0.00555 mol NaOH

Each mol of NaOH consumes a mol of aniline hydrochloride(HA) and produces an equal number of mol of aniline(A).

Recall that we began with 0.00500 mol of the acid (which we're representing as HA)

Substitution into the Ka expression yields:

$$\frac{[A^-][H_3O^+]}{[HA]} = 2.4 \times 10^{-5} \text{ and rearranging them } [H_3O^+] = 2.4 \times 10^{-5} \cdot \frac{[HA]}{[A]}$$

After 10 mL of base are added, 0.00185 mol HA are consumed, and 0.00185 mol A produced. The acid remaining is (000500 - 0.00185) mol.

Substituting into the rearranged equation:

$$[H_3O^+] = 2.4 \times 10^{-5} \cdot \frac{[0.00315 \text{ mol}]}{[0.00185 \text{ mol}]} = 4.1 \times 10^{-5} \text{ and pH} = -log (4.1 \times 10^{-5}) = 4.39$$

After 20 mL of base are added, 0.00370 mol HA are consumed, and 0.00370 mol A produced. The acid remaining is (000500 - 0.00370)mol.

Substituting into the rearranged equation:

$$[H_3O^+] = 2.4 \times 10^{-5} \cdot \frac{[0.00130 \text{ mol}]}{[0.00370 \text{ mol}]} = 8.4 \times 10^{-6} \text{ and pH} = -log (8.4 \times 10^{-6}) = 5.07$$

After 30 mL of base are added, all the acid is consumed, and excess strong base is present (0.00555 mol – 0.00500)mol. We can treat this solution as one of a strong base. The volume of the solution is (50.0 + 30.0 or 80.0 mL.).

The concentration of the NaOH is:

0.00055 mol/ 0.080 L = 0.006875 M so pOH = -log(0.006875) and pOH=2.16, the pH=11.84.

(f) The approximate titration curve:

Vol base	pH
0.0	2.81
10.0	4.39
13.5	4.62
20.0	5.07
27.0	8.72
30.0	11.84

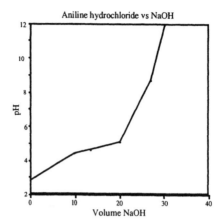

81. Make a buffer of pH= 2.50 from 100. mL of 0.230 M H_3PO_4 and 0.150 M NaOH.

For the first ionization of the acid, K_a= 7.5 x 10^{-3} (Appendix H)

The pK_a of the acid = 2.12. The Henderson-Hasselbalch equation will help to determine the ratio of the conjugate pair to make the pH 2.50.

We'll assume that (since the second K_a is approximately 10^{-8}, that we're dealing only with the first ionization. This *is an approximation.*

$$pH = pK_a + \log \frac{[\text{conjugate base}]}{[\text{acid}]} \quad \text{so } 2.50 = 2.12 + \log \frac{[\text{conjugate base}]}{[\text{acid}]}$$

$$0.38 = \log \frac{[A]}{[HA]} \quad \text{and the ratio of } \frac{[A]}{[HA]} = 2.37$$

The amount of HA = (0.100 L • 0.230M HA) = 0.0230 mol HA.

Rearranging the ratio: [A] = 2.37[HA]. Knowing that the acid present is *either* the molecular acid, HA, or the conjugate base of the acid, A, we can write:

A + HA = 0.0230 mol and with A = 2.37 HA

2.37 HA + HA = 0.0230mol ; 3.37 HA = 0.0230 mol; and HA = 0.0068 mol.

We need to consume (0.0230 mol HA- 0.0068 mol HA) or 0.0162 mol HA.

This requires 0.0162 mol NaOH, and from 0.150M NaOH we need:

0.0162 mol NaOH = 0.150 mol/L • V and V = 0.108 L or 110 mL of NaOH (to 2 sf).

83. For the titration of 0.150 M ethylamine (K_b = 4.27 x 10^{-4}) with 0.100 M HCl:
   (a) pH of 50.0 mL of 0.0150 M $CH_3CH_2NH_2$:

   For the weak base, $CH_3CH_2NH_2$, the equilibrium in water is represented as:

   $$CH_3CH_2NH_2 \text{ (aq) } + H_2O \text{ (}\ell\text{) } \Leftrightarrow CH_3CH_2NH_3^+ \text{ (aq) } + OH^- \text{ (aq)}$$

   Note : The K_b for ethylamine is given as 4.27 x 10^{-4}.

The slight dissociation of $CH_3CH_2NH_2$ would form equimolar amounts of $CH_3CH_2NH_3^+$ and OH^- ions.

$$K_b = \frac{[CH_3CH_2NH_3^+][OH^-]}{[CH_3CH_2NH_2]} = \frac{(x)(x)}{0.150 - x} = 4.27 \times 10^{-4}$$

The quadratic equation will be needed to find an exact solution.

$$x^2 = 6.405 - 4.27 \times 10^{-4}x$$

Rearranging: $x^2 + 4.27 \times 10^{-4}x - 6.405 \times 10^{-5} = 0$

and solving for $x = 7.79 \times 10^{-3}$; $pOH = -\log(7.79 \times 10^{-3}) = 2.11$

and $pH = 11.89$

(b) pH at the halfway point of the titration:

The volume of acid added isn't that important since the amounts of base and conjugate acid will be equal. In part (d), we find that 75.0 mL of acid are required for the equivalence point, so the volume of acid at this point is (0.5 • 75.0 mL)

Substituting that fact into the equilibrium expression we obtain:

$$\frac{[CH_3CH_2NH_3^+][OH^-]}{[CH_3CH_2NH_2]} = \frac{(x)[OH^-]}{(x)} = 4.27 \times 10^{-4} \text{ so we can see that the}$$

$[OH^-] = 4.27 \times 10^{-4}$ and $pOH = 3.37$ and $pH = (14.00-3.37) = 10.63$

(c) pH when 75% of the required acid has been added:

Amount of base initially present is (0.050 L • 0.150 M) or 0.0075 moles base.

So 75% of this amount is 0.00563 mol. requiring 0.00563 mol of HCl.

The volume of 0.100 M HCl containing that # mol is

0.00563 mol HCl/0.100 M = 0.0563 L or 56.3 mL

So 0.001875 mol base remain. Substituting into the K_b expression:

$$\frac{[CH_3CH_2NH_3^+][OH^-]}{[CH_3CH_2NH_2]} = \frac{0.00563 \text{ mol} \bullet [OH^-]}{0.001875 \text{ mol}} = 4.27 \times 10^{-4}$$

solving for hydroxyl ion concentration yields: 1.42×10^{-4} M and $pOH = 3.85$

and $pH = 10.15$.

(d) pH at the equivalence point:

At the equivalence point, there are equal # of moles of acid and base. The number of moles of ethylamine = (0.050 L • 0.150 M) or 0.00750 moles ethylamine .

That amount of HCl would be:

$$7.50 \times 10^{-3} \text{ mol HCl} \bullet \frac{1 \text{ L}}{0.100 \text{ mol HCl}} = 0.0750 \text{ L (or 75.0 mL)}$$

This total amount of solution would be: $75.0 + 50.0 = 125.0$ mL

Since we have added equal amounts of acid and base, the reaction between the two will result in the existence of only the salt.

and the concentration of salt would be: $\dfrac{7.50 \times 10^{-3} \text{ mol}}{0.1250 \text{ L}} = 6.00 \times 10^{-2}$ M

Recall that the salt will act as a weak acid and we can calculate the K_a:

$$CH_3CH_2NH_3^+ \text{ (aq)} + H_2O \text{ (}\ell\text{)} \Leftrightarrow CH_3CH_2NH_2 \text{ (aq)} + H_3O^+ \text{ (aq)}$$

The equilibrium constant (K_a) would be $\dfrac{K_w}{K_b} = \dfrac{1.0 \times 10^{-14}}{4.27 \times 10^{-4}} = 2.3 \times 10^{-11}$

The equilibrium expression would be:

$$\frac{[CH_3CH_2NH_2][H_3O^+]}{[CH_3CH_2NH_3^+]} = 2.3 \times 10^{-11}$$

Given that the salt would hydrolyze to form equal amounts of ethylamine and hydronium ion (as shown by the equation above). If we represent the concentrations of those species as x, then we can write (using our usual approximation):

$$\frac{x^2}{6.00 \times 10^{-2}} = 2.3 \times 10^{-11} \text{ and solving for } x = 1.19 \times 10^{-6}$$

Since x represents $[H_3O^+]$, then pH $= -\log(1.19 \times 10^{-6}) = 5.93$

(e) pH after addition of 10.0 mL of HCl more than required:

Past the equivalence point, the excess strong acid controls the pH. To determine the pH, we need only calculate the concentration of HCl. In part (d) we found that the total volume at the equivalence point was 125.0 mL. The addition of 10.0 mL will bring that total volume to 135.0 mL of solution. The # of moles of excess HCl is:

(0.100 mol HCl/L • 0.0100 L = 0.00100 mol HCl contained within 135.0 mL, for a concentration of 7.41×10^{-3} M HCl. Since HCl is a strong acid, the

pH $= -\log(7.41 \times 10^{-3})$ or 2.13.

(f) The titration curve.

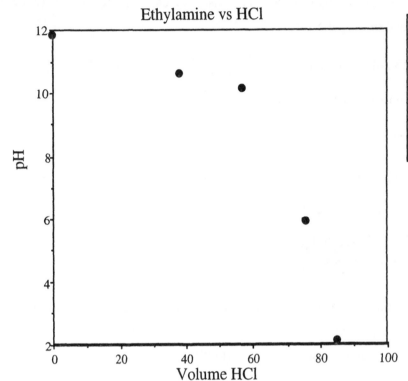

Volume HCl	pH
0.0	11.89
37.5	10.63
56.3	10.15
75.0	5.93
85.0	2.13

(g) Suitable indicator for endpoint: Alizarin or Bromcresol purple would be suitable indicators. (See Figure 18.10 for indicators.)

85. The effect on pH of:

(a) Adding $CH_3CO_2^-Na^+$ to 0.100 M CH_3CO_2H:

The equilibrium affected is: $CH_3CO_2H + H_2O \Leftrightarrow CH_3CO_2^- + H_3O^+$

Addition of sodium acetate increases the concentration of acetate ion. The equilibrium will shift to the left—reducing hydronium ion, and increase the pH.

(b) Adding $NaNO_3$ to 0.100 M HNO_3: No effect on pH. Since nitric acid is a strong acid, the equilibrium lies very far to the right. Addition of NO_3^- will not shift the equilibrium.

(c) The effects differ owing to the nature of the two acids. Nitric acid (a strong acid) exists as hydronium and nitrate ions. Acetic acid exists as the molecular acid and acetate ion. The addition of conjugate base of both acids ($CH_3CO_2^-$ and NO_3^-) will affect the acetic acid equilibrium but not the nitric acid system.

87. Buffer (200.0 mL) of Na_3PO_4 and Na_2HPO_4 has pH=12.00

   (a) Component present in the larger amount:

   The Henderson-Hasselbalch equation will be of use here:

   $$pH = pK_a + log \frac{[conjugate\ base]}{[acid]}$$

   The K_{a3} for the acid is 3.6 x 10^{-13} .

   The pKa is then $-log$ (3.6 x 10^{-13}) = 12.44. and solve for the log term.

   $$12.00 = 12.44 + log \frac{[PO_4^{3-}]}{[HPO_4^{2-}]} \quad so \quad -0.44 = log \frac{[PO_4^{3-}]}{[HPO_4^{2-}]}$$

   Solving for the ratio of the conjugates 10$^{-0.44}$ gives a value = 0.36, so the acidic component (HPO_4^{2-})is present in the greater amount (approximately 3 x the concentration of phosphate).

   (b) With a concentration of Na_3PO_4 of 0.400 M, the mass of Na_2HPO_4

   With the ratio of $0.36 = \dfrac{0.400\ M}{x}$ we can solve for the concentration of the Na_2HPO_4

   x = 0.400/0.36 = 1.11 M Na_2HPO_4 . The mass is then

   (1.11 mol/L • 0.200 L • 141.96 g/mol) or 32 g of Na_2HPO_4 (to 2 sf)

   (c) Substance needed to change pH to 12.25:

   Using Henderson-Hasselbalch as above we get:

   $$12.25 = 12.44 + log \frac{[PO_4^{3-}]}{[HPO_4^{2-}]} \quad so \quad -0.19 = log \frac{[PO_4^{3-}]}{[HPO_4^{2-}]}$$

   Solving for the ratio again, we get a ratio of phosphate to hydrogen phosphate of 0.646. Note that this is larger than the previous ratio, so we need to add phosphate ion to the solution to change the pH as desired.

   Note that [Na_2HPO_4] = 1.11 M (from (b) above) We can solve for the phosphate ion concentration needed to provide a ratio of 0.646.

   $$0.646 = \frac{x}{1.11\ M} \quad and \quad (0.646)(1.11) = 0.717\ M\ PO_4^{3-}$$

   The # of moles we need is (0.717 M PO_4^{3-})(0.2000 L) = 0.143 mol Na_3PO_4.

   We originally had 0.400 M phosphate in 200.0 mL, giving 0.0800 mol Na_3PO_4.

   The amount we need to add is: (0.143 mol Na_3PO_4 -0.0800 mol Na_3PO_4)= 0.063 mol

   The mass of sodium phosphate corresponding to this number of moles is:

   0.063 mol • 163.94 g Na_3PO_4/mol = 10.4 g Na_3PO_4 (or 10. g Na_3PO_4 to 2sf).

89. (a) These metal sulfates are 1:1 salts. The K_{sp} expression has the general form:

$$K_{sp} = [M^{2+}][SO_4{}^{2-}]$$

To determine the $[SO_4{}^{2-}]$ necessary to begin precipitation, we can divide the equation by the metal ion concentration to obtain:

$$\frac{K_{sp}}{[M^{2+}]} = [SO_4{}^{2-}]$$

The concentration of the three metal ions under consideration are each 0.10 M. Substitution of the appropriate K_{sp} for the sulfates and 0.10 M for the metal ion concentration yields the sulfate ion concentrations in the table below. As the soluble sulfate is added to the metal ion solution, the sulfate ion concentration increases from zero molarity. The lowest sulfate ion concentration is reached first, with higher concentrations reached later. The order of precipitation is listed in the last column of the table below.

Compound	K_{sp}	Maximum $[SO_4{}^{2-}]$	Order of Precipitation
$BaSO_4$	1.1×10^{-10}	1.1×10^{-9}	1
$SrSO_4$	3.4×10^{-7}	3.4×10^{-6}	2

(b) Concentration of Ba^{2+} when $SrSO_4$ begins to precipitate:

From (a) we note that $[SO_4{}^{2-}]$ will be 3.4×10^{-6} when $SrSO_4$ begins to precipitate.

So we calculate the concentration of Ba^{2+} by noting that since that barium salt has been precipitating, it is a **saturated solution of $BaSO_4$**.

$$Ksp = [Ba^{2+}][SO_4{}^{2-}] = 1.1 \times 10^{-10}. \text{ So}$$

$$[Ba^{2+}] = 1.1 \times 10^{-10}/3.4 \times 10^{-6} = 3.2 \times 10^{-5}.$$

91. The equation $AgCl(s) + I^-(aq) \Leftrightarrow AgI(aq) + Cl^-$

can be obtained by adding two equations:

1. $AgCl(s) \Leftrightarrow Ag^+(aq) + Cl^-(aq)$	$K_{sp1} = 1.8 \times 10^{-10}$
2. $Ag^+(aq) + I^-(aq) \Leftrightarrow AgI(s)$	$\dfrac{1}{K_{sp2}} = 1.2 \times 10^{16}$

The net equation: $AgCl(s) + I^-(aq) \Leftrightarrow AgI(aq) + Cl^-$

$$K_{net} = K_{sp1} \cdot \frac{1}{K_{sp2}} = (1.8 \times 10^{-10})(1.2 \times 10^{16}) = 2.2 \times 10^6$$

The equilibrium lies to the right. This indicates that *AgI will form* if I^- is added to a saturated solution of AgCl.

<u>93.</u> Regarding the solubility of barium and calcium fluorides:

(a) [F⁻] that will precipitate maximum amount of calcium ion:

BaF_2 begins to precipitate when the ion product just exceeds the K_{sp}.

$K_{sp} = [Ba^{2+}][F^-]^2 = 1.8 \times 10^{-7}$, and the concentration of barium ion = 0.10 M

$[F^-] = (1.8 \times 10^{-7}/0.10)^{0.5} = 1.3 \times 10^{-3}$ M

This would be the maximum fluoride concentration permissible.

(b) $[Ca^{2+}]$ remaining when BaF_2 just begins to precipitate:

The $[Ca^{2+}]$ ion remaining will be calculable from the K_{sp} expression for CaF_2.

$K_{sp} = [Ca^{2+}][F^-]^2 = 5.3 \times 10^{-11}$ and we know that the [F⁻] is 1.3×10^{-3} M.

So $\dfrac{5.3 \times 10^{-11}}{\left(1.3 \times 10^{-3}\right)^2} = 2.9 \times 10^{-5}$ M

<u>95.</u> Regarding the precipitation of $CaSO_4$ and $PbSO_4$ from a solution 0.010 M in each metal:

(a) Examine the K_{sp} of the two salts:

K_{sp} for $CaSO_4 = 4.9 \times 10^{-5}$ and for $PbSO_4 = 2.5 \times 10^{-8}$

These data indicate that $PbSO_4$ will begin to precipitate first—as it is the less soluble of the two salts. This can be found by examining the general Ksp expression for 1:1 salts:

$K_{sp} = [M^{2+}][SO_4^{2-}]$. With both metal concentrations at 0.010 M, the sulfate ion for the less soluble sulfate will be exceeded first.

(b) When the calcium salt just begins to precipitate, the $[Pb^{2-}] = ?$.

We know that the metal ion concentration is 0.010 M. So the sulfate ion concentration when the more soluble calcium sulfate begins to precipitate is:

$K_{sp} = [0.010][SO_4^{2-}] = 4.9 \times 10^{-5}$ and $[SO_4^{2-}] = (4.9 \times 10^{-5}/0.010) = 4.9 \times 10^{-3}$ M

At that point the $[Pb^{2+}]$ would be: $K_{sp} = [Pb^{2+}][SO_4^{2-}] = 2.5 \times 10^{-8}$

Then $[Pb^{2+}][4.9 \times 10^{-3}] = 2.5 \times 10^{-8}$ and $[Pb^{2+}] = (2.5 \times 10^{-8}/4.9 \times 10^{-3})$ or 5.1×10^{-6} M.

Summary and Conceptual Questions

<u>97.</u> To separate CuS and $Cu(OH)_2$:

The K_{sp} for CuS is 6×10^{-37} and that for $Cu(OH)_2$ is 2.2×10^{-20}.

The small size of the K_{sp} for CuS indicates that it is not very soluble. Addition of acid will cause the more soluble hydroxide to dissolve, leaving the CuS in the solid state.

99. How does a buffer solution control pH upon addition of a strong base. A buffer consisting of CH_3CO_2H and $CH_3CO_2^-$ contains both an acidic(CH_3CO_2H) and basic($CH_3CO_2^-$) substance. Upon addition of a strong base (OH^-), the following reaction occurs:

$CH_3CO_2H + OH^- \Leftrightarrow CH_3CO_2^- + H_2O$.

Note that the consumption of hydroxide ion has resulted in the formation of the much weaker base (acetate). The resulting equilibrium has a slightly reduced concentration of acetic acid, and a slightly increased concetration of acetate ion---with little change in pH.

101. Silver phosphate can be more soluble in water than calculated from Ksp data owing to competing reactions. The phosphate anion is also the conjugate base of a weak acid, and will undergo a reaction with water (hydrolysis), which lowers the $[PO_4^{3-}]$, and causes the solubility equilibrium to shift to the right, increasing the solubility of the phosphate salt. This is another example of LeChatelier's Principle in operation.

103. For the equilibrium between acetic acid and its conjugate base (acetate ion).

(a). One can understand the fraction of acetic acid present by viewing the Ka expression for the acid:

$Ka = \dfrac{[CH_3CO_2^-][H_3O^+]}{[CH_3CO_2H]}$, As pH increases, the concentration of H_3O^+ decreases. Since Ka remains constant, the "mathematical" result is that the "numerator term" ($[CH_3CO_2^-]$) must increase at the expense of the "denominator term" ($[CH_3CO_2H]$). Chemically, this means that the fraction of CH_3CO_2H decreases.

(b) Predominant species at pH = 4.0

This is easy to answer by rearranging the Ka expression for the acid.

$\dfrac{Ka}{[H_3O^+]} = \dfrac{[CH_3CO_2^-]}{[CH_3CO_2H]}$. Substituting values for Ka and the $[H_3O^+]$

$\dfrac{1.8 \times 10^{-5}}{1.0 \times 10^{-4}} = \dfrac{[CH_3CO_2^-]}{[CH_3CO_2H]} = 0.18$.

This means that the molecular acid is the predominant specie at pH = 4.

Predominant species at pH = 6.0?

Once again, examine the Ka expression, substituting 1×10^{-6} for the $[H_3O^+]$.

$\dfrac{1.8 \times 10^{-5}}{1.0 \times 10^{-6}} = \dfrac{[CH_3CO_2^-]}{[CH_3CO_2H]} = 18$. Now the conjugate base is present at a concentration level

almost 20 times that of the molecular acid.

(c) The pH at the equivalence point (when the fraction of acid is equal to the fraction of

conjugate base). Let's calculate the $[H_3O^+]$ at this point.

$1.8 \times 10^{-5} = \dfrac{[0.5][H_3O^+]}{[0.5]}$. Note that I simply chose 0.5 as the concentration for both

species—any number here will do. So $[H_3O^+] = 1.8 \times 10^{-5}$ and

$pH = -\log(1.8 \times 10^{-5})$ or 4.74.

105. For salicylic acid:

(a) Approximate values for the bond angles:

Bond	Angle
i	120
ii	120
iii	109
iv	120

(b) Hybridization of C in the ring: sp^2. With three atoms attached to any one carbon, the three
groups separate by an angle of approximately 120°. This is also true of the C in the
carboxylate group.

(c) What is Ka of acid?

Using SA to represent the molecular acid, the monoprotic acid would have a Ka expression
of:

$Ka = \dfrac{[H^+][A^-]}{[SA]}$. We calculate the concentration of SA:

$\dfrac{\dfrac{1.00 \text{ g SA}}{122.12 \text{ g/mol SA}}}{0.460 \text{ L}} = 1.78 \times 10^{-2}$ mol/L SA

Noting that pH = 2.4, lets us know that $[H^+] = 10^{-2.4}$ or 3.98×10^{-3} M.

Since SA is a monoprotic acid, the concentration of A^- is also 3.98×10^{-3} M.

The Ka is then: $\dfrac{[3.98 \times 10^{-3}][3.98 \times 10^{-3}]}{(1.78 \times 10^{-2}) - (3.98 \times 10^{-3})} = 1.15 \times 10^{-3}$ or 1×10^{-3} (to 1 sf)

(d) If pH = 2.0, the % of salicylate ion present:

$Ka = \dfrac{[H^+][A^-]}{[SA]} = 1 \times 10^{-3}$ and rearranging to isolate the salicylate ion: $\dfrac{[A^-]}{[SA]} = \dfrac{1 \times 10^{-3}}{1 \times 10^{-2}}$

At pH = 2.0, the salicylate ion is 10% of the acid form, with the molecular acid, SA, being 90% of the acid present.

(e) In a titration of 25.0 mL of a 0.024M solution of SA titrated with 0.010 M NaOH:

(i) What is pH at the halfway point ?

At the halfway point, the $[A^-] = [SA]$ so $[H^+] = 1 \times 10^{-3}$, so pH = 3.0

(ii) What is pH at the equivalence point?

At the equivalence point, only salt (A^-) remains. Begin by asking "What's the concentration of the salt?". The amount of SA present is (0.025L)(0.024 mol SA/L) = 0.00060 mol. The amount of 0.010 M NaOH that contains this number of moles of NaOH is (0.00060mol)/(0.010 mol NaOH/L) = 0.060 L or 60 mL. At the equivalence point, the total volume is then (60.0 + 25.0) or 85.0 mL. The concentration is then: 6.0×10^{-4} mol/0.085 L = 7.1×10^{-3} M. Substituting into the Kb expression for the salicylate anion reacting:

$$A^- (aq) + H_2O(\ell) \Leftrightarrow HA(aq) + OH^-(aq)$$

$Kb = \dfrac{[HA][OH^-]}{[A^-]} = \dfrac{1.0 \times 10^{-14}}{1.0 \times 10^{-3}}$

Since [HA] = [OH$^-$], we can substitute into the Kb expression to get:

$\dfrac{[OH^-]^2}{[7.1 \times 10^{-3}]} = 1.0 \times 10^{-11} = 7.1 \times 10^{-14}$ and $[OH^-] = 2.7 \times 10^{-7}$ M, so pOH = 6.7,

and pH = 7.3

107. The Ksp expression: $Ksp = [Pb^{2+}][Cl^-]^2$ Once equilibrium is achieved, the balance of lead(II) and chloride ions has been achieved (a dynamic steady state). Addition of chloride ion (according to our friend, LeChatelier) would shift the equilibrium " to the left"—forming MORE $PbCl_2$ solid. The macroscopic manifestation would be that "more precipitate" forms.

109. The increase in solubility of Co(OH)$_2$ as pH changes by a factor of 1 unit:

The Ksp expression for Co(OH)$_2$ is: $[Co^{2+}][OH^-]^2 = 2.5 \times 10^{-16}$. The composition of this salt is such that the molar solubility of the salt is numerically equal to the $[Co^{2+}]$. This is visible if you consider that for x mol/L of the salt that dissolve, x mol/L of Co^{2+} result. Rearrange the Ksp expression to isolate the $[Co^{2+}]$:

$[Co^{2+}] = \dfrac{2.5 \times 10^{-16}}{[OH^-]^2}$. Note that each time pH (and of course, pOH) changes by one unit, the

[OH$^-$] *changes* by a power of 10. The Ksp expression indicates that the hydroxide ion concentration should be *squared*, so the resulting $[Co^{2+}]$ would change by a factor of 100.

Chapter 19
Principles of Reactivity: Entropy and Free Energy

Practicing Skill

Entropy

1. Compound with the higher entropy:

   (a) CO_2 (s) at $-78°$ vs **CO_2 (g) at 0 °C**: Entropy increases with temperature.

   (b) H_2O (ℓ) at 25 °C vs **H_2O (ℓ) at 50 °C**: Entropy increases with temperature.

   (c) Al_2O_3 (s) (pure) vs **Al_2O_3 (s) (ruby)**: Entropy of a solution (even a solid one) is greater than that of a pure substance.

   (c) **1 mol N_2 (g) at 1bar** vs 1 mol N_2 (g) at 10 bar: With the increased P, molecules have greater order.

3. Compound with higher standard entropy:

   (a) O_2 (g) vs **CH_3OH (g)**: Entropy increases with molecular complexity.

   (a) HF (g) vs HCl (g) vs **HBr (g)**: Entropy increases with molecular size (mass)

   (a) NH_4Cl (s) vs **NH_4Cl (aq)**: Entropy of solutions is greater than that of the solid.

   (a) **HNO_3 (g)** vs HNO_3 (ℓ) vs HNO_3 (aq): Entropy of the gaseous state is very high. The liquid state has relatively lower entropy.

Predicting and Calculating Entropy Changes

5. Entropy changes:

   (a) KOH (s) $\rightarrow$ KOH(aq)

   $$\Delta S° = 91.6 \ \frac{J}{K•mol} \text{ (1 mol)} \ - 78.9 \ \frac{J}{K•mol} \text{(1 mol)} \ = +12.7 \ \frac{J}{K}$$

   The increase in entropy reflects the greater disorder of the solution state.

   (b) Na (g) $\rightarrow$ Na (s)

   $$\Delta S° = 51.21 \ \frac{J}{K•mol} \text{ (1 mol)} \ - 153.765 \ \frac{J}{K•mol} \text{ (1 mol)} \ = -102.55 \ \frac{J}{K}$$

   The lower entropy of the solid state is evidenced by the negative sign.

   (c) Br_2 (ℓ) $\rightarrow$ Br_2 (g)

   $$\Delta S° = 245.42 \ \frac{J}{K•mol} \text{ (1 mol)} \ - 152.2 \ \frac{J}{K•mol} \text{(1 mol)} \ = +93.2 \ \frac{J}{K}$$

   The increase in entropy is expected with the transition to the disordered state of a gas.

(d) HCl (g) → HCl (aq)

$$\Delta S° = 56.5 \frac{J}{K \cdot mol} (1 \text{ mol}) - 186.2 \frac{J}{K \cdot mol} (1 \text{ mol}) = -129.7 \frac{J}{K}$$

The lowered entropy reflects the greater order of the solution state over the gaseous state.

7. For the reaction: 2 C (graphite) + 3 H$_2$ (g) → C$_2$H$_6$ (g)

$$\Delta S° = 1 \cdot S° \text{ C}_2\text{H}_6 - [2 \cdot S° \text{ C (graphite)} + 3 \cdot S° \text{ H}_2 \text{ (g)}]$$

$$= (1 \text{ mol})(229.2 \frac{J}{K \cdot mol}) - [(2 \text{ mol})(5.6 \frac{J}{K \cdot mol}) + (3 \text{ mol})(130.7 \frac{J}{K \cdot mol})]$$

$$= -174.1 \frac{J}{K}$$

9. Standard Entropy change for compound formation from elements:
   (a) HCl (g): Cl$_2$ (g) + H$_2$ (g) → 2 HCl (g)

$$\Delta S° = 2 \cdot S° \text{ HCl (g)} - [1 \cdot S° \text{ Cl}_2 \text{ (g)} + 1 \cdot S° \text{ H}_2 \text{ (g)}]$$

$$= (2 \text{ mol})(186.2 \frac{J}{K \cdot mol}) - [(1 \text{ mol})(223.08 \frac{J}{K \cdot mol}) + (1 \text{ mol})(130.7 \frac{J}{K \cdot mol})]$$

$$= +18.6 \frac{J}{K}$$

   (b) Ca(OH)$_2$ (s): Ca(s) + O$_2$ (g) + H$_2$ (g) → Ca(OH)$_2$ (s)

$$\Delta S° = 1 \cdot S° \text{ Ca(OH)}_2 \text{ (s)} - [1 \cdot S° \text{ Ca (s)} + 1 \cdot S° \text{ O}_2 \text{ (g)} + 1 \cdot S° \text{ H}_2 \text{ (g)}]$$

$$= (1 \text{ mol})(83.39 \frac{J}{K \cdot mol}) - [(1 \text{ mol})(41.59 \frac{J}{K \cdot mol}) + (1 \text{ mol})(205.07 \frac{J}{K \cdot mol}) +$$

$$(1 \text{ mol})(130.7 \frac{J}{K \cdot mol})]$$

$$= -293.97 \frac{J}{K} \text{ (or } -294.0 \frac{J}{K} \text{ to 4 sf)}$$

11. Standard molar entropy changes for:
   (a) 2 Al (s) + 3 Cl$_2$ (g) → 2 AlCl$_3$ (s)

$$\Delta S° = 2 \cdot S° \text{ AlCl}_3 \text{ (s)} - [2 \cdot S° \text{ Al (s)} + 3 \cdot S° \text{ Cl}_2 \text{ (g)}]$$

$$= (2 \text{ mol})(109.29 \frac{J}{K \cdot mol}) - [(2 \text{ mol})(28.3 \frac{J}{K \cdot mol}) + (3 \text{ mol})(223.08 \frac{J}{K \cdot mol})]$$

$$= -507.3 \frac{J}{K}$$

   (b) 2 CH$_3$OH (ℓ) + 3 O$_2$ (g) → 2 CO$_2$ (g) + 4 H$_2$O (g)

$$\Delta S° = [2 \cdot S° \text{ CO}_2 \text{ (g)} + 4 \cdot S° \text{ H}_2\text{O (g)}] - [2 \cdot S° \text{ CH}_3\text{OH (ℓ)} + 3 \cdot S° \text{ O}_2 \text{ (g)}]$$

$$= \left[(2 \text{ mol})(213.74 \ \frac{J}{K \cdot mol}) + (4 \text{ mol})(188.84 \ \frac{J}{K \cdot mol})\right] -$$

$$\left[(2 \text{ mol})(127.19 \ \frac{J}{K \cdot mol}) + (3 \text{ mol})(205.07 \ \frac{J}{K \cdot mol})\right]$$

$$= + 313.25 \ \frac{J}{K}$$

In (a) we see the sign of ΔS as negative—expected since the reaction results in the decrease in the number of moles of gas. In (b), the sign of ΔS is positive—as expected, since the reaction produces a larger number of moles of gaseous products than gaseous reactants.

$\Delta S°$univ and Spontaneity

13. Is the reaction: $Si (s) + 2 Cl_2 (g) \rightarrow SiCl_4 (g)$ spontaneous?

$$\Delta S°_{sys} = 1 \cdot S° \ SiCl_4 (g) - [1 \cdot S° \ Si (s) + 2 \cdot S° \ Cl_2 (g)]$$

$$= (1 \text{ mol})(330.86 \ \frac{J}{K \cdot mol}) - [(1 \text{ mol})(18.82 \ \frac{J}{K \cdot mol}) + (2 \text{ mol})(223.08 \ \frac{J}{K \cdot mol})]$$

$$= -134.12 \ \frac{J}{K}$$

To calculate $\Delta S°_{surr}$, we calculate $\Delta H°_{sys}$:

$$\Delta H° = 1 \cdot \Delta H° \ SiCl_4 (g) - [1 \cdot \Delta H° \ Si (s) + 2 \cdot \Delta H° \ Cl_2 (g)]$$

$$= (1 \text{ mol})(-662.75 \ \frac{kJ}{mol}) - [(1 \text{ mol})(0 \ \frac{kJ}{mol}) + (2 \text{ mol})(0 \ \frac{kJ}{mol})] = -662.75 \ \frac{kJ}{mol}$$

$$\Delta S°_{surr} = -\Delta H°/T = (662.75 \times 10^3 \text{ J/mol})/298.15 \text{ K} = 2222.9 \ \frac{J}{K}$$

$$\Delta S°univ = \Delta S°_{sys} + \Delta S°_{surr} = -134.12 \ \frac{J}{K} + 2222.9 \ \frac{J}{K} = 2088.7 \ \frac{J}{K}$$

15. Is the reaction: $2 H_2O(\ell) \rightarrow 2 H_2 (g) + O_2 (g)$ spontaneous?

$$\Delta S°_{sys} = [2 \cdot S° \ H_2 (g) + 1 \cdot S° \ O_2 (g)] - 2 \cdot S° \ H_2O (\ell)$$

$$= [(2 \text{ mol})(130.7 \ \frac{J}{K \cdot mol}) + (1 \text{ mol})(205.07 \ \frac{J}{K \cdot mol})] - (2 \text{ mol})(69.95 \ \frac{J}{K \cdot mol})$$

$$= +326.57 \ \frac{J}{K} \text{ and for decomposition of 1 mol of water: } +326.57/2 = 163.3 \ \frac{J}{K}$$

To calculate $\Delta S°_{surr}$, we calculate $\Delta H°_{sys}$:

$$\Delta H°_{sys} = [2 \cdot \Delta H° \ H_2 (g) + 1 \cdot \Delta H° \ O_2 (g)] - 2 \cdot \Delta H° \ H_2O (\ell)$$

$$= [(2 \text{ mol})(0 \ \frac{kJ}{mol}) + (1 \text{ mol})(0 \ \frac{kJ}{mol})] - (2 \text{ mol})(-285.83 \ \frac{kJ}{mol})$$

$$= +571.66 \frac{kJ}{mol} \text{ and for decomposition of 1 mol of water: } +571.66/2 = 285.83 \frac{kJ}{mol}$$

$$\Delta S^\circ_{surr} = -\Delta H^\circ/T = -(285.83 \times 10^3 \text{ J/mol})/298.15 \text{ K} = -958.68 \frac{J}{K}$$

$$\Delta S^\circ_{univ} = \Delta S^\circ_{sys} + \Delta S^\circ_{surr} = 163.3 \frac{J}{K} + -958.68 \frac{J}{K} = -795.4 \frac{J}{K} \text{ Since this value is less}$$

than zero, the process is not spontaneous.

17. Using Table 19.2 classify each of the reactions:

(a) ΔH°system = -, ΔS°system = - Product-favored at lower T

(b) ΔH°system = +, ΔS°system = - Not product-favored under any conditions

Effect of Temperature on Reactions

19. For the decomposition of $MgCO_3$ (s)) $\rightarrow$ MgO (s) + CO_2 (g):

(a) ΔS°_{sys} = [1 • S° MgO (s) + 1 • S° CO_2 (g)] - 1 • S° $MgCO_3$ (s)

$$= [(1 \text{ mol})(26.85 \frac{J}{K\bullet mol}) + (1 \text{ mol})(213.74 \frac{J}{K\bullet mol})] - (1 \text{ mol})(65.84 \frac{J}{K\bullet mol})$$

$$= +174.75 \frac{J}{K}$$

To calculate ΔS°_{surr} we calculate ΔH°_{sys}:

ΔH°_{sys} = [1 • ΔH° MgO (s) + 1 • ΔH° CO_2 (g)] - 1 • ΔH° $MgCO_3$ (s)

$$= [(1 \text{ mol})(-601.24 \frac{kJ}{mol}) + (1 \text{ mol})(-393.509 \frac{kJ}{mol})] - (1 \text{ mol})(-1111.69 \frac{kJ}{mol})$$

$$= +116.94 \frac{kJ}{mol}$$

$$\Delta S^\circ_{surr} = -\Delta H^\circ/T = -(285.83 \times 10^3 \text{ J/mol})/298.15 \text{ K} = -392.4 \frac{J}{K}$$

(b) $\Delta S^\circ_{univ} = \Delta S^\circ_{sys} + \Delta S^\circ_{surr} = +174.75 \frac{J}{K} + -392.4 \frac{J}{K} = -217.67 \frac{J}{K}$ Since this value

is less than zero, the process is not spontaneous.

(c) From Table 19.2 we observe that this type of reaction (ΔH = + and ΔS = +) is spontaneous at higher T (product-favored).

Changes in Free Energy

21. Calculate ΔG°_{rxn} for :

(a) 2 Pb (s) + O_2 (g) $\rightarrow$ 2 PbO (s)

$$\Delta H°rxn = (2 \text{ mol})(-219 \frac{kJ}{mol}) - [0 + 0] = -438 \text{ kJ}$$

$$\Delta S° = (2 \text{ mol})(66.5 \frac{J}{K \cdot mol}) -$$

$$[(2 \text{ mol})(64.81 \frac{J}{K \cdot mol}) + (1 \text{ mol})(205.07 \frac{J}{K \cdot mol})] = -201.7 \text{ J/K}$$

$$\Delta G°rxn = \Delta H°_f - T\Delta S°$$

$$= -438 \text{ kJ} - (298.15 \text{ K})(-201.7 \text{ J/K})(\frac{1.000 \text{ kJ}}{1000. \text{ J}}) = -378 \text{ kJ}$$

Reaction is product-favored since $\Delta G < 0$. With the very large negative ΔH, the process is enthalpy driven.

(b) $NH_3(g) + HNO_3(aq) \rightarrow NH_4NO_3 (aq)$

$$\Delta H°rxn = (1 \text{ mol})(-339.87 \frac{kJ}{mol}) - [(1 \text{ mol})(-45.90 \frac{kJ}{mol}) + (1 \text{ mol})(-207.36 \frac{kJ}{mol}] = -86.61 \text{ kJ}$$

$$\Delta S° = (1 \text{ mol})(259.8 \frac{J}{K \cdot mol}) -$$

$$[(1 \text{ mol})(192.77 \frac{J}{K \cdot mol}) + (1 \text{ mol})(146.4 \frac{J}{K \cdot mol})] = -79.4 \text{ J/K}$$

$$\Delta G°_{rxn} = \Delta H°_{rxn} - T\Delta S° = -86.61 \text{ kJ} - (298.15 \text{ K})(-79.4 \text{ J/K})(\frac{1.000 \text{ kJ}}{1000. \text{ J}}) = -62.9 \text{ kJ}$$

Reaction is product-favored since $\Delta G < 0$. With the very large negative ΔH, the process is enthalpy driven.

23. Calculate the molar free energies of formation for:

(a) CS_2 (g) The reaction is: C (graphite) + 2 S (s,rhombic) $\rightarrow$ CS_2 (g)

$$\Delta H°rxn = (1 \text{ mol})(116.7 \frac{kJ}{mol}) - [0 + 0] = +116.7 \text{ kJ}$$

$$\Delta S° = (1 \text{ mol})(237.8 \frac{J}{K \cdot mol}) - [(1 \text{ mol})(5.6 \frac{J}{K \cdot mol}) + (2 \text{ mol})(32.1 \frac{J}{K \cdot mol})]$$

$$= +168.0 \text{ J/K}$$

$$\Delta G°_f = \Delta H°_f - T\Delta S° = (116.7 \text{ kJ}) - (298.15 \text{ K})(168.0 \frac{J}{K})(\frac{1.000 \text{ kJ}}{1000 \text{ J}})$$

$$= +66.6 \text{ kJ} \qquad \text{Appendix value: 66.61 kJ/mol}$$

(b) NaOH (s)   The reaction is: Na (s) + $\frac{1}{2}$ O$_2$(g) + $\frac{1}{2}$ H$_2$(g) → NaOH (s)

$\Delta H°_f$ = (1 mol)(-425.93 $\frac{kJ}{mol}$) - [0 + 0+ 0] = -425.93 kJ

$\Delta S°$ = (1 mol)(64.46 $\frac{J}{K \cdot mol}$) -

  [(1 mol)(51.21 $\frac{J}{K \cdot mol}$) +($\frac{1}{2}$ mol)(205.07 $\frac{J}{K \cdot mol}$) + ($\frac{1}{2}$ mol)(130.7 $\frac{J}{K \cdot mol}$)]

  = -154.6 J/K

$\Delta G°_f$ = $\Delta H°_f$ - $T\Delta S°$ = (-425.93 kJ) - (298.15 K)(-154.6 J/K)($\frac{1.000 \text{ kJ}}{1000. \text{ J}}$)

    = -379.82 kJ   Appendix value: -379.75 kJ/mol

(c) ICl (g) The reaction is: $\frac{1}{2}$ I$_2$ (g) + $\frac{1}{2}$ Cl$_2$ (g) → ICl (g)

$\Delta H°_f$ = (1 mol)(+17.51 $\frac{kJ}{mol}$) - [0 + 0] = +17.51 kJ

$\Delta S°$ = (1 mol)(247.56 $\frac{J}{K \cdot mol}$) -

    [($\frac{1}{2}$mol)(116.135 $\frac{J}{K \cdot mol}$) + ($\frac{1}{2}$ mol)(223.08 $\frac{J}{K \cdot mol}$)]

  = +77.95 J/K

$\Delta G°_f$ = $\Delta H°_f$ - $T\Delta S°$ = (+17.51 kJ)- (298.15 K)(+77.95 J/K)($\frac{1.000 \text{ kJ}}{1000. \text{ J}}$)

    = - 5.72 kJ   Appendix value: - 5.73 kJ/mol

Free Energy of Formation

25. Calculate $\Delta G°_{rxn}$ for the following equations. Are they product-favored?

(a) 2 K (s) + Cl$_2$ (g) → 2 KCl (s)

$\Delta G°_{rxn}$ = [2 •$\Delta G°_f$ KCl(s)] – [1 •$\Delta G°_f$ Cl$_2$(g) + 2• $\Delta G°_f$ K (s)]

  = [(2 mol)(- 408.77 $\frac{kJ}{mol}$] - [(1 mol)(0 $\frac{kJ}{mol}$) +(2 mol)(0 $\frac{kJ}{mol}$)]

  = - 817.54 kJ and per mol of KCl = -408.77 kJ/mol

With a $\Delta G < 0$, the reaction is product-favored.

(b) 2 CuO (s) → 2 Cu (s) + O$_2$ (g)

$\Delta G°_{rxn}$ = [2 •$\Delta G°_f$ Cu(s) + $\Delta G°_f$ O$_2$ (g)] – [2• $\Delta G°_f$ CuO (s)]

$$\Delta G°_{rxn} = [(2 \text{ mol})(0\ \frac{kJ}{mol}) + (1 \text{ mol})(0\ \frac{kJ}{mol})] - [(2\text{mol})(-128.3\ \frac{kJ}{mol})$$

$$= + 256.6 \text{ kJ and } + 128.3 \text{ kJ/mol CuO}$$

With a $\Delta G > 0$, the reaction is not product-favored.

(c) $4 \text{ NH}_3 \text{ (g)} + 7 \text{ O}_2 \text{ (g)} \rightarrow 4 \text{ NO}_2 \text{ (g)} + 6 \text{ H}_2\text{O (g)}$

$$\Delta G°_{rxn} = [4 •\Delta G°_f \text{ NO}_2 \text{ (g)} + 6• \Delta G°_f \text{ H}_2\text{O (g)}] - [4• \Delta G°_f \text{ NH}_3 \text{ (g)} + 7 •\Delta G°_f \text{ O}_2 \text{ (g)}]$$

$$\Delta G°_{rxn} = [(4\text{mol})(+51.23\ \frac{kJ}{mol}) + (6 \text{ mol})(-228.59\ \frac{kJ}{mol}] -$$

$$[(4 \text{ mol})(-16.37\ \frac{kJ}{mol}) + (7 \text{ mol})(0\ \frac{kJ}{mol})] = -1101.14 \text{ kJ}$$

With a $\Delta G < 0$, the reaction is product-favored.

27. Value for $\Delta G°_f$ of $BaCO_3(s)$:

$$\Delta G°_{rxn} = [\Delta G°_f \text{ BaO(s)} + \Delta G°_f \text{ CO}_2\text{(g)}] - [\Delta G°_f \text{ BaCO3(s)}]$$

$$+219.7 \text{ kJ} = [(1 \text{ mol})(-520.38\ \frac{kJ}{mol}) + (1 \text{ mol})(-394.359\ \frac{kJ}{mol})] - \Delta G°_f \text{ BaCO3(s)}$$

$$+219.7 \text{ kJ} = -914.74 \text{ kJ} - \Delta G°_f \text{ BaCO3(s)}$$

$$-1134.4 \text{ kJ/mol} = + \Delta G°_f \text{ BaCO3(s)}$$

Effect of Temperature on ΔG

29. Entropy-favored or Enthalpy-favored reactions?

(a) $N_2 \text{ (g)} + 2 \text{ O}_2 \text{ (g)} \rightarrow 2 \text{ NO}_2 \text{ (g)}$

$$\Delta H° = (2 \text{ mol})(+33.1\ \frac{kJ}{mol}) - [0 + 0] = +66.2 \text{ kJ}$$

$$\Delta S° = (2 \text{ mol})(+240.04\ \frac{J}{K•mol}) -$$

$$[(1\text{mol})(191.56\ \frac{J}{K•mol}) + (2 \text{ mol})(+205.07\ \frac{J}{K•mol})] = -121.62\ \frac{J}{K}$$

$$\Delta G° = (2 \text{ mol})(51.23\ \frac{kJ}{mol}) - [(1 \text{ mol})(0\frac{kJ}{mol}) + (1 \text{ mol })(0\frac{kJ}{mol})] = 102.5 \text{ kJ}$$

The reaction is **not** entropy OR enthalpy favored. There is **no** T at which $\Delta G < 0$.

(b) $2 \text{ C (s)} + \text{O}_2 \text{ (g)} \rightarrow 2 \text{ CO (g)}$

$$\Delta H° = (2 \text{ mol})(-110.525\ \frac{kJ}{mol}) - [0 + 0] = - 221.05 \text{ kJ}$$

$$\Delta S^\circ = (2 \text{ mol})(+197.674 \frac{J}{K \bullet mol}) -$$

$$[(2 \text{ mol})(+5.6 \frac{J}{K \bullet mol}) + (1 \text{mol})(+205.07 \frac{J}{K \bullet mol})] = +179.1 \frac{J}{K}$$

$$\Delta G^\circ = (2 \text{ mol})(-137.168 \frac{kJ}{mol}) - [(1 \text{ mol})(0 \frac{kJ}{mol}) + (1 \text{ mol})(0 \frac{kJ}{mol})]$$

$$= -274.336 \text{ kJ}$$

This reaction is **both** entropy- and enthalpy-favored *at all temperatures.*

(c) $CaO \text{ (s)} + CO_2 \text{ (g)} \rightarrow CaCO_3 \text{ (s)}$

$$\Delta H^\circ = (1 \text{ mol})(-1207.6 \frac{kJ}{mol}) - [(1 \text{mol})(-635.0 \frac{kJ}{mol}) + (1 \text{mol})(-393.509 \frac{kJ}{mol})]$$

$$= -179.0 \text{ kJ}$$

$$\Delta S^\circ = (1 \text{ mol})(+91.7 \frac{J}{K \bullet mol}) -$$

$$[(1 \text{mol})(38.2 \frac{J}{K \bullet mol}) + (1 \text{mol})(+213.74 \frac{J}{K \bullet mol})] = -160.2 \frac{J}{K}$$

$$\Delta G^\circ = (1 \text{ mol})(-1129.16 \frac{kJ}{mol}) -$$

$$[(1 \text{mol})(-603.42 \frac{kJ}{mol}) + (1 \text{mol})(-394.359 \frac{kJ}{mol})] = -131.4 \text{ kJ}$$

This reaction is *enthalpy-favored* and will be spontaneous at low temperatures.

(d) $2 \text{ NaCl (s)} \rightarrow 2 \text{ Na (s)} + Cl_2 \text{ (g)}$

$$\Delta H^\circ = [(2 \text{ mol})(0 \frac{kJ}{mol}) + (1 \text{mol})(0 \frac{kJ}{mol})] - (2 \text{mol})(-411.12 \frac{kJ}{mol})] = +822.24 \text{ kJ}$$

$$\Delta S^\circ = [(2 \text{ mol})(+51.21 \frac{J}{K \bullet mol}) + (1 \text{mol})(+223.08 \frac{J}{K \bullet mol})] -$$

$$(2 \text{mol})(+72.11 \frac{J}{K \bullet mol})] = +181.28 \frac{J}{K}$$

$$\Delta G^\circ = [(2 \text{ mol})(0 \frac{kJ}{mol}) + (1 \text{mol})(0 \frac{kJ}{mol})] - (2 \text{mol})(-384.04 \frac{kJ}{mol})] = +768.08 \text{ kJ}$$

This reaction is *entropy-favored* and will be spontaneous at high temperatures.

31. Estimate the temperature to decompose HgS(s) into Hg(ℓ) and S(g):

For this process, we need to find the temperature at which the ΔG° becomes < 0

Calculate ΔH for the process: $HgS \text{ (s)} \rightarrow Hg \text{ (ℓ)} + S \text{ (g)}$

317

$$\Delta H°_{rxn} = [\Delta H°_f \, Hg \, (\ell) + \Delta H°_f \, S \, (g)] - \Delta H°_f \, HgS \, (s)]$$

$$\Delta H°_{rxn} = [0 + (1mol)(+ 278.98 \frac{kJ}{mol})] - (1 \, mol)(-58.2 \frac{kJ}{mol}] = +337.2 \, kJ$$

$$\Delta S°_{rxn} = [(1mol)(76.02 \frac{J}{K \cdot mol}) + (1mol)(167.83 \frac{J}{K \cdot mol})] -$$

$$(1 \, mol)(82.4 \frac{J}{K \cdot mol}] = +161.5 \frac{J}{K}$$

The process becomes spontaneous when ΔG just becomes negative. So calculate the T at which $\Delta G = 0$.

$\Delta G = \Delta H - T\Delta S$ and if $\Delta G = 0$ then $\Delta H = T\Delta S$

$337.2 \, kJ(1000 \, J/kJ) = T \cdot 161.5 \frac{J}{K}$ so $T = 2088 \, K$ (Wow, that's hot!)

Free Energy and Equilibrium Constants

33. Calculate K_p for the reaction:

$$\frac{1}{2} N_2(g) + \frac{1}{2} O_2(g) \rightarrow NO \, (g) \qquad \Delta G°_f = + 86.58 \, kJ/mol \, NO$$

$\Delta G°_{rxn} = - RT \ln K_p$ so $86.58 \times 10^3 \, J/mol = - (8.3145 \frac{J}{K \cdot mol})(298.15 \, K) \ln K_p$

$- 34.926 = \ln K_p$ and $6.8 \times 10^{-16} = K_p$

Note that the + value of $\Delta G°_f$ results in a value of K_p which is small--reactants are favored. A negative value would result in a large K_p -- a process in which the products were favored.

35. From the $\Delta G°$ and Kp, determine if the hydrogenation of ethylene is product-favored:

Using $\Delta G°$ data from the Appendix: $C_2H_4 \, (g) + H_2 \, (g) \rightarrow C_2H_6 \, (g)$

$$\Delta G° = (1 \, mol)(-31.89 \frac{kJ}{mol}) - [(1mol)(68.35 \frac{kJ}{mol}) + (1mol)(0 \frac{kJ}{mol})] = -100.24 \, kJ$$

and since $\Delta G° = - RT \ln K_p$

$$-100.24 \times 10^3 \, J/mol = - (8.3145 \frac{J}{K \cdot mol})(298.15 \, K) \ln K_p$$

$40.436 = \ln K_p$ and $K_p = 3.64 \times 10^{17}$

The negative value of ΔG means the reaction is product-favored. The large value of ΔG means that Kp is very large.

General Questions on Thermodynamics

37. Calculate the ΔS for (1) $C(s) + 2\,H_2\,(g) \rightarrow CH_4\,(g)$

$$\Delta S_1^\circ = (1\text{ mol})(+186.26\,\frac{J}{K\bullet mol}\,) -$$

$$[(1\text{mol})(+5.6\,\frac{J}{K\bullet mol}) + (2\text{mol})(+130.7\,\frac{J}{K\bullet mol})] = -80.7\,\frac{J}{K}$$

Calculate the ΔS for (2) $CH_4(g) + \frac{1}{2}\,O_2\,(g) \rightarrow CH_3OH(\ell)$

$$\Delta S_2^\circ = (1\text{ mol})(+127.19\,\frac{J}{K\bullet mol}) -$$

$$[(1\text{mol})(+186.26\,\frac{J}{K\bullet mol}) + (\frac{1}{2}\text{mol})(+205.07\,\frac{J}{K\bullet mol})] = -161.60\,\frac{J}{K}$$

Calculate the ΔS for (3) $C(s) + 2\,H_2\,(g) + \frac{1}{2}\,O_2\,(g) \rightarrow CH_3OH(\ell)$

$$\Delta S_3^\circ = (1\text{ mol})(+127.19\,\frac{J}{K\bullet mol}) -$$

$$[(1\text{mol})(+5.6\,\frac{J}{K\bullet mol}) + (2\text{mol})(+130.7\,\frac{J}{K\bullet mol}) + (\frac{1}{2}\text{mol})(+205.07\,\frac{J}{K\bullet mol})]$$

$$= -242.3\,\frac{J}{K}$$

So $\Delta S_1^\circ + \Delta S_2^\circ = (-80.7\,\frac{J}{K}) + (-161.60\,\frac{J}{K}) = -242.3\,\frac{J}{K}$

39. Calculate ΔH°_{rxn} and ΔS°_{rxn} for the combustion of ethane:

	$C_2H_6\,(g)$	+	$7/2\,O_2\,(g) \rightarrow$	$2\,CO_2(g)$	+	$3\,H_2O\,(g)$
ΔH°_f (kJ/mol)	-83.85		0	-393.509		-241.83
S° (J/K•mol)	+229.2		+205.07	+213.74		+188.84

$\Delta H^\circ_{rxn} = [2 \bullet \Delta H^\circ_f\,CO_2(g) + 3 \bullet \Delta H^\circ_f\,H_2O\,(g)] - [1 \bullet \Delta H^\circ_f\,C_2H_6\,(g) + 7/2 \bullet \Delta H^\circ_f\,O_2\,(g)]$

$$= [(2\text{mol})(-393.509\,\frac{kJ}{mol}) + (3\text{mol})(-241.83\,\frac{kJ}{mol})] - [(1\text{mol})(-83.85\text{ kJ/mol}) + 0]$$

$$= -1428.66\text{ kJ}$$

$\Delta S^\circ_{rxn} = [2 \bullet S^\circ\,CO_2(g) + 3 \bullet S^\circ\,H_2O\,(g)] - [1 \bullet S^\circ\,C_2H_6\,(g) + 7/2 \bullet S^\circ\,O_2\,(g)]$

$$= [2\text{ mol})(213.74\,\frac{J}{K\bullet mol}) + (3\text{ mol})(188.84\,\frac{J}{K\bullet mol})] -$$

$$[(1\text{ mol})(229.2\,\frac{J}{K\bullet mol}) + (7/2\text{ mol})(205.07\,\frac{J}{K\bullet mol})] = +47.1\text{ J/K}$$

$$\Delta S°_{surroundings} \quad = \frac{-\Delta H_{rxn}}{T} \quad \text{(Assuming we're at 298 K)}$$

$$= \frac{1428.66 \text{ kJ}}{298.15 \text{ K}} \cdot \frac{1000 \text{ J}}{1 \text{ kJ}} \quad = 4791.7 \text{ J/K}$$

$$\text{so } \Delta S°_{system} + \Delta S°_{surroundings} \quad = +47.1 \text{ J/K} + 4791.7 \text{ J/K}$$

$$= 4838.8 \text{ J/K}$$

Since $\Delta H° = -$ and $\Delta S° = +$, the process is product-favored.

This calculation is consistent with our expectations. We know that hydrocarbons burn completely (in the presence of sufficient oxygen) to produce carbon dioxide and water.

41. (a) Calculate $\Delta G°_{rxn}$ for NH_3 (g) + HCl (g) $\rightarrow$ NH_4Cl (s)

S° (J/K•mol)	192.77	186.2	94.85
$\Delta H°_f$ (kJ/mol)	– 45.90	- 92.31	- 314..55

$\Delta S°_{rxn} = 1 \cdot S° NH_4Cl$ (s) - $[1 \cdot S° NH_3$ (g) $+ 1 \cdot S° HCl$ (g)]

$$= (1 \text{ mol})(94.85 \text{ J/K} \cdot \text{mol}) - [(1 \text{ mol})(192.77 \text{ J/K} \cdot \text{mol}) +$$
$$(1 \text{ mol})(186.2 \text{ J/K} \cdot \text{mol})] = - 284.1 \text{ J/K}$$

$\Delta H°_{rxn} = 1 \cdot \Delta H°_f NH_4Cl$ (s) - $[1 \cdot \Delta H°_f NH_3$ (g) $+ 1 \cdot \Delta H°_f HCl$ (g)]

$$= (1 \text{ mol})(- 314.55 \text{ kJ/mol}) - [(1 \text{ mol})(- 45.90 \text{ kJ/mol}) + (1 \text{ mol})(- 92.31 \text{ kJ/mol})]$$

$$= - 176.34 \text{ kJ}$$

$\Delta G°_{rxn} = \Delta H°_f - T \Delta S°_{rxn}$

$$= - 176.34 \text{ kJ} - (298.15 \text{ K})(- 284.1 \text{ J/K})(\frac{1.000 \text{ kJ}}{1000 \text{ J}})$$

$$= - 176.01 \text{ kJ} + 84.67 \text{ kJ} = - 91.64 \text{ kJ}$$

$$\Delta S°_{surroundings} \quad = \frac{-\Delta H_{rxn}}{T} \quad \text{(Assuming we're at 298 K)}$$

$$= \frac{176.34 \text{ kJ}}{298.15 \text{ K}} \cdot \frac{1000 \text{ J}}{1 \text{ kJ}} \quad = 591.45 \text{ J/K}$$

$\text{so } \Delta S°_{system} + \Delta S°_{surroundings} \quad = - 284.12 \text{ J/K} + 591.45 \text{ J/K} = +307.3 \text{ J/K}$

The values for $\Delta G°$ for the equation is negative, indicating that is product-favored. The reaction is enthalpy driven ($\Delta H < 0$).

(b) Calculate K_p for the reaction:

$\Delta G°_{rxn} = - RT \ln K_p$ so -91.64×10^3 J/mol = $- (8.3145 \frac{J}{K \cdot mol})(298.15 \text{ K}) \ln K_p$

$36.97 = \ln K_p$ and $1.13 \times 10^{16} = K_p$

43. Calculate K_p for the formation of methanol from its' elements:

Begin by calculating a ΔG_{rxn}

$$\Delta G_{rxn} = \Delta G_{CH3OH(\ell)} - [\Delta G_{C(graphite)} + 1/2 \cdot \Delta G_{O2(g)} + 2 \cdot \Delta G_{H2(g)}]$$

$$\Delta G_{rxn} = -166.14 \text{ kJ} - (0 \text{ kJ} + 0 \text{ kJ} + 0 \text{ kJ}) = -166.14 \text{ kJ}$$

$$\Delta G°_{rxn} = - RT \ln K_p$$

$$-166.1 \times 10^3 \text{ J/mol} = - (8.3145 \frac{J}{K \cdot mol})(298.15 \text{ K}) \ln K_p$$

$$67.00 = \ln K_p \text{ and } 1.3 \times 10^{29} = K_p$$

The large value of K_p indicates that this process is product-favored at 298 K. Judging by the relative numbers of gaseous particles, (without doing a calculation), one can see that ΔS for the reaction is < 0, so higher temperatures would reduce the value of K_p.

Regarding the connection between $\Delta G°$ and K, the more negative the value of $\Delta G°$, the larger the value of K.

45. Calculate the $\Delta S°$ for the vaporization of ethanol at 78.0 °C.

$$\text{The } \Delta S = \frac{\Delta H_{vap}}{T} = \frac{39.3 \times 10^3 \text{ J}}{351 \text{ K}} = 112 \text{ J/K}$$

47. Estimate the normal boiling point of ethanol:

Ethanol boils when the transition $(\ell) \rightarrow (g)$ is spontaneous. That is signaled by a ΔG that makes the transition from + values to – values. Calculate the point at which $\Delta G = 0$.

$$\Delta S_{sys} = S \text{ CH}_3\text{CH}_2\text{OH (g)} - S \text{ CH}_3\text{CH}_2\text{OH } (\ell)$$

$$= (1 \text{ mol}) \cdot 282.70 \frac{J}{K \cdot mol} - (1 \text{ mol}) \cdot 160.7 \frac{J}{K \cdot mol} = 122.0 \text{ J/K}$$

$$\Delta H = \Delta H_f \text{ CH}_3\text{CH}_2\text{OH (g)} - \Delta H_f \text{ CH}_3\text{CH}_2\text{OH } (\ell)$$

$$= (1 \text{mol})(-235.3 \frac{kJ}{mol}) - (1 \text{mol})(-277.0 \frac{kJ}{mol}) = 41.7 \text{ kJ}$$

At $\Delta G = 0$, $\Delta H = T \cdot \Delta S$ so 41.7 kJ(1000J/kJ) = T $\cdot$122.0 J/K and solving for T:

T = 341.8 K (or 341.8 – 273.15 = 68.7 °C) (Compared to literature value of 78 °C)

49. For the decomposition of phosgene:

$$\Delta H°_{rxn} = [\Delta H°_f \text{ CO (g)} + \Delta H°_f \text{ Cl}_2 \text{ (g)}] - [\Delta H°_f \text{ COCl}_2(g)]$$

$$\Delta H°_{rxn} = [(1 \text{ mol.})(-110.525 \frac{kJ}{mol}) + (1 mol)(0 \frac{kJ}{mol})] - [(1 \text{ mol})(-218.8 \frac{kJ}{mol})] = 108.275 \frac{kJ}{mol}$$

$$\Delta S°_{rxn} = [S° \ CO (g) + S° \ Cl_2 (g)] - [S° \ COCl_2(g)]$$

$$\Delta S°_{rxn} = [(1 \text{ mol})(197.674 \frac{J}{K \cdot mol}) + (1 \text{ mol})(223.07 \frac{J}{K \cdot mol})] -$$

$$[(1 \text{ mol})(283.53 \frac{J}{K \cdot mol})] = 137.2 \text{ J/K}$$

Using the ΔS data, we can see that raising the temperature will favor the endothermic decomposition of this substance.

51. (a)The reaction: $2 \ HgO(s) \rightarrow 2 \ Hg(\ell) + O_2(g)$

S° (J/K•mol) 70.29	76.02	205.07
$\Delta H°_f$ (kJ/mol) – 90.83	0	0

$$\Delta S°_{rxn} = [2 \cdot S° \ Hg(\ell) + 1 \cdot S° \ O_2(g)] - 2 \cdot S° \ HgO(s)]$$

$$= (2 \text{ mol})(76.02 \text{ J/K} \cdot mol) + [(1 \text{ mol})(205.07 \text{ J/K} \cdot mol) - (2 \text{ mol})(70.29 \text{ J/K} \cdot mol)]$$

$= 216.53$ J/K. This positive value is expected, with the production of liquid & gas.

$$\Delta H°_{rxn} = [2 \cdot \Delta H°_f \ Hg(\ell) + 1 \cdot \Delta H°_f \ O_2(g)] - 2 \cdot \Delta H°_f \ HgO(s)]$$

$$= [(2 \text{ mol})(0 \text{ kJ/mol}) + (1 \text{ mol})(0 \text{ kJ/mol})] - (2 \text{ mol})(-90.83 \text{ kJ/mol})] = 181.66 \text{ kJ}$$

The positive value isn't that surprising given that we're forming elements from a compound.

$$\Delta G°_{rxn} = \Delta H°_f - T \ \Delta S°_{rxn}$$

$$= 181.66 \text{ kJ} - (298.15 \text{ K})(216.53 \text{ J/K})(\frac{1.000 \text{ kJ}}{1000 \text{ J}})$$

$= 181.66$ kJ $- 64.56$ kJ $= 117.10$ kJ The positive value is consistent with the expectation that the reaction would not occur spontaneously.

$$\Delta S°_{surroundings} = \frac{-\Delta H_{rxn}}{T} \quad \text{(Assuming we're at 298 K)}$$

$$= \frac{-181.66 \text{ kJ}}{298.15 \text{ K}} \cdot \frac{1000 \text{ J}}{1 \text{ kJ}} = -609.29 \text{ J/K}$$

so $\Delta S°_{system} + \Delta S°_{surroundings} = 216.53$ J/K $+ -609.29$ J/K $= -392.76$ J/K

(b) Calculate K_p for the reaction:

$$\Delta G°_{rxn} = -RT \ln K_p \text{ so } 117.10 \times 10^3 \text{ J/mol} = -(8.3145 \frac{J}{K \cdot mol})(298.15 \text{ K}) \ln K_p$$

$-47.24 = \ln K_p$ and $3.06 \times 10^{-21} = K_p$

53. For the reaction of sodium with water: Na (s) + H$_2$O(ℓ) $\rightarrow$ NaOH(aq) + $\frac{1}{2}$ H$_2$ (g)

Predict signs for $\Delta H°$ and $\Delta S°$:

This one seems easy !! The reaction of sodium with water gives off heat, and the heat frequently ignites the hydrogen gas that is concomitantly evolved. $\Delta H° = -$.

Regarding entropy, the system changes from one with a solid (low entropy) and a liquid (higher entropy) to a solution (*frequently* higher entropy than liquid) and a gas (high entropy). So we would predict that the entropy would increase, i.e. $\Delta S° = +$.

Now for the calculation:

$$\Delta H°_{rxn} = [1 \bullet \Delta H°_f \text{ NaOH(aq)} + \frac{1}{2} \bullet \Delta H°_f \text{ H}_2\text{(g)}] - [1 \bullet \Delta H°_f \text{ Na(s)} + 1 \bullet \Delta H°_f \text{ H}_2\text{O}(\ell)]$$

$$= [(1 \text{ mol})(-469.15 \frac{\text{kJ}}{\text{mol}}) + (\frac{1}{2}\text{mol})(0)] - [(1 \text{ mol})(0) + (1 \text{ mol})(-285.83 \frac{\text{kJ}}{\text{mol}})]$$

$$= -183.32 \text{ kJ}$$

$$\Delta S°_{rxn} = [1 \bullet S° \text{ NaOH(aq)} + \frac{1}{2} \bullet S° \text{ H}_2\text{(g)}] - [1 \bullet S° \text{ Na(s)} + 1 \bullet S° \text{ H}_2\text{O}(\ell)]$$

$$= [(1 \text{ mol})(48.1 \frac{\text{J}}{\text{K} \bullet \text{mol}}) + (\frac{1}{2} \text{ mol})(130.7 \frac{\text{J}}{\text{K} \bullet \text{mol}})] -$$

$$[(1 \text{ mol})(51.21 \frac{\text{J}}{\text{K} \bullet \text{mol}}) + (1 \text{ mol})(69.95 \frac{\text{J}}{\text{K} \bullet \text{mol}})]$$

$$= -7.7 \frac{\text{J}}{\text{K} \bullet \text{mol}}$$

As expected, the $\Delta H°_{rxn}$ for the reaction is negative! The surprise comes in the calculation for $\Delta S°_{rxn}$. While we anticipate the sign to be positive, we find a slightly negative number—reflecting the order (hence a decrease in entropy) that can occur as solutions occur.

55. For the reaction :

	BCl$_3$(g) +	3/2 H$_2$(g) $\rightarrow$	B(s) +	3HCl(g)
S° ($\frac{\text{J}}{\text{K}\bullet\text{mol}}$)	290.17	130.7	5.86	186.2
$\Delta H°$($\frac{\text{kJ}}{\text{mol}}$)	-402.96	0	0	-92.31

$$\Delta H°_{rxn} = [3 \bullet \Delta H°_f \text{ HCl(g)} + 1 \bullet \Delta H°_f \text{ B(s)}] - [1 \bullet \Delta H°_f \text{ BCl}_3\text{(g)} + 3/2 \Delta H°_f \text{ H}_2\text{(g)}]$$

$$\Delta H°_{rxn} = [(3 \text{ mol})(-92.31 \frac{\text{kJ}}{\text{mol}}) + (1 \text{ mol })(0)] - [(1 \text{ mol})(-402.96 \frac{\text{kJ}}{\text{mol}}) + (3/2 \text{ mol})(0)]$$

$$= 126.03 \text{ kJ}$$

$$\Delta S°_{rxn} = [3 \bullet S° \text{ HCl(g)} + 1 \bullet S° \text{ B(s)}] - [1 \bullet S° \text{ BCl}_3\text{(g)} + 3/2 S° \text{ H}_2\text{(g)}]$$

$$\Delta S^{\circ}_{rxn} = [(3 \text{ mol})(186.2 \frac{J}{K \bullet mol}) + (1 \text{ mol})(5.86 \frac{J}{K \bullet mol})] - [(1 \text{mol})(290.17 \frac{J}{K \bullet mol})$$

$$+ (3/2 \text{ mol})(130.7 \frac{J}{K \bullet mol})]$$

$$= 78.2 \text{ J/K}$$

$$\Delta G^{\circ}_{rxn} = \Delta H^{\circ}_{f} - T \Delta S^{\circ}_{rxn} = 126.03 \text{ kJ} - (298.15 \text{ K})(78.2 \frac{J}{K})(\frac{1.000 \text{ kJ}}{1000 \text{ J}}) = 103 \text{ kJ}$$

The reaction is not product-favored.

57. The equilibrium constant for the reduction of iron(II) oxide is 0.422 at 700 °C.
Estimate ΔG°_{rxn}.

Given that $\Delta G^{\circ}_{rxn} = - RT \ln K_p$

$$\Delta G^{\circ}_{rxn} = - (8.3145 \frac{J}{K \bullet mol})(973 \text{ K}) \ln 0.422$$

$$\Delta G^{\circ}_{rxn} = 6.98 \text{ kJ}$$

59. For the reaction $C_6H_6 (\ell) + 3 H_2(g) \rightarrow C_6H_{12}(\ell)$, $\Delta H^{\circ} = - 206.7$ kJ, and $\Delta S^{\circ} = -361.5$ J/K

Is the reaction spontaneous under standard conditions? Calculating ΔG° will answer this
question.

$$\Delta G^{\circ}_{rxn} = \Delta H^{\circ}_{rxn} - T \Delta S^{\circ}_{rxn}$$

$$= -206.7 \text{ kJ} - (298.15 \text{ K})(-361.5 \frac{J}{K})(\frac{1.000 \text{ kJ}}{1000 \text{ J}}) = \text{ kJ}$$

$$= -206.7 \text{ kJ} - (-107.8 \text{ kJ}) = -98.9 \text{ kJ}$$

The negative value for ΔG°_{rxn} tells us that the reaction would be spontaneous (product-favored)
under standard conditions. The negative value for ΔH° tells us that the reaction is enthalpy-
driven.

61. For the reaction: $2 SO_3 (g) \rightarrow 2 SO_2 (g) + O_2 (g)$

$$\Delta H^{\circ}_{rxn} = [2 \bullet \Delta H^{\circ}_{f} SO_2 (g) + 1 \bullet \Delta H^{\circ}_{f} O_2 (g)] - [2 \bullet \Delta H^{\circ}_{f} SO_3 (g)]$$

$$= [(2 \text{ mol})(- 296.84 \frac{kJ}{mol}) + 0] - [(2 \text{ mol})(-395.77 \frac{kJ}{mol})] = 197.86 \text{ kJ}$$

$$\Delta S^{\circ}_{rxn} = [2 \bullet S^{\circ} SO_2 (g) + 1 \bullet S^{\circ} O_2 (g)] - [2 \bullet S^{\circ} SO_3 (g)]$$

$$= [(2 \text{ mol})(248.21 \frac{J}{K \bullet mol}) + (1 \text{ mol})(205.07 \frac{J}{K \bullet mol})] - [(2 \text{ mol})(256.77 \frac{J}{K \bullet mol})]$$

$$= 187.95 \text{ J/K}$$

(a) Is the reaction product-favored at 25 °C ?:

$\Delta G°_{rxn} = \Delta H°_{rxn} - T \Delta S°_{rxn}$

$= 197.86 \text{ kJ} - (298.15 \text{ K})(187.95 \frac{J}{K})(\frac{1.000 \text{ kJ}}{1000 \text{ J}}) = 141.82 \text{ kJ}$

The reaction is not product-favored.

(b) The reaction can become product-favored if there is some T at which $\Delta G°_{rxn} < 0$. To see if such a T is feasible, let's set $\Delta G°_{rxn} = 0$ and solve for T! *Remember that the units of energy must be the same, so let's convert units of J (for the entropy term) into units of kJ*

$\Delta G°_{rxn} = \Delta H°_{rxn} - T \Delta S°_{rxn}$

$0 = 197.86 \text{ kJ} - T(0.18795 \frac{\text{kJ}}{K})$

$T = \dfrac{197.86 \text{ kJ}}{0.18795 \dfrac{\text{kJ}}{K}} = 1052.7 \text{ K}$ or $(1052.7 - 273.1) = 779.6 \text{ °C}$

(c) The equilibrium constant for the reaction at 1500 °C. Since we know that

$\Delta G°_{rxn} = \Delta H°_{rxn} - T\Delta S°_{rxn} = - RT\ln K$, we can solve for K if we know $\Delta G°_{rxn}$ at 1500 °C

$\Delta G°_{rxn} = \Delta H°_{rxn} - T \Delta S°_{rxn} = - RT \ln K$

$= 197.86 \text{ kJ} - (1773 \text{ K})(187.95 \text{ J/K})(\frac{1.000 \text{ kJ}}{1000 \text{ J}})$

And solving for $\Delta G°_{rxn}$ gives –135.4 kJ. Substitute into the last part of the equation:

$- 135.4 \text{ kJ} = - 8.314 \dfrac{J}{K \cdot mol} \cdot \dfrac{1 \text{ kJ}}{1000 \text{ J}} \cdot 1773 \text{ K} \cdot \ln K$

$K = 9.7 \times 10^3$ or 1×10^4 (to 1 sf)

63. Reaction: H_2S (g) + 2 O_2 (g) → H_2SO_4 (ℓ)

$\Delta H°_f$ (kJ/mol)	-20.63	0	- 814
S° (J/K • mol)	205.79	205.07	156.9

$\Delta H°_{rxn} = [(1 \text{ mol})(- 814 \frac{\text{kJ}}{mol})] - [(1 \text{ mol})(-20.63 \frac{\text{kJ}}{mol}) + 0] = - 793 \text{ kJ}$

$\Delta S°_{rxn} = [(1 \text{ mol})(156.9 \frac{J}{K \bullet mol})] - [(1 \text{ mol})(205.79 \frac{J}{K \bullet mol}) + (2 \text{ mol})(205.07 \frac{J}{K \bullet mol})]$

$= -459.0 \text{ J/K}$

$\Delta G°_{rxn} = \Delta H°_f - T \Delta S°_{rxn}$

$= - 793 \text{ kJ} - (298.15 \text{ K})(- 459.0 \frac{J}{K})(\frac{1.000 \text{ kJ}}{1000 \text{ J}}) = - 657 \text{ kJ}$

The reaction is product-favored at 25 °C ($\Delta G° < 0$) and enthalpy-driven ($\Delta H°_{rxn} < 0$)

65. Calculate the $\Delta G°$ for the transition of S_8 (rhombic) $\rightarrow$ S_8 (monoclinic)

(a) At 80 °C $\Delta G°_{rxn}$ $= \Delta H° - T \Delta S°$

$$\Delta G°_{rxn} = 3.213 \text{ kJ} - (353 \text{ K})(0.0087 \frac{\text{kJ}}{\text{K}})$$

$$= 0.14 \text{ kJ}$$

At 110 °C $\Delta G°_{rxn}$ $= \Delta H° - T \Delta S°$

$$\Delta G°_{rxn} = 3.213 \text{ kJ} - (383 \text{ K})(0.0087 \frac{\text{kJ}}{\text{K}})$$

$$= -0.12 \text{ kJ}$$

The rhombic form of sulfur is the more stable at lower temperature, while the monoclinic form is the more stable at higher temperature. The transition to monoclinic form is product-favored at temperatures above 110 degrees C.

(b) The temperature at which $\Delta G°_{rxn} = 0$:

$$\Delta G°_{rxn} = 3.213 \text{ kJ} - (T)(0.0087 \frac{\text{kJ}}{\text{K}})$$

$$0 = 3.213 \text{ kJ} - (T)(0.0087 \frac{\text{kJ}}{\text{K}})$$

$$T = \frac{3.213 \text{ kJ}}{0.0087 \frac{\text{kJ}}{\text{K}}} = 370 \text{ K} = 96 °C$$

67. (a) Calculate K_p at 25 °C for the reaction:

	$N_2(g)$	+ $O_2(g)$	$\rightarrow$	2 NO (g)
$\Delta G°_f$ kJ/mol	0	0		86.58
$\Delta H°_f$ kJ/mol	0	0		90.29
$\Delta S°_f$ J/K•mol	191.56	205.07		210.76

$$\Delta H°_{rxn} = [(2 \text{ mol})(90.29 \frac{\text{kJ}}{\text{mol}})] - [(1 \text{ mol})(0) + (1 \text{ mol})(0)] = 180.58 \text{ kJ}$$

$$\Delta S°_{rxn} = [(2 \text{ mol})(210.76 \frac{\text{J}}{\text{K} \bullet \text{mol}})] -$$

$$[(1 \text{ mol})(191.56 \frac{\text{J}}{\text{K} \bullet \text{mol}}) + (1 \text{ mol})(205.07 \frac{\text{J}}{\text{K} \bullet \text{mol}})] = 24.89 \text{J/K}$$

$\Delta G°_{rxn} = [(2 \text{ mol})(86.58 \text{ kJ/mol})] - [0 + 0] = 173.16 \text{ kJ}$

Calculating K_p: $\Delta G°_{rxn} = -RT \ln K_p$

$$173.16 \times 10^3 \text{ J/mol} = -(8.3145 \frac{\text{J}}{\text{K} \bullet \text{mol}})(298.15 \text{ K}) \ln K_p$$

$$-69.85 = \ln K_p \quad \text{and} \quad 4.62 \times 10^{-31} = K_p$$

Note that the + value of ΔG°_f results in a value of K_p which is small--reactants are favored. A negative value would result in a large K_p -- a process in which the products were favored.

(b) ΔG°_{rxn} at 973K: $\Delta G^\circ = \Delta H^\circ - T\Delta S^\circ = 180.58$ kJ $- (973$ K$)(24.89$J/K$)(1$kJ/1000 J$)$

$$= 156.4 \text{ kJ}$$

The estimate for Kp at 700 °C: 156.4×10^3 J/mol $= -(8.3145 \dfrac{J}{K \bullet mol})(973$ K$) \ln K_p$

$\ln K_p = -19.33$ and $K_p = 4 \times 10^{-9}$

The reaction is still reactant favored, although K is much greater than at 25°C.

(c) With $K_p = 4 \times 10^{-9}$, equilibrium pressures of N_2, O_2 and NO:

$$\frac{P_{NO}^2}{P_{N_2} \bullet P_{O_2}} = 4 \times 10^{-9}$$

If the $P(N_2) = P(O_2) = 1.00$ bar initially, and we allow some amount, x, to form NO, then equilibrium concentrations would be: $P(N_2) = P(O_2) = (1.00-x)$ bar and $P(NO) = x$.

Substituting into the equilibrium expression:

$$\frac{x^2}{(1.00 - x)(1.00 - x)} = 4 \times 10^{-9} \text{ and taking the square root of both sides:}$$

$\dfrac{x}{1.00 - x} = 6.32 \times 10^{-5}$ Solving for $x = 6 \times 10^{-5}$ and the equilibrium concentrations are:

$P(N_2) = P(O_2) = 1$ bar (essentially unchanged) $P(NO) = 6 \times 10^{-5}$ bar

69. Calculate ΔG°_f for HI(g) at 350 °C, given the following equilibrium partial pressures:

P(H$_2$)= 0.132 bar, P(I$_2$) = 0.295 bar, and P(HI) =1.61 bar. At 350 °C ,1 bar, I_2 is a gas.

$$1/2 \text{ H}_2(g) + 1/2 \text{ I}_2(g) \Leftrightarrow \text{HI}(g)$$

Calculate Kp: $\dfrac{P_{HI}}{P_{H_2}^{1/2} \bullet P_{I_2}^{1/2}} = \dfrac{1.61}{0.363 \bullet 0.543} = 8.16$

Knowing that $\Delta G^\circ = -RT\ln K$, we can solve:

$\Delta G^\circ = -RT\ln K = -(8.3145 \dfrac{J}{K \bullet mol})(623.15K)\ln 8.16 = -10{,}873$J or -10.9 kJ/mol

71. An examination of the equation Hg(ℓ) $\rightarrow$ Hg(g) shows that the equilibrium constant expression would be $K_p = P_{Hg(g)}$. So to find the temperature at which $K_p = 1.00$ bar and 1/760 bar, we need only to find the temperature at which the vapor pressure of mercury is 1.00 bar and 1/760 bar, respectively.

We can calculate the T for Kp = 1.00 bar easily. At the equilibrium point, we can calculate T at which Kp = 1.00bar if we know ΔG°. Since at equilibrium, ΔG°= 0, we can rewrite the equation: $\Delta G° = \Delta H° - T\Delta S°$ to read. $\Delta H°/\Delta S = T$

$$\Delta H°rxn = (1\ mol)(61.38\ \frac{kJ}{mol}) - (1\ mol)(0) = 61.38\ kJ\ \text{and for entropy:}$$

$$\Delta S°rxn = (1\ mol)(174.97\ \frac{J}{K\bullet mol}) - (1\ mol)(76.02\ \frac{J}{K\bullet mol}) = 98.95\ \frac{J}{K}$$

Substituting into the equation:

$$\Delta H°/\Delta S = T = (61.38\ kJ \bullet 1000\ J/kJ)/\ 98.95\ \frac{J}{K} = 620.3\ K.\ \text{or}\ 347.2\ °C.$$

(b) Temperature at which Kp = 1/760   Using the Clausius-Clapeyron equation:

$$\ln\left(\frac{P_2}{P_1}\right) = \frac{\Delta H}{R}\left(\frac{1}{T_1} - \frac{1}{T_2}\right) \text{ and } \ln\left(\frac{1}{760}\right) = \frac{61.38 \times 10^3\ \frac{J}{mol}}{8.3145\ \frac{J}{K\bullet mol}}\left(\frac{1}{620.3K} - \frac{1}{T_2}\right)$$

$$\frac{8.3145\ \frac{J}{K\bullet mol} \bullet -6.633}{61.38 \times 10^3\ \frac{J}{mol}} = \left(\frac{1}{620.3K} - \frac{1}{T_2}\right) \text{ so } -8.98 \times 10^{-4}\ \frac{1}{K} = \left(\frac{1}{620.3K} - \frac{1}{T_2}\right)$$

$- 1/T_2 = - 8.98 \times 10^{-4}\ 1/K - 1.61 \times 10^{-3} = -2.51 \times 10^{-3}$ and $T_2 = 398.3\ K$ or $125.2\ °C$.

So in summary, Kp = 1 at 347.2 °C and is 1/760 at 125.2 °C.

Summary and Conceptual Questions

73. Processes product-favored or reactant-favored under standard conditions:

   (a) $Hg(\ell) \rightarrow Hg(s)$. Since Hg exists at room T as a liquid, we predict reactant-favored.

   (b) $H_2O(g) \rightarrow H_2O(\ell)$ Water exists mainly as a liquid at room T, and water vapor condenses spontaneously, so product-favored.

   (c) $2\ HgO(s) \rightarrow Hg(\ell) + O_2(g)$ Reactant-favored, since energy is required to carry out this process

   (d) $C(s) + O_2(g) \rightarrow CO_2(g)$ Product-favored since carbon will burn readily to form CO_2.

   (e) $NaCl(s) \rightarrow NaCl(aq)$ Table salt spontaneously dissolves in water, so product-favored.

   (f) $CaCO_3(s) \rightarrow Ca^{2+}(aq) + CO_3^{2-}(aq)$ Calcium carbonate does not dissolve to any appreciable extent, so reactant-favored.

75. Following statements false or true?

(a) The entropy of a liquid increases on going from the liquid to the vapor state at any temperature. **True**. For a given substance, the entropy of the vapor state of that substance is greater than for the liquid state.

(b) An exothermic reaction will always be spontaneous. **False**. While exothermic reactions are *almost always spontaneous*, the entropy does play a role, and,should the entropy increase greatly enough, cause the reaction to be non-spontaneous (i.e. reactant-favored).

(c) Reactions with a $+ \Delta H$ and a $+\Delta S$ can never be product-favored. **False**. At very high temperatures, such reactions can be product-favored ($\Delta G < 0$).

(d) If ΔG_{rxn} is < 0, the reaction will have an equilibrium constant greater than 1. **True**. Since ΔG and K are related by the expression $\Delta G = $ -RT lnK, if $\Delta G < 0$, then mathematically K will be greater than 1.

77. If we dissolve a solid (e.g. table salt), the process proceeds spontaneously ($\Delta G° < 0$).
If $\Delta H° = 0$, we can write: $\Delta G° = \Delta H° - T\Delta S°$ and $(-) = 0 - (+)\Delta S°$.
The only mathematical condition for which this equation is true is if $\Delta S° = +$, hence the process is entropy driven.

79. For the reaction: $2 C_2H_6(g) + 7 O_2(g) \rightarrow 4 CO_2(g) + 6 H_2O(g)$

(a) Predict whether signs of $\Delta S°_{sys}$, $\Delta S°_{surr}$, $\Delta S°_{univ}$ are greater than, equal to, or less than 0.
$\Delta S°_{sys}$ will be > 0, since 9 mol of gas form 10 mol of gas as the reaction proceeds.
$\Delta S°_{surr}$ will be > 0, since the reaction liberates heat, and would increase the entropy of the surroundings. With both $\Delta S°_{sys}$ and $\Delta S°_{surr}$ increasing, $\Delta S°_{univ}$ would also increase.

(b) Predict signs of $\Delta H°$, and $\Delta G°$: Since the reaction is exothermic, $\Delta H°$ would be "-". With a negative $\Delta H°$ and an increasing entropy, $\Delta G°$ would be "-" as well.

(c) Will value of Kp be very large, very small, or nearly 1? With the relatively large number of moles of carbon dioxide and water being formed, $\Delta H°$ for the reaction will be *large and negative*, and with the increasing entropy $\Delta G°$ will also be relatively *large and negative*. Since $\Delta G = $ -RT lnK, we anticipate Kp being **very large**.
Will Kp be larger or smaller at temperatures greater than 298 K? Rearrange the expression $\dfrac{\Delta G}{RT} = -\ln K$. As T increases, the term on the left will decrease, resulting in a smaller value of K.

81. Explain why the entropy of the system increases on dissolving NaCl(s) in water.

The calculation for the process: $NaCl(s) \rightarrow NaCl(aq)$ is:

$$\Delta S_{sys} = S\ NaCl(aq) - S\ NaCl(s) = (1\ mol)\cdot 115.5\ \frac{J}{K\cdot mol} - (1\ mol)\cdot 72.11\ \frac{J}{K\cdot mol} = 43.4\ J/K$$

Obviously the numbers tell us that the entropy of system increases. Moreover, from a qualitative standpoint, as the solid (with the ions arranged in the crystal lattice) dissolves, the order of those ions in the aqueous state with definitely decrease, with the concommitant **increase** in entropy.

83. The key phrase needed to answer the question: "What is the sign......" is "Iodine dissolves readily...." . This phrase tell us that $\Delta G°$ is negative.

Enthalpy-driven processes are exothermic. The "neutrality" of the ΔH for this reaction tells us that the process is NOT enthalpy-driven. Since the iodine goes from the solid state to the "solution" state, we anticipate an increase in entropy, and would therefore state that the process is entropy-driven.

85. (a) Confirm that $Mg(s) + 2\ H_2O(\ell) \rightarrow Mg(OH)_2\ (s) + H_2(g)$ is a spontaneous reaction.

$\Delta H°_f$ (kJ/mol)   0   -285.83   -924.54   0

$S°$ (J/K • mol) 32.67   69.95   63.18   130.7

$$\Delta H°_{rxn} = [(1\ mol)(-924.54\ \frac{kJ}{mol}) + (1mol)(0)] - [(1mol)0 + (2\ mol)(-285.83\ \frac{kJ}{mol})\ ]$$

$$= -352.88\ kJ$$

$$\Delta S°_{rxn}\ [(1\ mol)(63.18\ \frac{J}{K\cdot mol}) + (1mol)(130.7\ \frac{J}{K\cdot mol})] -$$

$$[(1mol)(32.67\ \frac{J}{K\cdot mol}) + (2\ mol)(69.95\ \frac{J}{K\cdot mol})\ ]$$

$$= 21.31 J/K$$

$$\Delta G°_{rxn} = \Delta H°_f - T\ \Delta S°_{rxn}$$

$$= -352.88\ kJ - (298.15\ K)(21.31\ \frac{J}{K})(\frac{1.000\ kJ}{1000\ J}) = -359.23\ kJ$$

With a negative $\Delta G°$, we anticipate the reaction to be spontaneous.

(b) Mass of Mg to produce sufficient energy to heat 225 mL of water (D = 0.996 g/mL) from 25°C to the boiling point (100 °C)? [100-25 = 75 °C or 75K]

Heat required: 225 mL • 0.996 g/mL • 4.184 J/g•K • 75 K. = 70,322.58 J or 70.3 kJ

The $\Delta H°_{rxn}$ =-352.88 kJ for 1 mol of Mg.

$$\frac{70.3 kJ}{1} \cdot \frac{1\ mol\ Mg}{352.88\ kJ} = 0.2\ mol\ Mg\ or\ 24.3\ g/mol • 0.2\ mol = 4.84\ g\ Mg$$

87. (a) Equation for the reaction of hydrazine and oxygen.

$N_2H_4 (\ell) + O_2(g) \rightarrow 2\ H_2O(\ell) + N_2(g)$

Oxygen is the **oxidizing agent,** and hydrazine is the **reducing agent.** There are several ways to assess this. Note that the oxidation state for O_2 is 0 (as reactant) and –2 (as product)—it has been reduced (by hydrazine). Note that hydrazine **loses H**—in going from reactant to product—a definition for being oxidized.

(b) Calculate $\Delta H°$, $\Delta S°$, and $\Delta G°$:

$\Delta H°_{rxn} = [2 \cdot \Delta H°_f H_2O\ (\ell) + 1 \cdot \Delta H°_f\ N_2\ (g)] - [1\cdot \Delta H°_f\ N_2H_4\ (\ell) + 1\cdot \Delta H°_f\ O_2\ (g)]$

$= [(2\ mol)(- 285.830\ \frac{kJ}{mol}) + 0] - [(1\ mol)(50.63\ \frac{kJ}{mol}) + 0] = - 622.29\ kJ$

$\Delta S°_{rxn} = [2 \cdot S°H_2O\ (\ell) + 1 \cdot S°\ N_2\ (g)] - [1\cdot S°\ N_2H_4\ (\ell) + 1\cdot S°\ O_2\ (g)]$

$= [(2\ mol)(69.95\ \frac{J}{K \cdot mol}) + (1\ mol)(191.56\ \frac{J}{K \cdot mol})] -$

$[(1\ mol)(121.52\ \frac{J}{K \cdot mol}) + (1\ mol)(205.07\ \frac{J}{K \cdot mol})]$

$= 4.87\ J/K$

$\Delta G°_{rxn} = \Delta H°_f - T\ \Delta S°_{rxn}$

$= - 622.29\ kJ - (298\ K)(4.87\ \frac{J}{K})(\frac{1.000\ kJ}{1000.\ J}) = - 623.74\ kJ/mol$

(c) T change of 5.5×10^4 L of water (assuming 1 mole of N_2H_4 reacts:

1 mol of hydrazine releases –622.29 kJ,

Heat = m $\cdot$ c $\cdot$ Δt [Assume D of water = 0.996 g/mL]

622.29×10^3 J = 5.5×10^4 L $\cdot$ 996 g/L $\cdot$ 4.184 J/g $\cdot$K $\cdot$ Δt.

$\dfrac{6.2229 \times 10^5\ J}{\left(5.5\ \times 10^4 L \cdot 996\ \frac{g}{L} \cdot 4.184 \frac{J}{g \cdot K}\right)} = \Delta t.\ 2.7 \times 10^{-3}$ K.

(d) Solubility of O_2 = 0.000434 g O_2/100g water.

5.5×10^4 L $\cdot$ 996 g/L $\cdot$ 4.34×10^{-4} g O_2/100g water $\cdot$ 1 mol O_2/32.00 g O_2

7.4 mol O_2 (or approximately 240 g)

(e) If hydrazine is present in 5% solution, what mass of hydrazine solution is needed to consume the O_2 present?

$\dfrac{7.4\ mol\ O_2}{1} \cdot \dfrac{1\ mol\ N_2H_4}{1\ mol\ O_2} \cdot \dfrac{32.05\ g\ N_2H_4}{1\ mol\ N_2H_4} \cdot \dfrac{100\ g\ solution}{5.00\ g\ N_2H_4} = 4.8 \times 10^3$ g solution

(to 2 sf)

(f) Assuming N_2 escapes as gas, calculate V of N_2 at STP,

The balanced equation tells us that 7.4 mol of O_2 will liberate 7.4 mol of N_2. At STP, 7.4 mol of this gas will occupy (7.4 mol • 22.4 L/mol) or 170 L (to 2 sf)

89. From the CD-ROM,

(a) the $\Delta S^{\circ}{}_{sys}$ at 400K = -60.49 J/K

(b) No, $\Delta S^{\circ}{}_{sys}$ is not a function of temperature.

(c) Yes, $\Delta S^{\circ}{}_{surr}$ does change with temperature.

(d) Yes, $\Delta S^{\circ}{}_{surr}$ always changes with temperature.

(e) Exothermic reactions always give positive values for ΔS_{surr}.

(f) If one examines the $\Delta S^{\circ}{}_{univ}$ for the two temperatures mentioned, the $\Delta S^{\circ}{}_{univ}$ is positive at 400 K, but is negative at 700 K—so the reaction is spontaneous at 400K, but NOT at 700K.

91. (a) For the reaction of Fe_2O_3 with C:ΔG decreases (becomes less positive) with increasing T. The temperature at which the reaction becomes spontaneous is that T at which ΔG becomes negative (approximately 835K)

(b) For the reaction of HCl and Na_2CO_3, is there a T at which the reaction is NOT spontaneous? No. With a negative ΔH° and a positive ΔS°, this reaction is spontaneous at ALL T.

(c) The spontaneity of a reaction is driven by ΔG, which is a function of ΔH, T, and ΔS. For a given system, ΔH and ΔS are fixed for the system. Depending on the value of ΔS, there may be a temperature at which the $T\Delta S^{\circ}$ term becomes large enough to drive ΔG positive. Hence T plays an important role in the spontaneity of certain reactions.

Chapter 20
Principles of Reactivity: Electron Transfer Reactions

Practicing Skills

Balancing Equations for Oxidation-Reduction Reactions

1. Balance the following:

	reactant is	overall process is
(a) $Cr\ (s)\ \rightarrow\ Cr^{3+}\ (aq) + 3\ e^-$	reducing agent	oxidation
(b) $AsH_3\ (g) \rightarrow\ As\ (s) + 3\ H^+\ (aq) + 3\ e^-$	reducing agent	oxidation
(c) $VO_3^-\ (aq) + 6\ H^+\ (aq) + 3\ e^-\ \rightarrow$	oxidizing agent	reduction
$V^{2+}\ (aq) + 3\ H_2O\ (\ell)$	reducing agent	oxidation

(d) $2\ Ag\ (s) + 2\ OH^-\ (aq)\ \rightarrow Ag_2O(s) + H_2O(l) + 2\ e^-$

Note: e^- are used to balance charge; H^+ balances only H atoms; H_2O (or OH^- in base) balances both H and O atoms.

3. Balance the equations (in acidic solutions):

Balancing redox equations in neutral or acidic solutions may be accomplished in several steps. They are:

1. Separating the equation into two equations which represent reduction and oxidation
2. Balancing mass of elements (other than H or O)
3. Balancing mass of O by adding H_2O
4. Balancing mass of H by adding H^+
5. Balancing charge by adding electrons
6. Balancing electron gain (in the reduction half-equation) with electron loss (in the oxidation half-equation)
7. Combining the two half equations

For the parts of this problem, each step will be identified with a number corresponding to the list above. In addition, the physical states of all species will be omitted in all but the final step. While this omission is <u>not generally recommended</u>, it should increase the clarity of the steps involved. In addition when a step leaves a half equation unchanged from the previous step, we have omitted the half equation.

(a) $Ag\ (s) + NO_3^-(aq) \rightarrow NO_2\ (g) + Ag^+\ (aq)$

Oxidation half-equation	Reduction half-equation	Step
$Ag \rightarrow Ag^+$	$NO_3^- \rightarrow NO_2$	1 & 2
	$NO_3^- \rightarrow NO_2 + H_2O$	3
	$2\ H^+ + NO_3^- \rightarrow NO_2 + H_2O$	4
$Ag \rightarrow Ag^+ + 1e^-$	$2\ H^+ + NO_3^- + 1e^- \rightarrow NO_2 + H_2O$	5 & 6
$2\ H^+(aq) + NO_3^-\ (aq) + Ag\ (s) \rightarrow Ag^+(aq) + NO_2\ (g) + H_2O(\ell)$		7

(b) $MnO_4^-\ (aq) + HSO_3^-\ (aq) \rightarrow Mn^{2+}(aq) + SO_4^{2-}\ (aq)$

Oxidation half-equation	Reduction half-equation	Step
$HSO_3^- \rightarrow SO_4^{2-}$	$MnO_4^- \rightarrow Mn^{2+}$	1 & 2
$H_2O + HSO_3^- \rightarrow SO_4^{2-}$	$MnO_4^- \rightarrow Mn^{2+} + 4H_2O$	3
$H_2O + HSO_3^- \rightarrow SO_4^{2-} + 3\ H^+$	$8\ H^+ + MnO_4^- \rightarrow Mn^{2+} + 4H_2O$	4
$H_2O + HSO_3^- \rightarrow SO_4^{2-} + 3\ H^+ + 2e^-$	$8\ H^+ + MnO_4^- + 5e^- \rightarrow Mn^{2+} + 4H_2O$	5
$5\ H_2O + 5\ HSO_3^- \rightarrow 5\ SO_4^{2-} + 15\ H^+ + 10e^-$	$16\ H^+ + 2MnO_4^- + 10e^- \rightarrow 2\ Mn^{2+} + 8H_2O$	6
$5\ HSO_3^-\ (aq) + H^+\ (aq) + 2\ MnO_4^-\ (aq) \rightarrow 5\ SO_4^{2-}\ (aq) + 2\ Mn^{2+}\ (aq) + 3\ H_2O\ (\ell)$		7

(c) $Zn\ (s) + NO_3^-(aq) \rightarrow Zn^{2+}\ (aq) + N_2O\ (g)$

Oxidation half-equation	Reduction half-equation	Step
$Zn \rightarrow Zn^{2+}$	$NO_3^- \rightarrow N_2O$	1
	$2\ NO_3^- \rightarrow N_2O$	2
	$2\ NO_3^- \rightarrow N_2O + 5\ H_2O$	3
	$10\ H^+ + 2\ NO_3^- \rightarrow N_2O + 5\ H_2O$	4
$Zn \rightarrow Zn^{2+} + 2\ e^-$	$10\ H^+ + 2\ NO_3^- + 8\ e^- \rightarrow N_2O + 5\ H_2O$	5
$4\ Zn \rightarrow 4\ Zn^{2+} + 8\ e^-$		6
$10\ H^+\ (aq) + 2\ NO_3^-\ (aq) + 4\ Zn\ (s) \rightarrow 4\ Zn^{2+}(aq) + N_2O\ (g) + 5\ H_2O\ (\ell)$		7

(d) $Cr(s) + NO_3^- (aq) \rightarrow Cr^{3+} (aq) + NO (g)$

Oxidation half-equation	Reduction half-equation	Step
$Cr \rightarrow Cr^{3+}$	$NO_3^- \rightarrow NO$	1 & 2
	$NO_3^- \rightarrow NO + 2 H_2O$	3
	$4 H^+ + NO_3^- \rightarrow NO + 2 H_2O$	4
$Cr \rightarrow Cr^{3+} + 3 e^-$	$4 H^+ + NO_3^- + 3 e^- \rightarrow NO + 2 H_2O$	5 & 6
$Cr (s) + 4 H^+(aq) + NO_3^- (aq) \rightarrow NO(g) + 2 H_2O(\ell) + Cr^{3+}(aq)$		7

5. Balancing redox equations in basic solutions may be accomplished in several steps. There is only a *slight change* from the "acidic solution".

1. Separating the equation into two equations which represent reduction and oxidation
2. Balancing mass of elements (other than H or O)
3. Balancing mass of O by adding H_2O
4. Balancing mass of H by adding H^+
5. Balancing charge by adding electrons
6. Balancing electron gain (in the reduction half-equation) with electron loss (in t oxidation half-equation)
7. Combining the two half equations (removing any redundancies).
8. *Add as many OH^- to both sides of the equation as there are H^+ ions, to form water.*
9. *Remove any redundancies in H_2O molecules.*

As before, each step will be identified with a number corresponding to the list above, and physical states of all species will be omitted in all but the final step.

(a) $Al (s) + OH^- (aq) \rightarrow Al(OH)_4^- (aq) + H_2 (g)$

Oxidation half-equation	Reduction half-equation	Step
$Al \rightarrow Al(OH)_4^-$	$OH^- \rightarrow H_2$	1 & 2
$Al + 4 H_2O \rightarrow Al(OH)_4^-$	$OH^- \rightarrow H_2 + H_2O$	3
$Al + 4 H_2O \rightarrow Al(OH)_4^- + 4H^+$	$3 H^+ + OH^- \rightarrow H_2 + H_2O$	4
$Al + 4 H_2O \rightarrow Al(OH)_4^- + 4 H^+ + 3 e^-$	$3 H^+ + OH^- + 2 e^- \rightarrow H_2 + H_2O$	5
$2Al + 8 H_2O \rightarrow 2 Al(OH)_4^- + 8 H^+ + 6 e^-$	$9 H^+ + 3 OH^- + 6 e^- \rightarrow 3 H_2 + 3 H_2O$	6
$2 Al + 5 H_2O + H^+ + 3 OH^- \rightarrow 2 Al(OH)_4^- + 3 H_2$ (combine H^+ with 1 OH^- on left side)		7 & 8
$2Al + 5 H_2O + (H_2O) + 2 OH^- \rightarrow 2 Al(OH)_4^- + 3 H_2$		8
$2Al (s) + 6 H_2O (\ell) + 2 OH^-(aq) \rightarrow 2 Al(OH)_4^-(aq) + 3 H_2(g)$		9

(b) CrO_4^{2-} (aq) + SO_3^{2-} (aq) → $Cr(OH)_3$ (s) + SO_4^{2-} (aq)

Oxidation half-equation	Reduction half-equation	Step
SO_3^{2-} → SO_4^{2-}	CrO_4^{2-} → $Cr(OH)_3$	1 & 2
H_2O + SO_3^{2-} → SO_4^{2-}	CrO_4^{2-} → $Cr(OH)_3$ + H_2O	3
H_2O + SO_3^{2-} → SO_4^{2-} + 2 H^+	5 H^+ + CrO_4^{2-} → $Cr(OH)_3$ + H_2O	4
H_2O + SO_3^{2-} → SO_4^{2-} + 2 H^+ +2 e^-	5 H^+ + CrO_4^{2-} + 3 e^- → $Cr(OH)_3$ + H_2O	5
3 H_2O +3 SO_3^{2-} → 3 SO_4^{2-} + 6 H^+ + 6 e^-	$10H^+$ + $2CrO_4^{2-}$ + $6e^-$ → $2Cr(OH)_3$ + $2H_2O$	6
$H_2O(\ell)$ + 3 SO_3^{2-} + 4 H^+ + 2 CrO_4^{2-} → 2 $Cr(OH)_3$ + 3 SO_4^{2-}		7
$H_2O(\ell)$ + 3 SO_3^{2-} + 4 H^+ + *4 OH*⁻ + 2 CrO_4^{2-} → 2 $Cr(OH)_3$ + 3 SO_4^{2-} + *4 OH*⁻		8
5 $H_2O(\ell)$ + 3 SO_3^{2-}(aq) + 2 CrO_4^{2-} (aq) → 2 $Cr(OH)_3$ (s) + 3 SO_4^{2-}(aq) + *4 OH*⁻(aq)		9

(c) Zn(s) + $Cu(OH)_2$ (s) → $Zn(OH)_4^{2-}$ (aq) + Cu (s)

Oxidation half-equation	Reduction half-equation	Step
Zn → $Zn(OH)_4^{2-}$	$Cu(OH)_2$ → Cu	1 & 2
4 H_2O + Zn → $Zn(OH)_4^{2-}$	$Cu(OH)_2$ → Cu + 2 H_2O	3
4 H_2O + Zn → $Zn(OH)_4^{2-}$ + 4 H^+	2 H^+ + $Cu(OH)_2$ → Cu + 2 H_2O	4
4 H_2O + Zn → $Zn(OH)_4^{2-}$ + 4 H^+ + 2 e^-	2 H^+ + $Cu(OH)_2$ + 2 e^- → Cu + 2 H_2O	5 & 6
2 H_2O (ℓ) + Zn (s) + $Cu(OH)_2$(s) → $Zn(OH)_4^{2-}$(aq) + 2 H^+ (aq) + Cu(s)		7
$2H_2O(\ell)$ + Zn(s) + $Cu(OH)_2$(s) +*2OH⁻(aq)* → $Zn(OH)_4^{2-}$(aq) + $2H^+$(aq) + *2OH⁻(aq)* + Cu(s)		8
Zn(s) + $Cu(OH)_2$(s) + *2 OH⁻(aq)* → $Zn(OH)_4^{2-}$(aq) + Cu(s)		9

(d) HS^- (aq) + ClO_3^- (aq) → S(s) + Cl^- (aq)

Oxidation half-equation	Reduction half-equation	Step
$HS^- → S$	$ClO_3^- → Cl^-$	1 & 2
	$ClO_3^- → Cl^- + 3 H_2O$	3
$HS^- → S + H^+$	$6 H^+ + ClO_3^- → Cl^- + 3 H_2O$	4
$HS^- → S + H^+ + 2 e^-$	$6 e^- + 6 H^+ + ClO_3^- → Cl^- + 3 H_2O$	5
$3 HS^- → 3 S + 3 H^+ + 6 e^-$		6
$3 HS^- + 3 H^+ + ClO_3^- → Cl^- + 3 H_2O + 3 S$		7
$3 HS^- + 3 H^+ + 3 OH^- + ClO_3^- → Cl^- + 3 H_2O + 3 S + 3 OH^-$		8
$3 HS^-(aq) + ClO_3^-(aq) → Cl^- (aq) + 3 S (s) + 3 OH^-(aq)$		9

Constructing Voltaic Cells

7. For the reaction : $2 Cr (s) + 3 Fe^{2+}$ (aq) → $2 Cr^{3+}$ (aq) + 3 Fe (s) :

Electrons in the external circuit flow from the <u>Cr</u> electrode to the <u>Fe</u> electrode. Negative ions move in the salt bridge from the <u>iron</u> half-cell to the <u>chromium</u> half-cell. The half-reaction at the anode is <u>$Cr(s) → Cr^{3+}(aq) + 3 e^-$</u> and that at the cathode is <u>$Fe^{2+}(aq) + 2 e^- → Fe(s)$</u>.

Note that the reaction shows that Cr is being oxidized (to Cr^{3+}) by Fe^{2+} which is being reduced (to Fe). The electrons leave the anode (and head to the cathode via the external circuit (wire). With reduction occurring in the cathode half-cell , a net deficit of positive ions accumulates, necessitating the assistance of + ions from the salt bridge. See SQ20.2 and 20.3 for additional questions of this type.

9. Like SQ20.7, we can complete the paragraph in part (c), by deciding on the spontaneous or product-favored reaction between the iron and oxygen half cells. The reduction potentials are:

$$O_2 (g) + 4 H^+ (aq) + 4e^- (aq) → 2 H_2O \qquad E° = 1.229V$$

$$Fe^{2+} (aq) + 2 e^- → Fe (s) \qquad E° = -0.44V$$

(a) Oxidation half-reaction: $2 Fe (s) → 2 Fe^{2+}$ (aq) + 4 e^-

Reduction half-reaction: $O_2 (g) + 4 H^+$ (aq) + 4e^- (aq) → 2 H_2O

Net cell reaction: $2 Fe (s) + O_2 (g) + 4 H^+$ (aq) → $2 H_2O (ℓ) + 2 Fe^{2+}(aq)$

How to decide which half-reaction occurs as oxidation? An examination of Table 20.1 (or Appendix M) shows O_2 as a stronger oxidizing agent than Fe^{2+}.

Alternatively, Fe is a stronger reducing agent than H_2O. Either of these conclusions points to the direction of reaction which is product-favored.

(b) The anode half-reaction is the oxidation half-reaction: $2 \text{ Fe (s)} \rightarrow 2 \text{ Fe}^{2+} \text{ (aq)} + 4 \text{ e}^-$

At the cathode, the reduction half-reaction: $O_2(g) + 4 \text{ H}^+ \text{ (aq)} + 4 \text{ e}^-\text{(aq)} \rightarrow 2 \text{ H}_2O$

(c) Electrons in the external circuit flow from the <u>Fe</u> electrode to the <u>O_2</u> electrode. Negative ions move in the salt bridge from the <u>oxygen</u> half-cell to the <u>iron</u> half-cell.

Commercial Cells

11. Similarities and differences between dry cells, alkaline batteries, and Ni-cad batteries:
The first two types of cells are non-rechargeable batteries—also called primary batteries. They also share the common anode, Zinc. Ni-cad batteries are rechargeable. Alkaline and Ni-cad batteries are in a basic environment, whereas dry cells are in an acidic environment.

Standard Electrochemical Potentials

13. Calculate E° for each of the following, and decided if it is product-favored as written:
$$E°_{cell} = E°_{cathode} - E°_{anode}$$

(a) $2 \text{ I}^- \text{ (aq)} + \text{Zn}^{2+}\text{(aq)} \rightarrow I_2 \text{ (g)} + \text{Zn (s)}$

Cathode reaction: $\text{Zn}^{2+}\text{(aq)} + 2 \text{ e}^- \rightarrow \text{Zn (s)}$	E° = -0.763 V
Anode reaction: $2 \text{ I}^- \text{ (aq)} \rightarrow I_2 \text{ (g)} + 2 \text{ e}^-$	<u>E° = +0.535 V</u>
Cell voltage:	E° = -1.298 V(not product-favored)

(b) $\text{Zn}^{2+} \text{ (aq)} + \text{Ni (s)} \rightarrow \text{Zn (s)} + \text{Ni}^{2+} \text{ (aq)}$

Cathode reaction: $\text{Zn}^{2+} \text{ (aq)} + 2 \text{ e}^- \rightarrow \text{Zn (s)}$	E° = -0.763 V
Anode reaction: $\text{Ni (s)} \rightarrow \text{Ni}^{2+} \text{ (aq)} + 2 \text{ e}^-$	<u>E° = -0.25 V</u>
Cell voltage:	E° = -0.51 V(not product-favored)

(c) $2 \text{ Cl}^- \text{ (aq)} + \text{Cu}^{2+} \text{ (aq)} \rightarrow \text{Cu(s)} + \text{Cl}_2 \text{ (g)}$

Cathode reaction: $\text{Cu}^{2+} \text{ (aq)} + 2 \text{ e}^- \rightarrow \text{Cu(s)}$	E° = +0.337 V
Anode reaction: $2 \text{ Cl}^- \text{ (aq)} \rightarrow \text{Cl}_2 \text{ (g)} + 2 \text{ e}^-$	<u>E° = +1.360 V</u>
Cell voltage:	E° = -1.023 V(not product-favored)

(d) $\text{Fe}^{2+} \text{ (aq)} + \text{Ag}^+ \text{ (aq)} \rightarrow \text{Fe}^{3+} \text{ (aq)} + \text{Ag (s)}$

Cathode reaction: $\text{Ag}^+ \text{ (aq)} + \text{ e}^- \rightarrow \text{Ag (s)}$	E° = +0.80 V
Anode reaction: $\text{Fe}^{2+}\text{(aq)} \rightarrow \text{Fe}^{3+} \text{ (aq)} + \text{ e}^-$	<u>E° = +0.771 V</u>
Cell voltage:	E° = 0.029 V (product-favored)

15. (a) Sn^{2+} (aq) + 2 Ag (s) → Sn (s) + 2 Ag^+ (aq)

   Cathode reaction: Sn^{2+} (aq) + 2e⁻ → Sn (s)   E° = -0.14 V

   Anode reaction: 2 Ag (s) → 2 Ag^+ (aq) + 2e⁻   <u>E° = +0.80 V</u>

   Cell voltage: E° = - 0.94 V (not product-favored)

(b) 2 Al (s) + 3 Sn^{4+} (aq) → 3 Sn^{2+} (aq) + 2 Al^{3+} (aq)

   Cathode reaction: 3 Sn^{4+} (aq) + 6e⁻ →3 Sn^{2+} (aq)   E° = +0.15 V

   Anode reaction: 2 Al(s) → 2 Al^{3+} (aq) + 6e⁻   <u>E° = - 1.66 V</u>

   Cell voltage: E° = 1.81 V (product-favored)

(c) ClO_3^- (aq) + 6 Ce^{3+} (aq) + 6 H^+(aq) → Cl^- (aq) + 6 Ce^{4+} (aq) + 3 H_2O(ℓ)

   Cathode reaction: ClO_3^- (aq) + 6 e⁻ +6 H^+(aq)   E° = +0.62 V

   → Cl^-(aq)+ 3 H_2O(ℓ)

   Anode reaction: 6 Ce^{3+} (aq) → 6 Ce^{4+} (aq)   <u>E° = +1.61 V</u>

   Cell voltage: E° = -0.99 V

(not product-favored)

(d) 3 Cu (s) + 2 NO_3^-(aq) + 8 H^+(aq) → 3 Cu^{2+}(aq) + 2 NO (g) + 4 H_2O(ℓ)

   Cathode reaction: 2 NO_3^-(aq) + 8 H^+(aq) + 6 e⁻   E° = +0.96 V

   → 2 NO (g) + 4 H_2O (ℓ)

   Anode reaction: 3 Cu (s) → 3 Cu^{2+}(aq) + 6e⁻   <u>E° = +0.337 V</u>

   . Cell voltage: E° = 0.62 V   (product-favored)

17. From the following half-reactions:

(a) The metal most easily oxidized:

   From the list **Al** is the most easily oxidized metal. Having the most negative reduction potential, Al is the strongest reducing agent of the group, and reducing agents are oxidized as they perform their task.

(b) Metals on the list capable of reducing Fe^{2+} to Fe

   Zn and **Al** both have more negative reduction potentials than Fe, hence are stronger reducing agents, and can reduce Fe^{2+} to Fe.

(c) A balanced equation for the reaction of Fe^{2+} with Sn. Is the reaction product-favored?

   Fe^{2+} (aq) + Sn (s) → Fe (s) + Sn^{2+} (aq); Since Fe is a stronger reducing agent than Sn, this reaction would not have an E° > 0, and the reaction would be reactant-favored.

(d) A balanced equation for the reaction of Zn^{2+} with Sn. Is the reaction product-favored?

Zn^{2+} (aq) + Sn (s) → Zn (s) + Sn^{2+} (aq);Zn is a stronger reducing agent than Sn, this reaction would not have an E° > 0, and the reaction would be reactant-favored.

Ranking Oxidizing and Reducing Agents

19. Best element from the group that is the best reducing agent?

Zn has the most negative standard reduction potential of the group. Recall that the more negative the reduction potential, the stronger a substance is as a reducing agent.

21. Ion from the group that is most easily reduced?

The specie with the most positive reduction potential is the strongest oxidizing agent of the list, and with that role, becomes the most easily reduced. **Ag^+** fits that role from this list.

23. Regarding the halogens:

(a)The halogen most easily reduced is the one with the most positive reduction potential **F_2** has the most negative of the halogens.

(b)MnO_2 has a reduction potential of 1.23V. Both F_2 and Cl_2 have more positive reduction potentials than MnO_2, and are better oxidizing agents than MnO_2.

Electrochemical Cells Under Nonstandard Conditions

25. The Voltage of a cell that has dissolved species at 0.025 M:

Calculate the standard voltage of the cell:

(1) 2 $H_2O(\ell)$ + 2e⁻ → H_2 (g) + 2 OH⁻ (aq)   E° = -0.8277 V

(2) $[Zn(OH)_4]^{2-}$ (aq) + 2 e⁻ → Zn(s) + 4 OH⁻ (aq)   E° = -1.22 V

The net equation is given as: Zn(s) + 2 H_2O (ℓ) + 2 OH⁻ (aq) →Zn(OH)$_4^-$ (aq) + H_2 (g)

The equilibrium expression (Q) would be: $\dfrac{\left[Zn(OH)_4^-\right]P_{H2}}{\left[OH\text{-}\right]^2}$

We'll assume a hydrogen pressure of 1 atm, and we know the concentrations of the other terms: 0.025 M. Note that n (in the Nernst equation) corresponds to 2, since the balanced overall equation indicates that 2 moles of electrons are lost and 2 moles of electrons are gained. The cell equation indicates Zn as a reactant, meaning that reaction (2) runs in the reverse direction, making that process an oxidation, so the E°anode = -1.22 V.

Calculate E° $_{cell}$ = E°$_{cathode}$ − E°$_{anode}$ = -0.8277 − (-1.22) = 0.3923 V

The Nernst equation $\qquad E_{cell} = E°_{cell} - \dfrac{0.0257}{n} \ln \dfrac{\left[Zn(OH)_4^-\right]P_{H2}}{\left[OH-\right]^2}$

$E_{cell} = 0.3923 - \dfrac{0.0257}{2} \ln \dfrac{[0.025] \cdot 1}{[0.025]^2} = 0.3923 - (0.0257 \cdot 3.69/2) = 0.345$ V

27. The voltage of a cell that has Ag in a 0.25 M solution of Ag^+ and Zn electrode in 0.010 M Zn^{2+}: Calculate the standard voltage of the cell:

(1) Zn^{2+} (aq) + 2e⁻ → Zn (s) $\qquad\qquad\qquad\qquad$ E° = -0.763 V

(2) Ag^+ (aq) + e⁻ → Ag(s) $\qquad\qquad\qquad\qquad\quad$ E° = +0.80 V

The net equation is given as: 2 Ag^+(aq) + Zn (s) → Zn^{2+} (aq) + 2 Ag(s).

The equilibrium expression (Q) would be: $\dfrac{\left[Zn^{2+}\right]}{\left[Ag^+\right]^2}$

The cell will run in the direction that is product favored, so we can calculate

$\quad$ E° cell = E°cathode – E°anode = +0.80 – (-0.763) = 1.563 V

We also note from the balanced equation that 2 moles(n) of electrons are transferred.

Using the Nernst equation $\quad E_{cell} = E°_{cell} - \dfrac{0.0257}{n} \ln \dfrac{\left[Zn^{2+}\right]}{\left[Ag^+\right]^2}$

$E_{cell} = 1.563 - \dfrac{0.0257}{2} \ln \dfrac{[0.010]}{[0.25]^2} = 1.563 - (0.0257/2) \cdot (-1.83) = 1.563 + 0.0235 = 1.58$ V

29. The voltage of a cell that has Ag in a ? M solution of Ag^+ and Zn electrode in 1.0 M Zn^{2+}: Calculate the standard voltage of the cell:

(1) Zn^{2+} (aq) + 2e⁻ → Zn (s) $\qquad\qquad\qquad\qquad$ E° = -0.763 V

(2) Ag^+ (aq) + e⁻ → Ag(s) $\qquad\qquad\qquad\qquad\quad$ E° = +0.80 V

The net equation is given as: 2Ag^+(aq) + Zn (s) → Zn^{2+} (aq) + 2 Ag(s).

The equilibrium expression (Q) would be: $\dfrac{\left[Zn^{2+}\right]}{\left[Ag^+\right]^2}$

The cell will run in the direction that is product favored, so we can calculate

E° cell = E°cathode – E°anode = +0.80 – (-0.763) = 1.563 V

We also note from the balanced equation that 2 moles(n) of electrons are transferred.

Using the Nernst equation $E_{cell} = E°_{cell} - \dfrac{0.0257}{n} \ln \dfrac{[Zn^{2+}]}{[Ag^+]^2}$

Given the Ecell = 1.48V , we should be able to calculate the value of the "ln term".

Recall that we **know** the concentration of $[Zn^{2+}]$, but **don't know** the $[Ag^+]$.

$1.48 = 1.563 - \dfrac{0.0257}{2} \ln \dfrac{[Zn^{2+}]}{[Ag^+]^2}$ or $1.48 = 1.563 - (0.0257/2) \bullet \ln \dfrac{[Zn^{2+}]}{[Ag^+]^2}$

$\dfrac{-(1.48 - 1.563)2}{0.0257} = \ln \dfrac{[Zn^{2+}]}{[Ag^+]^2}$. or $6.459 = \ln \dfrac{[Zn^{2+}]}{[Ag^+]^2}$; and $\dfrac{[Zn^{2+}]}{[Ag^+]^2} = 638.5$

So $\dfrac{[1.0]}{638.5} = [Ag^+]^2$ and $[Ag^+] = 0.040$ M

Electrochemistry, Thermodynamics, and Equilibrium

31. Calculate $\Delta G°$ and K for the reactions:

(a) $2 Fe^{3+}(aq) + 2 I^- (aq) \rightarrow 2 Fe^{2+} (aq) + 2 I_2 (aq)$

Using the potentials: $Fe^{3+}(aq) + e^- \rightarrow 2 Fe^{2+}$ $E° = 0.771V$ and

$2I^- (aq) \rightarrow 2 I_2 (s) + 2 e^-$ $E° = 0.535V$, we calculate an $E°_{cell}$.

$E°_{cell} = (0.771V - 0.535V) = 0.236V$

The relationship between $\Delta G°$ and $E°$ is: $\Delta G° = -nFE°$ so.

$\Delta G° = -(2 \text{ mol e})(96{,}500 \text{ C/mol e})(0.236 \text{ V})$ and $1V = 1J/C$ so

$\Delta G° = -(2 \text{ mol e})(96{,}500 \text{ C/mol e})(0.236 \text{ J/C}) = -45548$ J or -45.5 kJ

$\Delta G° = - RT\ln K$ so $- \Delta G°/RT = \ln K$

$\dfrac{45548J}{8.314 \dfrac{J}{K \bullet mol} \bullet 298K} = \ln K$ so $18.38 = \ln K$ and $K = 9 \times 10^7$

(b) $I_2 (aq) + 2Br^- (aq) \rightarrow 2 I^- (aq) + Br_2 (\ell)$

Using the potentials: $Br_2 (\ell) + 2 e^- \rightarrow 2Br^- (aq)$ $E° = 1.08$ V   and

$2 I_2 (s) + 2 e^- \rightarrow 2I^- (aq)$ $E° = 0.535V$

We calculate an $E°_{cell}$, noting that in the cell reaction given, molecular iodine is being

reduced (the cathode) and bromide ion is being oxidized (the anode).

$E°_{cell} = (0.535V - 1.08V) = -0.545V$

The relationship between $\Delta G°$ and $E°$ is: $\Delta G° = -nFE°$ so.

$\Delta G° = -(2 \text{ mol e})(96,500 \text{ C/mol e})(-0.545V)$ and $1V = 1J/C$ so

$\Delta G° = -(2 \text{ mol e})(96,500 \text{ C/mol e})(-0.545J/C) = +105,185 \text{ J}$ or $+110 \text{ kJ}$ (to 2sf)

$\Delta G° = -RT\ln K$ so $-\Delta G°/RT = \ln K$

$$\frac{-110,000J}{8.314 \frac{J}{K \bullet mol} \bullet 298.15 \text{ K}} = \ln K \text{ so } -42.45 = \ln K \text{ and } K = 4 \times 10^{-19}$$

33. Calculate K_{sp} for AgBr using the following reactions:

(1) $AgBr (s) + 1 e^- \rightarrow Ag(s) + Br^- (aq)$ $E° = 0.0713 \text{ V}$

(2) $Ag^+ (aq) + 1e^- \rightarrow Ag(s)$ $E° = 0.7994 \text{ V}$

Write the Ksp expression for AgBr; $AgBr(s) \Leftrightarrow Ag^+ (aq) + Br^- (aq)$

Note that we can accomplish this as an overall reaction, by reversing equation (2) and adding that to equation (1). Equation (1) is presently written as a reduction (naturally) and the *reverse* of Equation (2) would be an oxidation—so the roles for our "cell" are defined.

$E°cell = E°cathode = E°anode = 0.0713 \text{ V} - 0.7994 \text{ V} = -0.7281V$

Using our $\Delta G°$ relationships: $\Delta G° = -nFE° = -RT\ln K$ so $\ln K = \dfrac{nFE°}{RT}$

$$\text{So } \ln K = \frac{1 \text{ mol e} \bullet 96,500 \dfrac{C}{mol \, e} \bullet -0.7281\dfrac{J°}{C}}{8.314 \dfrac{J}{K \bullet mol} \bullet 298.15 \text{ K}} = -28.36 \text{ and } K = 4.9 \times 10^{-13}$$

35. Calculate the $K_{formation}$ for $AuCl_4^- (aq)$

(1) $AuCl_4^- (aq) + 3e^- \rightarrow Au(s) + 4 Cl^- (aq)$ $E° = 1.00 \text{ V}$

(2) $Au^{3+} (aq) + 3e^- \rightarrow Au (s)$ $E° = 1.50 \text{ V}$

The formation reaction for the complex is: $Au^{3+} (aq) + 4 Cl^- (aq) \Leftrightarrow AuCl_4^- (aq)$

To achieve this reaction as a net reaction, we need to reverse equation (1) and add it to equation (2).

The $E°cell$ for that process would be: $E°cathode - E°anode = 1.50 - 1.00 = 0.50 \text{ V}$

Using our $\Delta G°$ relationships: $\Delta G° = -nFE° = -RT\ln K$ so $\ln K = \dfrac{nFE°}{RT}$

$$\text{So } \ln K = \frac{3 \text{ mol e} \cdot 96{,}500 \, \dfrac{C}{\text{mol e}} \cdot 0.50 \, \dfrac{J}{C}}{8.314 \, \dfrac{J}{K \cdot \text{mol}} \cdot 298.15 \text{ K}} = 58.42 \quad \text{and } K = 2 \times 10^{25}$$

37. For the disproportionation of Iron(II):

   (a) The half-reactions that make up the disproportionation reaction:

   $$Fe^{2+}(aq) + 2 \, e^- \rightarrow Fe \, (s) \qquad\qquad E° = -0.44 \text{ V} \quad \text{(cathode)}$$

   $$\underline{2 \, Fe^{2+}(aq) \rightarrow 2 \, Fe^{3+}(aq) + 2 \, e^- \qquad\qquad E° = 0.771 \text{ V (anode)}}$$

   $$3 \, Fe^{2+}(aq) \rightarrow 2 \, Fe^{3+}(aq) + Fe \, (s) \quad E° = -0.44 \text{ V} - 0.771 \text{ V} = -1.21 \text{V}$$

   (b) Given the voltage calculated, the disproportionation reaction is not product-favored.

   $$\text{(c) } \ln K = \frac{2 \text{ mol e} \cdot 96{,}500 \, \dfrac{C}{\text{mol e}} \cdot -1.211 \, \dfrac{J}{C}}{8.314 \, \dfrac{J}{K \cdot \text{mol}} \cdot 298.15 \text{ K}} = -94.34 \quad \text{and } K_{formation} = 1 \times 10^{-41}$$

Electrolysis

39. Diagram of an electrolysis apparatus for molten NaCl:

Downs Cell Schematic

$$2 \, Na^+ + e^- \longrightarrow Na \, (l) \quad \text{(cathode, reduction)}$$

$$2 \, Cl^- \longrightarrow Cl_2 + 2e^- \text{ (anode, oxidation)}$$

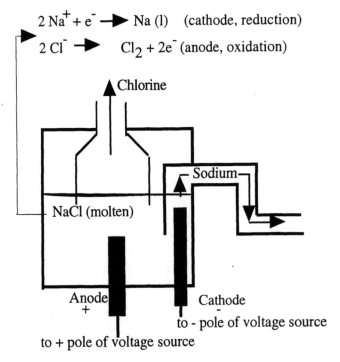

41. For the electrolysis of a solution of KF(aq), what product is expected at the anode: O_2 or F_2

Example 20.11 provides additional help with this concept. The **bottom line** is that the process that occurs in an electrolysis is the one requiring the smaller applied potential.

For an electrolysis: $E°cell = E°cathode - E°anode$. For aqueous KF, those voltages are:

$2 F^- (aq) \rightarrow F_2 (g) + 2e^-$ $E° = +2.87$ V

$2 H_2O (\ell) \rightarrow O_2 (g) + 4 H^+ (aq) + 4 e^-$ $E° = +1.23$ V

At the cathode: $2 H_2O(\ell) + 2 e^- \rightarrow H_2(g) + 2 OH^-(aq)$ $E° = -0.83$ V

[K has such a large negative reduction potential, that it *will not be reduced*.]

So the two choices are: $E°cell = E°cathode - E°anode$

For fluorine oxidation: $E°cell = (-0.83 - +2.87) = -3.7$ V

For oxygen oxidation: $E°cell = (-0.83 - +1.23) = -2.06$ V

Oxygen oxidation will require the lower applied potential, and will be produced at the anode.

43. For the electrolysis of KBr (aq):

(a) The reaction occurring at the cathode: $2 H_2O(\ell) + 2 e^- \rightarrow H_2(g) + 2 OH^-(aq)$

(b) The reaction occurring at the anode: $2 Br^- (aq) \rightarrow Br_2 (\ell) + 2e^-$

Counting Electrons

45. Solutions to problems of this sort are best solved by beginning with a factor containing the desired units. Connecting this factor to data provided usually gives a direct path to the answer.

units desired

$\downarrow$

$$\frac{58.69 \text{ g Ni}}{1 \text{ mol Ni}} \cdot \frac{1 \text{ mol Ni}}{2 \text{ mol e}^-} \cdot \frac{1 \text{ mol e}^-}{9.65 \times 10^4 \text{ C}} \cdot \frac{1 \text{ C}}{1 \text{ amp} \cdot \text{s}}$$

$$\cdot \frac{0.150 \text{amps}}{1} \cdot \frac{60\text{s}}{1 \text{ min}} \cdot \frac{12.2 \text{ min}}{1} = 0.0334 \text{ g Ni}$$

The second factor ($\frac{1 \text{ mol Ni}}{2 \text{ mol e}^-}$) is arrived at by looking at the reduction half-reaction:
$Ni^{2+} (aq) + 2 e^- \rightarrow Ni(s)$

All other factors are either data or common unity factors (e.g. $\frac{60\text{s}}{1 \text{ min}}$).

47. Follow the pattern from SQ20.45, in starting with a unit that has the units of the desired answer—in this case units of *time*.

$$\frac{1\ amp \cdot s}{1\ C} \cdot \frac{1}{0.66\ amp} \cdot \frac{96500\ C}{1\ mol\ e} \cdot \frac{2\ mol\ e}{1\ mol\ Cu} \cdot \frac{1\ mol\ Cu}{63.546\ g\ Cu} \cdot \frac{0.50\ g\ Cu}{1} = 2300\ s.$$

49. Once again, since the requested answer has units of *hours*, start with a unit that has hours in it.

$$\frac{1\ hr}{3600\ s} \cdot \frac{1\ amp \cdot s}{1\ C} \cdot \frac{1}{1.0\ amp} \cdot \frac{96500\ C}{1\ mol\ e} \cdot \frac{3\ mol\ e}{1\ mol\ Al} \cdot \frac{1\ mol\ Al}{26.98\ g\ Al} \cdot \frac{84\ g\ Al}{1} = 250\ hr$$

General Questions

51. Balanced equations for the following half-reactions:

(a) $UO_2^+(aq) \rightarrow U^{4+}(aq)$

Reduction half-equation	
$UO_2^+ (aq) \rightarrow U^{4+}(aq)$	Balance all non H,O
$UO_2^+ (aq) \rightarrow U^{4+}(aq) + 2H_2O$	Balance O with H_2O
$4\ H^+(aq) + UO_2^+(aq) \rightarrow U^{4+}(aq) + 2H_2O\ (\ell)$	Balance H with H^+
$4\ H^+(aq) + UO_2^+(aq) + 1\ e^- \rightarrow U^{4+}(aq) + 2\ H_2O\ (\ell)$	Balance charge with e^-

(b) $ClO_3^- (aq) \rightarrow Cl^-(aq)$

Reduction half-equation	
$ClO_3^- (aq) \rightarrow Cl^-(aq)$	Balance all non H,O
$ClO_3^- (aq) \rightarrow Cl^-(aq) + 3H_2O(\ell)$	Balance O with H_2O
$6\ H^+(aq) + ClO_3^- (aq) \rightarrow Cl^-(aq) + 3H_2O(\ell)$	Balance H with H^+
$6\ H^+(aq) + ClO_3^- (aq) + 6\ e^- \rightarrow Cl^-(aq)$ $+3H_2O(\ell)$	Balance charge with e^-

(c) $N_2H_4 (aq) \rightarrow N_2 (g)$

Oxidation half-equation	
$N_2H_4 (aq) \rightarrow N_2 (g)$	Balance all non H,O
$N_2H_4 (aq) \rightarrow N_2 (g) + 4\ H_2O(\ell)$	Balance H with H_2O
$4\ OH^-(aq) + N_2H_4 (aq) \rightarrow N_2 (g) + 4\ H_2O(\ell)$	Balance O with OH^-
$4\ OH^-(aq) + N_2H_4 (aq) \rightarrow N_2 (g) + 4\ H_2O(\ell)$ $+ 4e^-$	Balance charge with e^-

(d) $OCl^-(aq) \rightarrow Cl^-(aq)$

Reduction half-equation	
$OCl^-(aq) \rightarrow Cl^-(aq)$	Balance all non H,O
$OCl^-(aq) \rightarrow Cl^-(aq) + OH^- (aq)$	Balance O with OH^-
$H_2O(\ell) + OCl^-(aq) \rightarrow Cl^-(aq) + OH^- (aq) + OH^- (aq)$	Balance H with H_2O
	rebalance H with OH^-
$H_2O(\ell) + OCl^-(aq) + 2 e^- \rightarrow Cl^-(aq) + 2 OH^-(aq)$	Balance charge with e^-

53. For the electrochemical cell involving Mg and Ag:

(a) Parts of the cell:

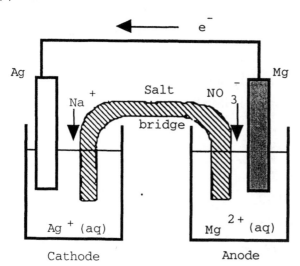

(b) Anode (oxidation):
$$Mg \rightarrow Mg^{2+} + 2e^-$$

Cathode (reduction):
$$2 Ag^+ + 2 e^- \rightarrow 2 Ag$$

Adding gives the net reaction:
$$2 Ag^+ + Mg \rightarrow Mg^{2+} + 2 Ag$$

(c) The flow of electrons in the outer circuit, as described on the diagram above is from the Mg electrode to the Ag electrode. The ion flow in the salt bridge is also shown on the diagram above. The salt bridge is necessary to negate the charge differential that would grow as the cell operates. For example, in the Ag half-cell, the compartment that originally contained equal amounts of + and – charges (from Ag^+ and NO_3^- ions respectively) would accrue a net "–" charge, as the "+" silver ions are reduced. The Na^+ ions from the salt bridge would flow into the Ag compartment, neutralizing that growing negative charge.

55. (a) Half-cells that might be used in conjunction wih the Ag+/Ag half-cell to produce a cell with voltage close to 1.7V. Consider cells in which the silver half-cell could function either as cathode or anode. $E° Ag^+ = 0.7994V$

Recall that $E° \text{ cell} = E°\text{cathode} – E° \text{anode}$.

With $E°$ cell desired to be 1.7V, we can substitute the Ag/Ag^+ half cell first as cathode, then as anode.

$E°$ cell = $E°$cathode – $E°$ anode   1.7V = 0.7994V – $E°$ anode, and $E°$anode = 0.7994-1.7 = -0.90 V (so the Chromium half-cell , with $E°$ = -0.91 V would be an appropriate half-cell.)

And $E°$ cell = $E°$cathode – $E°$ anode   1.7V = $E°$cathode – 0.7994V and $E°$cathode = 0.7994 +1.7 = 2.5V so fluorine ($E°$ = 2.87V is a reasonable suggestion as the other half-cell.

(b) Half-cells that might be used in conjunction wih the Ag+/Ag half-cell to produce a cell with voltage close to 0.5V. Consider cells in which the silver half-cell could function either as cathode or anode. $E°$ Ag$^+$ = 0.7994V. As in (a) above, we can substitute the Ag/Ag$^+$ half cell first as cathode, then as anode—for a $E°$ cell = 0.5V

$E°$ cell = $E°$cathode – $E°$ anode = 0.5V = 0.7994V – $E°$ anode, and $E°$anode = 0.7994 - 0.5 = -0.299 V (so the Cu couple with $E°$ = 0.337V would be a possible half-cell.)

Substituting with the Ag couple as anode:

$E°$ cell = 0.5V = $E°$ cathode - 0.7994V , and $E°$cathode = 0.5 + 0.7994 = 1.30 V (so the Cl couple with $E°$ = 1.36V would be a possible half-cell.)

57. Examine the reduction potentials for: $\underline{E°}$

$$Au^+ (aq) + 1\ e^- \rightarrow Au\ (s) \qquad +1.68$$
$$Ag^+ (aq) + 1\ e^- \rightarrow Ag\ (s) \qquad +0.80$$
$$Cu^{2+} (aq) + 2\ e^- \rightarrow Cu\ (s) \qquad +0.337$$
$$Sn^{2+} (aq) + 2e^- \rightarrow Sn\ (s) \qquad -0.14$$
$$Co^{2+} (aq) + 2\ e^- \rightarrow Co\ (s) \qquad -0.28$$
$$Zn^{2+} (aq) + 2\ e^- \rightarrow Zn\ (s) \qquad -0.763$$

To clarify the trends, I have listed them in the descending reduction potential typical of Reduction Potential Charts, as in Table 20.1.

(a) Metal ion that is the weakest oxidizing agent? — Zn^{2+} Strength of oxidizing agents increase with the *increasing* $+ E°$.

(b) Metal ion that is the strongest oxidizing agent? –Au^+

(c) Metal that is the strongest reducing agent? –Zn (s) (since it's oxidized partner is the weakest oxidizing agent, the metal is the strongest reducing agent.

(d) Metal that is the weakest reducing agent?—Au (s)

(e) Will Sn(s) reduce Cu^{2+} to Cu (s)? – Yes. Use the "NW to SE rule to see that this will give a positive $E°$cell.

(f) Will Ag (s) reduce Co^{2+}(aq) to Co (s)? No. See part (e)

348

(g) Which metal ions from the list can be reduced by Sn(s). Using the logic as in part (e) Au^+, Ag^+, and Cu^{2+} can be reduced by Sn.

(h) Metals that can be oxidized by Ag^+(aq) –Any metal "below" Ag: Cu, Sn, Co, Zn

59. Examine the following reductions $E°$

$$Cu^{2+} (aq) + 2\ e^- \rightarrow Cu\ (s) \qquad +0.337$$
$$Fe^{2+} (aq) + 2\ e^- \rightarrow Fe\ (s) \qquad -0.44$$
$$Cr^{3+} (aq) + 3\ e^- \rightarrow Cr\ (s) \qquad -0.74$$
$$Mg^{2+} (aq) + 2\ e^- \rightarrow Mg\ (s) \qquad -2.37$$

(a) In which of the voltaic cells would the S.H.E. be the cathode?

$E°_{cell}$ must be positive, and the E° S.H.E.= 0.00V, so for $E°_{cell}$ = 0.00V – ($E°_{anode}$) to be positive, the $E°_{anode}$ would have to be *negative*, so the iron, chromium, and magnesium half-cells would fit this description.

(b) Voltaic cell with the highest and lowest voltages?

For $E°_{cell}$ = 0.00V – ($E°_{anode}$) to have the largest value the $E°_{anode}$ would have to be the most negative (Magnesium)

For $E°_{cell}$ = 0.00V – ($E°_{anode}$) to have the smallest value the $E°_{anode}$ would have to be the least negative (Iron). While Cu^{+2} is a tempting choice, note that the Cu couple, when paired with the S.H.E. would give a negative voltage!

61. Mass of Al metal produced from electrolysis of Al^{3+} salt with 5.0V and 1.0×10^5A in 24 hr.

$$\frac{26.98\ g\ Al}{1\ mol\ Al} \cdot \frac{1 mol\ Al}{3\ mol\ e^-} \cdot \frac{1\ mol\ e^-}{9.65\ x\ 10^4\ C} \cdot \frac{1\ C}{1\ amp \cdot s} \cdot \frac{1.0\ x\ 10^5\ amp}{1} \cdot \frac{3600\ s}{1\ hr} \cdot \frac{24\ hr}{1}$$

$$= 8.1\ x\ 10^5\ g\ Al$$

63. (a) $E°_{cell}$ = - 0.142V = E°cathode – E° anode, so - 0.142V = -0.126 – E° anode

[Note that the value, -0.126V, is arrived at by examining the Reduction Potential Table]

Rearranging to solve: E° anode = -0.126 + 0.142 = 0.016V

(b) Given these data, estimate Ksp for $PbCl_2$.

The expression for this salt is: Ksp =$[Pb^{2+}][Cl^-]^2$

From equation 20.3, we can calculate the equilibrium constant (Q):

$$Ecell\ =\ E°cell\ -\frac{0.0257}{n}lnQ$$

If we are at equilibrium, Ecell = 0, and Q = Ksp.

$$0.0 = -0.142 - \frac{0.0257}{2} \ln Ksp \quad \text{and rearranging:} \quad \frac{-(0.0 + 0.142) \cdot 2}{0.0257} = \ln Ksp$$

$-11.1 = \ln Ksp$ and $Ksp = 2 \times 10^{-5}$

65. For a cell with $E° = +2.91$, the $\Delta G°$ is:

$$\Delta G° = -nFE° = -2 \text{ mole} \cdot 9.65 \times 10^4 \frac{J}{\text{volt} \cdot \text{mol}} \cdot 2.91 \text{ volt} \cdot \frac{1.000 \text{ kJ}}{1000 \text{ J}} = -562 \text{ kJ}$$

67. Mass of Cl_2 and Na produced by 7.0V with a current of 4.0×10^4A, flowing for 1 day?

$$\frac{70.91 \text{ g } Cl_2}{1 \text{ mol } Cl_2} \cdot \frac{1 \text{ mol } Cl_2}{2 \text{ mol } e^-} \cdot \frac{1 \text{mole}^-}{9.65 \times 10^4 C} \cdot$$

$$\frac{1C}{1 \text{amps}} \cdot \frac{4.0 \times 10^4 \text{ amp}}{1} \cdot \frac{3600s}{1hr} \cdot \frac{24 \text{ hr}}{1 \text{ day}} = 1.3 \times 10^6 \text{ g } Cl_2$$

and for sodium:

$$\frac{23.0 \text{gNa}}{1 \text{molNa}} \cdot \frac{1 \text{molNa}}{1 \text{mole}^-} \cdot \frac{1 \text{ mol } e^-}{9.65 \times 10^4 C} \cdot$$

$$\frac{1 C}{1 \text{ amp} \cdot s} \cdot \frac{4.0 \times 10^4 \text{ amp}}{1} \cdot \frac{3600 s}{1 \text{ hr}} \cdot \frac{24 \text{ hr}}{1 \text{ day}} = 8.2 \times 10^5 \text{ g Na}$$

The energy consumed:

$$\frac{1 \text{ kwh}}{3.6 \times 10^6 J} \cdot \frac{1 J}{1 V \cdot C} \cdot \frac{7.0 V}{1} \cdot \frac{1C}{1 \text{amp s}} \cdot \frac{4.0 \times 10^4 \text{ amp}}{1} \cdot \frac{3600s}{1hr} \cdot \frac{24hr}{1 \text{ day}} = 6720 \text{ kwh}$$

(6700 khw to 2 sf)

69. To calculate the charge on the Ru^{n+} ion, we need to know two things:

1. How many moles of elemental ruthenium are reduced?

2. How many moles of electrons caused that reduction?

Moles of ruthenium: $\quad \dfrac{0.345 \text{ gRu}}{1} \cdot \dfrac{1 \text{molRu}}{101.07 \text{gRu}} = 3.4 \times 10^{-3} \text{ mol Ru}$

Moles of electrons:

$$0.44 \text{ amp} \cdot \frac{1 C}{1 \text{ amp} \cdot s} \cdot \frac{3600 s}{1 \text{ hr}} \cdot \frac{(25/60) \text{ hr}}{1} \cdot \frac{1 \text{ mol } e^-}{9.65 \times 10^4 C} = 6.8 \times 10^{-3} \text{ mol } e^-$$

Recall that our general reduction reactions are written: $M^{+x} + x\, e^- \rightarrow M$

If we know the number of $\dfrac{\text{moles of electrons}}{\text{mol of metal}}$, we know the charge on the cation, hence for

the Rh^{x+} ion we have $\dfrac{6.8 \times 10^{-3} \text{mole}^-}{3.4 \times 10^{-3} \text{molRu}} = 2.0 \dfrac{\text{mole}^-}{\text{molRu}}$.

The ion is therefore the Ru^{2+} ion! The formula for the nitrate salt would be $Ru(NO_3)_2$.

71. Mass of Cl_2 produced by electrolysis with 3.0×10^5 amperes at 4.6V in a 24-hr day.

$$\frac{70.91 \text{ g } Cl_2}{1 \text{ mol } Cl_2} \cdot \frac{1 \text{ mol } Cl_2}{2 \text{ mol } e^-} \cdot \frac{1 \text{ mol } e^-}{96500 \text{ C}} \cdot \frac{1 \text{ C}}{1 \text{ amp} \bullet s} \cdot \frac{3.0 \times 10^5 \text{ amp}}{1} \cdot \frac{3600 \text{ s}}{1 \text{ hr}} \cdot \frac{24 \text{ hr}}{1} =$$

$$9.5 \times 10^6 \text{ g } Cl_2$$

73. What amount of Au will be deposited by the current that deposits 0.089g Ag in 10 minutes.

$$\frac{0.089 \text{ g Ag}}{1} \cdot \frac{1 \text{ mol Ag}}{107.9 \text{ g Ag}} \cdot \frac{1 \text{ mol } e^-}{1 \text{ mol Ag}} = 8.2 \times 10^{-4} \text{ mol } e^-$$

$$\frac{8.2 \times 10^{-4} \text{ mol } e^-}{1} \cdot \frac{1 \text{ mol Au}}{3 \text{ mol } e^-} \cdot \frac{197.0 \text{ g Au}}{1 \text{ mol Au}} = 0.054 \text{ g Au}$$

75. Balance the equations:

(a) Separating the equation into an oxidation and reduction half-reaction we get:

$Ag^+ + 1e^- \rightarrow Ag$ (reduction)

$C_6H_5CHO \rightarrow C_6H_5CO_2H$ (oxidation)

Using the method outlined earlier in this chapter, we balance the oxidation half-equation:
(Numbers in parentheses correspond to the steps listed in SQ20.3)

$C_6H_5CHO + H_2O \rightarrow C_6H_5CO_2H$ (3)

$C_6H_5CHO + H_2O \rightarrow C_6H_5CO_2H + 2 H^+$ (4)

$C_6H_5CHO + H_2O \rightarrow C_6H_5CO_2H + 2 H^+ + 2 e^-$ (5)

Note that the reduction half equation gains 1 electron/ silver ion. To balance electron gain with electron loss, we multiply the reduction equation by 2.

$2Ag^+ + 2e^- \rightarrow 2Ag$
$C_6H_5CHO + H_2O \rightarrow C_6H_5CO_2H + 2 H^+ + 2e^-$
$\overline{2Ag^+ + C_6H_5CHO + H_2O \rightarrow C_6H_5CO_2H + 2 H^+ + 2Ag}$

(b) Separating the equation into an oxidation and reduction half-equation we get:

$Cr_2O_7^{2-} \rightarrow Cr^{3+}$ (reduction)

$C_2H_5OH \rightarrow CH_3CO_2H$ (oxidation)

Performing the "steps" on each half-equation

$Cr_2O_7^{2-} \rightarrow Cr^{3+}$ $C_2H_5OH \rightarrow CH_3CO_2H$

$Cr_2O_7^{2-} \rightarrow 2 Cr^{3+}$ $C_2H_5OH \rightarrow CH_3CO_2H$ (2)

$Cr_2O_7^{2-} \rightarrow 2 Cr^{3+} + 7H_2O$ $C_2H_5OH + H_2O \rightarrow CH_3CO_2H$ (3)

$14H^+ + Cr_2O_7^{2-} \rightarrow 2 Cr^{3+} + 7H_2O$ $C_2H_5OH + H_2O \rightarrow CH_3CO_2H + 4H^+$ (4)

$14H^+ + Cr_2O_7^{2-} + 6e^- \rightarrow 2 Cr^{3+} + 7H_2O$

$C_2H_5OH + H_2O \rightarrow CH_3CO_2H + 4H^+ + 4e^-$ (5)

Multiplying the reduction half-equation by 2, and the oxidation half-equation by 3 will equalize electron gain with electron loss.

$28H^+ + 2\ Cr_2O_7^{2-} + 12e^- \rightarrow 4\ Cr^{3+} + 14H_2O$

$3C_2H_5OH + 3H_2O \rightarrow 3CH_3CO_2H + 12H^+ + 12e^-$

Adding the two equations, and removing any duplications:

$28H^+ + 2\ Cr_2O_7^{2-} + \cancel{12e^-} \rightarrow 4\ Cr^{3+} + 14H_2O$

$\underline{3C_2H_5OH + 3H_2O \rightarrow 3CH_3CO_2H + 12H^+ + \cancel{12e^-}}$

$16H^+ + 2\ Cr_2O_7^{2-} + 3C_2H_5OH \rightarrow 3CH_3CO_2H + 11H_2O + 4\ Cr^{3+}$

77. Comparing the silver/zinc battery with the lead storage battery:

(a) The stoichiometry of the silver/zinc battery indicates a reaction of one mole each of silver oxide, zinc, and water. The mass of one mole of each of these three substances is:

$$
\begin{array}{ll}
1\ mol\ Ag_2O & 231.7\ g \\
1\ mol\ Zn & 65.4\ g \\
1\ mol\ H_2O & \underline{18.0\ g} \\
& 315.1\ g
\end{array}
$$

The energy associated with the battery is:

$$\frac{1.59\ V}{1} \cdot \frac{96500\ C}{1\ mol\ e} \cdot \frac{1\ J}{1\ V \cdot C} \cdot \frac{2\ mol\ e}{1\ mol\ reactant} \cdot \frac{1\ mol\ reactant}{315.1\ g} = 973.88\ J/g\ or\ 0.974\ kJ/g$$

(b) Performing the same calculations for the lead storage battery, using a stoichiometric amount for the overall battery reaction:

$$
\begin{array}{ll}
1\ mol\ Pb & 207.2\ g \\
1\ mol\ PbO_2 & 239.2\ g \\
2\ mol\ H_2SO_4 & \underline{196.2\ g} \\
& 642.6\ g
\end{array}
$$

$$\frac{2.0\ V}{1} \cdot \frac{96500\ C}{1\ mol\ e} \cdot \frac{1\ J}{1\ V \cdot C} \cdot \frac{2\ mol\ e}{1\ mol\ reactant} \cdot \frac{1\ mol\ reactant}{642.6\ g} = 600.\ J/g\ or\ 0.60\ kJ/g\ (2sf)$$

(c) The silver/zinc battery produces more energy/gram.

Summary and Conceptual Questions

79. In the electrolysis of 150 g of CH_3SO_2F:

(a) The mass of HF required to electrolyze 150 g of CH_3SO_2F:

$$\frac{150\ g\ CH_3SO_2F}{1} \cdot \frac{1\ mol\ CH_3SO_2F}{98.10\ g\ CH_3SO_2F} \cdot \frac{3\ mol\ HF}{1\ mol\ CH_3SO_2F} \cdot \frac{20.01\ g\ HF}{1\ mol\ HF} = 92\ g\ HF$$

$$\frac{150\ g\ CH_3SO_2F}{1} \cdot \frac{1\ mol\ CH_3SO_2F}{98.10\ g\ CH_3SO_2F} \cdot \frac{1\ mol\ CF_3SO_2F}{1\ mol\ CH_3SO_2F} \cdot \frac{152.07\ g\ CF_3SO_2F}{1\ mol\ CF_3SO_2F} =$$

$$230\ g\ CF_3SO_2F$$

$$\frac{150 \text{ g CH}_3\text{SO}_2\text{F}}{1} \cdot \frac{1 \text{ mol CH}_3\text{SO}_2\text{F}}{98.10 \text{ g CH}_3\text{SO}_2\text{F}} \cdot \frac{3 \text{ mol H}_2}{1 \text{ mol CH}_3\text{SO}_2\text{F}} \cdot \frac{2.02 \text{ g H}_2}{1 \text{ mol H}_2} = 9.3 \text{ g H}_2$$

(b) H_2 produced at the anode or cathode? Since H is being reduced from +1 to 0, it will be produced at the cathode.

(c) Energy consumed: $\dfrac{1 \text{ kwh}}{3.60 \times 10^6 \text{ J}} \cdot \dfrac{1 \text{ J}}{1 \text{ V} \bullet \text{C}} \cdot \dfrac{8.0 \text{ V}}{1} \cdot \dfrac{250 \text{ C}}{1 \text{ s}} \cdot \dfrac{3600 \text{ s}}{1 \text{ hr}} \cdot \dfrac{24 \text{ hr}}{1} = 48 \text{kwh}$

81. (a) Cell diagram:

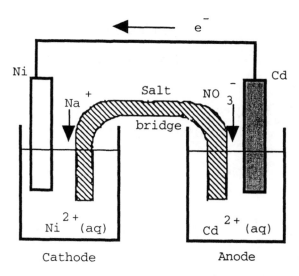

(b) As shown in part (a), Cd serves as the anode, so it must be oxidized to the 2+ cation. Likewise Ni ions would be reduced to elemental Ni.

The balanced equation is: $Ni^{2+}(aq) + Cd(s) \rightarrow Cd^{2+}(aq) + Ni(s)$

(c) The anode (Cd) serves as the source of electrons to the external circuit, and we label it "-". The cathode is labeled "+".

(d) The $E^{\circ}_{cell} = E^{\circ}_{cathode} - E^{\circ}_{anode} = (-0.205) - (-0.403) = +0.15$ V

(e) As shown on the diagram, electrons flow from Cd to Ni compartments.

(f) The direction of travel for the sodium and nitrate ions are shown on the diagram.

(g) The K for the reaction: $\Delta G^{\circ} = -nFE^{\circ}_{cell}$

$$= -(2 \text{ mol e})(96500 \text{ C/mol e})(+0.15 \text{ J/C}) = 28950 \text{J}$$

$\Delta G^{\circ} = -RT\ln K$ and $\ln K = \dfrac{nFE^{\cdot}}{RT}$

$$\text{So } \ln K = \frac{2 \text{ mol e} \bullet 96{,}500 \dfrac{C}{\text{mol e}} \bullet 0.15 \dfrac{J^{\cdot}}{C}}{8.314 \dfrac{J}{K \bullet \text{mol}} \bullet 298 \text{ K}} = 11.68 \quad \text{and } K = 1 \times 10^5$$

(h) If $[Cd^{2+}] = 0.010M$ and $[Ni^{2+}] = 1.0M$, the value for Ecell = ?

Using the Nernst equation $E_{cell} = E°_{cell} - \dfrac{0.0257}{n} \ln \dfrac{[Cd^{2+}]}{[Ni^{2+}]}$

$E_{cell} = 0.15 - \dfrac{0.0257}{2} \ln \dfrac{[0.010]}{[1.0]} = 0.15 - (0.0257/2)\cdot (-4.605) = 0.15 + 0.0592 = 0.21$ V

(i) Lifetime use of battery?

We have 1.0 L of each solution, we should determine the limiting reagent, if there is one.

The spontaneous reaction reduces Ni^{2+} and oxidizes Cd..

The Nickel solution contains: $1.0 L \cdot 1.0M = 1.0$ mol Ni^{2+} The cadmium electrode weighs 50.0 g, so we have: 50.0 g Cd $\cdot$ (1 mol Cd/112.41 g Cd) = 0.445 mol Cd, so Cd is the limiting reagent. Now we can calculate:

0.445 mol Cd $\cdot \dfrac{2 \text{ mol e}}{1 \text{ mol Cd}} \cdot \dfrac{96500 \text{ C}}{1 \text{ mol e}} \cdot \dfrac{1 \text{ amp} \bullet \text{s}}{1C} \cdot \dfrac{1}{0.050 \text{ A}} = 1.7$ x 106 s or 480 hr

83. Explain the reaction of Cu^{2+} with I^-:

The appropriate reduction potentials:

$$Br_2 + 2e^- \rightarrow 2Br^- \quad 1.08V$$
$$Cl_2 + 2e^- \rightarrow Cl^- \quad 1.36V$$
$$I_2 + 2e^- \rightarrow 2I^- \quad 0.535V$$

The relative values for the reduction potentials tell us that I^- is the strongest reducing agent, capable of reducing the Cu(II) ion, to Cu(I). The Cu^{+1} ion would react with the I^- to form the insoluble CuI. The equation: $2Cu^{2+}(aq) + 4 I^- (aq) \rightarrow 2CuI(s) + I_2(aq)$

85. Since the reaction depends on the oxidation of elemental hydrogen to water (2 mol e^- per mol H_2), we must determine the amount of H_2 present:

$$n = \dfrac{(200. \text{ atm})(1.0 \text{ L})}{(0.0821 \frac{\text{L} \bullet \text{atm}}{\text{K} \bullet \text{mol}})(298 \text{ K})} = 8.2 \text{ mol } H_2$$

The amount of time this can produce current:

8.2 mol $H_2 \cdot \dfrac{2 \text{ mol e}^-}{1 \text{ mol } H_2} \cdot \dfrac{9.65 \text{ x } 10^4 \text{ C}}{1 \text{ mol e}^-} \cdot \dfrac{1 \text{ A} \bullet \text{s}}{1 \text{ C}} \cdot \dfrac{1}{1.5 \text{ A}} = 1.1$ x 10^6 s   (290 hrs)

87. From the simulation screen 20.6 on the General Chemistry Now CD-ROM:

   (a) Species that can be reduced by Cu. In general any oxidizing agent (left hand column in the reduction potential table) that resides "above" Cu (a reducing agent) can be reduced by Cu. A screen capture from the CD-ROM Screen 20.6:

$Au^{3+}(aq) + 3\,e^-$ $\longrightarrow$	$Au(s)$	+1.50
$Br_2(l) + 2\,e^-$ $\longrightarrow$	$2\,Br^-(aq)$	+1.08
$Hg^{2+}(aq) + 2\,e^-$ $\longrightarrow$	$Hg(l)$	+0.855
$Ag^+(aq) + e^-$ $\longrightarrow$	$Ag(s)$	+0.80
$Hg_2^{2+}(aq) + 2\,e^-$ $\longrightarrow$	$2\,Hg(l)$	+0.789
$Cu^{2+}(aq) + 2\,e^-$ $\longrightarrow$	$Cu(s)$	+0.337

   As an example, suppose we let Cu react with Au^{3+}. The $E°_{cell}$ would be $+1.50 - 0.337 = 1.16V$. Note that the value is **positive**. This would be true for any specie above Cu in the list. Hence Au^{3+}, Br_2, Hg^{2+}, Ag^+, Hg_2^{2+} would all be capable of being reduced by Cu.

   (b) Metals that can be oxidized by Cd^{2+}:

   A screen capture from the CD-ROM Screen 20.6:

$Cd^{2+}(aq) + 2\,e^-$ $\longrightarrow$	$Cd(s)$	-0.40
$Al^{3+}(aq) + 3\,e^-$ $\longrightarrow$	$Al(s)$	-1.66
$Mg^{2+}(aq) + 2\,e^-$ $\longrightarrow$	$Mg(s)$	-2.37
$Li^+(aq) + e^-$ $\longrightarrow$	$Li(s)$	-3.045

   With Cd^{2+} acting as oxidizing agent, Li, Mg, and Al could be oxidized.

   As an example, suppose we let Li react with Cd^{2+}. The $E°_{cell}$ would be $-0.40 + 3.045 = 2.65V$. Note that the value is **positive**. This would be true for any specie below Cd^{2+} in the list.

Chapter 21
The Chemistry of the Main Group Elements

Practicing Skills

Properties of the Elements

1. Examples of two basic oxides, and equations showing the oxide formation from its elements:

$4 Li + O_2 \rightarrow 2 Li_2O$; is basic owing to the reaction: $Li_2O + H_2O \rightarrow 2LiOH$

$2 Ca + O_2 \rightarrow 2 CaO$; is basic owing to the reaction: $CaO + H_2O \rightarrow Ca(OH)_2$

3. Name and symbol with a valence configuration $ns^2 np^1$:

The configuration shown is characteristic of the elements in Group 3A (13)

B:Boron	Al:Aluminum	Ga:Gallium	In:Indium	Tl:Thallium

5. Select an alkali metal; Write the balanced equation with Cl_2:

$2 Na + Cl_2 \rightarrow 2 NaCl$

Reactions of alkali metals with halogens are exothermic—given the excellent ability of these metals to function as reducing agents, and the ability of Cl_2 to function as oxidizing agents.

Typically compounds formed from elements of greatly different electronegativities (a metal and a nonmetal) are ionic in nature.

7. Predict color, state of matter, water solubility of compound in SQ21.5:

Such compounds are typically colorless (white in the solid state); are high melting (a solid), and are water soluble.

9. Calcium is **not** expected to be found free in the earth's crust. Its great reactivity with oxygen and water would dispose the metal to exist as an oxide.

11. The oxides listed in order of increasing basicity:

Metal oxides are basic, nonmetal oxides are acidic. The three listed range from a nonmetal oxide (CO_2) to a metal oxide, SnO_2. So the order is: $CO_2 < SiO_2 < SnO_2$

13. Balanced equations for the following reactions:

(a) $2 Na (s) + Br_2 (\ell) \rightarrow 2 NaBr (s)$

(b) $2 Mg (s) + O_2 (g) \rightarrow 2 MgO (s)$

(c) $2 \, Al \, (s) + 3 \, F_2 \, (g) \rightarrow 2 \, AlF_3 \, (s)$

(d) $C(s) + O_2 \, (g) \rightarrow CO_2 \, (g)$

Hydrogen

15. Balanced chemical equation for hydrogen gas reacting with oxygen, chlorine, and nitrogen.

$2 \, H_2 \, (g) + O_2 \, (g) \rightarrow 2 \, H_2O \, (g)$

$H_2 \, (g) + Cl_2 \, (g) \rightarrow 2 \, HCl \, (g)$

$3 \, H_2 \, (g) + N_2 \, (g) \rightarrow 2 \, NH_3 \, (g)$

17. The reaction: $CH_4(g) + H_2O(g) \rightarrow CO(g) + 3 \, H_2(g)$

Data from Appendix L:

$\Delta H°$ (kJ/mol)	-74.87	-241.83	-110.525	0
$S°$ (J/K•mol)	+186.26	+188.84	+197.674	+130.7
$\Delta G°$ (kJ/mol)	-50.8	-228.59	-137.168	0

$\Delta H°_{rxn} = [(1mol•-110.525 \, kJ/mol) + (3 \, mol• \, 0)] -$

$[(1mol • -74.87 \, kJ/mol)+(1mol•-241.83 \, kJ/mol)]$

$= 206.18 \, kJ$

$\Delta S°_{rxn} = [(1mol• +197.674 \, J/K•mol) + (3 \, mol• +130.7 J/K•mol)] -$

$[(1mol • +186.26 J/K•mol)+(1mol• +188.84 J/K•mol)]$

$= + 214.7 \, J/K$

$\Delta G°_{rxn} = [(1mol•-137.168 \, kJ/mol) + (3 \, mol• \, 0)] -$

$[(1mol • -50.8 \, kJ/mol)+(1mol•-228.59 \, kJ/mol)]$

$= 142.2 \, kJ$

19. Prepare a balanced equation for each of the 3 steps, and show that the sum is the decomposition of water to form hydrogen and oxygen.

(a) $2 \, SO_2 \, (g) + 4 \, H_2O \, (\ell) + 2 \, I_2 \, (s) \rightarrow 2 \, H_2SO_4 \, (\ell) + 4 \, HI \, (g)$

(b) $2 \, H_2SO_4 \, (\ell) \rightarrow 2 \, H_2O \, (\ell) + 2 \, SO_2 \, (g) + O_2 \, (g)$

(c) $4 \, HI \, (g) \rightarrow 2 \, H_2 \, (g) + 2 \, I_2 \, (g)$

Net: $2 \, H_2O \, (\ell) \rightarrow 2 \, H_2 \, (g) + O_2 \, (g)$

Alkali Metals

21. Equations for the reaction of sodium

    with the halogens:

    $2 \, Na(s) \; + \; F_2(g) \rightarrow \; 2 \, NaF(s)$

    $2 \, Na(s) \; + \; Cl_2(g) \rightarrow 2 \, NaCl(s)$

    $2 \, Na(s) \; + \; Br_2(\ell) \rightarrow 2 \, NaBr(s)$

    $2 \, Na(s) \; + \; I_2(s) \rightarrow \; 2 \, NaI(s)$

> Physical properties of the alkaline metal halides: (1) ionic solids (2) high melting and boiling points (3) white color (4) water soluble

23. In the electrolysis of aqueous NaCl:

    (a) The balanced equation for the process:

    $$2 \, NaCl(aq) \; + \; 2 \, H_2O(\ell) \rightarrow 2 \, NaOH(aq) \; + \; Cl_2(g) + \; H_2(g)$$

    (b) Anticipated masses ratios:

    $$\frac{1 \, mol \, Cl_2}{2 \, mol \, NaOH} \; = \; \frac{70.9 \, g \, Cl_2}{80.0 \, g \, NaOH} = \; 0.88 \, g \, Cl_2/g \, NaOH$$

    $$\text{Actual:} \; \frac{1.14 \times 10^{10} \, kg \, Cl_2}{1.19 \times 10^{10} \, kg \, NaOH} \; = \; 0.96 \, \frac{kg \, Cl_2}{kg \, NaOH}$$

    The difference in ratios means that alternative methods of producing chlorine are used. One of these is the Kel-Chlor process which uses HCl, NOCl, and O_2. Other products are NO and H_2O.

Alkaline Earth Elements

25. Balanced equations for the reaction of magnesium with nitrogen and oxygen:

    $3 \, Mg(s) \; + \; N_2(g) \rightarrow \quad Mg_3N_2(s)$

    $2 \, Mg(s) \; + \; O_2(g) \rightarrow \quad 2 \, MgO(s)$

27. Uses of limestone:

    agricultural: to furnish Ca^{2+} to plants and neutralize acidic soils

    building: lime (CaO) is used in mortar and absorbs CO_2 to form $CaCO_3$

    steel-making: $CaCO_3$ furnishes lime (CaO) in the basic oxygen process. The lime reacts with gangue (SiO_2) to form calcium silicate.

    The balanced equation for the reaction of limestone with carbon dioxide in water:

    $$CaCO_3(s) \; + \; H_2O(\ell) \; + \; CO_2(g) \rightarrow Ca^{2+}(aq) \; + \; 2 \, HCO_3^-(aq)$$

    This reaction is important in the formation of "hard water" (not particularly a great happening for plumbing) and stalagmites and stalactites (aesthetically pleasing in caves).

29. The amount of SO_2 that could be removed by 1200 kg of CaO by the reaction:

$$CaO\ (s) + SO_2\ (g) \rightarrow\ CaSO_3\ (s)$$

$$1.2 \times 10^6\ g\ CaO\ \bullet\ \frac{1\ mol\ CaO}{56.079\ g\ CaO}\ \bullet\ \frac{1\ mol\ SO_2}{1\ mol\ CaO}\ \bullet\ \frac{64.059\ g\ SO_2}{1\ mol\ SO_2}\ = 1.4 \times 10^6\ g\ SO_2$$

Boron and Aluminum

31. Structure for the cyclic anion in the salt $K_3B_3O_6$, and the chain anion in $Ca_2B_2O_5$

33. For the reactions of boron hydrides:

(a) Balanced equation for the combusion of $B_5H_9(g)$ in air:

$$2\ B_5H_9(g) + 12\ O_2(g) \rightarrow 5\ B_2O_3\ (s) + 9\ H_2O(g)$$

(b) Heat of combustion for $B_5H_9(g)$:

$$\Delta H°_{comb} = [9\bullet \Delta H°_f\ (H_2O(g)) + 5\bullet \Delta H°_f(B_2O_3\ (s))]$$

$$- [2\bullet\Delta H°_f(B_5H_9(g)) + 12\bullet \Delta H°_f(O_2(g))]$$

$$\Delta H°_{comb} = [9mol\bullet\ (-241.83kJ/mol) + 5\bullet\ (-1271.9\ kJ/mol)] - [2\bullet(73.2\ kJ/mol)) + 12\bullet\ (0)]$$

$$= -8682.37\ kJ$$

So in kJ/mol: -8682.37kJ/2mol $B_5H_9(g)$ = -4341.2 kJ/mol

The text reports the Heat of Combustion for B_2H_6 is-2038 kJ/mol, so the Heat of

Combustion for B_5H_9 is more than twice as great than that of diborane's combustion.

(c) Compare Heats of Combustion for $C_2H_6(g)$ with $B_2H_6(g)$:

For B: $B_2H_6 + 3\ O_2 \rightarrow\ B_2O_3\ (s) + 3\ H_2O(\ell)$ $\Delta H°_{comb}$ = -2038 kJ/mol

So the energy/g is –2038 kJ/mol or for 27.67 g B_2H_6 (-73.7 kJ/g B_2H_6)

For C: $2C_2H_6 + 7\ O_2 \rightarrow\ 4\ CO_2\ (g) + 6\ H_2O(g)$

$$\Delta H°_{comb} = [6\bullet\ (-241.83kJ/mol)\ + 4\bullet\ (-393.509\ kJ/mol)] - [2\bullet(-83.85kJ/mol) + 7\bullet\ 0)]$$

$$\Delta H°_{comb} = -2857.32\ kJ\ and\ -1,428.66\ kJ/mol\ C_2H_6\ \ or\ for\ 30.07\ g\ C_2H_6\ (47.51\ kJ/g)$$

35. The equations for the reaction of aluminum with HCl, Cl_2 and O_2:

$$2 \, Al(s) + 6 \, HCl(aq) \rightarrow 2 \, Al^{3+}(aq) + 6 \, Cl^{-}(aq) + 3 \, H_2(g)$$

$$2 \, Al(s) + 3 \, Cl_{2(g)} \rightarrow 2 \, AlCl_3(s)$$

$$4 \, Al(s) + 3 \, O_{2(g)} \rightarrow 2 \, Al_2O_3(s)$$

37. The equation for the reaction of aluminum dissolving in aqueous NaOH:

$$2 \, Al \, (s) + 2 \, NaOH \, (aq) + 6 \, H_2O \, (\ell) \rightarrow 2 \, NaAl(OH)_4 \, (aq) + 3 \, H_2 \, (g)$$

Volume of H_2 (in mL) produced when 13.2 g of Al react:

$$13.2 \text{ g Al} \cdot \frac{1 \text{ mol Al}}{26.98 \text{ g Al}} \cdot \frac{3 \text{ mol } H_2}{2 \text{ mol Al}} = 0.734 \text{ mol } H_2$$

$$V = \frac{(0.734 \text{ mol } H_2)(0.082057 \, \frac{L \cdot atm}{K \cdot mol})(295.7 \text{ K})}{735 \text{ mm Hg} \cdot \frac{1 \text{ atm}}{760 \text{ mm Hg}}} = 18.4 \text{ L}$$

39. The equation for the reaction of aluminum oxide with sulfuric acid:

$$Al_2O_3(s) + 3 \, H_2SO_4(aq) \rightarrow Al_2(SO_4)_3 \, (s) + 3 \, H_2O(\ell)$$

$$1.00 \times 10^3 \text{ g } Al_2(SO_4)_3 \cdot \frac{1 \text{ mol } Al_2(SO_4)_3}{342.1 \text{ g } Al_2(SO_4)_3} \cdot \frac{1 \text{ mol } Al_2O_3}{1 \text{ mol } Al_2(SO_4)_3} \cdot$$

$$\frac{102.1 \text{ g } Al_2O_3}{1 \text{ mol } Al_2O_3} \cdot \frac{1 \text{ kg}}{1 \times 10^3 \text{ g}} = 0.298 \text{ kg } Al_2O_3$$

$$1.00 \times 10^3 \text{ g } Al_2(SO_4)_3 \cdot \frac{1 \text{ mol } Al_2(SO_4)_3}{342.1 \text{ g } Al_2(SO_4)_3} \cdot \frac{3 \text{ mol } H_2SO_4}{1 \text{ mol } Al_2(SO_4)_3} \cdot$$

$$\frac{98.07 \text{ g } H_2SO_4}{1 \text{ mol } H_2SO_4} \cdot \frac{1 \text{ kg}}{1 \times 10^3 \text{ g}} = 0.860 \text{ kg } H_2SO_4$$

41. A Lewis electron dot structure for $AlCl_4^-$:

The $AlCl_4^-$ will have a tetrahedral geometry. The aluminum atom would utilize sp^3 hybridization.

Silicon

43. The structure of SiO_2 is tetrahedral with the silicon atom being surrounded by four oxygen atoms. The geometry around C in CO_2 is linear with two groups attached to the C atom.

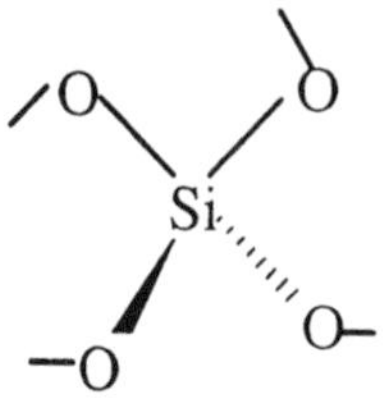

> The energies of four Si - O single bonds is greater than two Si=O double bonds. CO_2, on the other hand, forms discrete C=O double bonds.

The high melting point of SiO_2 is due to the network structure of the SiO_4 tetrahedron, while CO_2 exists as discrete nonpolar molecules—and as a gas at room temperature.

45. In the synthesis of dichlorodimethylsilane:

   (a) The balanced equation: $Si\ (s)\ +\ 2\ CH_3Cl\ (g) \rightarrow (CH_3)_2SiCl_2\ (\ell)$

   (b) Stoichiometric amount of CH_3Cl to react with 2.65 g silicon:

$$2.65\ g\ Si\ \bullet\ \frac{1\ mol\ Si}{28.09\ g\ Si}\ \bullet\ \frac{2\ mol\ CH_3Cl}{1\ mol\ Si}\ =\ 0.189\ mol\ CH_3Cl$$

$$P\ =\ \frac{\left(0.189\ mol\ CH_3Cl\right)\left(0.082057\ \frac{L\bullet atm}{K\bullet mol}\right)(297.7\ K)}{5.60\ L}\ =\ 0.823\ atm$$

   (c) Mass of $(CH_3)_2SiCl_2$ produced assuming 100 % yield:

$$0.0943\ mol\ Si\ \bullet\ \frac{1\ mol\ (CH_3)_2SiCl_2}{1\ mol\ Si}\ \bullet\ \frac{129.1\ g\ (CH_3)_2SiCl_2}{1\ mol\ (CH_3)_2SiCl_2}\ =12.2\ g\ (CH_3)_2SiCl_2$$

Nitrogen and Phosphorus

47. The $\Delta G°_f$ data from Appendix L are shown below:

compound:	NO(g)	NO₂(g)	N₂O(g)	N₂O₄(g)
$\Delta G°\ (\frac{kJ}{mol})$	+86.58	+51.23	+104.20	+97.73

To ask if the oxide is stable with respect to decomposition is to ask about the $\Delta G°$ for the

process: $N_xO_y\ \rightarrow\ \frac{x}{2}\ N_2\ (g)\ +\ \frac{y}{2}\ O_2(g)$

This reaction is *the reverse* of the ΔG_f for each of these oxides. The ΔG for such a process would have an *opposite sign* from that data given above. Since that sign would be negative, the process is product-favored. Hence **all** of the oxides shown above are unstable with respect to decomposition.

49. Calculate $\Delta H°_{rxn}$ for the reaction: $2\,NO(g)$ + $O_2(g)$ $\rightarrow$ $2\,NO_2(g)$

$$\Delta H°\,(\tfrac{kJ}{mol})\quad +90.29\qquad\quad 0\qquad\qquad +33.1$$

$$\Delta H°_{rxn} = [(2\text{ mol})(+33.1\,\tfrac{kJ}{mol})] - [(2\text{ mol})(+90.29\,\tfrac{kJ}{mol})] = -114.4\text{ kJ}$$

The reaction is exothermic.

51. (a) The reaction of hydrazine with dissolved oxygen:

   $$N_2H_4\,(aq) + O_2\,(g) \rightarrow N_2\,(g) + 2\,H_2O(\ell)$$

   (b) Mass of hydrazine to consume the oxygen in 3.00×10^4 L of water:

   $$3.00 \times 10^4\text{ L} \cdot \frac{3.08\text{ cm}^3\,O_2}{0.100\text{ L}} \cdot \frac{1\text{ mol }O_2}{22400\text{ cm}^3} \cdot \frac{1\text{ mol }N_2H_4}{1\text{ mol }O_2} \cdot \frac{32.05\text{g}N_2H_4}{1\text{ mol }N_2H_4}$$

   $$\uparrow \text{ at STP}$$

   $$= 1.32 \times 10^3\text{ g }N_2H_4$$

53. The half equations:

   $$N_2H_5^+\,(aq) \rightarrow N_2\,(g) + 5\,H^+\,(aq) + 4\,e^-$$

   $$IO_3^-\,(aq) \rightarrow I_2\,(s)$$

   are balanced according to the procedure in Chapter 5 to give:

   $$5\,N_2H_5^+\,(aq) + 4\,IO_3^-\,(aq) \rightarrow 5\,N_2\,(g) + 2\,I_2\,(aq) + 12\,H_2O(\ell) + H^+(aq)$$

   E° for the reaction is: E° for the hydrazine equation (oxidation)   = – 0.23 V

   E° for the iodate equation (reduction)   = + 1.195 V

   $E°_{net}$ = + 1.43 V

55. For phosphonic acid, $H_4P_2O_5$:

   (a) oxidation number of P atom:

   $4(H) + 2(P) + 5(O) = 0$ where the symbol represents the oxidation number of the atom.

   $4(+1) + 2(P) + 5(-2) = 0$; $2(P) -6 = 0$; $2(P) =+6$ so P $=+3$

   (b) The sructure of diphosphonic acid, $H_4P_2O_5$, and the maximum number of protons that can

   dissociate in this acid:

   The "acidic" H atoms are those attached to
   the O atoms, and NOT those directly bonded
   to the P atoms—so 2 protons can dissociate.

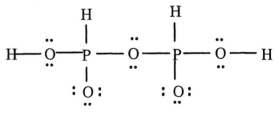

Oxygen and Sulfur

57. (a) Allowable release of SO_2: (0.30 %)

$$1.80 \times 10^6 \text{ kg H}_2\text{SO}_4 \cdot \frac{1 \text{ mol H}_2\text{SO}_4}{98.08 \text{ g H}_2\text{SO}_4} \cdot \frac{1 \text{ mol SO}_2}{1 \text{ mol H}_2\text{SO}_4} \cdot \frac{64.06 \text{ g SO}_2}{1 \text{ mol SO}_2}$$

$$\cdot \frac{0.0030 \text{ kg SO}_2 \text{ released}}{1.00 \text{ kg SO}_2 \text{ produced}} = 3.5 \times 10^3 \text{ kg SO}_2$$

(b) Mass of $Ca(OH)_2$ to remove 3.53×10^3 kg SO_2:

$$3.53 \times 10^3 \text{ kg SO}_2 \cdot \frac{1 \text{ mol SO}_2}{64.06 \text{ g SO}_2} \cdot \frac{1 \text{ mol Ca(OH)}_2}{1 \text{ mol SO}_2} \cdot \frac{74.09 \text{ g Ca(OH)}_2}{1 \text{ mol Ca(OH)}_2}$$

$$= 4.1 \times 10^3 \text{ kg Ca(OH)}_2$$

59. The disulfide ion S_2^{2-} can be pictured as below:

$$\left[\; \ddot{\underset{\cdot\cdot}{S}} \; \vdots \; \ddot{\underset{\cdot\cdot}{S}} \; \right]^{2-}$$

Fluorine and Chlorine

61. Calculate the equivalent net cell potential for the oxidation:

2 [$Mn^{2+}(aq)$ + 4 $H_2O(\ell)$ $\rightarrow$ $MnO_4^-(aq)$ + 8 $H^+(aq)$ + 5 e^-]   -1.51 V

2 $BrO_3^-(aq)$ + 12 $H^+(aq)$ + 10 e^- $\rightarrow$ $Br_2(aq)$ + 6 $H_2O(\ell)$]   +1.44 V

net 2 $Mn^{2+}(aq)$ + 2 $H_2O(\ell)$ + 2 $BrO_3^-(aq)$ $\rightarrow$ 2 $MnO_4^-(aq)$ -0.07 V

+ 4 $H^+(aq)$ + $Br_2 (aq)$

NOTE: The first equation has been reversed from its usual "reduction form", indicating that we want to have this process occur as an oxidation. Recall that the task is to ask if BrO_3^- can oxidize Mn^{2+}! So we arrange the two half-equations to give us a specific net equation, and then use that format to ask, "Will it go this way?" If E°_{cell} is > 0, the answer is **Yes**. If E°cell is < 0, the answer is **No**. Given that we calculate E°_{cell} by *subtracting* the E°_{anode} from the $E^\circ_{cathode}$, I have given the reduction potential for the manganese half-reaction a sign that is *opposite* the normal sign. If we add the values, this is the mathematical equivalent of our usual procedure. The negative net potential indicates that this process doesn't favor products with 1.0 M bromate ion.

63. The balanced equation for the reaction of Cl_2 with Br^-

$$2\,Br^-\,(aq)\quad\rightarrow\quad Br_2\,(\ell)+2\,e^-\qquad\qquad -1.08\text{ V}$$

$$Cl_2\,(g)+2\,e^-\rightarrow 2\,Cl^-\,(aq)\qquad\qquad +1.36\text{ V}$$

net: $\;Cl_2\,(g)\;+2\,Br^-\,(aq)\rightarrow 2\,Cl^-\,(aq)+Br_2\,(\ell)\;\;+0.28\text{ V}$

- Note that bromide ions are losing electrons (and donating them to chlorine), causing chlorine to be reduced—so bromide is the reducing agent and chlorine, removing the electrons from bromide ions, is causing the bromide ions to be oxidized—so chlorine is the oxidizing agent.
- Note also that the voltage for the cell we "constructed" is positive, making this process a "product-favored" reaction.

65. Mass of F_2 that can be proeuced per 24hr by a current of 5.00×10^3 amps (at 10.0V)

$$\frac{38.00\text{ g }F_2}{1\text{ mol }F_2}\cdot\frac{1\text{ mol }F_2}{2\text{ mol }e^-}\cdot\frac{1\text{ mol }e^-}{9.65\times10^4\text{ C}}\cdot\frac{1\text{ C}}{1\text{ amp}\cdot s}$$

$$\cdot\frac{5.00\times10^3\text{ amps}}{1}\cdot\frac{3600\text{ s}}{1\text{ hr}}\cdot\frac{24\text{ hr}}{1}=8.51\times10^4\text{ g }F_2$$

General Questions

67. Describe the elements in the third period:

Atomic No.	Element	(a) Type	(b) Color	(c) State
11	Sodium	metal	grey, shiny	solid
12	Magnesium	metal	grey, shiny	solid
13	Aluminum	metal	grey,shiny	solid
14	Silicon	metalloid	grey,shiny	solid
15	Phosphorus	nonmetal	red,white,black	solid
16	Sulfur	nonmetal	yellow	solid
17	Chlorine	nonmetal	pale green	gas
18	Argon	nonmetal	colorless	gas

69. Reactions of Na, Mg, Al, Si, P, S

   (a) Balanced equations of the elements with elemental chlorine:

   $2\,Na\,(s) + Cl_2\,(g) \rightarrow 2\,NaCl\,(s)$ $Si\,(s) + 2\,Cl_2\,(g) \rightarrow SiCl_4\,(\ell)$

   $Mg\,(s) + Cl_2\,(g) \rightarrow MgCl_2\,(s)$ $P_4\,(s) + 10\,Cl_2\,(g) \rightarrow 4\,PCl_5\,(s)$

   $2\,Al\,(s) + 3\,Cl_2\,(g) \rightarrow 2\,AlCl_3\,(s)$ $S_8\,(s) + 8\,Cl_2\,(g) \rightarrow 8\,SCl_2(s)$

   (b) Bonding in the products:

   $NaCl$ and $MgCl_2$ are ionic; all others are covalent. This bonding is attributable—to a first approximation--to the differing electronegativities of the atoms bonded to each other.

   (c) Electron dot structures for the products; electron-pair geometry; molecular geometry

Compound	Electron-Pair geometry	Molecular geometry
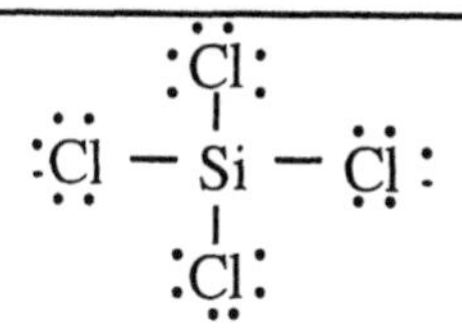	Tetrahedral	Tetrahedral
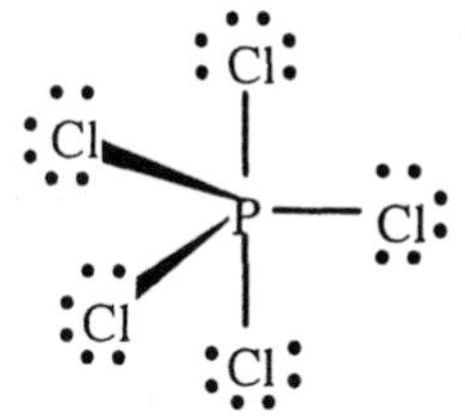	Trigonal bipyramidal	Trigonal bipyramidal

71. Complete and balance the equations:

   (a) $2\,KClO_3\,(s) + heat \rightarrow 2\,KCl(s) + 3\,O_2(g)$

   (b) $2\,H_2S(g) + 3\,O_2(g) \rightarrow 2\,H_2O(g) + 2\,SO_2(g)$

   (c) $2\,Na(s) + O_2\,(g) \rightarrow Na_2O_2\,(s)$

   (d) $P_4(s) + 3\,KOH(aq) + 3\,H_2O(\ell) \rightarrow PH_3(g) + 3\,KH_2PO_4(aq)$

   (e) $NH_4NO_3(s) \rightarrow N_2O(g) + 2\,H_2O(g)$

   (f) $2\,In(s) + 3\,Br_2(\ell) \rightarrow 2\,InBr_3(s)$

   (g) $SnCl_4(\ell) + 2\,H_2O(\ell) \rightarrow 4\,HCl(aq) + SnO_2(s)$

73. (a) Lewis dot structure for B_2Cl_4:

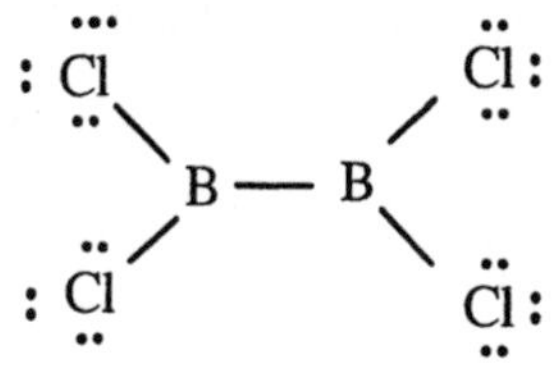

(b) Hybridization of B in the molecule and geometry around each B.

Noting that three pairs of electrons are gathered around each B, we would anticipate Cl-B-Cl bond angles of 120°--or trigonal planar geometry around each B. The three bonds would require three orbitals, so we anticipate sp^2 hybridization.

75. (a) Heating BaO in oxygen gives barium peroxide. The balanced equation:

$$2\ BaO(s) +\ O_2\ (g)\ \rightarrow\ 2\ BaO_2(s)$$

(b) Reaction of Fe with $BaO_2(s)$:

$$3\ BaO_2\ (s) +\ 2\ Fe\ (s) \rightarrow\ 3\ BaO(s) +\ Fe_2O_3(s)$$

77. The products expected when molten LiH is electrolyzed:

Molten LiH contains lithium (Li^+) and hydride (H^-) ions.

At the cathode, lithium ions are reduced: $Li^+\ (\ell) + e^- \rightarrow Li\ (\ell)$

At the anode, hydride ions are oxidized: $2\ H^-\ (\ell) \rightarrow H_2\ (g) + 2\ e^-$

79. Calculate $\Delta G°$ for the decomposition of the metal carbonates for Mg, Ca, and Ba

$$MCO_3(s) \longrightarrow \quad MO(s)\ +\ CO_2(g) \qquad\qquad \underline{M}$$

-1028.2		- 568.93	- 394.359	(Mg) $\Delta G°$ ($\frac{kJ}{mol}$)	0
- 1129.16		- 603.42	- 394.359	(Ca) $\Delta G°$	0
- 1134.41		- 520.38	- 394.359	(Ba) $\Delta G°$	0

$\Delta G°rxn\ =\ \Delta G°_f\ MO\ +\ \Delta G°_f\ CO_2\ -\ \Delta G°_f\ MCO_3$

$MgCO_3$ = (- 568.93 $\frac{kJ}{mol}$)(1 mol) + (-394.359 $\frac{kJ}{mol}$)(1 mol) - (-1028.2 $\frac{kJ}{mol}$)(1 mol)

= 64.9 kJ

$CaCO_3$ = (- 603.42 $\frac{kJ}{mol}$)(1 mol) + (-394.359 $\frac{kJ}{mol}$)(1 mol) - (- 1129.16 $\frac{kJ}{mol}$)(1 mol)

= 131.38 kJ

$BaCO_3$ = (- 520.38 $\frac{kJ}{mol}$)(1 mol) + (-394.359 $\frac{kJ}{mol}$)(1 mol) - (-1134.41 $\frac{kJ}{mol}$)(1 mol)

= 219.67 kJ

The relative tendency for decomposition is then $MgCO_3\ >\ CaCO_3\ >\ BaCO_3$.

81. (a) Since $\Delta G^\circ_{rxn} < 0$ for the reaction to be product-favored, calculate the value for

ΔG°_f MX that will make the ΔG°_{rxn} zero.

$\Delta G^\circ_{rxn} = \Delta G^\circ_f (MX_n) - n \Delta G^\circ_f (HX)$

for HCl $= \Delta G^\circ_f (MX_n) - n(-95.1 \dfrac{kJ}{mol})$

so if $n(-95.1 \ kJ) = \Delta G^\circ_f(MX_n)$ then $\Delta G^\circ = 0$

and if $n (-951 \ kJ) > \Delta G^\circ (MX_n)$ then $\Delta G^\circ_{rxn} < 0$.

(b) Examine ΔG°_f MX values for

metal:	Ba	Pb	Hg	Ti
ΔG°_f MX:	-810.4	-314.10	-178.6	-737.2
n:	2	2	2	4
n(-95.1):	-190.2	-190.2	-190.2	-380.4

For Barium, Lead, and Titanium, $n (-95.1) > \Delta G^\circ(MX)$; We expect these reactions to be spontaneous.

83. The average O-F bond energy in OF_2, given that the $\Delta H_f^\circ = +24.5$ kJ/mol:

The reaction is represented as: $O_2 (g) + 2F_2 (g) \rightarrow 2OF_2 (g)$

ΔH = Energy input – Energy released

Energy input: 1 mol O=O = 1 mol • 498 kJ/mol =   498 kJ

2 mol F-F = 2 mol • 155 kJ/mol =   310 kJ

Total input   =   808 kJ

Energy released : 4 mol O-F = 4 x (where x = O-F bond energy)

Total input   = 4x kJ

ΔH = Energy input – Energy released

+49.0 kJ = 808 kJ – 4x kJ so +49.0 kJ - 808 kJ = -4x and

-759 kJ = -4x   and 190 kJ = x. So the O-F bond energy is 190. kJ/mol

85. For the equation: $H_2NN(CH_3)_2(\ell) + 2 N_2O_4(\ell) \rightarrow 3 N_2(g) + 4 H_2O(g) + 2 CO_2(g)$

(a) The oxidizing and reducing agents: Noting that $N_2O_4(\ell)$ loses O as it reacts, the oxide is

reduced—making N_2O_4 the **oxidizing agent.** This means that $H_2NN(CH_3)_2(\ell)$ serves

as the **reducing agent**.

(b) What mass of $N_2O_4(\ell)$ was consumed if 4100kg of $H_2NN(CH_3)_2$ reacts? Noting that the mass of the substituted hydrazine was given in kg, we can express the answer in kg, and treat the mass *as if it were in grams*.

$$\left(\frac{4100 \text{ kg } H_2NN(CH_3)_2}{1} \cdot \frac{1 \text{ mol } H_2NN(CH_3)_2}{60.10 \text{ g } H_2NN(CH_3)_2}\right) \cdot \frac{2 \text{ mol } N_2O_4}{1 \text{ mol } H_2NN(CH_3)_2} \cdot \frac{92.01 \text{ g } N_2O_4}{1 \text{ mol } N_2O_4}$$

$$= 1.3 \times 10^4 \text{ kg } N_2O_4 \text{ (to 2 sf)}$$

What mass of N_2, H_2O, CO_2 will be formed?

$$\left(\frac{68.2 \text{ mol } H_2NN(CH_3)_2}{1}\right) \cdot \frac{3 \text{ mol } N_2}{1 \text{ mol } H_2NN(CH_3)_2} \cdot \frac{28.01 \text{ g } N_2O_4}{1 \text{ mol } N_2} = 5.7 \times 10^3 \text{ kg } N_2 \text{ (to 2sf)}$$

$$\left(\frac{68.2 \text{ mol } H_2NN(CH_3)_2}{1}\right) \cdot \frac{4 \text{ mol } H_2O}{1 \text{ mol } H_2NN(CH_3)_2} \cdot \frac{18.02 \text{ g } H_2O}{1 \text{ mol } H_2O} = 4.9 \times 10^3 \text{ kg } H_2O \text{ (to 2sf)}$$

$$\left(\frac{68.2 \text{ mol } H_2NN(CH_3)_2}{1}\right) \cdot \frac{2 \text{ mol } CO_2}{1 \text{ mol } H_2NN(CH_3)_2} \cdot \frac{44.01 \text{ g } CO_2}{1 \text{ mol } CO_2} = 6.0 \times 10^3 \text{ kg } CO_2 \text{ (to 2sf)}$$

87. For N_2O_3:

(a) Explain differences in bond lengths for the three N-O bonds:
The Lewis dot picture at right indicates that the "left" N has a
N=O, while the "right" N has two oxygens, each connected with
a N-O "one and a half" bond, owing to resonance. Naturally,
the N-O bonds are expected to be equal to one other and longer
than the N=O double bond.

(b) $\Delta H = +40.5$kJ/mol N_2O_3; $\Delta G = -1.59$ kJ/mol; What are ΔS and K for the reaction at 298K?

Given that $\Delta G° = \Delta H° - T\Delta S°$, $\Delta S° = -(\Delta G° - \Delta H°)/T = (\Delta H° - \Delta G°)/T$

(+40.5kJ/mol- -1.59 kJ/mol)/298K = (38.91x 10^3 J/mol)/298K = 131 J/K•mol

and K can be calculated via the relationship:

$\Delta G° = -RT\ln K$ so -1.59 kJ/mol = - (8.3145 J/K•mol)298K•lnK

$$\frac{-1.59 \times 10^3 \dfrac{J}{mol}}{- (8.3145 \text{ J/K} \cdot \text{mol}) \cdot 298K} = \ln K = 6.417 \times 10^{-1} \text{ and } K = 1.90$$

(c) Calculate $\Delta H_f°$ for $N_2O_3(g)$

The reaction: $N_2O_3(g) \rightarrow NO(g) + NO_2(g)$ has $\Delta H = +40.5$ kJ/mol N_2O_3.

$\Delta H_{rx} = [\Delta H_f° (NO(g)) + \Delta H_f° (NO_2(g))] - \Delta H_f°(N_2O_3(g)) = +40.5$ kJ

$\Delta H_{rx} = [(+90.29 \text{kJ/mol}) \cdot 1\text{mol} + (+33.1 \text{kJ/mol}) \cdot 1\text{mol}] - (1\text{mol}) \cdot \Delta H_f°(N_2O_3(g)) = +40.5$ kJ

$\Delta H_{rx} = 123.39\text{kJ} - \Delta H_f°(N_2O_3(g)) = +40.5$ kJ

$123.39\text{kJ} - 40.5 \text{ kJ} = \Delta H_f°(N_2O_3(g)) = +82.9$ kJ/mol

89. Examine the enthalpy change for the reaction:

$$2 N_2 (g) + 5 O_2 (g) + 2 H_2O (\ell) \rightarrow 4 HNO_3 (aq)$$

$\Delta H°_{rxn} = [(4 \text{ mol})(-207.36 \frac{kJ}{mol})] - [(2 \text{ mol})(0 \frac{kJ}{mol}) + (2 \text{ mol})(0 \frac{kJ}{mol}) +$

$(2 \text{ mol})(-285.83 \frac{kJ}{mol})] = -257.78$ kJ

The reaction is **exothermic** so it is a reasonable "first-guess" that this might be a way to "fix" nitrogen. The only way to be certain is to calculate the $\Delta G°_{rxn}$.

$\Delta G°_{rxn} = [(4 \text{ mol})(-111.25 \frac{kJ}{mol})] - [(2 \text{ mol})(0 \frac{kJ}{mol}) + (2 \text{ mol})(0 \frac{kJ}{mol}) +$

$(2 \text{ mol})(-237.15 \frac{kJ}{mol})] = 29.19$ kJ

The positive value for $\Delta G°_{rxn}$ tells us that the **reaction is not likely at 25 °C,** The decrease in the number of moles of gas (entropy decrease) is also a factor that works against this process at 25 °C. The favorable $\Delta H°$ does indicate that a lower temperature might be feasible, and below 268 K, the ΔG_{rxn} is favorable, but at this temperature water is a solid—not a good "sign".

91. $1.00 \text{ kg Cl}_2 \cdot \dfrac{1 \times 10^3 \text{ g Cl}_2}{1.0 \text{ kg Cl}_2} \cdot \dfrac{1 \text{ mol Cl}_2}{70.906 \text{ g Cl}_2} \cdot \dfrac{2 \text{ mol e}^-}{1 \text{ mol Cl}_2} \cdot \dfrac{9.6485 \times 10^4 \text{ C}}{1 \text{ mol e}^-} = 2,720 \text{ C}$

The power is then:

$\dfrac{2720 \text{ C}}{1} \cdot \dfrac{4.6 \text{ V}}{1} \cdot \dfrac{1 \text{ J}}{1 \text{ V} \cdot \text{C}} \cdot \dfrac{1 \text{ kwh}}{3.6 \times 10^6 \text{ J}} = 3.5 \text{ kwh.}$ (2 sf)

The factor, $\dfrac{1 \text{ kwh}}{3.6 \times 10^6 \text{ J}}$, is derived from the fact that a kwh = 1000 watts $\cdot$ 1 hr

and the relationship that 1 J = 1 watt $\cdot$ s or 1 J/s = 1 watt

$1 \text{ kwh} = 1000 \text{ watts} \cdot 1 \text{ hr} \cdot \dfrac{1 \text{J/s}}{1 \text{watt}} \cdot \dfrac{3600 \text{ s}}{1 \text{ hr}} = \dfrac{3.60 \times 10^6 \text{ J}}{1 \text{ kwh}}$

Summary and Conceptual Questions

93. A procedure to determine the % hydrogen in a mixture of argon and hydrogen in a 1.0-L flask. A simple solution is to convert the hydrogen to water (burn the mixture). Argon doesn't react (noble gas, right?). Quantitatively collecting the gaseous mixture, and chilling it until water liquefies, one can separate the water from the argon. The reduced pressure of the gas that remains is due to argon. The difference between the reduced pressure and the initial pressure (745 mmHg) would be attributable to the original H_2. Given the pressure, volume, and T, one could calculate the amount of hydrogen gas originally present in the mixture.

95. To extinguish a Na fire, addition of water is NOT ADVISABLE, since the reaction of the water with the Na would produce HEAT and HYDROGEN GAS—which would also burn!! The optimal solution would be to smother the fire—perhaps with sand.

97. Stoppered flask contains either H_2, N_2, or O_2. What experiment would identify the gas? Several possibilities exist. Chilling a small sample of the gas with liquid N_2(bp $-196°C$), would cause O_2 to liquefy—but not H_2, and only N_2 slowly. If the gas liquefies—it's Oxygen! If the gas doesn't liquefy at all, the gas is either H_2 (bp $-253°C$) or N_2(bp $-196°C$). Allowing a small sample of the gas to escape through a narrow opening (perhaps the tip of a plastic medicine dropper) while holding a smoldering splint at the tip of the opening would confirm the presence of H_2 (the escaping gas would burn), or N_2 (the escaping gas would **not** burn). *CAUTION: Burning H_2 in glass vessels is TO BE AVOIDED. Serious physical damage frequently results!*

99. The change to more positive values in **Reduction Potentials** as one descends the group (from Al to Tl) indicates a diminishing ability for the metals to act as a reducing agent as you descend the group. Another noticeable trend is the large **differences** between the reduction potentials of Al and Ga, and between In and Tl. The trend toward more positive reduction potentials also indicates the increased stability of +1 oxidation states (over higher oxidation states) for the heavier elements in the family.

<u>101</u>. Reaction scheme:

Clue:

1.   1.00 g A   +   heat $\longrightarrow$ B   +   gas (P = 209 mm; V = 450 mL; T = 298 K)
     white solid   white solid

2.   Gas (from 1) + $Ca(OH)_2$ (aq) $\longrightarrow$ C (s)

                                  white solid

3.   Aqueous solution of B is basic (turns red litmus paper blue)

4.   B (aq) + HCl (aq) + heat $\longrightarrow$ D

                          white solid

5.   Flame test for B: green flame

6.   B (aq) + H_2SO_4 (aq) $\rightarrow$ E

                    white solid

Clue 5 indicates that **B** is a barium salt.

Clue 2 suggests that the gas evolved in Clue 1 is CO_2, and that *C would be CaCO₃*.

Heating of carbonates liberates CO_2 (g).

Compound *B is a metal oxide (Clue 3), and probably has the formula BaO*.

Clues 3 and 5 suggest the oxide reacts with HCl and H_2SO_4 to form *BaCl₂ (compound D)* and

     BaSO₄ (compound E) respectively.

Since **B** is most likely BaO, *compound A must be BaCO₃.*

One gram of $BaCO_3$ (Molar mass 197) corresponds to 5.06×10^{-3} mol $BaCO_3$.

Compare this amount of substance to the amount of gas liberated when substance **A** is

     heated.Substitution of data from clue 1 yields:

$$n = \frac{(209 \text{ mm Hg}) (0.450 \text{ L})}{(62.4 \frac{\text{L} \cdot \text{mm Hg}}{\text{K} \cdot \text{mol}}) (298 \text{ K})} = 5.06 \times 10^{-3} \text{ mol gas}$$

This is the quantity of CO_2 anticipated from the thermal decomposition of 5.06×10^{-3} mol

     $BaCO_3$.

Practicing Skills

Properties of Transition Elements

1. (a) Cr^{3+} 3d [↑|↑|↑| |]   4s [ ]   paramagnetic

    (b) V^{2+} 3d [↑|↑|↑| |]   4s [ ]   paramagnetic

    (c) Ni^{2+} 3d [↑↓|↑↓|↑↓|↑|↑]   4s [ ]   paramagnetic

    (d) Cu^{+} 3d [↑↓|↑↓|↑↓|↑↓|↑↓]   4s [ ]   diamagnetic

3. Ions from first series transition metals that are isoelectronic with:

   (a) Fe^{3+} has 5 3d electrons so it is isoelectronic with Mn^{2+}

   (b) Zn^{2+} has 10 3d electrons so it is isoelectronic with Cu^{+}(see 1(d) above)

   (c) Fe^{2+} has 6 3d electrons so it is isoelectronic with Co^{3+}

   (d) Cr^{3+} has 3 3d electrons so it is isoelectronic with V^{2+} (see 1(b) above)

5. Balance:
   Each is balanced by the procedure shown in Chapter 5 of the Solutions Manual. Refer to that procedure. While each of these is a reduction-oxidation reaction, one can also use the "inspection method" to obtain the balanced equations:

   (a) $Cr_2O_3(s) + 2\,Al(s) \rightarrow Al_2O_3(s) + 2\,Cr(s)$

   (b) $TiCl_4(\ell) + 2\,Mg(s) \rightarrow Ti(s) + 2\,MgCl_2(s)$

   (c) $2\,[Ag(CN)_2]^-(aq) + Zn(s) \rightarrow 2\,Ag(s) + [Zn(CN)_4]^{2-}(aq)$

   (d) $3\,Mn_3O_4\,(s) + 8\,Al\,(s) \rightarrow 9\,Mn(s) + 4\,Al_2O_3\,(s)$

Formulas of Coordination Compounds

7. Classify each of the following as monodentate or multidentate:

   (a) CH_3NH_2 monodentate (lone pair on N)

   (b) $CH_3C\equiv N$ monodentate (lone pair on N)

   (c) N_3^- monodentate (lone pair on a N atom)

   (d) ethylenediamine   multidentate (lone pairs on terminal N atoms)

(e) Br⁻   monodentate (lone pair on Br)

(f) phenanthroline   multidentate (lone pairs on N atoms)

9.

Compound	Metal	Oxidation Number
(a) $[Mn(NH_3)_6]SO_4$	Mn	+2

Ammonia is a neutral ligand. SO_4^{2-} means that Mn has to have a +2 oxidation no.

(b) $K_3[Co(CN)_6]$	Co	+3

The 3 K^+ ions mean that the complex ion must have a -3 charge. Each CN^- has a –1 charge, so Co must have a +3 charge.

(c) $[Co(NH_3)_4Cl_2]Cl$	Co	+3

The chloride anion means that the complex ion must have a net +1 charge. While the ammonia ligand is neutral, each Cl has a –1 charge, so Co must have +3.

(d) $Cr(en)_2Cl_2$	Cr	+2

The ethylenediamine ligand is neutral, so with each Cl having a –1 charge, Cr must be +2.

11. Formula of complex: $[Ni(NH_3)_3(H_2O)en]^{2+}$. The complex has to have a 2+ charge, since each of the three types of ligands is neutral.

Naming Coordination Compounds
13. Formulas for:

   (a) dichlorobis(ethylenediamine)nickel(II) $Ni(en)_2Cl_2$

   (b) potassium tetrachloroplatinate(II) $K_2[PtCl_4]$

   (c) potassium dicyanocuprate(I) $K[Cu(CN)_2]$

   (d) tetraamminediaquairon(II) $[Fe(NH_3)_4(H_2O)_2]^{2+}$

15.

Formula	Name
(a) $[Ni(C_2O_4)_2(H_2O)_2]^{2-}$	diaquabis(oxalato)nickelate(II) ion
(b) $[Co(en)_2Br_2]^+$	dibromobis(ethylenediamine)cobalt(III) ion
(c) $[Co(en)_2(NH_3)Cl]^{2+}$	amminechlorobis(ethylenediamine)cobalt(III) ion
(d) $Pt(NH_3)_2(C_2O_4)$	diammineoxalatoplatinum(II)

17. The name or formula for the ions or compounds shown below:

    (a) $[Fe(H_2O)_5(OH)]^{2+}$ Pentaaquahydroxoiron(III) ion

    (b) $K_2[Ni(CN)_4]$ Potassium tetracyanonickelate(II)

    (c) $K[Cr(C_2O_4)_2(H_2O)_2]$ Potassium diaquabis(oxalato)chromate(III)

    (d) $(NH_4)_2[PtCl_4]$ Ammonium tetrachloroplatinate(IV)

Isomerism

19. Geometric Isomers of

    (a) $Fe(NH_3)_4Cl_2$

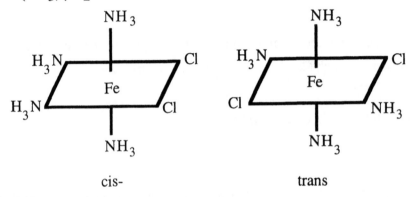

                cis-   trans

    (b) $Pt(NH_3)_2(SCN)(Br)$

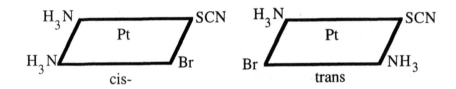

                 cis-   trans

    (c) $Co(NH_3)_3(NO_2)_3$

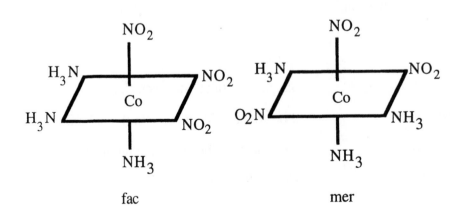

                 fac   mer

(d) [Co(en)Cl4]⁻

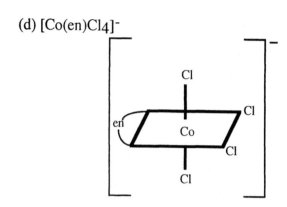

21. Which of the following species has a chiral center?

(a) [Fe(en)₃]²⁺   Yes

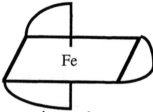   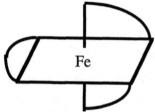

The two mirror images shown above are not superimposable and therefore possess a chiral center.

(b) trans-[Co(en)₂Br₂]⁺   No. The *cis* isomer has a chiral center.

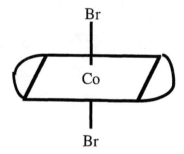

(c) fac-[Co(en)(H₂O)Cl₃]   No

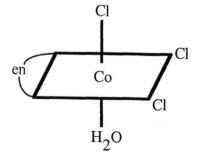   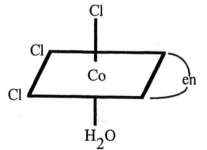

As you can see by the two structures above, a 180° rotation along the Cl-Co-H$_2$O axis, would make these two mirror images superimposable. The *mer* complex would also be superimposable, and therefore possess no chiral center.

(d) Pt(NH$_3$)(H$_2$O)Cl(NO$_2$)

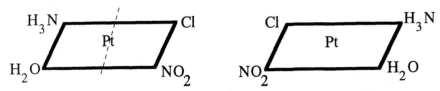

Above are two mirror images of the complex. Rotation of the first complex along the dotted axis by 180° results in the second complex. There are no nonsuperimposable isomers.

Magnetic Properties of Complexes

23. In the name of clarity, the counterion has been omitted. The counterions do not affect the magnetic behavior of the complex ion.

(a) [Mn(CN)$_6$]$^{4-}$

Mn^{2+} has a d^5 configuration

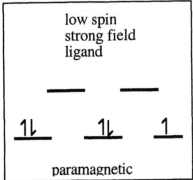

(b) [Co(NH$_3$)$_6$]$^{3+}$ Co^{3+} has a d^6 configuration

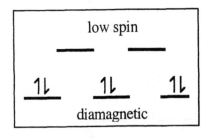

(c) [Fe(H$_2$O)$_6$]$^{3+}$

Fe^{3+} has a d^5 configuration (like Mn^{2+}), and 1 unpaired electron (see diagram in part (a).

(d) $[Cr(en)_3]^{2+}$

Cr^{2+} has the d^4
configuration

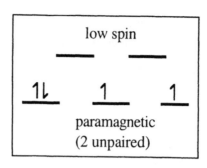

25. For the following (high spin) tetrahedral complexes, determine the number of unpaired
electrons:

(a) $[FeCl_4]^{2-}$ (d^6 ,paramagnetic)	(b) $[CoCl_4]^{2-}$ (d^7, paramagnetic)

↑ ↑ ↑
  ↑↓ ↑
(4 unpaired)
	↑ ↑ ↑
	  ↑↓ ↑↓
	(3 unpaired)

(c) $[MnCl_4]^{2-}$ (d^5, paramagnetic)	(d) $[ZnCl_4]^{2-}$ (d^{10}, diamagnetic)

↑ ↑ ↑
  ↑ ↑
(5 unpaired)
	↑↓ ↑↓ ↑↓
	  ↑↓ ↑↓
	(0 unpaired)

27. For $[Fe(H_2O)_6]^{2+}$

(a) The coordination number of iron is 6. 6 monodentate ligands are attached.

(b) The coordination geometry is octahedral. (Six groups attached to the central metal ion)

(c) The oxidation state of iron is 2+. The charge on the complex is 2+.
Water is a neutral ligand.

(d) Fe^{2+} is a d^6 case. Water is a weak-field ligand (high spin complex) there are 4 unpaired
electrons.

(e) The complex would be paramagnetic

29. Aqueous cobalt(III) sulfate is diamagnetic. H_2O provides a large enough Δ_0 to force

$[Co(H_2O)_6]^{3+}$ to have no unpaired electrons. An excess of F$^-$ results in the conversion of the

hexaaqua complex to the $[CoF_6]^{3-}$ ion. The weaker F$^-$ ligands don't separate the d orbital

energies as much as H_2O, resulting in four unpaired electrons—a paramagnetic complex.

Spectroscopy of Complexes

31. 500nm light is in the blue region of the spectrum (sidebar on page 1098 of your text. The transmitted light, and the color of the solution--is yellow.

General Questions

33. Describe an experiment to determine if:

Nickel in $K_2[NiCl_4]$ is in a square planar or tetrahedral environment.

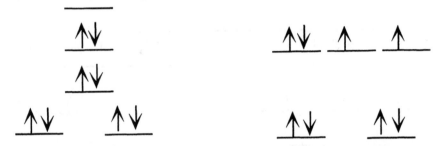

d- orbitals for square-planar complex   d- orbitals for tetrahedral complex

Square-planar nickel(II) is diamagnetic while tetrahedral nickel(II) is paramagnetic. Measuring the magnetic moment would discriminate between the two.

35. Number of unpaired electrons for high-spin and low-spin complexes of Fe^{2+}:

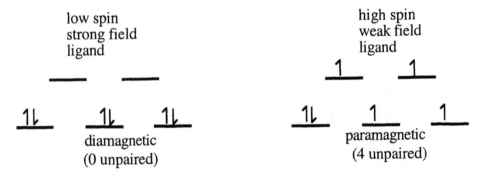

37. Which of the following complexes are square planar?

Of the four complexes, $[Ni(CN)_4]^{2-}$ and $[Pt(CN)_4]^{2-}$ are square planar, since such complexes have the geometry assumed by d^8 metal ions. The other complexes are tetrahedral.

39. Geometric isomers of $Pt(NH_3)(CN)Cl_2$:

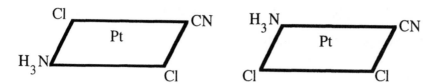

41. Complex absorbs 425-nm light, the blue-violet end of the visible spectrum. So red and green are transmitted , and the complex appears to be yellow.

43. For the high-spin complex $Mn(NH_3)_4Cl_2$ determine:

(a) The oxidation number of manganese is +2. With a neutral overall complex, and 2 chloride ions (each with a –1 charge) and 4 neutral ammonia ligands, manganese has to be +2.

(b) With six monodentate ligands, the coordination number for manganese is 6.

(c) With six monodentate ligands, the coordination geometry for manganese is octahedral.

(d) Number of unpaired electrons: 5 The ligands attached are weak-field ligands resulting in five unpaired electrons.

(e) With unpaired electrons, the complex is paramagnetic.

(f) Cis- and trans- geometric isomers are possible.

45. Structures for cis- and trans- isomers of $CoCl_3 \cdot 4NH_3$

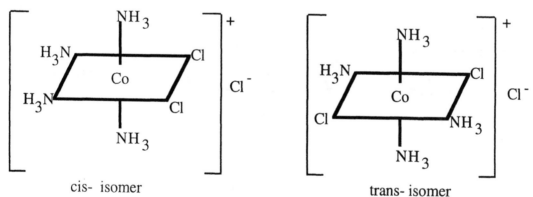

cis- isomer   trans- isomer

cis-tetraamminedichlorocobalt(III) chloride   *trans*-tetraamminedichlorocobalt(III) chloride

47. Formula of a complex containing a Co^{3+} ion, two ethylenediamine molecules, one water molecule, and one chloride ion: $[Co(en)_2(H_2O)Cl]^{2+}$

The ethylenediamine molecules and the water molecule are neutral, so with a (3+) and (1-) charge, the net charge on the ion is 2+.

49. (a) Structures for the fac- and mer- isomers of Cr(dien)Cl3. In these diagrams the curved lines represent the H2N-CH2CH2-NH-CH2CH2- NH2 ligand--with attachments to the metal ion through the electron pair on the N atoms.

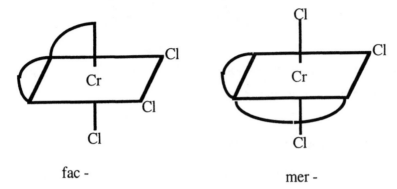

        fac -   mer -

(b) Two different isomers of mer-Cr(dien)BrCl2:

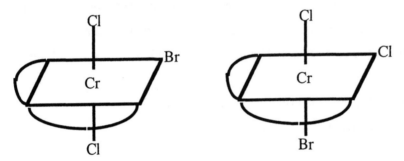

(c) The geometric isomers for isomers of [Cr(dien)2]$^{3+}$

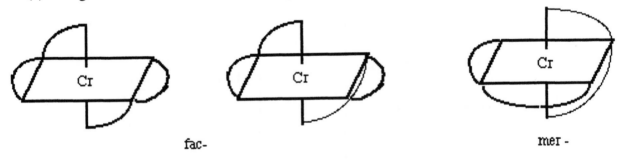

        fac-   mer -

51. Three geometric isomers of [Co(en)(NH3)2(H2O)2]$^{3+}$]:

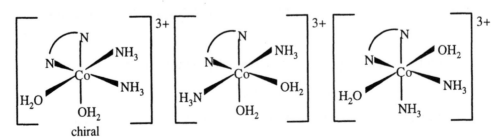

53. For the two Mn^{2+} complexes:

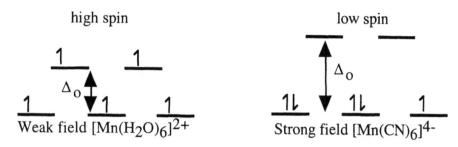

high spin   low spin

Weak field $[Mn(H_2O)_6]^{2+}$ Strong field $[Mn(CN)_6]^{4-}$

The cyano ligand is a *strong field* ligand, so the difference in energies of the doubly vs triply degenerate d orbitals is greater than with the *weak field* aqua ligand. The differences in these energy levels are referred to as Δ_o, and are shown in the diagram above.

55.   A + $BaCl_2$ → ppt $(BaSO_4)$ ⇒ A = $[Co(NH_3)_5Br]SO_4$

B + $BaCl_2$ → no ppt   ⇒ B = $[Co(NH_3)_5SO_4]Br$

Complex A has the sulfate ion in the outer sphere of the transition metal compound. As such, this compound (like many ionic compounds) dissolves in water—and dissociates, liberating sulfate ions. Barium ions react with the sulfate ions to produce the precipitate.

Complex B has the sulfate ion as a part of the inner sphere of the compound. This ion is bound tightly to the cobalt ion, and not available to the barium ions—hence no precipitate.

The reaction between A and $BaCl_2$:

$[Co(NH_3)_5Br]SO_4(aq) + BaCl_2(aq) \rightarrow BaSO_4 (s) + [Co(NH_3)_5Br]^{2+}(aq) + 2\ Cl^-(aq)$

57. For the complex ion $[Co(CO_3)_3]^{3-}$:

(a) Color of complex: blue or cyan—absorbing the red light (at 640 nm), the blue or cyan color would be observed.

(b) Place the CO_3^{2-} ion in the spectrochemical series:

$Cl^-, Br^-, I^- < CO_3^{2-} < C_2O_4^{2-} < H_2O < NH_3,\ en < phen < CN^-$

(c) Is the complex para- or diamagnetic?

If the carbonate ion gives rise to a small Δ, the electrons would tend to occupy as many d orbitals as possible—since the pairing energy would be greater than the Δ. With the d electrons occupying as many orbitals as possible—unpaired electrons would result and the complex would be paramagnetic.

59. The five geometric isomers of $Cu(H_2NCH_2CO_2)_2(H_2O)_2$:

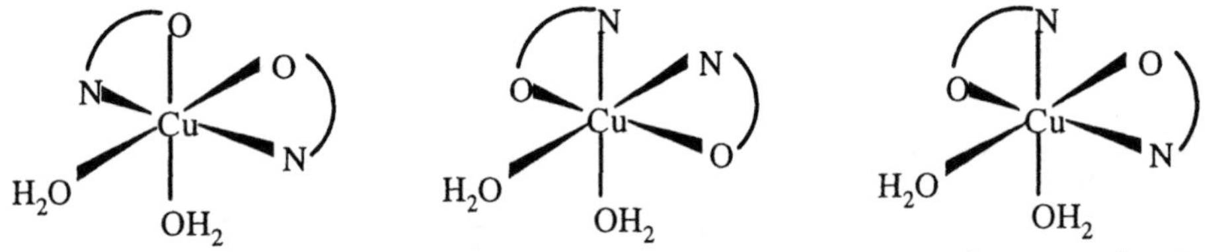

The three isomers above each have a nonsuperimposable mirror image—that is they contain a chiral center. The two isomers shown below have no chiral center.

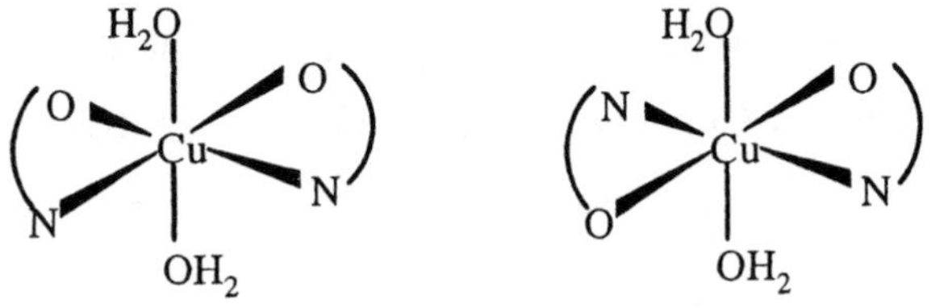

61. Volume of 18.0 M H_2SO_4 need to react with 1.00 kg of ilmenite:

$$\frac{1000.\ g\ FeTiO_3}{1} \cdot \frac{1\ mol\ FeTiO_3}{151.745\ g\ FeTiO_3} \cdot \frac{3\ mol\ H_2SO_4}{1\ mol\ FeTiO_3} \cdot \frac{1\ L}{18.0\ mol\ H_2SO_4} = 1.10\ L$$

Mass of TiO_2 obtainable:

This is straightforward if one calculates the % of TiO_2 in $FeTiO_3$:

$$\frac{1000.\ g\ FeTiO_3}{1} \cdot \frac{79.88\ g\ TiO_2}{151.73\ g\ FeTiO_3} = 526\ g\ TiO_2$$

63. A 0.213 g sample of $UO_2(NO_3)_2$ contains xxxx mol of $UO_2(NO_3)_2$.

$$\frac{0.213\ g\ UO_2(NO_3)_2}{1} \cdot \frac{1\ mol\ g\ UO_2(NO_3)_2}{394.038\ g\ UO_2(NO_3)_2} = 5.41\ x\ 10^{-4}\ mol\ of\ UO_2(NO_3)_2$$

(a) If MnO_4^- is reduced to Mn^{2+}, each mol of MnO_4^- gains 5 mol of electrons (as Mn^{7+} is reduced to Mn^{2+}. The number of mol of MnO_4^- is:

$$\frac{0.0173\ mol\ MnO_4^-}{1\ L} \cdot \frac{0.01247\ L}{1} = 2.15731\ x\ 10^{-4}\ mol\ MnO_4^-.$$ With this information, we

know that 5 mol e^-/mol MnO_4^- • 2.15731 x 10^{-4} mol MnO_4^-. = 1.08 x10^{-3} mol e^-

Knowing that we have 5.41 x 10^{-4} mol of $UO_2(NO_3)_2$ and 1.08 x10^{-3} mol e^-, we calculate

the number of electrons per mol of uranium salt.

$$\frac{1.08\ x10^{-3}\ mol\ electrons}{5.41\ x\ 10^{-4}\ mol\ UO_2(NO_3)_2} = 2\ mol\ electrons/mol\ uranium\ salt.$$

So with U having an oxidation state of +6 (in UO_2^{2+}), and each mol of U gaining 2 mol

electrons, the U would be reduced to +4. So n = 4.

(b) Balanced net ionic equation for reduction of UO_2^{2+} by zinc:

We know that Zn will be oxidized to 2+ (the ion that Zn forms):

$Zn \rightarrow Zn^{2+}$ and the reduction equation: $UO_2^{2+} \rightarrow U^{4+}$ (with a 2 electron change)

The net reduction equation: $UO_2^{2+} + 4 H^+ + 2e^- \rightarrow U^{4+} + 2 H_2O$, with an overall

equation: $UO_2^{2+}(aq) + 4H^+(aq) + Zn(s) \rightarrow U^{4+}(aq) + 2 H_2O(\ell) + Zn^{2+}(aq)$

(c) The net ionic equation for the oxidation of U^{4+} to UO_2^{2+} by MnO_4^-.

The oxidation equation is: $U^{4+} \rightarrow UO_2^{2+}$. The reduction equation is: $MnO_4^- \rightarrow Mn^{2+}$

In acid, the oxidation equation is: $U^{4+} + 2 H_2O \rightarrow UO_2^{2+} + 4H^+ + 2e^-$.

The reduction equation (in acid): $MnO_4^- + 8 H^+ + 5 e^- \rightarrow Mn^{2+} + 4 H_2O$

To equalize electron gain and loss we multiply the oxidation equation by 5, and the reduction equation by 2 to obtain:

$$5 U^{4+} + 10 H_2O \rightarrow 5 UO_2^{2+} + 20 H^+ + 10 e^-$$

$$2 MnO_4^- + 16 H^+ + 10 e^- \rightarrow 2 Mn^{2+} + 8 H_2O$$

Adding the two equations, and removing any redundancies:

$2 MnO_4^-(aq) + 5 U^{4+}(aq) + 2 H_2O (\ell) \rightarrow 5 UO_2^{2+}(aq) + 4 H^+(aq) + 2 Mn^{2+}(aq)$.

Summary and Conceptual Questions

65. The relative stabilities of the hexaammine complexes with Co^{2+}, Ni^{2+}, Cu^{2+}, and Zn^{2+}:
    From Appendix K, the data are:

$Zn^{2+}(aq) + 4 NH_3 (aq) \Leftrightarrow [Zn(NH_3)_4]^{2+}$ $K_f = 2.9 \times 10^9$

$Cu^{2+}(aq) + 4 NH_3 (aq) \Leftrightarrow [Cu(NH_3)_4]^{2+}$ $K_f = 6.8 \times 10^{12}$

$Ni^{2+}(aq) + 6 NH_3 (aq) \Leftrightarrow [Ni(NH_3)_6]^{2+}$ $K_f = 5.6 \times 10^8$

$Co^{2+}(aq) + 6 NH_3 (aq) \Leftrightarrow [Co(NH_3)_6]^{2+}$ $K_f = 7.7 \times 10^4$

While the general order is followed, the copper tetraammine complex has the largest K_f of this

series, a trend that is generally followed for many transition metal complexes.

Practicing Skills

Nuclear Reactions

12. Balance the following nuclear equations, supplying the missing particle.

[The missing particle is emboldened.]

(a) $^{54}_{26}\text{Fe} + {}^{4}_{2}\text{He} \rightarrow 2\,{}^{1}_{0}\text{n} + {}^{\mathbf{56}}_{\mathbf{28}}\text{Ni}$

(b) $^{27}_{13}\text{Al} + {}^{4}_{2}\text{He} \rightarrow {}^{30}_{15}\text{P} + {}^{\mathbf{1}}_{\mathbf{0}}\text{n}$

(c) $^{32}_{16}\text{S} + {}^{1}_{0}\text{n} \rightarrow {}^{1}_{1}\text{H} + {}^{\mathbf{32}}_{\mathbf{15}}\text{P}$

(d) $^{96}_{42}\text{Mo} + {}^{2}_{1}\text{H} \rightarrow {}^{1}_{0}\text{n} + {}^{\mathbf{97}}_{\mathbf{43}}\text{Tc}$

(e) $^{98}_{42}\text{Mo} + {}^{1}_{0}\text{n} \rightarrow {}^{99}_{43}\text{Tc} + {}^{\mathbf{0}}_{\mathbf{-1}}\text{e}$

(f) $^{18}_{9}\text{F} \rightarrow {}^{18}_{8}\text{O} + {}^{\mathbf{0}}_{\mathbf{1}}\text{n}$

14. Balance the following nuclear equations, supplying the missing particle.

[The missing particle is emboldened.]

(a) $^{111}_{47}\text{Ag} \rightarrow {}^{111}_{48}\text{Cd} + {}^{\mathbf{0}}_{\mathbf{-1}}\text{e}$

(b) $^{87}_{36}\text{Kr} \rightarrow {}^{0}_{-1}\text{e} + {}^{\mathbf{87}}_{\mathbf{37}}\text{Rb}$

(c) $^{231}_{91}\text{Pa} \rightarrow {}^{227}_{89}\text{Ac} + {}^{\mathbf{4}}_{\mathbf{2}}\text{He}$

(d) $^{230}_{90}\text{Th} \rightarrow {}^{4}_{2}\text{He} + {}^{\mathbf{226}}_{\mathbf{88}}\text{Ra}$

(e) $^{82}_{35}\text{Br} \rightarrow {}^{82}_{36}\text{Kr} + {}^{\mathbf{0}}_{\mathbf{-1}}\text{e}$

(f) $^{\mathbf{24}}_{\mathbf{11}}\text{Na} \rightarrow {}^{24}_{12}\text{Mg} + {}^{0}_{-1}\text{e}$

16. $^{235}_{92}\text{U} \rightarrow {}^{231}_{90}\text{Th} \rightarrow {}^{231}_{91}\text{Pa} \rightarrow {}^{227}_{89}\text{Ac} \rightarrow {}^{227}_{90}\text{Th} \rightarrow {}^{223}_{88}\text{Ra} \rightarrow {}^{219}_{86}\text{Rn} \rightarrow {}^{215}_{84}\text{Po}$

$+ \quad + \quad + \quad + \quad + \quad + \quad +$

$^{4}_{2}\text{He} \quad {}^{0}_{-1}\text{e} \quad {}^{4}_{2}\text{He} \quad {}^{0}_{-1}\text{e}\ {}^{4}_{2}\text{He} \quad {}^{4}_{2}\text{He} \quad {}^{4}_{2}\text{He}$

and continuing (from Po-215) we have:

$$\begin{array}{ccccc}
^{215}_{84}\text{Po} \rightarrow & ^{211}_{82}\text{Pb} \rightarrow & ^{211}_{83}\text{Bi} \rightarrow & ^{211}_{84}\text{Po} \rightarrow & ^{207}_{82}\text{Pb} \\
+ & + & + & + & \\
^{4}_{2}\text{He} & ^{0}_{-1}\text{e} & ^{0}_{-1}\text{e} & ^{4}_{2}\text{He} &
\end{array}$$

Nuclear Stability and Nuclear Decay

18. The particle emitted in the following reactions: [The missing particle is emboldened.]

(a) $^{198}_{79}\text{Au} \rightarrow ^{198}_{80}\text{Hg} + \mathbf{^{0}_{-1}e}$

(b) $^{222}_{86}\text{Rn} \rightarrow ^{218}_{84}\text{Po} + \mathbf{^{4}_{2}He}$

(c) $^{137}_{55}\text{Cs} \rightarrow ^{137}_{56}\text{Ba} + \mathbf{^{0}_{-1}e}$

(d) $^{110}_{49}\text{In} \rightarrow ^{110}_{48}\text{Cd} + \mathbf{^{0}_{1}n}$

20. Predict the probable mode of decay for each of the following:

(a) $\mathbf{^{80}_{35}}\text{Br}$ (large number of neutrons /proton—beta emission) $\mathbf{^{80}_{35}}\text{Br} \rightarrow \mathbf{^{80}_{36}}\text{Kr} + \mathbf{^{0}_{-1}e}$

(b) $\mathbf{^{240}_{98}}\text{Cf}$ (large isotope- alpha emission) $\mathbf{^{240}_{98}}\text{Cf} \rightarrow \mathbf{^{236}_{96}}\text{Cm} + \mathbf{^{4}_{2}He}$

(c) $\mathbf{^{61}_{27}}\text{Co}$ (mass # > atomic number—beta emission) $\mathbf{^{61}_{27}}\text{Co} \rightarrow \mathbf{^{61}_{28}}\text{Ni} + \mathbf{^{0}_{-1}e}$

(d) $\mathbf{^{11}_{6}}\text{C}$ (more protons than neutrons—positron emission or K-capture)

$$\mathbf{^{11}_{6}}\text{C} \rightarrow \mathbf{^{11}_{5}}\text{B} + \mathbf{^{0}_{-1}}\text{e} \quad \text{or} \quad \mathbf{^{11}_{6}}\text{C} + \mathbf{^{0}_{-1}}\text{e} \rightarrow \mathbf{^{11}_{5}}\text{B}$$

22. Beta particle and positron emission:

(a) Beta particle emission occurs (usually) when the ratio of neutrons/protons is high

Hydrogen-3 has 1 proton and 2 neutrons—**beta particle emission (forms $^{3}_{2}\text{He}$)**

Oxygen-16 has 8 protons and 8 neutrons—not expected

Fluorine-20 has 9 protons and 11 neutrons-- **beta particle emission (forms $^{20}_{10}\text{Ne}$)**

Nitrogen-13 has 7 protons and 6 neutrons – not expected

(b) Position emission occurs when the neutron/proton ratio is too low:

Uranium-238 has 92 protons and 146 neutrons—not expected

Fluorine-19 has 9 protons and 10 neutrons—not expected

Sodium-22 has 11 protons and 11 neutrons—positron emission expected (forms)

Sodium-24 has 11 protons and 13 neutrons—not expected

24. The change in mass (Δ m) for ^{10}B is:

$$\Delta m = 10.01294 - [5(1.00783) + 5(1.00867)]$$

$$= 10.01294 - 10.0825 = -0.06956 \text{ g/mol}$$

Binding energy is: $\Delta mc^2 = (6.956 \times 10^{-5} \text{ kg/mol})(3.00 \times 10^8 \text{ m/s})^2 \left(\dfrac{1 \text{ J}}{1 \text{ kg} \cdot \text{m}^2 \cdot \text{s}^{-2}} \right)$

$$= 6.26 \times 10^{12} \text{ J/mol}$$

The **binding energy per nucleon** : $\dfrac{6.26 \times 10^9 \text{ kJ}}{10 \text{ mol nucleons}} = 6.26 \times 10^8 \dfrac{\text{kJ}}{\text{nucleon}}$

The mass change for ^{11}B is:

$$\Delta m = 11.00931 - [5(1.00783) + 6(1.00867)]$$

$$= 11.00931 - 11.09117 = -0.08186 \text{ g/mol}$$

Binding energy is: $\Delta mc^2 = (8.186 \times 10^{-5} \text{ kg/mol})(3.00 \times 10^8 \text{ m/s})^2 \left(\dfrac{1 \text{ J}}{1 \text{ kg} \cdot \text{m}^2 \cdot \text{s}^{-2}} \right)$

$$= 7.367 \times 10^{12} \text{ J/mol}$$

The **binding energy per nucleon**: $\dfrac{7.37 \times 10^9 \text{ kJ}}{11 \text{ mol nucleons}} = 6.70 \times 10^8 \dfrac{\text{kJ}}{\text{nucleon}}$

26. The binding energy per nucleon for calcium-40:

The change in mass (Δ m) for ^{40}Ca is:

$$\Delta m = 39.96259 - [20(1.00783) + 20(1.00867)]$$

$$= 39.96259 - 40.3300$$

$$= -0.3674 \text{ g/mol}$$

The energy change is then:

$$\Delta E = (3.674 \times 10^{-4} \text{ kg/mol})(3.00 \times 10^8 \text{ m/s})^2 \left(\dfrac{1 \text{ J}}{1 \text{ kg} \cdot \text{m}^2 \cdot \text{s}^{-2}} \right)$$

$$= 3.307 \times 10^{13} \text{ J/mol or } 3.307 \times 10^{10} \text{ kJ/mol}$$

This energy can be converted into the **binding energy per nucleon:**

$$\dfrac{3.307 \times 10^{10} \text{ kJ}}{40 \text{ mol nucleons}} = 8.27 \times 10^8 \dfrac{\text{kJ}}{\text{nucleon}}$$

28. Binding energy per nucleon for Oxygen-16

The change in mass (Δ m) for ^{16}O is:

Δ m   = 15.99492 - [8(1.00783) + 8(1.00867)]

   = -0.13708 g/mol

The energy change is then:

$$\Delta E = (1.3708 \times 10^{-4} \text{ kg/mol})(3.00 \times 10^8 \text{ m/s})^2 \left(\frac{1 \text{ J}}{1 \text{ kg} \cdot \text{m}^2 \cdot \text{s}^{-2}} \right)$$

   = 1.234 $\times 10^{13}$ J/mol or 1.234 $\times 10^{10}$ kJ/mol

This energy can be converted into the **binding energy per nucleon:**

$$\frac{1.234 \times 10^{10} \text{ kJ/mol}}{16 \text{ mol nucleons}} = 7.71 \times 10^8 \text{ kJ/nucleon}$$

Rates of Radioactive Decay

30. For ^{64} Cu, $t_{1/2}$ =12.7 hr

The fraction remaining as ^{64} Cu following n half-lives is equal to $\left(\frac{1}{2}\right)^n$.

Note that 64 hours corresponds to exactly **five** half-lives.

The <u>fraction</u> remaining as ^{64} Cu is $\left(\frac{1}{2}\right)^5$ or $\frac{1}{32}$ or 0.03125.

The mass remaining is: (0.03125)(25.0 μg) = 0.781 μg.

32. (a) The equation for β–decay of ^{131}I is: $^{131}_{53}\text{I} \rightarrow {}^{0}_{-1}\text{e} + {}^{131}_{54}\text{Xe}$

(b) The amount of ^{131}I remaining after 40.2 days:

For ^{131}I, $t_{1/2}$ is 8.04 days--so 40.2 days is exactly **five** half-lives:

The fraction of ^{131}I remaining is $\left(\frac{1}{2}\right)^5$ or $\frac{1}{32}$ or 0.03125.

The amount of the original 2.4 μg remaining will be : (0.03125)(2.4 μg) = 0.075 μg

34. To determine the mass of Gallium-67 left after 13 days, determine the number of half-lives corresponding to 13 days.

$$\frac{13 \text{ days}}{1} \cdot \frac{24 \text{ hrs}}{1 \text{ day}} = 312 \text{ hours}$$

The rate constant is : $k = \dfrac{0.693}{t\frac{1}{2}} = \dfrac{0.693}{78.25 \text{ hr}} = 0.00886 \text{ hrs}^{-1}$

The fraction remaining can be calculated: $\ln(x) = -(0.00886 \text{ hrs}^{-1}) \cdot (312 \text{ hrs})$ and solving for x yields 0.06309 (where x represents the fraction of Ga-67 remaining.

The amount of Gallium-67 remaining is then $(0.06309)(0.015 \text{ mg}) = 9.5 \times 10^{-4} \text{ mg}$

36. For the decomposition of Radon-222:

   (a) The balanced equation for the decomposition of Rn-222 with α particle emission.

   $$^{222}_{86}\text{Rn} \rightarrow \,^{4}_{2}\text{He} + \,^{218}_{84}\text{Po}$$

   (b) Time required for the sample to decrease to 20.0 % of its original activity:

   Since this decay follows 1st order kinetics, we can calculate a rate constant:

   $$k = \dfrac{0.693}{t\frac{1}{2}} = \dfrac{0.693}{3.82 \text{ days}} = 0.181 \text{ days}^{-1}$$

   With this rate constant , using the 1st order integrated rate equation, we can calculate the time required:

   $$\ln\left(\dfrac{20.0}{100}\right) = -(0.181 \text{ days}^{-1}) \cdot t \quad \text{and} \quad \dfrac{-1.609}{-0.181 \text{ days}^{-1}} = t = 8.87 \text{ days}$$

38. The age of the fragment can be determined if:

   (1st) we calculate a rate constant and,

   (2nd) we use the 1st order integrated rate equation (much as we did in question 32b above)

   $$(1\text{st}) \ \ k = \dfrac{0.693}{t\frac{1}{2}} = \dfrac{0.693}{5730 \text{ yr}} = 1.21 \times 10^{-4} \text{ yr}^{-1}$$

   (2nd) Now we can calculate the time required for the Carbon-14 : Carbon-12 to decay to 72% of that ratio in living organisms.

   $$\ln\left(\dfrac{72}{100}\right) = -(1.21 \times 10^{-4} \text{ yr}^{-1}) \cdot t$$

   $$\dfrac{-0.3285}{-1.21 \times 10^{-4} \text{ yr}^{-1}} = t = 2700 \text{ years (to 2 significant figures)}$$

40. For the decay of Cobalt-60, $t_{1/2}$ is 5.27 yrs:

   (a) Time for Co-60 to decrease to 1/8 of its original activity:

   Following the methodology of questions 34,36 and 38, determine the rate constant:

   $$k = \dfrac{0.693}{t\frac{1}{2}} = \dfrac{0.693}{5.27 \text{ yr}} = 0.131 \ \text{ yr}^{-1}$$

Substituting into the equation:

$$\ln(\tfrac{1}{8}) = -0.131 \, yr^{-1} \cdot t \quad \text{and}$$

$$\ln(0.125) = -0.131 \, yr^{-1} \cdot t \quad \text{and solving for } t = 15.8 \, yrs$$

A " short-cut" is available here if you notice that 1/8 corresponds to $(\tfrac{1}{2})^3$. Said another way, one-eighth of the Co-60 will remain after **three half-lives** have passed, so
$3 \cdot 5.27 \, yrs = 15.8 \, years \, !!$

(b) Fraction of Co-60 remaining as Co-60 after 1.0 years:

Now we can solve for the fraction on the "left-hand side" of the rate equation:

$$\ln (\text{fraction remaining}) = -k \cdot t = -0.131 \, yr^{-1} \cdot 1.0 \, yr$$

$$\ln (\text{fraction remaining}) = -0.131 \text{ and } e^{-0.131} = \text{fraction remaining}$$

fraction remaining $= 0.877$, so 88% remains after 1.0 years.

42. Graph for P-31 of disintegrations per minute as a function of time for a period of one year

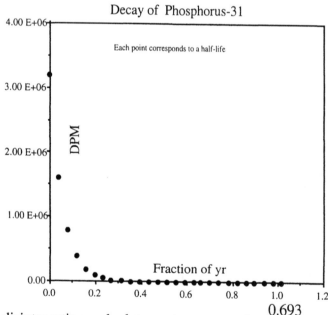

For the disintegration, calculate a rate constant: $k = \dfrac{0.693}{t_{1/2}} = \dfrac{0.693}{14.28 \text{ days}} = 0.04853 \text{ days}^{-1}$

On the graph above is plotted the results of the calculation:

$\ln \dfrac{x}{3.2 \times 10^6} = -(0.04853 \text{ days}^{-1})t$ I have substituted multiples of 14.28 days, so each data

point corresponds to a half-life.

44. To determine the half-life of polonium-210, we will plot ln (dpm) vs time.

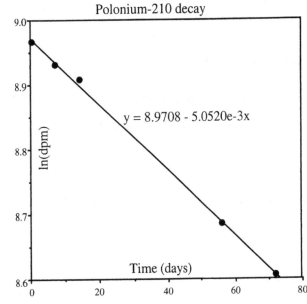

Polonium-210 decay

$y = 8.9708 - 5.0520e\text{-}3x$

The data show a slope of 5.05×10^{-3}. This corresponds to the rate constant, k. The half-life is

$$k = \frac{0.693}{t_{1/2}} \quad \text{or rearranged}$$

$$\frac{0.693}{k} = t_{1/2} \quad \text{so}$$

$$\frac{0.693}{5.05 \times 10^{-3}} = 140 \text{ days}$$

Nuclear Reactions

46. For the decay of Americium-241: $^{239}_{94}\text{Pu} + ^{4}_{2}\text{He} \rightarrow ^{240}_{95}\text{Am} + ^{1}_{1}\text{H} + 2\ ^{1}_{0}\textbf{n}$

48. Synthesis of Element 114: $^{242}_{94}\text{Pu} + ^{48}_{20}\text{Ca} \rightarrow ^{287}_{114}\text{Uuq} + 3\ ^{1}_{0}\text{n}$

50. Complete the following equations using deuterium bombardment:

[The missing particle is emboldened.]

(a) $^{114}_{48}\text{Cd} + ^{2}_{1}\text{D} \rightarrow ^{\textbf{115}}_{48}\textbf{Cd} + ^{1}_{1}\text{H}$

(c) $^{40}_{20}\text{Ca} + ^{2}_{1}\text{D} \rightarrow ^{38}_{19}\text{K} + ^{\textbf{4}}_{\textbf{2}}\textbf{He}$

(b) $^{6}_{3}\text{Li} + ^{2}_{1}\text{D} \rightarrow ^{7}_{\textbf{4}}\textbf{Be} + ^{1}_{0}\text{n}$

(d) $^{\textbf{63}}_{\textbf{29}}\textbf{Cu} + ^{2}_{1}\text{D} \rightarrow ^{65}_{30}\text{Zn} + \gamma$

52. The equation for the bombardment of Boron-10 with a neutron, and the subsequent release of an alpha particle: $^{10}_{5}\text{B} + ^{1}_{0}\text{n} \rightarrow ^{7}_{3}\text{Li} + ^{4}_{2}\text{He}$

General Questions

54. The rate constant, $k = \dfrac{0.693}{t_{1/2}} = \dfrac{0.693}{4.8 \times 10^{10}} = 1.44375 \times 10^{-11} \text{ yr}^{-1}$

At some time ,t, we have 1.8×10^{-3} mol Rb. We also have 1.6×10^{-3} mol Sr, which resulted from the decay of an equal amount of Rb. So the initial amount of Rb = $(1.6 + 1.8) \times 10^{-3}$.

Substituting into the first-order equation we get: $\ln\dfrac{1.8}{3.4} = -(1.44 \times 10^{-11} \text{ yr}^{-1})t$

$$\dfrac{\ln(0.529)}{-1.44 \times 10^{-11}\text{yr}^{-1}} = \dfrac{-0.636}{-1.44 \times 10^{-11}\text{yr}^{-1}} = t = 4.4 \times 10^{10} \text{ yr}$$

56. If the ratio of $\dfrac{\text{Pb-206}}{\text{U-238}}$ is 0.33, what we know is that 1/3 of the uranium has undergone decay. This would imply 2/3 of the uranium remains as U-238. We can then calculate a rate constant and use the 1st order rate equation to calculate the time required for this to occur:

$k = \dfrac{0.693}{4.5 \times 10^9 \text{ yr}} = 1.54 \times 10^{-10} \text{ yr}^{-1}$ and $\ln(\dfrac{67}{100}) = -1.54 \times 10^{-10} \text{ yr}^{-1} \bullet t$

$-0.400 = -1.54 \times 10^{-10} \text{ yr}^{-1} \bullet t$ and $t = \dfrac{-0.400}{-1.54 \times 10^{-10} \text{ yr}^{-1}} = 2.6 \times 10^9 \text{ yr}$

58. The decay of Uranium-238 to produce Plutonium-239:

(a) $^{238}_{92}\text{U} + ^{1}_{0}\text{n} \rightarrow ^{239}_{92}\text{U} + \gamma$

(b) $^{239}_{92}\text{U} \rightarrow ^{239}_{93}\text{Np} + ^{0}_{-1}\text{e}$

(c) $^{239}_{93}\text{Np} \rightarrow ^{239}_{94}\text{Pu} + ^{0}_{-1}\text{e}$

(d) $^{239}_{94}\text{Pu} + ^{1}_{0}\text{n} \rightarrow 2^{1}_{0}\text{n} + $ other nuclei + energy

60. The energy liberated by one pound of ^{235}U:

$$\dfrac{2.1 \times 10^{10} \text{ kJ}}{1 \text{ mol } ^{235}\text{U}} \bullet \dfrac{1 \text{ mol } ^{235}\text{U}}{235 \text{ g U}} \bullet \dfrac{453.6 \text{ g U}}{1.000 \text{ lb}} = \dfrac{4.1 \times 10^{10} \text{ kJ}}{1.000 \text{ lb U}}$$

comparing this amount of energy to the energy per ton of coal yields:

$$\dfrac{4.1 \times 10^{10} \text{ kJ}}{1.000 \text{ lb U}} \bullet \dfrac{1 \text{ ton coal}}{2.6 \times 10^7 \text{ kJ}} = 1.6 \times 10^3 \text{ tons coal/lb U}$$

62. This "dilution" problem may be solved using the equation which is useful for solutions:

$$M_c \times V_c = M_d \times V_d$$

where c and d represent the concentrated and diluted states, respectively. We'll use the number of disintegrations/second as our "molarity."

$$2.0 \times 10^6 \text{ dps} \bullet 1.0 \text{ mL} = 1.5 \times 10^4 \text{ dps} \bullet V_d$$

$$130 \text{ mL} = V_d$$

64. If we assume that the catch represents a homogeneous sample of the tagged fish, the problem is rather simple. The percentage of tagged fish in the sample is $\dfrac{27}{5250} = 0.00514$ or 0.51%.

If our 1000 fish represent 0.51% of the fish in the lake, the number is approximately:

$\dfrac{1000}{0.0051} = 190{,}000$ fish.

66. For the radioactive decay series of U-238:

(a) Why can the masses be expressed as $m = 4n + 2$?

These masses correlate in this fashion since the **principle** mode of decay in the U-238 series is **alpha particle** emission. The mass of an alpha particle is **4** (since it is basically 2 neutrons + 2 protons). Hence the loss of an alpha particle (say from U-238 to Th-234) results in a **mass loss of 4 units**. The series results in a total loss of 8 alpha particles and 6 beta particles. Since the beta particles (electrons) have an insignificant mass (compared to a neutron or a proton), the loss of a beta particle--or for that matter 6 of them--does not significantly affect the masses of the daughter products. These masses correspond to n values: n=(51 -> 59).

(b) Equations corresponding to the decay series for U-235 and Th-232 :

The U-235 series corresponds to the equation **$m = 4n+3$** with n values: (51 -> 58). The Th-232 series corresponds to the equation **$m = 4n$**

From an empirical standpoint, the masses of the most massive isotopes in the three series differ by 3 (U-238 $\rightarrow$ U–235) and (U–235 $\rightarrow$ Th-232. So given that the algorithm for the U-238 series is 4n+2, subtracting 3 gave 4n-1 (which also equals 4n+3—with n reduced by 1; subtracting 3 more gives 4n.

The isotopic masses for these 3 series are summarized in the table below:

n	= 4n+2	= 4n+3	= 4n
51	206		
52	210	207	208
53	214	211	212
54	218	215	216
55	222	219	220
56	226	223	224
57	230	227	228
58	234	231	232
59	238	235	

(c) Identify the series to which each of the following isotopes belong:

Isotope	series
226-Ra:	U-238 (4n + 2)
215-At:	U-235 (4n +3)
228-Th:	Th-232(4n)
210- Bi:	U-238 (4n + 2)

(d) Why is the series "4n+1" missing in the earth's crust?

To occur in the earth's crust, an element must be very stable—that is have a very long half-life or be non-radioactive! From hydrogen to lawrencium, with the exception of two isotopes of hydrogen (protium and tritium), every isotope of every element has a nucleus containing at least one neutron for every proton. With the mass of the neutron and proton being 1, the change in mass number would have to change by a factor of 2—so **4n+1** would not lead to a long-lived isotope and would not be found in the earth's crust.

68. For Protactinium:

(a) The series containing Protactinium-231 is the U-235 series. It corresponds to the (4n+3) series—See question 66.

(b) A series of reaction to produce Pa-231:

$$^{235}_{92}U \rightarrow {}^{231}_{90}Th + {}^{4}_{2}He \text{ followed by the decay } {}^{231}_{90}Th \rightarrow {}^{0}_{-1}e + {}^{231}_{91}Pa.$$

(c) Quantity of ore to provide 1.0 g of Pa-231 assuming 100% yield:

Stow your calculators! If the ore is 1 part per million, and you want 1.0 g, then you need 1,000,000 g of ore, check?

(d) Decay for Pa-231: $^{231}_{91}Pa \rightarrow {}^{227}_{89}Ac + {}^{4}_{2}He$

Energy Interchapter
The Chemistry of Fuels and Energy Sources

1. Mass of H_2 expected from the reaction of steam with 100. g of methane (CH_4), petroleum(CH_2) and coal (C):

$$CH_4(g) + H_2O(g) \quad \rightarrow \quad CO(g) \; + \quad 3H_2(g)$$

$$CH_2(g) + H_2O(g) \quad \rightarrow \quad CO(g) \; + \quad 2H_2(g)$$

$$C(s) + H_2O(g) \quad \quad \rightarrow \quad CO(g) \; + \quad H_2(g)$$

Based on the equations above, we convert the mass to moles of the C containing specie, use the stoichiometric ratio indicated by the balanced equation, and convert to mass of H_2, using the molar mass of H_2.

$$\frac{100.\text{ g }CH_4}{1} \cdot \frac{1\text{ mol }CH_4}{16.04\text{ g }CH_4} \cdot \frac{3\text{ mol }H_2}{1\text{ mol }CH_4} \cdot \frac{2.02\text{g }H_2}{1\text{mol }H_2} = 37.7 \text{ g } H_2$$

$$\frac{100.\text{ g }CH_2}{1} \cdot \frac{1\text{ mol }CH_2}{14.03\text{ g }CH_2} \cdot \frac{4\text{ mol }H_2}{1\text{ mol }CH_2} \cdot \frac{2.02\text{g }H_2}{1\text{mol }H_2} = 28.7 \text{ g } H_2$$

$$\frac{100.\text{ g }C}{1} \cdot \frac{1\text{ mol }C}{12.01\text{ g }C} \cdot \frac{1\text{ mol }H_2}{1\text{ mol }C} \cdot \frac{2.02\text{g }H_2}{1\text{mol }H_2} = 16.8 \text{ g } H_2$$

3. Calculate the energy evolved (in kJ) when 70. lb of coal is burned. We make the assumption that coal can be represented as carbon (C).

$$\frac{70.\text{ lb }C}{1} \cdot \frac{454\text{ g }C}{1\text{ lb }C} \cdot \frac{33\text{ kJ}}{1\text{ g }C} = 1.0 \times 10^6 \text{ kJ}$$

5. Energy consumption in U.S, is: 7.0 gallons of oil (or 70. lb coal) per person per day. Compare the energy produced by 7.0 gallons of oil with that of 70. lb of coal (see problem 3 above). From Table 2, the energy released per gram of crude petroleum (45 kJ/g) is found.

$$\frac{7.0\text{ gal oil}}{1} \cdot \frac{4\text{ qts}}{1\text{ gal}} \cdot \frac{1000\text{ mL}}{1.0567\text{ qt}} \cdot \frac{0.8\text{ g oil}}{1\text{ mL oil}} \cdot \frac{45\text{ kJ}}{1\text{ g oil}} = 9.5 \times 10^5 \text{ kJ}.$$

The energy released is slightly less for petroleum than for coal, but within the same order of magnitude.

7. Heat evolved per gram and per liter of isooctane:

$$\frac{5.45 \times 10^3 \text{ kJ}}{1 \text{ mol C}_8\text{H}_{18}} \cdot \frac{1 \text{ mol C}_8\text{H}_{18}}{114.23 \text{ g C}_8\text{H}_{18}} = 47.7 \text{ kJ/g C}_8\text{H}_{18}$$

The heat evolved per liter is calculated by converting the mass of isooctane to a volume (D= 0.688 g/mL)

$$\frac{5.45 \times 10^3 \text{ kJ}}{1 \text{ mol C}_8\text{H}_{18}} \cdot \frac{1 \text{ mol C}_8\text{H}_{18}}{114.23 \text{ g C}_8\text{H}_{18}} \cdot \frac{0.688 \text{ g C}_8\text{H}_{18}}{1.0 \text{ mL C}_8\text{H}_{18}} \cdot \frac{1000 \text{ mL}}{1 \text{ L}} = 3.28 \times 10^4 \text{ kJ/L}$$

9. Express 940 kilowatt-hours in units of kJ:

$$\frac{940 \text{ kwh}}{1} \cdot \frac{1000 \text{ watt} \bullet \text{hours}}{1 \text{ kwh}} \cdot \frac{1 \text{ J/s}}{1 \text{ watt}} \cdot \frac{3600 \text{ s}}{1 \text{ hour}} \cdot \frac{1 \text{ kJ}}{1000 \text{ J}} = 3.4 \times 10^6 \text{ kJ}$$

11. Confirm that oxidation of 1.0 L of CH_3OH to form $CO_2(g)$ and $H_2O(l)$ in a fuel cell provides at least 5.0 kwh of energy (D of $CH_3OH = 0.787$ g/mL).

We need to convert 1.0 L of methanol into mass and moles:

$$\frac{1.0 \text{ L CH}_3\text{OH}}{1} \cdot \frac{787 \text{ g CH}_3\text{OH}}{1.0 \text{ L}} \cdot \frac{1 \text{ mol CH}_3\text{OH}}{32.04 \text{ g CH}_3\text{OH}} = 25 \text{ mol CH}_3\text{OH}$$

Now the question is what energy change occurs when methanol burns. That question is answered by using thermodynamic data (from Applendix L)

$$2 \text{ CH}_3\text{OH (}\ell\text{) + 3 O}_2 \text{ (g)} \rightarrow 2 \text{ CO}_2\text{(g) + 4 H}_2\text{O(}\ell\text{)}$$

$\Delta Hrx = [2 \Delta Hf°CO_2(g) + 4 \Delta Hf°H_2O(\ell)] - [2 \Delta Hf°CH_3OH(\ell) + 3 \Delta Hf°O_2(g)]$

$\Delta Hrx = [2\text{mol} \bullet (-393.509 \text{ kJ/mol}) + 4\text{mol} \bullet (-285.83 \text{ kJ/mol})]$

$\qquad\qquad\qquad\qquad - [2\text{mol} \bullet (-238.4 \text{ kJ/mol}) + 3 \text{ mol} \bullet 0)]$

$\qquad = -1,453.5 \text{ kJ}$

This energy change is associated with 2 mol of CH_3OH (balanced equation).

The energy change per mol is then multiplied by the # of moles of methanol associated with 1.0 L of CH_3OH

$$\frac{-1,453.5 \text{ kJ}}{2 \text{ mol CH}_3\text{OH}} \cdot \frac{25 \text{ mol CH}_3\text{OH}}{1} = -1.8 \times 10^4 \text{ kJ (2 sf)}$$

Now converting that energy change to units of kwh:

$$\frac{1.8 \times 10^4 \text{ kJ}}{1} \cdot \frac{1000 \text{ J}}{1 \text{ kJ}} \cdot \frac{1 \text{ watt}}{1 \text{ J/s}} \cdot \frac{1 \text{ hr}}{3600 \text{ s}} \cdot \frac{1 \text{ kw}}{1000 \text{ watt}} = 5.0 \text{ kwh (to 2sf)}$$

13. The parking lot is 325m long and 50.0 m wide, or a surface area of 16250 m^2

    If the solar radiation is 2.6 x 10^7 J/m^2 (per day) , the parking lot would receive:

    2.6 x 10^7 J/m^2 • 16250 m^2 = 4.2 x 10^{11} J (per day).

15. The formula of the substance formed when Pd absorbs H_2.

    If 1.0 cm^3 absorbs 0.084 g of H_2, let's calculate the # of moles of each of the two substances:

    $$\frac{1.0\ cm^3\ Pd}{1} \bullet \frac{12.0\ g\ Pd}{1\ cm^3\ Pd} \bullet \frac{1\ mol\ Pd}{106.42\ g\ Pd} = 0.11\ mol\ Pd$$

    $$\frac{0.084\ g\ H_2}{1} \bullet \frac{1\ mol\ H_2}{2.02\ g\ H_2} \bullet \frac{2 mol H}{1\ mol\ H_2} = 0.083\ mol\ H$$

    So an approximate formula is $Pd_{0.11}H_{0.0832}$ or roughly $Pd_{1.35}H$

17. Energy consumed to drive 1.00 mile by a car rated at 55.0 mpg. The density of gasoline is

    0.737 g/cm^3 ; gasoline produces 48.0 kJ/g.

    $$\frac{48.0\ kJ}{1\ g\ gasoline} \bullet \frac{0.737\ g\ gasoline}{1\ cm^3\ gasoline} \bullet \frac{1000\ cm^3}{1\ L} \bullet \frac{1\ L}{1.0567\ qt} \bullet \frac{4\ qt}{1\ gal} \bullet \frac{1\ gal}{55.0\ mile} \bullet \frac{1.00\ mile}{1} =$$

    2.43 x 10^3 kJ

Biochemistry Interchapter
The Chemistry of Life

1. Lewis structures for:

   (a) Valine with amino and carboxyl groups in un-ionized form

   where CH_3 = H - C - H (with H above and H below the C)

   (b) Valine in zwitterionic form

   (c) At physiological pH, the zwitterion predominates since the carboxylic acid group and the amino group are ionized.

3. Two different ways in which glycine and alanine may be combined:

   glycine   alanine   alanine   glycine

   The differences result from varying which "amino group" is bonded to which "carboxyl group".

5. Two Lewis structures for the dipeptide alanine-isoleucine that show the resonance structures of the amide linkage.

Resonance structures are a result of shifting pairs of electrons, in this case, the lone pair of electrons from N to create the C=N bond. The right-hand structure shows the shifts. A pair of electrons from C is then shifted to a "lone pair" position. The resulting O bears a "-" charge, while the N bears a "+" charge.

7. Information given in a description of the quaternary structure of reverse transcriptase: Quaternary structure tells one about the arrangement of the subunits of the protein. The two subunits that compose reverse transcriptase are shown on p542 of your textbook.

9. The tetranucleotide AUGC is shown to the right. Note that as with all nucleotides, there are three parts, an organic base, a sugar, and a phosphoric acid linking group. The braces in this picture indicate the ENTIRE nucleotide. It is important to note that the letters A,U,G,C are specifically used to designate the bases Adenine, Uracil, Guanine, and Cytosine respectively.

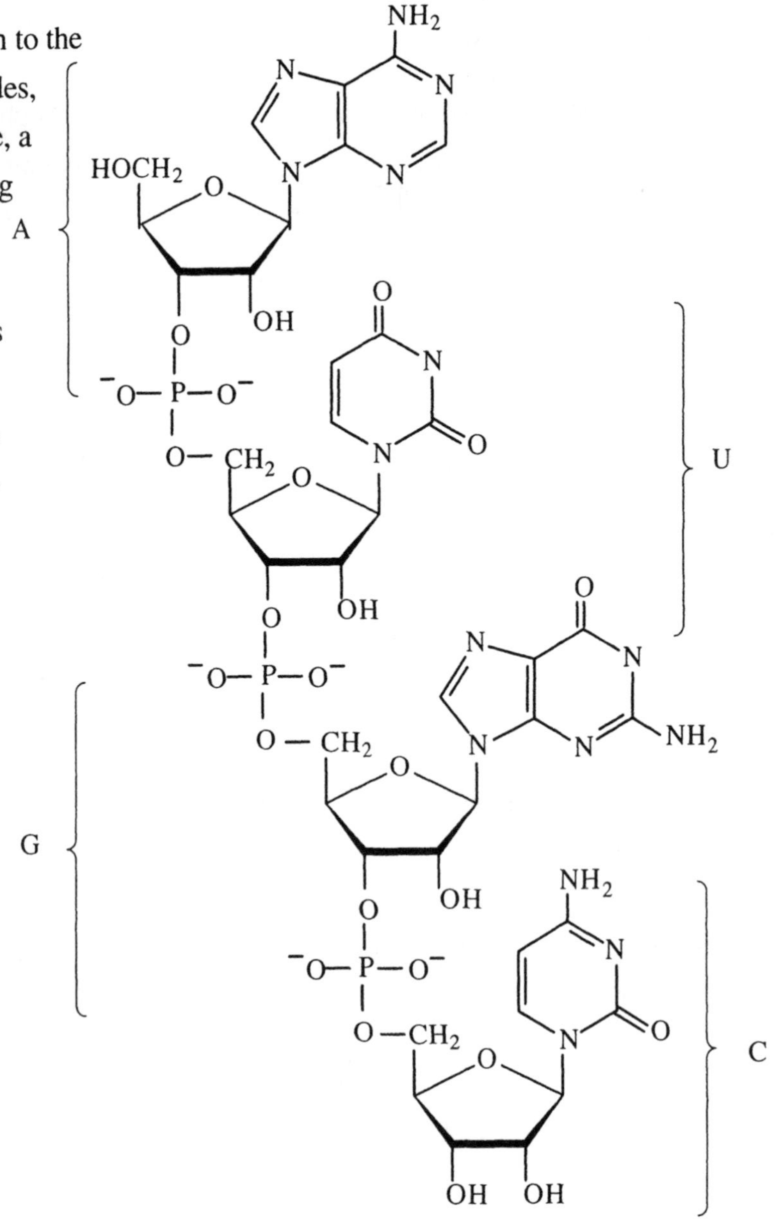

11. Base pairs proposed by Watson and Crick in their structure for DNA: As noted on page 538 of your textbook, Watson and Crick proposed A-T base pairs and C-G base pairs in DNA.

13. For the nucleotide sequence in DNA: 5'-ACGCGATTC-3':

(a) The sequence of the complementary strand would be found by nothing that every A has a T as a complement; every C a G. Recall that the 5' end of a DNA strand pairs with the 3' end of its complement. So if we write the complement to the sequence above, we'd get: 3'-TGCGCTAAG-5'. Since the two strands of DNA have 5' ends paired with 3' ends, we write the complement as the reverse of the listing above, namely: 5'-GAATCGCGT-3'.

(b) Write the sequence 5' to 3' for a strand of m-RNA to complement the original strand of DNA. For m-RNA Tyrosine (in DNA) is replaced with Uracil (in RNA). Hence taking the complementary strand from part (a), we replace every T with U: 5'-GAAUCGCGU-3'

(c) Beginning with the m-RNA strand shown in part (b), we now code each "letter" with its complement, recalling that we "flip" the 5' and 3' ends of the segment: so GAA becomes CUU, UCG becomes AGC, and CGU becomes GCA---and swapping the ends yields: CUU → UUC; AGC →CGA; and GCA →ACG so the three anticodons would be written: 5'- UUC-3'; 5'- CGA -3'; 5'- ACG -3'

(c) The sequence of amino acids coded by the three codons on m-RNA (part(b):

GAA codes for glu (glutamic acid)

UCG codes for ser (serine)

CGU codes for arg (arginine)

15. (a) Describe what occurs in the process of transcription:

The information contained in DNA is "transcribed" by the process of transcription. The process is NOT a straight copy, but "an exchange", in which complementary nitrogen bases appear in the "product"---m-RNA. So when DNA has an "A", that "A" is transcribed into m-RNA as U, a "C" is transcribed as "G", a "G" as "C: So a sequence in DNA that is CGCAA is transcribed into m-RNA as GCGUU.

(b) Describe what occurs in the process of translation: The information that m-RNA contains is "decoded" by t-RNA . That decoding results in the formation of an amino acid sequence (much as we did in 13(c) above). See Table 1 in the Biochemistry Interchapter (p 539)for the amino acids that result from this translation.

17. Which of the following statements are true?

(a) Breaking the P-O bond in ATP is exothermic. FALSE. The cleavage of P-O bonds in ATP requires energy—so the process is ENDOTHERMIC, not exothermic.

(b) Making a new bond between the P atoms in the phosphate group being cleaved off and the OH group of water is an exothermic process. TRUE- As noted on p 542 of your text, the formation of ADP from ATP releases energy. While the cleavage of a P-O bond requires energy, as does the cleavage of the O-H bond, the energy released upon formation of the P-OH bond releases more energy that is input in the cleavage of the P-O and O-H bonds, resulting in a net EXOTHERMIC process.

(c) Breaking bonds is an ENDOTHERMIC process. TRUE—To cleave bonds, energy must be put in to the process.

(d) Energy released in hydrolysis of ATP may be used to run endothermic reactions in a cell. TRUE. As noted on p 542 of the text (and in part (b) above), the hydrolysis of ATP requires an input of energy (breaking a P-O bond and a O-H bond). It also releases energy as the P-OH bond is formed—more energy than is required for the breaking—with the NET RELEASE of energy that can be channeled into driving otherwise endothermic processes in a cell.

19. (a) The enthalpy change for the production of one mole of glucoes by photosynthesis at 25°C:

The process may be represented: $6\ CO_2(g) + 6\ H_2O(\ell) \rightarrow C_6H_{12}O_6(s) + 6\ O_2(g)$

The energy change is:

$\Delta Hrx = (\Delta H_f^{\circ}\ C_6H_{12}O_6(s) + 6 \cdot \Delta H_f^{\circ}\ O_2(g)) - (6 \cdot \Delta H_f^{\circ}\ CO_2(g) + 6 \cdot \Delta H_f^{\circ}\ H_2O(\ell)\ )$

$\Delta Hrx = (1mol \cdot (-1273.3\ kJ/mol) + 6 \cdot (0))$

$\qquad\qquad\qquad - (6mol \cdot (-393.509\ kJ/mol) + 6\ mol \cdot (-285.83\ kJ/mol))$

$\Delta Hrx = 2803\ kJ$ for 1 mol of glucose.

(b) The enthalpy change for 1 molecule of glucose is found by dividing the energy change found in (a) by Avogadro's number:

$$\frac{2803\ kJ}{1\ mol\ glucose} \cdot \frac{1\ mol\ glucose}{6.022\ x\ 10^{23}\ molecules\ glucose} = 4.655\ x\ 10^{-21}\ kJ/molecule$$

(c) What is the energy of a photon of light with a wavelength of 650 nm?

To calculate the energy of one photon of light with 650 nm wavelength, we need to first calculate the frequency of the radiation:

$$frequency = \frac{speed\ of\ light}{wavelength} = \frac{2.9979x10^8\ m/s}{6.5\ x10^2\ nm} \cdot \frac{1.00\ x\ 10^9\ nm}{1.00\ m}$$

$$= 4.61\ x\ 10^{14}\ s^{-1}$$

And the energy is then $E = h\upsilon$ or $(6.626\ x\ 10^{-34}\ J \cdot s \cdot photons^{-1})(4.61\ x\ 10^{14}\ s^{-1}\ )$

$$= 3.06\ x\ 10^{-19}\ J \cdot photons^{-1}$$

(d) One molecule of glucose requires 4.655×10^{-21} kJ or 4.655×10^{-18} kJ . One photon with wavelength 650 nm has 3.06×10^{-19} J of energy, one photon will be **insufficient** to cause the production of one molecule of glucose.

Materials Interchapter
The Chemistry of Modern Materials

1. Maximum wavelength of light that can excite an electronic transition across the GaAs band gap:

   Region of specturm to which this corresponds.

   As noted on p 647 of your text, the band gap for GaAs is 140 kJ/mol

   Recall the correlation between energy and wavelength: $E = hc/\lambda$ and rearranging: $\lambda = hc/E$

   Where λ = wavelength,

   h = Planck's constant (6.626×10^{-34} J•s);

   c = speed of light (2.9979×10^{8} m/s)

   Noting that Planck's constant has units of J (not kJ), and noting that the energy is in kJ/mol, we need to (a) convert kJ to J and (b) convert mol electrons to electrons (Avogadro's number)

$$\lambda = \frac{6.626 \times 10^{-34} \, J \bullet s \bullet 2.9979 \times 10^{8} \, \frac{m}{s}}{140 \, \frac{kJ}{mol \, e} \bullet \frac{1 mol \, e}{6.022 \times 10^{23} \, e} \bullet \frac{1000 \, J}{1 \, kJ}} = 8.54 \times 10^{-7} m = 854 \times 10^{-9} \, m = 854 \, nm$$

   Noting that this wavelength is slightly longer than that associated with the "red" portion of the visible spectrum, this wavelength is in the IR region of the electromagnetic spectrum.

3. Energy produced/minute by a 1.0 cm^2 solar cell operating at 25% efficiency, if 925 watts/m^2 are striking the earth's surface.

   The energy striking the cell would be: $\dfrac{925 \, watts}{1 \, m^2} \bullet \dfrac{(1m)^2}{(100cm)^2} \bullet \dfrac{1 cm^2}{1} = 0.0925 \, watts$

   Since the cell operates at 25% efficiency, the power absorbed would be 0.023125 watts (or 1/4 of 0.0925 watts).

   Now we convert between power (watts) and energy (Joules), with the factor 1 watt = 1J/s

$$0.023125 \, watts \bullet \frac{1 \, \frac{J}{s}}{1 \, watt} \bullet \frac{60 \, s}{1 \, min} = 1.4 \, J/minute \; (\text{to 2sf})$$

5. From Figure 12, estimate the diameter of the gears in the photo. Compare with the diameter of a typical human blood cell.

According to the legend in Figure 12, a spider mite is about 0.8 mm long and 0.4 mm across. The diagram shows three gears that are of a length that compares to the width of the spider mite, so the average diameter of the gears shown is about 1/3(0.4mm) or 0.13mm (or 130 μm).

The diameter of a red blood cell is about 6.5μm (Reference: Anthony Carpi, www.visionlearning.com/library)

7. An approximate value for the density of aerogel which is, by volume, 99% air and 1% SiO_2.

Calculate the mass of a 1.0 cm^3 piece of aerogel.

Air has a density of 1.2929 kg/m^3 and SiO_2 has a density ranging from 2.32-2.66 g/cm^3 (61st Edition, CRC Handbook of Chemistry & Physics)—depending on the crystalline modification of the oxide. Assume that we use SiO_2 with a density 2.32 g/cm^3.

The density of the aerogel would be :

$(0.99 \times 1.29 \times 10^{-3}$ g/cm$^3) + (0.01 \times 2.32$ g/cm$^3) = 0.024$ g/cm^3 .

The mass of a 1.0-cm^3 piece of aerogel would be 0.024 g.

Environmental Interchapter
The Chemistry of The Environment

1. Calculate K for the reaction: $Ca(OH)_2(s) + Mg^{2+}(aq) \Leftrightarrow Ca^{2+}(aq) + Mg(OH)_2(s)$

   The K for this equation can be easily calculated if one recalls two facts: (1) Knet for an equation can be calculated if one can compose two or more equations that, when added, give the net reaction, and (2) These two processes involve the hydroxides of calcium and magnesium dissolving (e.g. the Ksp for the two hydroxides). Those equilibria have the form:

   (a) $Ca(OH)_2(s) \Leftrightarrow Ca^{2+}(aq) + 2\ OH^-\ (aq)$ Ksp = 5.5 x 10^{-5}

   (b) $Mg(OH)_2(s) \Leftrightarrow Mg^{2+}(aq) + 2\ OH^-(aq)$ Ksp = 5.6 x 10^{-12}

   Note that to obtain our "net" reaction, we need to reverse equation (b), and "add" the two:

   $Ca(OH)_2(s) \Leftrightarrow Ca^{2+}(aq) + 2\ \cancel{OH^-}\ (aq)$ Ksp = 5.5 x 10^{-5}

   $Mg^{2+}(aq) + 2\ \cancel{OH^-}(aq) \Leftrightarrow Mg(OH)_2(s)$ K = (1/ Ksp) = (1/5.6 x 10^{-12})

   $Ca(OH)_2(s) + Mg^{2+}(aq) \Leftrightarrow Ca^{2+}(aq) + Mg(OH)_2(s)$ K = 5.5 x 10^{-5} • (1/5.6 x 10^{-12})

   Multiplying: $\dfrac{5.5 \times 10^{-5}}{5.6 \times 10^{-12}}$ = 9.8 x 10^6 = K

3. Mass of CaO that must be added to 1.5 x 10^6 L of water to precipitate Ca^{2+} ions as CaCO$_3$ if the [Ca^{2+}] = 175 mg/L

   Begin by expression the calcium ion concentration as mol/L:

   $[Ca^{2+}] = \dfrac{175\ mg\ Ca^{2+}}{1\ L} \bullet \dfrac{1\ g\ Ca^{2+}}{1000\ mg\ Ca^{2+}} \dfrac{1\ mol\ Ca^{2+}}{40.08g}$ = 4.37 x 10^{-3}M

   The total number of moles of calcium ion in 1.5 x 10^6 L would be:

   $\dfrac{4.37 \times 10^{-3}\ mol\ Ca^{2+}}{1\ L} \bullet \dfrac{1.5 \times 10^6\ L}{1}$ = 6.5 x 10^3 mol Ca^{2+}

   The equation responsible for the precipitation of CaCO$_3$ (as noted on page 1001 of your text)

   is: $Ca^{2+}(aq) + 2\ HCO_3^-(aq) + CaO(s) \rightarrow 2\ CaCO_3(s) + H_2O(\ell)$

   This stoichiometry indicates that equal numbers of moles of CaO and moles of Ca^{2+} are required.

   $\dfrac{6.5 \times 10^3\ mol\ Ca^{2+}}{1} \bullet \dfrac{1\ mol\ CaO}{1\ mol\ Ca^{2+}} \bullet \dfrac{56.08\ g\ CaO}{1\ mol\ CaO}$ = 3.7 x 10^5 g CaO

5. Classify the reaction types: (1) formation of chloramines in water treatment and

(2) formation of ammonium salts that contribute to haze

as acid-base, oxidation-reduction, combustion, or gas-forming reactions.

(1) $HClO(aq) + NH_3(aq) \rightarrow NH_2Cl(aq) + H_2O(\ell)$

(1) $HClO(aq) + NH_2Cl(aq) \rightarrow NHCl_2 (aq) + H_2O(\ell)$

(1) $HClO(aq) + NHCl_2(aq) \rightarrow NCl_3 (aq) + H_2O(\ell)$

(2) $H_2SO_4(aq) + NH_3(aq) \rightarrow (NH_4)_2SO_4 (aq)$

(2) $HNO_3(aq) + NH_3(aq) \rightarrow NH_4NO_3 (aq)$

Note that all five equations are acid-base reactions, e.g. $HClO$ (acid) reacting with NH_3 (base) The first three equations (indicated as 1) are ALSO oxidation-reduction (recall that oxidation-reduction reactions can be classified as those in which H is lost/gained.

7. The Environmental Protectional Agency monitors several standards—air quality being one of them. The EPA web page is found at: http://www.epa.gov/

9. Estimate the PM_{10} and $PM_{2.5}$ index, from the data given in the graph.

The particulate mass index (PM) is defined as the mass of particulate matter in a given volume of air. The PM_{10} is the collection of particles with diameters smaller than $10\mu m$. [Note: Some documents include the number (e.g. particles with a diameter of $10\mu m$) with the distribution] Using $9.99\mu m$ as the upper limit for the particles for PM_{10} shown in the graph, we have:

$0.015 \ \mu g/m^3 + 0.010\mu g/m^3 + 0.005\mu g/m^3 + 0.001\mu g/m^3 = 0.031 \ \mu g/m^3$

For the $PM_{2.5}$ index, we have: $0.001\mu g/m^3$